Revised Edition

The Living Environment Biology

Rick Hallman
Principal
Benjamin N. Cardozo High School
Bayside, New York

AMSCO

AMSCO SCHOOL PUBLICATIONS, INC.,
a division of Perfection Learning®

The author wishes to acknowledge the helpful contributions of the following consultants in the preparation of this book:

Bart Bookman
Assistant Principal, Science Supervision, Retired
Adlai E. Stevenson High School, Bronx, NY

Mary P. Colvard
Biology/Science Research Teacher
Cobleskill-Richmondville H.S., Cobleskill, NY

Jay W. Costanza
Biology Teacher
School of the Arts in Rochester, Rochester, NY

Alan R. Doctor
Biology and Earth Science Teacher
The Michael J. Petrides School, Staten Island, NY

Sandra M. Latourelle
Instructor of Biological Sciences
Plattsburgh State Univ. of NY, Plattsburgh, NY

Laverne Lewis
Science Teacher
Adlai E. Stevenson High School, Bronx, NY

Barbara Nicolato
Biology Teacher
Newburgh Free Academy, Newburgh, NY

Dr. Jeffrey Piekarsky
Biology Teacher
Health Opportunities High School, Bronx, NY

Maureen Reilly-Gasior
Science Teacher
Hillcrest High School, Queens, NY

Michael F. Renna
Assistant Principal, Science Supervision
Hillcrest High School, Queens, NY

Suzanne Tansey
Science Teacher
John Ehret High School, Marrero, LA

Robert Zottoli, Ph.D.
Professor of Biology
Fitchburg State College, Fitchburg, MA

Edited by Carol Davidson Hagarty
Composition by Brad Walrod/High Text Graphics, Inc.
Cover and Text Design by Howard Petlack/A Good Thing, Inc.
Artwork by Hadel Studio
Photo Research by Tobi Zausner

When ordering this book, please specify:
470301 *or* THE LIVING ENVIRONMENT: BIOLOGY, REVISED EDITION, PAPERBACK *or*
470306 *or* THE LIVING ENVIRONMENT: BIOLOGY, REVISED EDITION, HARDBOUND
ISBN 978-0-87720-944-7 (Paperback)
ISBN 978-0-87720-943-0 (Hardbound)

Also available: *The Living Environment: Biology, Revised Edition, Teacher's Manual with Answers*
When ordering the *Teacher's Manual with Answers*, please specify **14659** *or* ISBN 978-1-62974-384-4

Please visit our Web sites at: ***www.amscopub.com*** and ***www.perfectionlearning.com***

Printed in the United States of America

5 6 7 8 9 10 EBM 18 17

To the Student

About the Living Environment: Biology

What causes goose bumps? How are dogs and bears related? Why is DNA called the most important substance on Earth? As you read this book, you will find answers to small and big questions such as these—questions about life, your life, and the living environment. You will be part of a process that has been going on for hundreds of years. By thinking like scientists—observing, questioning, testing, and explaining—men and women have made great progress in learning about life, about biology. And yet, much in biology remains to be learned. How can diseases be prevented? What is life like deep in the oceans? How did the human species originate? These are just some of the questions still being investigated.

Why study biology? Put simply, having an understanding of living organisms, including yourself, and of the living environment around you, makes you feel good. Also, throughout your life, you will have to make important decisions, for example, about your health and that of people close to you. To make these decisions, you will need the information you learned by studying biology. Finally, your study of biology is an invitation to join, in your own particular way, the amazing scientific journey to try to understand the gift that we all share: the gift of life.

While many facts are described in this text, the important point is that they are all centered around six main ideas, or *themes*, of biology. The six themes are divided into 28 chapters, each of which includes the following: a brief *Introduction* to the main topic; numerous diagrams and illustrations to highlight and enhance important ideas; a mid-chapter *Check Your Understanding* question; a boxed *Feature* on a related topic of interest; a *Laboratory Investigation*; and a *Chapter Review* section that includes multiple-choice, constructed-response, and reading-comprehension questions. Important science terms are boldfaced and defined in-text, and are presented again in the Vocabulary list in each Chapter Review. Some Laboratory Investigations will take more than one day to complete, so they are presented as extended investigations. The book contains a *Glossary* of all boldfaced vocabulary terms, followed by an *Index*.

Throughout your study of biology, you will learn about the six main themes in this book. You will be able to talk about them in your own words. Have a good time! Treasure what you learn. It can last a lifetime.

About the Part D Laboratory Activities

The Regents examination in Living Environment includes multiple-choice and open-ended questions based on several required laboratory activities to be completed during the school year. Part D of the Living Environment examinations in January, June, and August will test at least three of the four laboratory activities that are required for that year. There are now seven different laboratory activities scheduled for implementation and testing from 2004 through 2007. The first four activities (1, 2, 3, and 5) are required for the June 2004 and August 2004 exams.

While completing the laboratory activities, you will record your results and answers to questions in the Student Laboratory Packets given to you by your teacher. You are to keep these sheets for review before taking the Regents examination. You will also transfer your answers to separate Student Answer Packets, which will be used and kept by the school as evidence of your completion of the laboratory requirement for the Living Environment Regents exam. All directions to the teachers and printed materials for the students have been prepared and distributed by the New York State Education Department.

Required Laboratory Activity #1: Relationships and Biodiversity This activity is a simulation that consists of six tests done in the lab as well as a seventh task, a reading assignment. Your goal is to collect and analyze data on several different plant species in order to determine which of the species is most closely related to a valuable but endangered species. The evidence, both structural and molecular, is used to develop a hypothesis about the evolutionary relationships between the plant species. The final task of the activity, the reading passage, focuses on the importance of preserving biodiversity.

This laboratory activity is most closely correlated with topics covered in Themes I, V, and VI. The lab requires that you have an understanding of DNA and protein synthesis.

Required Laboratory Activity #2: Making Connections The purpose of this activity is to help you learn how to design and use a controlled experiment in order to draw a conclusion. In particular, you are to determine which of two conflicting claims is supported by your experimentation. The laboratory activity consists of two parts. In Part A, you will practice two simple techniques—taking a person's pulse and measuring muscle fatigue by squeezing a clothespin. Part B is the main portion of the activity. You will design your own investigation after reviewing guidelines for conducting a controlled experiment. You will use your experiment to determine which claim is supported, namely that a person can squeeze a clothespin more times by exercising first or more times by *not* exercising first. Results are put in writing and some students make oral presentations of their reports to the class for peer review.

This laboratory activity is most closely correlated with topics of human physiology covered in Chapters 8 and 9 within Theme II. However, a minimum of content knowledge is required for the activity. The main focus is to understand the concepts involved in experimental design.

Required Laboratory Activity #3: The Beaks of Finches The purpose of this activity is to use a simulation to study how structural differences affect the survival rate of members within a species. The lab is based on the observations of the many finch species on the Galápagos Islands that Charles Darwin used in support of the process of natural selection. You will work in pairs to represent a finch. Each pair is randomly assigned a grasping tool—such as forceps, tongs, pliers, or tweezers—that represents a type of beak to be used to pick up seeds of different sizes. The efficiency of the tool-beaks at picking up small seeds such as lentils determines whether the "finches" survive and stay in the same "environment" with these seeds or "migrate" in search of food to a different environment with larger seeds such as lima beans. The survivors in the activity now compete with others to continue to explore the efficiency of their "beaks."

This laboratory activity is most closely correlated with topics covered in Chapter 1 within Theme I. You will need to be familiar with the concepts of adaptation, variation, and natural selection.

Required Laboratory Activity #5: Diffusion through a Membrane In this laboratory activity, you will study the process of diffusion by using a model "cell" to test selective permeability of the cell membrane. The "cell" is made of dialysis tubing or a plastic bag that contains a glucose and starch solution, which is immersed in water for a period of time. The water in the beaker is then tested for the presence of glucose and starch. In the second part of the laboratory activity, you will use a microscope to observe the effects of salt water and distilled water on red onion cells. You will see the effects of the diffusion (osmosis) of water out of the cells when they are surrounded by salt water.

This laboratory activity is most closely correlated with sections on cell processes found in Chapter 6 within Theme II, as well as topics in Chapters 10 and 13 within Theme III.

Contents

THEME II
ENERGY, MATTER, AND ORGANIZATION

THEME III
MAINTAINING A DYNAMIC EQUILIBRIUM

THEME V
GENETICS AND MOLECULAR BIOLOGY

THEME VI
INTERACTION AND INTERDEPENDENCE

Evolution

The Process of Evolution

After you have finished reading this chapter, you should be able to:

Discuss Darwin's contributions to our understanding of how life on Earth has changed over time.

Explain how genetic mutations and variations in traits are important for natural selection to occur.

Describe how the theory of evolution explains the great variety and connectedness of all life-forms on Earth.

There is grandeur in this view of life . . . whilst this planet has gone cycling on according to the fixed laws of gravity, from so simple a beginning endless forms most beautiful and wonderful have been, and are being evolved.

Charles Darwin, *On the Origin of Species*

Introduction

Have you ever noticed a building wall or a tree trunk covered with ivy? If you look very closely at the ivy, you will notice groups of fibrous rootlike structures about every 10 centimeters along the stem. You may already know from past experiences that a plant's roots normally grow in the ground. There they anchor the plant in the soil and take in water and nutrients for the plant. Ivy plants have roots that grow in the ground, but they also produce roots along their stems. These roots are called aerial roots. You may wonder what function these roots have. The aerial roots of an ivy plant anchor, or hold, the ivy to a wall or tree trunk. In fact, it is the aerial roots that enable the ivy to climb up from the ground. (See Figure 1-1 on page 4.)

Some questions may come to mind at this point. How did ivy come to produce this special type of root? How did ivy plants develop an ability to climb? Does the ability to climb help an ivy plant survive?

Figure 1-1 The ivy has aerial roots that enable it to climb up walls or tree trunks.

Scientists know that ivy plants were once not able to climb. In the past, ivy lived close to the ground, as do many other kinds of plants. Over many thousands of years, the way that ivy plants grow has gradually changed. In this case, suppose a single plant developed a simple kind of aerial root and was able to climb, even a little bit, above other ivy plants. Since it was higher than the other plants, more sunlight would strike its leaves, allowing this plant to grow better than its relatives creeping along the ground. In time, the climbing ivy would produce more plants like itself, and these plants too would have an advantage over the ground-dwelling ivies. Eventually, these climbing ivies would look very different from their ancestors. Scientists would say that the new climbing ivy was so different from the ground-dwelling ivy that a new **species** (a group of related organisms that can breed and produce fertile offspring) was formed.

Remember that it was not one plant by itself that changed dramatically to produce a new species. It was a group of plants that changed over time, finally producing plants whose characteristics were so different from earlier forms that a new species of ivy was produced. You may wonder if any kind of change could produce a new species. It may, but in most cases a change that is harmful results in an organism that does not survive to reproduce. It is only those changes that increase an organism's ability to survive that may produce a new species. The explanation for the changes that gradually occurred in ivy is found in a theory that was proposed by the English scientist Charles Darwin in the middle of the nineteenth century. It is known as the **theory of evolution**. (See Figure 1-2.)

Figure 1-2 Charles Darwin proposed the theory of evolution by natural selection.

The theory of evolution explains how the immense variety of living things on Earth has developed from ancestral forms during the past 3.5 billion years of Earth's history. (Although Earth is considerably older, the best available evidence suggests that life began to appear on Earth about 3.5 billion years ago.) The theory of evolution is considered to be the most important idea in biology. It offers an explanation, based on a huge amount of fossil and other scientific evidence, about how Earth came to be populated by the millions of species now alive. There were also many species that once lived on Earth that are no longer living. These species are said to be **extinct**.

The theory of evolution answers some other important questions. For example:

- Why are the cells that make up all living things so similar?
- Why are the chemicals found in all cells so similar?
- Why have so many kinds of plants and animals disappeared from Earth?
- Why were many of the animals that disappeared in the past so different from the animals that live today?
- Why are organisms so well suited to their particular environment?

THE ORIGIN OF DARWIN'S THEORY OF EVOLUTION

Long before Charles Darwin's theory was published, some scientists suggested that the characteristics of living things changed over a long period of time. However, it was not until 1859 that the currently accepted explanation for *how* these changes occur was proposed. In that year, Charles Darwin's book, *On the Origin of Species*, was published.

Figure 1-3
H.M.S. *Beagle.*

In his youth, Darwin developed an intense curiosity about the natural world. This curiosity lasted throughout his life. Instead of becoming a physician like his father, Charles Darwin spent his life studying all sorts of living things, from worms to beetles to orchids. Darwin wrote, "One day, on tearing off some old bark, I saw two rare beetles and seized one in each hand; then I saw a third and new kind, which I could not bear to lose, so that I popped the one which I held in my right hand into my mouth. Alas, it ejected some intensely acrid fluid, which burnt my tongue so that I was forced to spit the beetle out, which was lost, as well as the third one."

Darwin's special knowledge and intense curiosity about insects and plants resulted in his being offered a position as "naturalist" on H.M.S. *Beagle,* a ship that was to sail around South America and into the Pacific Ocean. (See Figure 1-3.) Its mission was to map the coasts along its route and to collect plant and animal specimens from the countries it visited. During the long and difficult voyage, Darwin made countless observations and collected countless organisms. It was these observations that helped form his thoughts about the changes that occurred over time in the kinds of life on Earth.

As the *Beagle* sailed up the west coast of South America, Darwin visited the Galápagos Islands. These islands are located in the Pacific Ocean off the coast of Ecuador. (See Figure 1-4.) The location of the islands was important to Darwin. They are far enough from South America to make it difficult, but not impossible, for organisms to get there from the mainland. Also, since the islands are separated from each other, the organisms on each island are mostly isolated.

Darwin was fascinated by some species of finches on the Galápagos

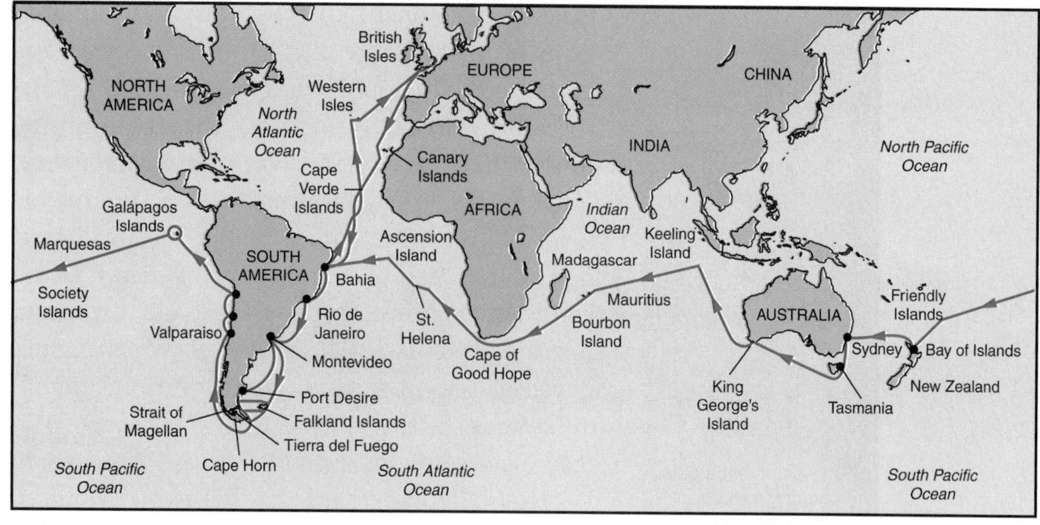

Figure 1-4 The *Beagle's* voyage, 1831–1836.

Islands. These small birds resembled each other in many characteristics, but they were very different from each other in the shape of their beaks and in the ways they lived. Years later, Darwin concluded that the birds had a common finch ancestor that had arrived on the islands many years before. The ancestors probably flew, with the help of strong winds, from the coast of South America. In time, in order to survive, the birds found different food sources and different places to live on the islands. Some ate seeds; others ate insects. Some ate fruit; others pecked the base of other birds' tails and drank blood as it dripped from the wound. Today, if you examine the beaks of the different species of finches, you will observe that the beaks are shaped in ways that deal with the particular food choices of the finches. There are 13 different finch species on the Galápagos Islands. Discovering and studying these different finch species had a profound effect on Darwin. (See Figure 1-5 on page 8.)

The work of other scientists also sparked Darwin's thoughts. Darwin read a geology textbook by Charles Lyell. Geology is the study of planet Earth and the changes that have occurred in it over time. In this textbook, Darwin learned that Earth was much older than previously thought. Lyell described the long geological processes that are involved in producing canyons and forming mountains. These and other geological processes occur over millions of years. Lyell supported the idea that Earth was actually many millions of years old, not just thousands of years old as others had proposed.

The observations and the work of Lyell and other scientists helped

Ground finch eats seeds

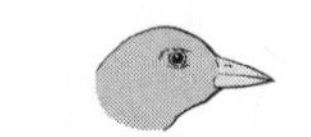

Catches insects with beak

Tool-using finch digs insects from under bark with thorn

Figure 1-5 The beaks of Galápagos finches differ.

Darwin develop his theory. Darwin's explanations of how change in organisms occurred was presented to the public on November 24, 1859—the date his book was published. It should be noted that Darwin was very modest and did much of his research for his own satisfaction. He probably would not have published his book if he had not been encouraged by his friends—and by a timely development that had occurred. Another scientist, Alfred Russel Wallace, had developed his own theory of evolution remarkably similar to the one proposed by Darwin. Wallace had sent his essay to Darwin in 1858. Darwin's friends, afraid that his work would go unrecognized, persuaded him to present his theory, along with Wallace's essay, at a scientific meeting. Darwin himself described the meeting as "exciting very little attention." However, this lack of attention was to change when Darwin's book was published. The entire printing of his book sold out in one day, testifying to the interest that had developed in one year's time.

A STRUGGLE FOR EXISTENCE

The elephant takes a long time to mature sexually and reach the age at which it is able to reproduce. The average female elephant begins to produce offspring at about 30 years of age. During its lifetime, a female elephant may produce only six babies. Using these facts, Darwin calculated that in 750 years, nearly 19 million elephants would descend from one pair of elephant parents. (These calculations are based on the assumption that all of the elephants survived to produce this number of baby elephants!)

Similarly, if a plant produced only two seeds a year (a very low number of seeds for most plants), and if the plants that grew from these seeds also produced two seeds a year, in 20 years there would be a million new plants descended from the original parent plant.

It should be very clear that Earth would run out of room for increases in populations such as those described above. Shortly after his return from the voyage of the *Beagle*, Darwin read an essay by the economist Thomas Malthus about the increase in size of the human population. Malthus showed, by mathematical calculations, that the human population had the potential to reach huge numbers in a relatively short time period. But this, in fact, does not happen—in elephants, plants, or people. The fact that Earth cannot support huge increases in populations, and that populations do not increase in as dramatic a manner as described, led Darwin to conclude that there is a "struggle for existence." In fact, only a relatively small number of offspring of any species survive to produce their own offspring.

Why are large numbers of organisms produced, yet relatively few survive? Lack of food and space are two factors that limit an organism's chances to survive and reproduce. Plants must have space and light to grow. Plant seeds must fall on good ground and have sufficient water to survive. If a large tree in a forest dies and falls, a space is created for a new tree to grow. Of the many tree seedlings that sprout, only one will have enough room to grow into an adult tree. Animals are often dependent on plants for their survival. The seeds, berries, insects, and honey that are an important part of the diet of a bear may be plentiful at one time of the year and scarce during another season. A bear must live in an area that is large enough to provide food for it and its family. If another bear lives nearby, they may compete for food. In times when food is plentiful, both bears may survive. In times when food is scarce, both bears or only one bear *may* survive. Food and space are resources provided by the environment that are used by organisms to survive. To this list can be added other needs such as light, water, minerals, and oxygen.

Competition exists for available resources among individuals of the same species and among different species living in an area. Many other factors add to the struggle for existence. Organisms must cope with **predators**, animals that feed on other animals. A lion is an example of a predator that survives by killing and eating other organisms. They must also cope with **parasites**, organisms that live on or within another organism. A tapeworm is an example of a parasite that lives within another organism and gets nourishment from it. Many diseases are due to microorganisms that are parasites. Extremes of climate, such as heat, cold, floods, and drought, may also affect an organism's chances for survival.

Check Your Understanding

How does the "struggle for existence" among organisms affect the survival of members of a species?

GENETIC VARIATION

Imagine a large family gathering. Sooner or later, grandparents, aunts, uncles, cousins, parents, sisters, and brothers are all arranged for the traditional family photograph. Some are standing, some sitting, and some kneeling on the ground. You can probably tell who married into the family and who is an original family member. You can see evidence in such features as facial structure, hair color, and eye color of the people in the photograph. However, what is most remarkable is that, even though they are all related to each other, they are not the same. There are many differences within this single group of related people. One person may be shorter, another has a slightly different skin tone. One person has straight hair, another tight little curls. Why are there such differences among individuals in a group of relatives?

Through the process of **reproduction**, characteristics, or traits, are passed on from parents to their offspring. The resulting offspring resemble their parents. However, as a family photograph shows, the offspring are not identical. Even identical twins usually show enough differences to be told apart by their parents, even if a person outside the family cannot. Traits that are passed on from one generation to the next are hereditary. The differences among offspring are called **genetic variations**.

One type of reproduction—asexual reproduction—occurs when a single parent organism splits in two to produce two organisms. The other type of reproduction—sexual reproduction—occurs when two parents mate and produce offspring. Sexual reproduction involves the combining of genetic material from each parent. The traits in the offspring are the result of a new assortment of traits inherited from both parents. Sexual reproduction produces greater genetic variation among offspring than does asexual reproduction.

In addition to the combination of genetic information that occurs during sexual reproduction, genetic variation can arise from mutations. A **mutation** is a sudden change that occurs in the genetic material of an organism. These changes occur spontaneously. A mutation may produce a small change in the resulting offspring, or a major effect in the offspring. Many mutations produce no noticeable effects in the offspring at all.

Some types of genetic variation are:

- Structural or anatomical changes—differences in the body of the organism, such as size, muscular strength, thickness of the fur of an animal, height of a plant, or the size of leaves.

- Physiological changes—differences in the functioning of the organism, such as resistance to disease, ability to withstand changes in temperature, ability to digest different types of food, or the amount of sunlight a plant needs to grow.
- Behavioral changes—differences in the behavior of an organism, such as the type of mating display, whether or not an organism migrates, the marking of territory with scent glands, or the time of day an organism feeds.
- Chromosomal and gene changes—differences in the arrangement of the genetic information along the **chromosomes**, the threadlike structures of genetic material in the cells. Each species has a specific number and arrangement of chromosomes. (See Figure 1-6.) Sometimes individual members of a species have an unusual number of chromosomes or have traits arranged differently on their chromosomes.

Figure 1-6 Chromosomes within female and male fruit fly cells.

- Molecular changes—differences in some of the molecules that make up an organism and allow it to function. For example, the enzymes necessary for breaking down food in the digestive system of humans are nearly identical in all people. Yet there may be small but important differences in individual people, such as some individuals who lack the enzyme that enables them to digest milk.

Chromosomal and molecular differences are the most important types of variations that occur because these variations are the causes of all other variations. All genetic differences among offspring of the same parents occur by chance. There is no special purpose for these differences. They just happen. It is through the process of evolution that these differences can take on special meaning.

Darwin recognized that there is variation among individuals of the same type produced by the same parents. The question he wanted to answer was: How do these differences within a group of related organisms lead to differences among groups? He asked this question about the species of finches he observed on the Galápagos Islands. Darwin realized that there were small differences among the individual birds that arrived on the islands from the mainland. How did these differences eventually lead to the significant differences that now exist in the finches? How did the finches develop into such distinct types that they can no longer mate with other similar finches? In other words, how did they develop into separate species?

Darwin came to recognize two clear facts:

- There is a struggle for existence, which limits the number of survivors.
- There are differences among offspring due to individual variations.

These facts must somehow play a role in the development of new species.

Perhaps more important, Darwin posed a question: What determines which individuals survive to reproduce and thus become the parents of the next generation of offspring? His answer to this question forms the basis for his theory of evolution. This answer revolutionized our understanding of how the various forms of life on Earth came to be.

NATURAL SELECTION

Have you ever seen a woodpecker in the forest? Many species of woodpeckers inhabit forests and other woody areas throughout the United States. The woodpecker is named accurately. It spends most of its day pecking at tree trunks in search of food. This bird's hearing is so acute that it can detect insects moving beneath a tree's bark. Woodpeckers have a sharp, strong beak—a bill strong enough to make a hole in tree bark. The bird's long, sticky tongue can probe the hole and trap any insects within reach. Its feet have special characteristics for climbing and holding onto a tree trunk. Two of its toes face forward, while the other two face backward. Each toe has a sharp claw to help the bird hold fast. Stiff tail feathers help support the woodpecker on a tree as it hammers away in search of food. It even has padding at the base of its bill that acts as a shock absorber as it drills into the wood. These special characteristics that make an organism well suited to its environment are called **adaptations.** How do organisms evolve the adaptations that enable them to survive so well in a particular environment? (See Figure 1-7.)

Darwin attempted to answer this question. He developed an answer by combining what he knew about the inheritance of traits with what he observed about an organism's struggle for existence. He concluded that whatever slight variations an organism had that gave it an advantage over other individuals in that environment would make it more likely to survive. An organism that was more likely to survive would pass on its genetic variations to future offspring. Those individuals that did not have characteristics that helped them to survive would be less likely to reproduce and pass on their characteristics. This process, whereby the conditions of life determine which organisms have survival advantages and are more likely to reproduce, Darwin called **natural selection**. He also used the phrase "survival of the fittest" to describe the process.

Figure 1-7
Woodpeckers are well adapted to their environment.

Now we can apply these concepts to explain how woodpeckers evolved into the birds we see today. At one time in the distant past, birds without the adaptations shown by modern woodpeckers moved into a forest area. Individual birds of the same species showed slight variations. They may have had a slightly sharper bill, or a slightly longer tongue, or more acute hearing. These birds were better able to take advantage of the food source of insects that until then lived safely beneath the bark of trees. By making use of a new, previously untapped food source, these birds survived and passed on their characteristics to their offspring. This process was repeated generation after generation, with those birds that by chance had favorable characteristics surviving to pass on those traits. The "fittest" birds, those most suited to exploiting the food source under the tree bark, survived and reproduced. In time, the characteristics of populations of these birds changed. The various species of woodpeckers evolved. Darwin called this process "descent with modification." It is this natural process, proceeding quietly, inevitably, and with astounding results in every corner of Earth, that has produced the millions of species of microorganisms, plants, fungi, and animals that live or have ever lived on this planet.

ANTIBIOTIC RESISTANCE IN BACTERIA—A CASE STUDY

In 1928, British researcher Sir Alexander Fleming made an important and accidental discovery. He found that a common fungus made a chemical that killed bacteria in the laboratory. He named the chemical penicillin. It was later determined that penicillin could kill some of the kinds of bacteria that cause diseases in humans, while leaving human cells unaffected. Penicillin was the first **antibiotic**. Later, many more antibiotics with the

Charles Darwin and His Saltwater Seeds

The main purpose of science is to look at events, occurrences, and patterns in nature and develop explanations for them. These explanations can always be changed as new observations are made and new evidence is found. A possible explanation of a natural event or pattern is called a hypothesis. Charles Darwin, in his own words, showed why he was a true scientist:

> From my early youth I have had the strongest desire to understand or explain whatever I observed—that is, to group all facts under some general laws. I have steadily endeavored to keep my mind free, so as to give up any hypothesis, however much beloved (and I cannot resist forming one on every subject), as soon as facts are shown to be opposed to it. I followed a golden rule that whenever a published fact, a new observation or thought came across me, which was opposed to my general results, to make a memorandum of it without fail and at once. During some part of the day I wrote my Journal, and took much pains in describing carefully and vividly all that I had seen; and this was good practice . . . and this habit of mind was continued during the five years of the voyage. I feel sure that it was this training which has enabled me to do whatever I have done in science.

On the voyage of the *Beagle*, Darwin saw seeds that had washed ashore on a small island near Australia. He wondered whether seeds could travel long distances in the ocean and still be able to grow. Back in England, he enthusiastically filled his home with bottles of seawater. Darwin soaked many different kinds of seeds in the salty water, for short and long periods of time. He used a variety of crop seeds—such as cabbage seeds, lettuce seeds, and celery seeds—23 kinds in all. He then tried to grow them in soil. Darwin's experiments showed exactly what the scientific method is:

- State the problem;
- Collect information;
- Form a hypothesis;
- Perform an experiment:

  Use an experimental group with a variable;

  Use a control group without the variable;
- Record observations and data;
- Draw a conclusion;
- Share your results.

See how the scientific method works for you as you carry out Extended Laboratory Investigation 1 to test Darwin's hypothesis that seeds can be soaked in seawater and still be able to grow when planted later.

ability to kill many different kinds of bacteria were discovered. Antibiotics quickly became widely used to cure diseases that were caused by bacterial infections.

In time, other scientists noticed that some strains of bacteria once killed by antibiotics were no longer affected. They had developed a *resistance* to penicillin and some of the other antibiotics.

Several questions needed to be answered. How did bacteria develop this resistance? Did the bacteria change in some way after being exposed to the antibiotic? Were there genetic variations within the bacterial populations? Did genetic variations mean that some bacteria were resistant to the antibiotic naturally, without having been exposed to it?

If there were bacteria that were resistant to the antibiotic from the beginning, then the addition of the antibiotic to the environment the bacteria lived in gave resistant bacteria an advantage. By killing off nonresistant bacteria, the antibiotic decreased the competition for food that existed in the original population. The resistant cells could survive, grow, and reproduce. This is an example of the process of natural selection at work. The result would be strains of bacteria with resistance to antibiotics.

In the early 1950s, an ingenious set of experiments was conducted by Esther and Joshua Lederberg. They used the common intestinal bacteria *Escherichia coli*. The Lederbergs prepared culture dishes of living bacterial colonies. A bacterial colony is a mass of bacterial cells that contains so many cells that it is visible to the eye. Bacterial cells reproduce by splitting in half. Thus, each colony developed over time from a single bacterium. Using a piece of **sterile** cloth, they were able to transfer a few cells from each colony to a new culture dish that contained an antibiotic. (See Figure 1-8.) Almost all of the transferred cells were unable to grow due to the antibiotic; but occasionally a colony did develop from the transferred cells. These cells were resistant to the antibiotic. The Lederbergs were able to identify the original bacterial colony that produced the resistant cells. Then they performed an experiment to determine whether the resistant bacteria developed their resistance after being transferred or already had the resistance to the antibiotic because

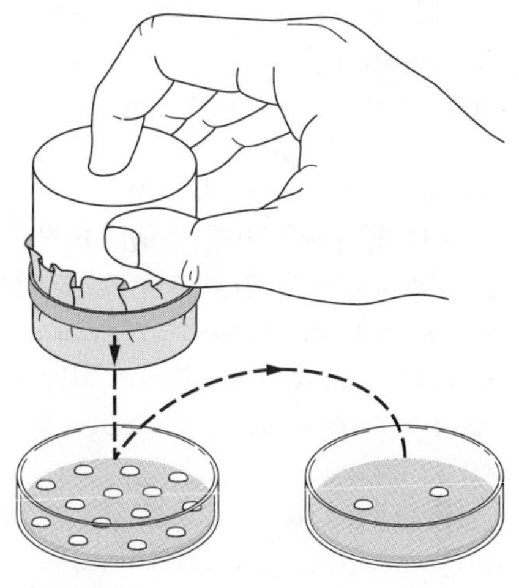

Figure 1-8 The Lederberg experiment.

of their genetic makeup. The Lederbergs tested the bacteria in the original colony and found that the cells in the colony had a natural genetic resistance to the antibiotic. Natural selection of a randomly occurring trait had taken place.

ARTIFICIAL SELECTION

More than 200 breeds, or varieties, of dogs can be identified. Each of these breeds, from the short-legged dachshund to the tall Great Dane, belong to one species, *Canis familiaris*. All dogs, regardless of their appearance, have certain common characteristics such as the same number and types of teeth, the same number and arrangement of toes, and the tendency to be covered by thick hair over their body. (There are one or two breeds of hairless dogs, which, as in other areas of science, tend to work against making generalizations!) Why is there such an enormous variety of dog breeds?

Evidence collected from ancient human habitations shows that dogs have been companions of people for at least 10,000 years. Humans have used dogs for many different purposes: as guards, as hunting companions, for pulling sleds and carts, as pets, and for chasing other animals from holes. For each purpose, a dog with particular characteristics was most desirable.

People who raised dogs to perform these tasks actually selected those pups with characteristics that best suited their intended function. For example, a dog that would be used to chase rats from a hole would ideally have short legs and a fearless temperament. Pups with short legs would be selected and cared for and would likely survive to reproduce. In nature, short legs might be a disadvantage. Dogs with short legs might not survive and reproduce. But the artificial situation resulting from human intervention might ensure that a short-legged dog would survive, if this was a characteristic that its human breeders wanted to preserve. In time, dogs with shorter legs could be mated with each other, producing a strain of puppies with short legs, since leg length is an inheritable trait. In this example, having short legs was a desirable trait and having long legs was an undesirable one. Eventually a breed of short-legged dogs was developed. However, they are not a separate species, since they can still breed with other dogs.

This selection by humans of organisms with specific, desired characteristics is known as **artificial selection**. It is a process similar to natural selection. However, humans—not the natural environment—select the organisms with certain traits. Humans decide which organisms will survive and pass on their inheritable traits to their offspring. The enormous variety of different dog breeds that exists shows the power of artificial selection over a relatively short period of time—10,000 to 20,000 years.

Plants are also artificially selected by humans. The large golden-yellow ears of corn that you can buy at the market were produced on plants that grow more than 2 meters tall. These plants were developed from an ancient corn plant that was short and produced small ears. Once again, through the process of artificial selection, humans have selected certain traits in an organism and developed a new variety that does not closely resemble its ancestors.

Artificial selection has been practiced by farmers and by plant and animal breeders for centuries. Many different plants and animals are the result. Artificial selection is evidence that within a population of organisms, there exists a tremendous potential for change—change that can be enhanced and selected for over many generations. Both artificial and natural selection, acting on genetic differences, produce the changes that develop in populations over time. Desirable traits are selected for, while undesirable traits are selected against. These changes are responsible for the production of new species.

EXTENDED LABORATORY INVESTIGATION 1

Can Seeds Survive Long Periods of Time in Ocean Water?

INTRODUCTION

Darwin made many observations during his five years as naturalist for the ship, the *Beagle*. These observations provided him with years of further scientific study, research, and experimentation after the voyage. One of his questions was how plant species from a mainland (continent) could colonize islands. He wondered whether seeds could survive being carried by ocean currents. In this investigation, you will recreate some of Darwin's experiments to test the hypothesis that seeds can be immersed in seawater and still germinate.

MATERIALS

Fast-growing seeds, including different varieties of the same species (radish, lettuce, cabbage, beans, mustard, carrot); tap water; seawater (or use instant-ocean mix from an aquarium store or use a solution of 35g table salt (NaCl) per liter of water); glass containers; growing pots or trays; sterile potting soil

PROCEDURE

1. Work in teams to design and conduct a controlled experiment that tests the effects of saltwater immersion on seed germination. Some factors that teams should consider are types of seeds; number of days seeds will be immersed; a control group of seeds; planting and growing of seeds; watering schedule; water temperature; exposure to sunlight; and so on.
2. After waiting for the germination time of the particular seed type, analyze the data and draw conclusions. Create a visual display of the experiment(s) and present your results to the class.

INTERPRETIVE QUESTIONS

1. Which plant species survived saltwater immersion the best?
2. Did the length of time of immersion affect the seeds?
3. Did different varieties of the same species all respond the same?
4. Does the class data help explain the colonization of islands by plant species? Why or why not?
5. Is there any evidence that contradicts the hypothesis that seeds are carried by ocean currents to islands?

CHAPTER 1 REVIEW

Answer these questions on a separate sheet of paper.

VOCABULARY

The following list contains all of the boldfaced terms in this chapter. Define each of these terms in your own words.

adaptations, antibiotic, artificial selection, competition, control group, chromosomes, experimental group, extinct, genetic variations, hypothesis, mutation, natural selection, parasites, predators, reproduction, species, sterile, theory of evolution

PART A—MULTIPLE CHOICE

Choose the response that best completes the sentence or answers the question.

1. The finches that Darwin observed in the Galápagos Islands had *a.* lost the ability to fly *b.* extremely similar beak shapes and habits *c.* adapted to eating seeds with thick, tough coats *d.* evolved from a common ancestor.
2. According to modern evolutionary theory, new traits that help a species survive in a particular environment will usually *a.* increase in frequency *b.* decrease in frequency *c.* not change in frequency *d.* increase and decrease unpredictably in frequency.
3. The theory of evolution is most closely associated with the scientist *a.* Thomas Malthus *b.* Charles Darwin *c.* Alexander Fleming *d.* Charles Lyell.
4. A substance that kills bacteria is known as an *a.* antibiotic *b.* antihistamine *c.* abiogenesis *d.* adaptation.
5. A sudden change that occurs in the genetic material of an organism is *a.* an antibiotic *b.* a mutation *c.* an adaptation *d.* a variation.
6. Which of the following organisms is best described as a parasite? *a.* wolf *b.* sparrow *c.* tick *d.* mushroom
7. Experiments conducted by the Lederbergs in the early 1950s demonstrated the *a.* evolution of antibiotic resistance in bacteria *b.* development of beak forms in Galápagos finches *c.* exponential growth of populations *d.* antibacterial properties of substances produced by certain fungi.

8. Competition for food in a given area is most likely to exist between *a.* lions and cheetahs *b.* foxes and hares *c.* mice and fleas *d.* oak trees and squirrels.
9. Species that once lived on Earth but are no longer living are said to be *a.* endangered *b.* archaic *c.* extinct *d.* evolved.
10. Which organism is best described as a predator? *a.* dandelion *b.* cat *c.* cow *d.* ivy
11. Because the gray-and-white Syrian bear and the enormous brown grizzly bear evolved from a common ancestor, the differences between them most likely arose from *a.* mutations and sexual reproduction *b.* mutations and asexual reproduction *c.* the inheritance of acquired characteristics *d.* genetic variation and artificial selection.
12. Which of the following parts occurs earliest in the scientific method? *a.* conclusion *b.* control group *c.* hypothesis *d.* experimental group.
13. The evolution of insect pests that are resistant to chemical insecticides illustrates *a.* genetic variation leading to behavioral changes *b.* natural selection *c.* purposeful change *d.* negative adaptations.
14. Most mutations *a.* have only a minor effect on the offspring *b.* greatly decrease the offspring's ability to survive *c.* greatly increase the offspring's ability to survive *d.* have no noticeable effect on the offspring.
15. Which statement is *not* included as part of the modern theory of evolution? *a.* Sexual reproduction and mutations provide variation among organisms. *b.* Traits are transmitted from parent to offspring by genes and chromosomes. *c.* New organs arise when they are needed. *d.* More offspring are produced than can possibly survive.

PART B—CONSTRUCTED RESPONSE

Use the information in the chapter to respond to these items.

16. Compare and contrast artificial selection and natural selection.
17. How can natural selection, genetic variation, and isolation result in the development of a new species?
18. Describe an example of competition, a predator, and a parasite in an environment with which you are familiar, such as a city park, a nearby pond, or your backyard.

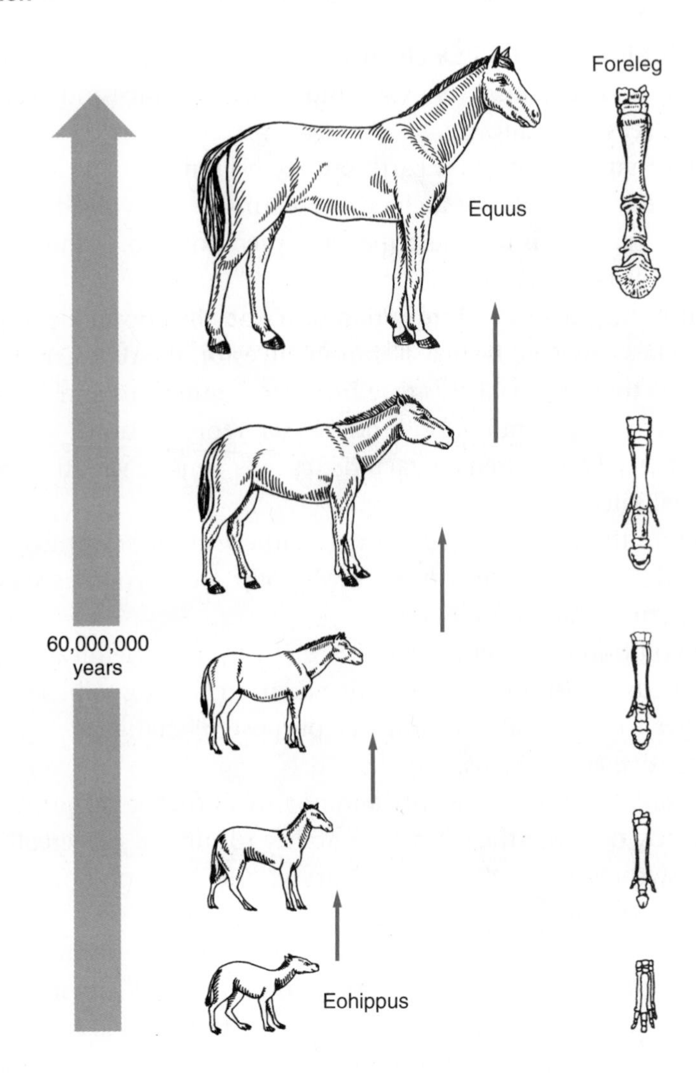

19. What process is illustrated in the above diagram?
20. How do the horses' feet and overall size change over time? How are these changes adaptations to living on open plains and to running?
21. Describe the main steps of the scientific method. Use an example of a research project to explain your answer.

PART C—READING COMPREHENSION

Base your answers to questions 22 through 24 on the information below and on your knowledge of biology. Source: *Science News* (May 10, 2003): vol. 157, p. 302.

Ancestral Split in Africa, [and] China

Homo erectus, the species usually regarded as the precursor of *Homo sapiens*, developed markedly different forms of behavior and social organization in Africa and China, says David E. Hopwood of the State University of New York at Binghamton.

In eastern Africa, *H. erectus* fashioned increasingly complex and diverse stone tools from around 1.8 million to 300,000 years ago, Hopwood contends. Occupation sites grew more numerous throughout that time. Many of them were eventually separated by a distance of only 1 to 2 miles, reflecting the social networking that was needed to organize travels to distant outcroppings to retrieve stones suitable for tools, Hopwood says.

In contrast, *H. erectus* artifacts found at sites in China from the same time span exhibit no substantial changes in toolmaking. Chinese sites are far more distantly spaced on the landscape, indicating little contact between inhabitants of different locations.

Intense competition for food with large predators, in an environment subject to frequent changes, prompted African *H. erectus* groups to invent new tools and forge cooperative bonds, Hopwood theorizes. Relatively stable environmental conditions in China encouraged more consistency in toolmaking among groups that had no need for regular interaction.

22. State two pieces of information about the behavior and social organization of *Homo erectus* in Africa.
23. Explain what the locations of occupation sites in China tell us about *Homo erectus* in that region.
24. Explain how the environments in Africa and China may have resulted in the differences between the *Homo erectus* groups in the two places.

2 Evidence for Evolution

After you have finished reading this chapter, you should be able to:

Explain how homologous structures, fossils, comparative anatomy, embryology, and biochemistry provide evidence for evolution.

Describe the differences between relative dating and absolute dating of fossils.

Compare and contrast microevolution and macroevolution.

Nothing in biology makes sense except in the light of evolution.

Theodosius Dobzhansky

Introduction

In the mid-19th century, Charles Darwin first proposed a theory to explain the ways species change through natural selection. Since that time, a tremendous body of evidence that supports this theory has been developed. Now, in the 21st century, the theory of evolution by natural selection is accepted by biologists in much the same way that other scientists accept the theory of gravitation and the atomic theory as valid explanations for events that occur in the natural world.

No single idea explains the enormous diversity and complexity of life on Earth more powerfully than the theory of evolution proposed by Darwin. Evidence that supports this theory includes fossils (see Figure 2-1), the shapes and structures of living organisms, the chemicals all living things are made of, and the distribution of species on Earth today.

CREATING A FAMILY TREE

Biologists try to understand the relationships that exist among different species. Often, simple diagrams can be used to represent these relation-

Figure 2-1 Fossilized trilobites.

ships. (See Figure 2-2.) Suppose that A, B, C, and D represent four living species. The letters E, F, and G represent ancestral forms of these species that are most likely extinct. In this case, organisms B and C are more closely related because they evolved from their common ancestor, E, most recently. B and C are both equally related to A, with a more distant common ancestor, F. Organism D is the least closely related to the others because it evolved from their common ancestor G the longest time ago.

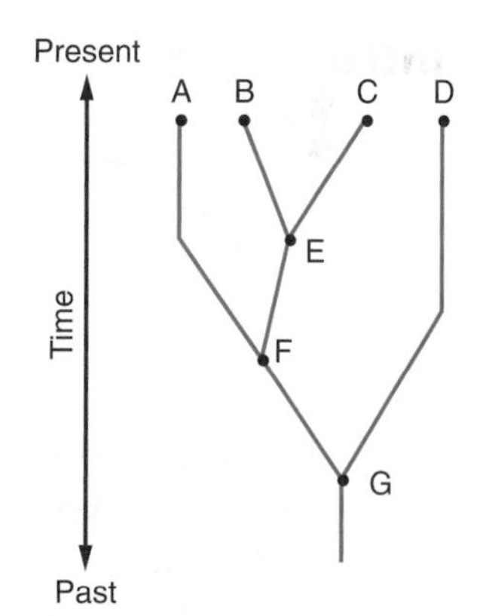

Figure 2-2 A diagram used to represent the relationships among different species.

An example might make this clearer. Suppose a group of biologists is studying the relationships that exist among horses, cats, whales, and humans. The biologists want to develop a family tree for these animals. They need to determine which of these animals evolved from common ancestors and at what time in the past this occurred. This task requires the biologists to study the similarities and differences that exist in these animals. The biologists would have to learn as much as possible about the animals' ancestors, which are usually, but not always, extinct species.

HOMOLOGOUS AND ANALOGOUS STRUCTURES

On way to determine the relationships among organisms is to find a characteristic they share that they inherited from a common ancestor. Such a feature is called a **homologous structure**. For example, similarities exist in the bones of the forelimbs of five very different animals. The wing of a bat, the flipper of a whale, the leg of a cat, the arm of a human, and the wing of a bird—although they appear to be quite different—are all made up of the same types of bones. These bones are attached to each other and to other bones in similar ways. The forelimbs of these five animals are homologous structures and indicate that, long ago, these different animals evolved from a common ancestor. (See Figure 2-3.)

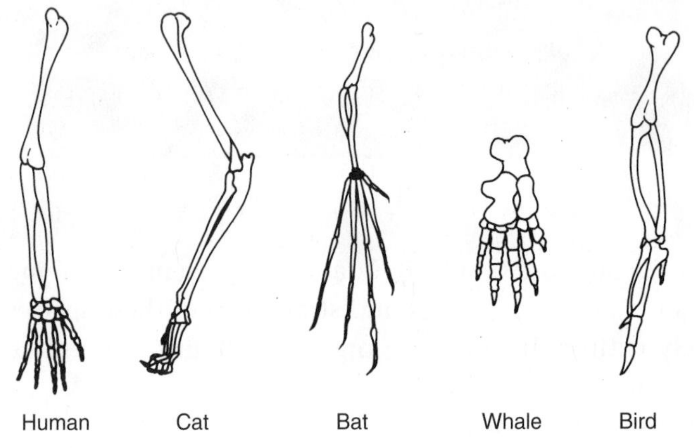

Figure 2-3 The bones in each forelimb are homologous.

Using similarities alone to conclude that organisms evolved from a common ancestor can sometimes lead to mistakes. For example, the penguin, shark, and sea lion are all excellent swimmers. (See Figure 2-4.) These very different animals have fins or flippers and streamlined bodies that enable them to propel themselves through the water. However, other evidence shows that these animals have very different ancestors. Because, over time, their ancestors became adapted to living in water, the living species appear similar in ways, even though they are not closely related. This process is called **convergent evolution**. The features shared by organisms through convergent evolution are called **analogous structures**.

Determining if organisms are related requires looking at other evidence besides homologous structures. For example, the first vertebrates (animals with backbones) to live on land had five digits on their fore-

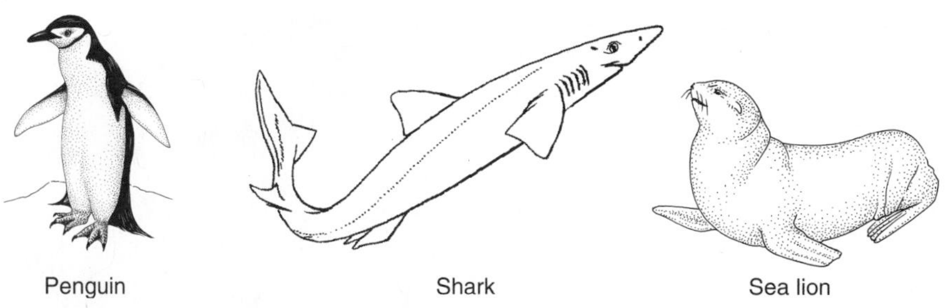

Figure 2-4 The similarity in the streamlined bodies of the penguin, shark, and sea lion is an example of convergent evolution.

limbs. Today, terrestrial vertebrates such as humans, crocodiles, and toads also have five digits that they inherited from a common ancestor. However, the feet of pigs, camels, and deer have only two digits. But we know that humans are more closely related to pigs, camels, and deer than to crocodiles or toads. This means that having five digits, like the more distant ancestors of terrestrial vertebrates, is a **primitive trait**. Over time, vertebrates that had fewer digits evolved, and this modified trait found in pigs, camels, and deer is called an **advanced trait**.

Learning which traits are primitive and which are advanced is important for biologists as they construct family trees that show the evolutionary relationships among organisms.

EVIDENCE FROM FOSSILS

Fossils are traces or remains of dead organisms that have been preserved by natural processes. There are many ways fossils can be formed. Although it is usually only the hard parts of organisms—such as bones, shells, or teeth—that become fossils, occasionally the complete remains of an organism are found. In ancient times, organisms were sometimes trapped in the sticky plant sap that oozed from certain kinds of trees. Later, the sap turned into clear yellow-orange amber, a substance that looks like rock. Amber samples collected from around the world contain mosquitoes, spiders, leaves, and even small vertebrates such as lizards that are many millions of years old. Ice has preserved the remains of woolly mammoths, elephantlike animals that have been extinct for thousands of years. (See Figure 2-5 on page 28.)

A common way fossils are formed is through the gradual replacement of the organism's remains by other substances. In this method of fossil formation, the organism is usually buried in sediments, its tissues slowly

Figure 2-5 Woolly mammoth.

replaced by minerals dissolved in underground water. In time, these minerals harden to form an exact copy of the original organism. In fact, the remains of dinosaurs found by scientists are usually not the dinosaur's actual bones but instead are copies of them made from minerals in this manner. (See Figure 2-6.) Tree trunks, too, have been turned into petrified wood by this process.

Fossils can also be formed if the original plant or animal creates an impression in soft mud or clay. The mud hardens and the remains of the organism disappear, leaving an empty mold. Later, when this mold fills with minerals, it forms a cast. This process shows only the original external shape of the organism and not the internal structure, as do some other methods of fossil formation.

A similar method is the creation of fossil imprints of animal footprints in clay or mud. Today, you can find dinosaur tracks in rock that was once the soft mud in which the animals walked. More than 3.5 million years ago, in eastern Africa, a volcano spread a layer of ash over the land. A

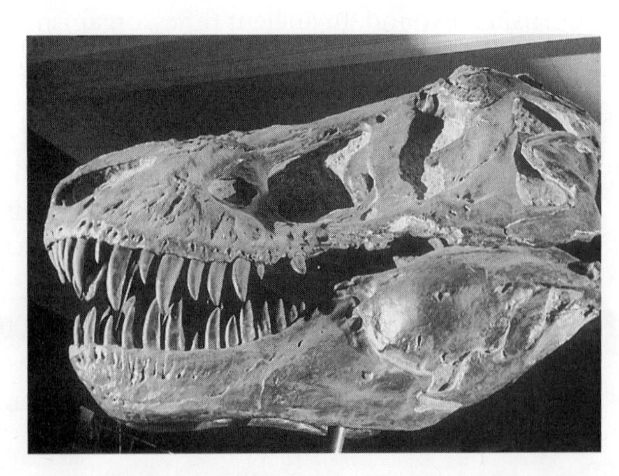

Figure 2-6 Dinosaur skull —a type of fossil.

light rain dampened the ash. At least 20 types of animals, including giraffes, a horse, birds, a wild cat, and at least three ancient humans, walked through the ash. The ash dried, baked by the heat of the sun. In time, the ash hardened into stone. The animals left footprints behind that today provide evidence of their presence long ago. These footprints were discovered by accident. A group of scientists working in the area were relaxing by throwing dried elephant dung at each other. When one scientist slipped and fell, he got a close-up view of the strange marks on the ground. After further investigation, the team of scientists made one of the most important discoveries of our time—the fossil footprints of early human ancestors, two adults and a youngster, who walked side by side, 3,600,000 years ago.

THE AGE OF FOSSILS

Fossils can be used as evidence for evolution only if the age of the fossils can be determined. For many years, scientists could determine only whether one fossil was older or younger than another—that is, their age relative to one another. Relative age is determined by observing the layers of the rock in which the fossils are found. Remains of plants and animals are normally found in **sedimentary rock**. This type of rock forms, often at the bottom of lakes or seas, when layers of sediments such as sand, mud, or clay are deposited. Eventually the lower layers are compressed by the layers settling down above them. The compressed lower layers become cemented into rock. (See Figure 2-7.) In a series of undisturbed layers, the oldest are found at the bottom and the most recent at the top. Therefore, the ages of fossils found in the layers can be related to the ages of the layers. Because this method determines only if one fossil is younger or older than another, it is called **relative dating**. (See Figure 2-8 on page 30.)

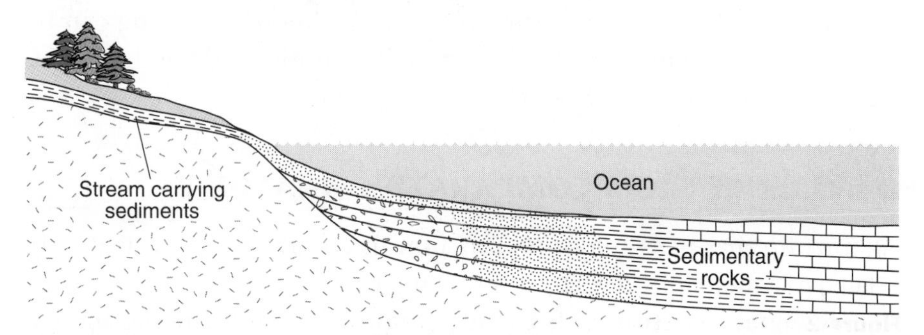

Figure 2-7 Formation of sedimentary rocks.

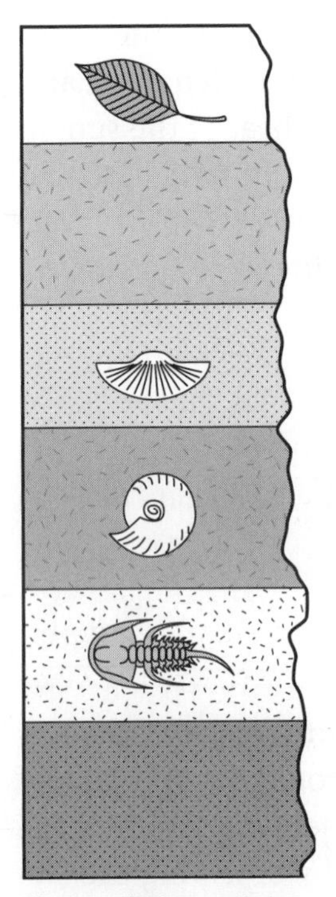

Figure 2-8 Fossils in a series of layers of sedimentary rock.

Today, scientists can determine the age of fossils much more accurately than by the position of rock layers. Very precise methods have been developed that determine when the fossil was formed. This method of dating is called **absolute dating**. Absolute dating techniques usually depend on **radioactive isotopes**, atoms of elements that release energy in the form of radiation at a uniform rate. As energy is released in a process called radioactive decay, the radioactive isotope changes into a different element. The process of radioactive decay continues. Over time, fewer atoms of the original isotope and more of the new element, called the **decay product**, are present. By measuring the amount of each element present in a sample, the length of time since the fossil was formed can be calculated.

Several methods use radioactive elements to measure the age of rocks and the fossils they contain. One method uses the ratio of the radioactive element uranium to its decay product, lead, in a sample. Another method calculates the ratio of radioactive potassium to its decay product, argon. These methods can date fossils and rocks that formed many millions of years ago.

Another method of absolute dating uses isotopes of carbon. This method measures the ratio of radioactive carbon-14 to carbon-12. When an organism is alive, it takes in fixed amounts of both carbon-12 and carbon-14. When it dies, the carbon-14 in the organism begins to change to its decay product, nitrogen, while the carbon-12 remains unchanged. Thus, the ratio of carbon-14 to carbon-12 changes. How long ago an organism died can be determined by finding the ratio of carbon-14 to carbon-12 in a sample from the organism. Radiocarbon dating can be used to date the remains of living things such as logs and bones that are less than 50,000 years old.

EVIDENCE FROM COMPARATIVE ANATOMY

A structure that has little or no function in an organism, but is clearly related to a more fully developed structure in another organism, is known as a **vestigial structure**. It is thought that such a structure did serve a

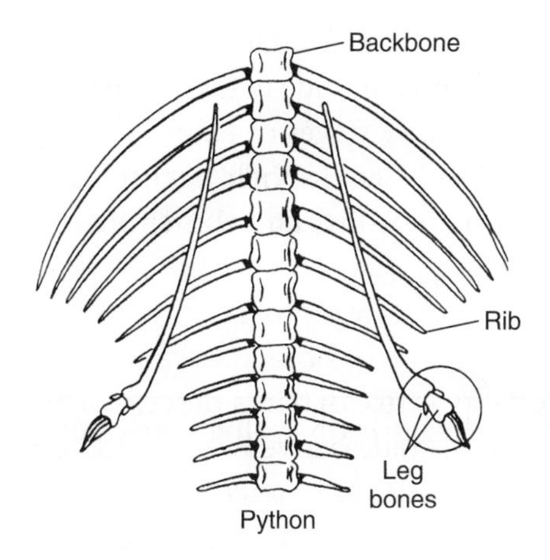

Figure 2-9 The skeletons of some snake species have vestigial leg bones.

function in an ancestral organism. The appendix, a vestigial structure present in humans, is a small sac attached to the place where the small and large intestines meet. In appearance, the appendix is a smaller copy of the cecum, which is a large pouch found in plant-eating mammals such as rabbits. In a rabbit, the cecum contains microorganisms that digest the plant materials the animal eats. The fact that a similar organ in one species is still useful is evidence that humans evolved from an ancestor that once had this larger, functional structure. There are many examples of vestigial structures found in organisms alive today. These structures support the idea that species have evolved over time as a result of natural selection. For example, although snakes have no legs, there are tiny leg bones in the skeletons of some snake species. These vestigial structures show that snakes evolved from animals that had legs. (See Figure 2-9.)

EVIDENCE FROM COMPARATIVE EMBRYOLOGY

Figure 2-10 illustrates five animals at very early stages in their development. Although all of these **embryos** resemble one another, you might be

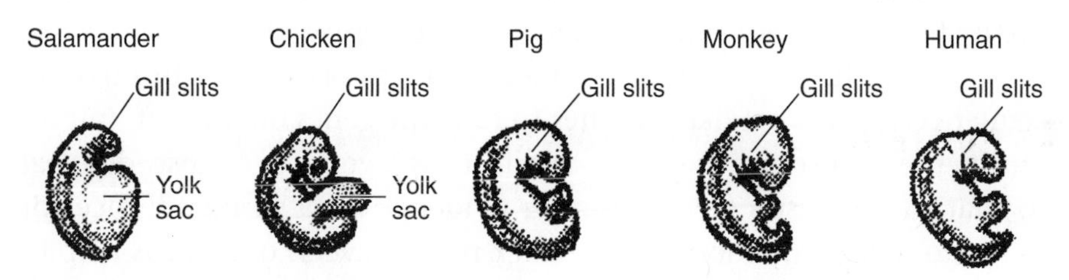

Figure 2-10 The early embryos of these five animals show some similarities.

surprised to learn that these show the embryos of a salamander, a chicken, a pig, a monkey, and a human. These diagrams provide evidence that all vertebrates follow a common plan in their early stages of embryological development. We are now learning that this is due to their having similar sets of **genes**, the molecular instructions for life; and this similarity comes from having common ancestors.

Check Your Understanding

How do organisms' similarities in anatomy and embryology provide evidence for evolution?

EVIDENCE FROM COMPARATIVE BIOCHEMISTRY

The chemistry of living things, called **biochemistry**, provides some of the strongest evidence that organisms evolved from common ancestors long ago. All organisms store the genetic information that is passed from one generation to the next in DNA molecules in a manner that is almost exactly the same. This genetic code shows that all organisms are related in fundamental ways.

Biochemistry also shows how different species are related to each other. Proteins, a type of molecule in all living things, are made up of smaller units called *amino acids*. Cytochrome c is a protein that helps organisms get energy. The 100 amino acids that make up the cytochrome c of humans and chimpanzees are identical. There are six differences in the amino acids that make up the cytochrome c of humans and monkeys. The cytochrome c of humans and kangaroos differs in 14 amino acids. Biologists now recognize that a small number of amino acid differences means that the two species are more closely related in evolutionary terms. On the other hand, a large number of differences means the two species are more distantly related.

More evidence for evolution has been discovered by molecular biologists who have uncovered the specific genes that direct the body development of very diverse animals. Called *homeotic genes*, these genes are like molecular architects. Within them are stretches of DNA, or **homeobox sequences**, which act like switches to turn the genes on and off. Importantly, the homeobox sequence in a fruit fly's genes is almost identical to that in a mouse's genes—further evidence for a common ancestor long ago. These common homeobox genes in diverse organisms help to explain the similarities in early embryonic development, as shown in Figure 2-10.

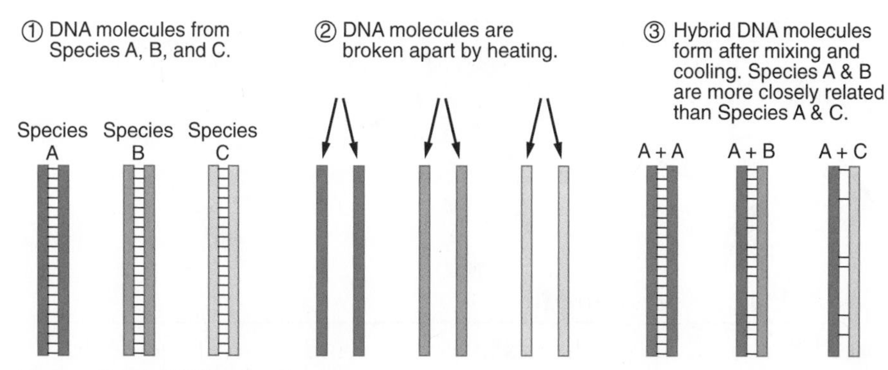

Figure 2-11 DNA hybridization.

DNA molecules store a species' genetic information in long sequences of subunits called nucleotides. For example, the DNA in each human cell that contains all the genetic information for a human has about one billion nucleotides in a specific order. The nucleotides that are present, and the order in which they are found, identify the species. By matching a DNA sequence from one individual with the DNA sequence from another, scientists can determine if the sequences belong to the same, closely related, or distantly related species. Again, the greater the similarity, the closer the species are related.

The most common method used to compare DNA sequences is called **DNA hybridization**. (See Figure 2-11.) DNA molecules from two different individual organisms are mixed together to form "hybrid" DNA. Wherever the molecules from the two individuals are similar, they attach to each other. Heat can be used to separate the hybrid DNA molecules. The greater the similarity, the greater the numbers of attachments and the more difficult it is to once again separate the two molecules. Biologists therefore measure the temperature required to separate the hybrid DNA molecules. The higher the temperature, the greater the number of attachments; hence the more closely related the species are.

Many evolutionary puzzles have now been solved using this technique of DNA hybridization. For example, *ornithologists*, scientists who study birds, disagreed on whether the flamingo was more closely related to storks or to ducks and geese. DNA hybridization studies have shown that flamingos are indeed closely related to storks, both having evolved most recently from a common ancestor. (See Figure 2-12 on page 34.)

A more famous controversy involved the relationship of the giant panda to the red panda. (See Figure 2-13 on page 35.) It has been known for a long time that the smaller red panda is a member of the raccoon family. Was the giant panda, whose anatomy and behavior closely resemble

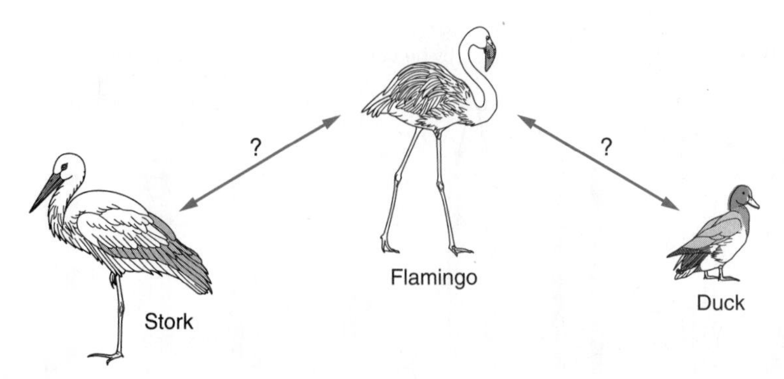

Figure 2-12 DNA hybridization shows that flamingos are closely related to storks.

that of the red panda, also a member of the same family and not a bear, as was commonly thought? From 1869, when the giant panda first became known to the Western world, until recently, this question remained unanswered. DNA hybridization has now shown that the giant panda is indeed a bear. The two pandas have developed by convergent evolution to resemble one another in some ways. This makes sense, because they are both native to the same mountainous environment in China, and both eat similar plants.

EVIDENCE FOR MICROEVOLUTION

Evolution occurs all the time within a species—on a small scale—as inheritable traits in a population change through natural selection from one generation to another. This kind of change is known as **microevolution**. When it occurs over a short period of time, this process does not produce significant enough change to result in a new species. However, by studying examples of microevolution, biologists can learn about the process of evolution that does lead to significant changes, new species, and eventually entirely new groups of organisms. Change on a large scale over a long enough period of time is known as **macroevolution**.

You may have observed moths fluttering around a bright light outdoors on a summer night. One species of moth, the peppered moth, has been carefully studied in England for more than a century. This study provides one of the best-known examples of microevolution. When the peppered moth was first studied, most were a light color. The moths were well camouflaged when they rested on trees and rocks covered with light-colored lichens. (See Figure 2-14 on page 36.) In 1845, a dark-colored peppered moth was observed for the first time, in the northern English city of Manchester. At that time, soot and smoke produced by coal-burning

Figure 2-13 DNA hybridization has shown that the red panda (above) and the giant panda are similar as a result of convergent evolution.

factories had begun to pollute the air. The trees and rocks became dark with soot, and the lichens began to die. The light-colored moths were easily seen against the new, darker backgrounds on which they rested. Increasingly, the light-colored moths became the easy prey of insect-eating birds. By 1900, most of the peppered moths in Manchester were dark-colored. Why did this happen? The darker moths were better camouflaged against the tree trunks and rocks blackened by the soot from the factories. The process of dark-colored moths replacing light-colored moths is known as **industrial melanism**. It has been observed with many species

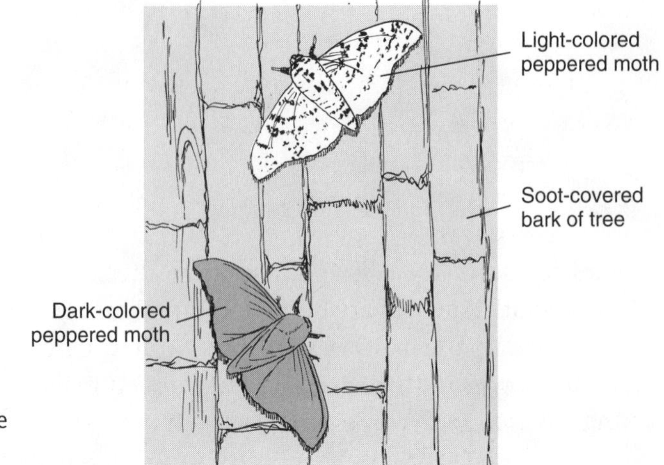

Figure 2-14 The light-colored and dark-colored peppered moth populations vary due to microevolution.

since then, not only in Manchester but also in similar industrial cities in the United States, such as Pittsburgh, Pennsylvania. It is important to understand that the light-colored moths did not change color; they were replaced by dark-colored moths through natural selection. The peppered moths with a dark color were the ones least likely to be preyed upon and so most likely to survive to pass on their genetic traits. It has been discovered that the dark color was a mutation that occurred infrequently in the population, due to the genetic variation that occurs naturally within every species. Natural selection, acting on the peppered moth, in the process of microevolution, changed the moth population from the light-colored to the dark-colored variety. Interestingly, but not surprisingly, as air pollution controls have been implemented over the years, the number of light-colored moths is once again increasing. Another example of microevolution that you can now recognize is the appearance of drug-resistant bacteria, as described in Chapter 1.

EVIDENCE FOR MACROEVOLUTION

The large-scale changes of macroevolution led to the development of new adaptive features, new species, and new groups of species. Examples of new features that resulted from macroevolution are the legs on amphibians (such as a salamander), the egg of a turtle, and the large brain of primates. An entirely new group of species that arose through macroevolution was the flowering plants. Macroevolution has also led to the disappearance of species, for example, the extinction of dinosaurs.

Naturalists living in Europe and England were accustomed to observing relatively small numbers of species. People thought that each species always existed exactly as it currently appeared, living in the same place it was currently living. However, as a result of wide-ranging explorations around the globe, such as Darwin's voyage on the *Beagle,* it came to be realized that many more species of organisms existed than anyone could have possibly imagined. For example, tropical rain forests revealed a diversity of life previously unimagined. In just one hectare (10,000 square meters) of a tropical rain forest, more than 100 species of trees and up to 40,000 species of insects can be found! The rain forests that surround the Amazon River in South America are home to more than 1600 bird species. As Darwin traveled up the west coast of South America, he was impressed by the gradual changes in the kinds of plants and animals that live there. These changes corresponded to the gradually changing environments. This wide diversity of species strongly supported the idea that over a very long period of time, organisms had evolved, adapting to many different environments in many different ways. It was becoming difficult to argue that each of these many different species had remained unchanged from its original form.

Another observation led Darwin to his theory of evolution. Darwin realized that there was a relationship between the types of plants and animals and the regions they live in. The study of the relationship of living things to their distribution in particular areas is called **biogeography**. Biogeography raises many questions that can logically be answered only by the theory of evolution. For example:

- Why are rabbitlike animals distributed throughout the world? (See Figure 2-15.) The European rabbit, the Patagonian hare of South America (related to rats), and the Australian banded hare (related to kangaroos) are similar in appearance and behavior. It is most likely that these animals evolved, from unrelated ancestors, over millions of years in

Figure 2-15 Rabbitlike animals are found throughout the world.

Mitochondrial Eve Meets Mungo Man

In 1987, a group of scientists thought they could use a better method to study human evolution. Instead of studying bits and pieces of fossil remains, they decided to study the genes contained within the cells of living people. These genes, passed from generation to generation, have stored within them a history of our origins. The molecular biologists decided to examine the DNA located in our cells' mitochondria. Unlike ordinary DNA—the genetic material in the nuclei of our cells that we get from both parents—mitochondrial DNA (mtDNA) in our cells comes only from our mother. From one generation to the next, mtDNA never gets mixed with the DNA in the genes we get from our father.

The researchers collected mtDNA from women living in many parts of the world. By studying the similarities and differences in mtDNA in these women, the researchers were able to look back in time to study the origins of human history. Their startling conclusion was that the molecular evidence indicated that all humans alive today are the descendants of a single female who lived in Africa about 200,000 year ago. Some people began to call this person "Mitochondrial Eve."

Since 1987, scientists have disagreed widely on the results. Some scientists claim that the computer program the researchers used for their analysis was not used correctly. Others think that Mitochondrial Eve lived only 150,000 years ago. Still others point to evidence showing that modern humans evolved much earlier—and in several parts of the world, not only in Africa.

In 2001, scientists determined that mtDNA from a human fossil found years earlier in Mungo, Australia, showed no linkage to any humans living today. Therefore, "Mungo Man," as the fossil is called, could not have descended from Mitochondrial Eve. "Put the gloves on, Mitochondrial Eve, because Mungo Man has stepped into the ring," began a recent article on the topic, showing how the debate still continues. This kind of open discussion is what science is all about—questions are asked and answered; and then, when more evidence is found, even more questions arise. For now, the answer to this question about our ancestry remains undecided.

similar environments but in different areas of Earth. They adapted in some of the same ways to fill similar roles in their respective environments. Hence, today these animals superficially resemble each other. (This is, in fact, another example of convergent evolution.)

- Why are the species found in South American rain forests more closely related to those species that live in South American deserts than to species found in Asian or African rain forests? Could it be that all the South American animals evolved from common ancestors, adapting to the rain forest or desert environments in their own particular ways, while the African and Asian species evolved independently, adapting to their own regions?

- Darwin's questions about biogeography were most clearly answered when he investigated the Galápagos Islands off the coast of Ecuador. (See Figure 2-16.) Why did each of these 13 islands have its own unique species of birds and reptiles, very similar to those on the other islands and to those on the mainland? If each island were to be given its own unique species in some special process, why weren't they fundamentally different from each other? It is much more logical to accept that ancestors of these animals arrived from the South American mainland after the islands—which had formed from volcanoes on the ocean's bottom—appeared above the water's surface. On each isolated island, the ancestral birds and reptiles evolved independently, in response to the slightly different environments they lived in.

These questions, and many others, can be explained only by recognizing that organisms have changed over time in response to particular environmental pressures.

One final type of evidence was used by Darwin to support his theory

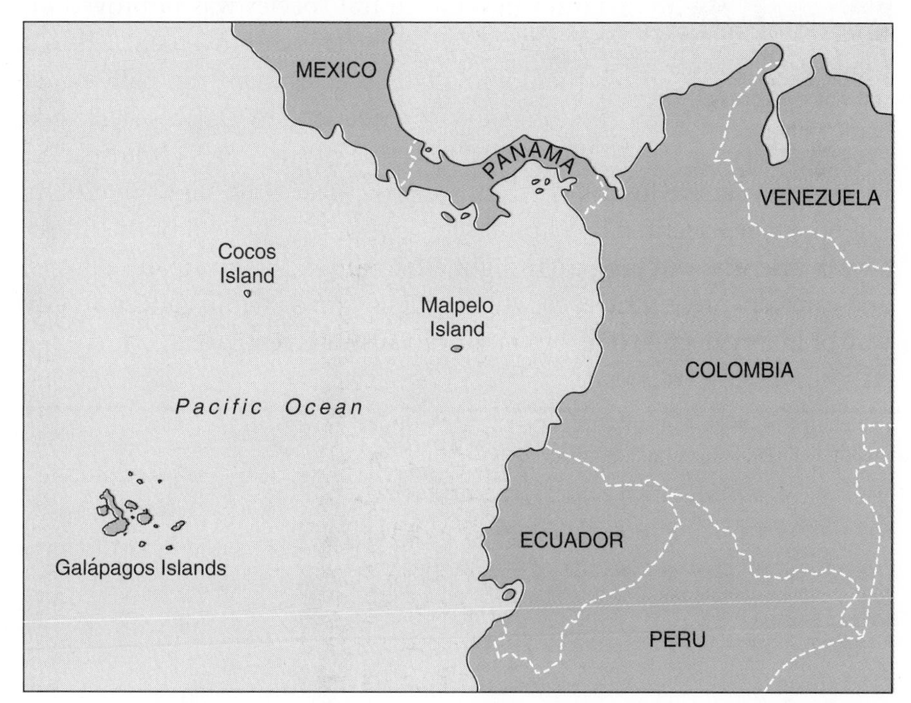

Figure 2-16 On the Galápagos Islands, the ancestral birds and reptiles evolved independently in response to their environments.

of evolution. Darwin argued that a particular adaptation that exists not only in its most advanced form but also in a wide range of intermediate forms suggests that evolution has occurred. An example he used in this argument—the eye—is also one that those who object to the theory of evolution sometimes use. They have argued: How could an organ as remarkable and highly developed as the eye be the product of a long series of small evolutionary changes? How could an eye have evolved from something else?

In *On the Origin of Species,* Darwin showed that there are many animals that have organs sensitive to light in a variety of ways. For example, a sea star has small gel-filled depressions that may concentrate light at a particular spot on its body. (See Figure 2-17.) These depressions cannot form an actual picture of a sea star's surroundings. Yet for the sea star, this type of "eye" is sufficient to aid the animal's survival and gives it an advantage over animals that do not have even this structure. (Likewise, other invertebrates have a variety of "eyes" that have differing abilities to see or perceive different levels of light.) On the other hand, it is possible that this type of adaptation in an ancestral species was improved on by natural selection. This improvement led to a structure in a later species that eventually could "see," as our eye does today. Darwin described this gradual selection of structures ("transparent layers") to perceive light. "We must suppose," Darwin stated, "that there is a power, represented by natural selection or the survival of the fittest, always intently watching each slight alteration in the transparent layers; and carefully preserving each which, under varied circumstances, in any way or in any degree, tends to produce a distinctive image."

Figure 2-17 The sea star has eyespots at the tip of each arm.

LABORATORY INVESTIGATION 2
How Have Animals Evolved Over Time?

INTRODUCTION

Much evidence to support the theory of evolution is found in Earth's fossil record. In this investigation, you will make a model that shows how a series of imaginary fossil organisms may have evolved.

MATERIALS

"Creature Sheet" handout (from the *Teacher's Manual*), scissors, glue or tape, unlined paper

PROCEDURE

1. Cut out each imaginary creature from the Creature Sheet.
2. Select one creature that seems to represent the oldest fossil, the ancestor of all the others. (*Hint:* Its form will be simpler than the others.)
3. Determine which creature appears to have evolved next. Continue arranging the drawings to show a pattern of evolution from one form to the next. There are several possible ways to arrange them. You may arrange your creatures in a straight line or in a set of branched lines similar to the branches of a tree. The pattern can branch into two or more directions. You should be able to support the order you place the organisms in with some logical reasons.
4. Glue or tape the creatures to the unlined paper in the order in which you have arranged them. Draw arrows that show the direction of evolutionary change.

INTERPRETIVE QUESTIONS

1. Did you choose a linear or branching diagram, or a combination of the two types? Explain your choice.
2. What shared characteristics did you use to determine the main sequences or branches of your pattern?
3. Imagine what kind of environment the organisms at the end of each branch or sequence of the pattern would live in. How might they move? What might they eat?

CHAPTER 2 REVIEW

Answer these questions on a separate sheet of paper.

VOCABULARY

The following list contains all of the boldfaced terms in this chapter. Define each of these terms in your own words.

absolute dating, advanced trait, analogous structures, biochemistry, biogeography, convergent evolution, decay product, DNA hybridization, embryos, fossils, genes, homeobox sequence, homologous structure, industrial melanism, macroevolution, microevolution, primitive trait, radioactive isotopes, relative dating, sedimentary rock, vestigial structure

PART A—MULTIPLE CHOICE

Choose the response that best completes the sentence or answers the question.

1. Which organism has the forelimb with the most "primitive" trait? *a.* bird *b.* human *c.* horse *d.* whale
2. Examples of homologous structures are *a.* a dog's tail and a fish's tail *b.* a bee's eye and a fish's eye *c.* a porpoise's flipper and a hawk's wing *d.* a crab's claw and a penguin's wing.
3. Evidence for evolution is shown by *a.* homologous structures *b.* fossils *c.* biochemistry *d.* all of these.
4. Scientists who study birds are called *a.* ornithologists *b.* ichthyologists *c.* aviaries *d.* biogeographers.
5. The streamlined bodies and powerful tails of sharks, dolphins, and ichthyosaurs (ocean-dwelling relatives of the dinosaurs) illustrate *a.* divergent evolution *b.* convergent evolution *c.* retention of ancestral traits *d.* development of vestigial organs.
6. Which of the following is *not* compared in order to clarify the evolutionary relationships among related organisms? *a.* analogous structures *b.* embryo development *c.* genes *d.* DNA hybridization
7. Fossils are usually found in *a.* radioactive isotopes *b.* igneous rock *c.* trap rock *d.* sedimentary rock.
8. DNA from a giant panda hybridizes more readily with DNA from a brown bear than with DNA from a red panda. This indicates that a giant panda

a. shares numerous analogous structures with a brown bear
b. is more closely related to a brown bear than to a red panda
c. is more closely related to a red panda than to a brown bear
d. lives in a similar environment to a brown bear.

9. Which of the following statements is false? *a.* In late nineteenth-century Manchester, natural selection resulted in the peppered moths turning from light gray to dark gray. *b.* Light-colored moths had dark-colored offspring because the moths needed to blend in with sooty tree bark. *c.* The technical name for the color change from light to dark as a result of pollution is industrial melanism. *d.* The peppered moths' change in color is an example of microevolution.
10. The older a specimen of sedimentary rock, *a.* the greater the amount of radioactive isotopes in it *b.* the closer to the surface it is found in undisturbed rock layers *c.* the greater the amount of decay products present *d.* the more likely it is to contain mold-and-cast fossils.
11. Examples of vestigial structures include *a.* a seal's flipper and a penguin's wing *b.* a duckbill platypus's bill and a duck's bill *c.* a bat's foot and a human's foot *d.* a human's appendix and a whale's hind limbs.
12. In three layers of undisturbed sedimentary rock, *a.* the middle layer contains the oldest fossils *b.* the top layer contains the youngest fossils *c.* the fossils in the bottom layer are younger than the fossils in the middle layer *d.* the fossils in the bottom layer have usually been destroyed by geologic activity.
13. Homeobox gene sequences provide evidence in support of evolution because *a.* they evolved only recently *b.* mammals have these genes but no other animals do *c.* similar homeobox gene sequences are found in very different organisms *d.* fruit flies have a very ancient form of these genes.
14. The study of the relationship of plants and animals to their regional distribution is called *a.* ornithology *b.* paleontology *c.* biogeography *d.* melanism.
15. Sea urchins and jellyfish are both round, have no head or tail, and live in the ocean; but their embryos develop in radically different ways. They probably *a.* are descended from the same ancestor *b.* should be placed in the same group *c.* have numerous vestigial structures *d.* are not very closely related.

PART B—CONSTRUCTED RESPONSE

Use the information in the chapter to respond to these items.

16. Scientists analyzed the protein cytochrome c in different organisms in order to construct this evolutionary diagram. Explain how such biochemical data provide evidence for evolution.

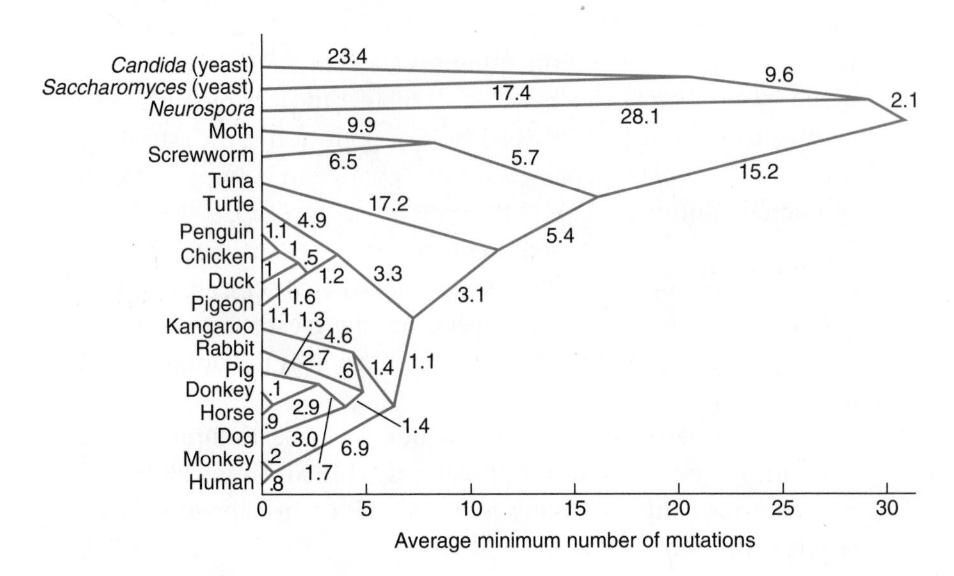

17. Use the diagram to answer the following questions.
    a. Is a turtle most closely related to a dog, tuna, penguin, horse, or silkworm moth?
    b. What organism is most closely related to a horse?
    c. Which group of organisms is least closely related to the other groups—fungi, insects, fish, reptiles, birds, or mammals?
    d. What can you conclude about the turtle's placement on this "family tree"?
18. Describe three ways in which fossils are formed.
19. Describe and give specific examples of evidence for evolution from fossils, comparative embryology, comparative biochemistry, comparative anatomy, microevolution, and biogeography.
20. Compare and contrast relative dating and absolute dating.

PART C—READING COMPREHENSION

Base your answers to questions 21 through 23 on the information below and on your knowledge of biology. Source: *Science News* (February 22, 2003): vol. 163, p. 126.

Farming Sprouted in Ancient Ecuador

People living in the lowlands of what's now southwestern Ecuador began to grow squash between 10,000 and 9,000 years ago, about the same time that residents of Mexico's southern highlands domesticated the vegetable (*SN:* 5/24/97, p. 322), according to a study in the Feb. 14 *Science.*

The comparably ancient roots of plant cultivation in these two regions indicate that "in South America, there was no single center of agricultural origins," conclude Dolores R. Piperno of the Smithsonian Tropical Research Institute in Balboa, Panama, and Karen E. Stothert of the University of Texas at San Antonio.

In the soil of two prehistoric sites in Ecuador, the scientists isolated and studied microscopic crystals from squash rinds that had been uncovered there. These ancient crystals were the same size as those in the squash's modern domesticated form, but not those in its present-day wild counterpart. Piperno and Stothert were able to date tiny bits of carbon that were trapped inside the ancient crystals as they formed.

21. State what has been learned about farming that went on in Ecuador and Mexico approximately 10,000 years ago.
22. Explain what conclusion can be made about agriculture in South America.
23. Explain how scientists have learned about plants that were grown by people in Ecuador long ago.

3 The Origin and Extinction of Species

After you have finished reading this chapter, you should be able to:

Discuss how an organism's morphological, behavioral, and physiological adaptations are suited to its environment.

Define what a species is; describe ways speciation occurs; and distinguish between gradualism and punctuated equilibrium.

Explain how species can become extinct as a result of natural causes or human activities.

Happy is he who could learn the causes of things.

Virgil, *Georgics*

Introduction

Today we identify about 15 bear species. (See Figure 3-1.) They vary in appearance and size (when standing) from the 1.2-meter-tall sun bear to

Figure 3-1 The polar bear is one of 15 bear species.

the 3.3-meter-tall brown bear and polar bear. The study of evolution has shown that, about 60 million years ago, the early ancestors of bears were fierce catlike carnivores. About 37 million years ago, one particular group evolved from these animals. This group contained the earliest known ancestors of dogs—animals the size of foxes with long tails and sharp teeth able to slice through meat. Finally, about 20 million years ago, the ancestors of modern bears evolved. Today, some bear species live in frigid Arctic regions, while other species live in the hot tropics. How did each of these bear species come to exist? The formation of new species is called **speciation**, the topic of this chapter.

ADAPTATION TO THE ENVIRONMENT

Every species lives in a particular place. Every place on Earth has specific conditions, including air temperature, average monthly rainfall, kinds of minerals in the soil, and the direction and strength of wind that blows. Darwin's theory states that those organisms that are best suited to tolerate the specific conditions of a particular environment will be most likely to survive and pass on their inherited traits to their offspring.

An adaptation has either a positive, negative, or neutral value, as determined by the specific conditions of the environment. In the pre-industrial environment of England, the coloration of the rare dark peppered moth had a negative adaptive value, because it made this moth more visible on the light lichen-covered tree trunks and rocks. The same dark coloration had a positive adaptive value when pollution killed the lichens and turned trees and rocks dark. If the main predator of moths hunted by smell and not by sight, the coloration of the moths would have had neutral adaptive value—providing neither a survival advantage nor disadvantage to the organism.

The most common types of adaptations are those that involve the **morphology**, that is, the shape and structure, of organisms. Observe leaves of as many different plants as possible as you travel home from school. You might see the leaves of a maple tree, an evergreen, an oak, or of the grasses and weeds in a lawn. You might even see a cactus plant on a windowsill in an apartment. Leaves are morphological adaptations of plants that serve two purposes:

- Leaves provide a surface area that collects light energy from the sun to make food.
- Leaves have openings that allow the gases carbon dioxide (CO_2) and oxygen (O_2) to move in and out of the plant while controlling the amount of water that is lost.

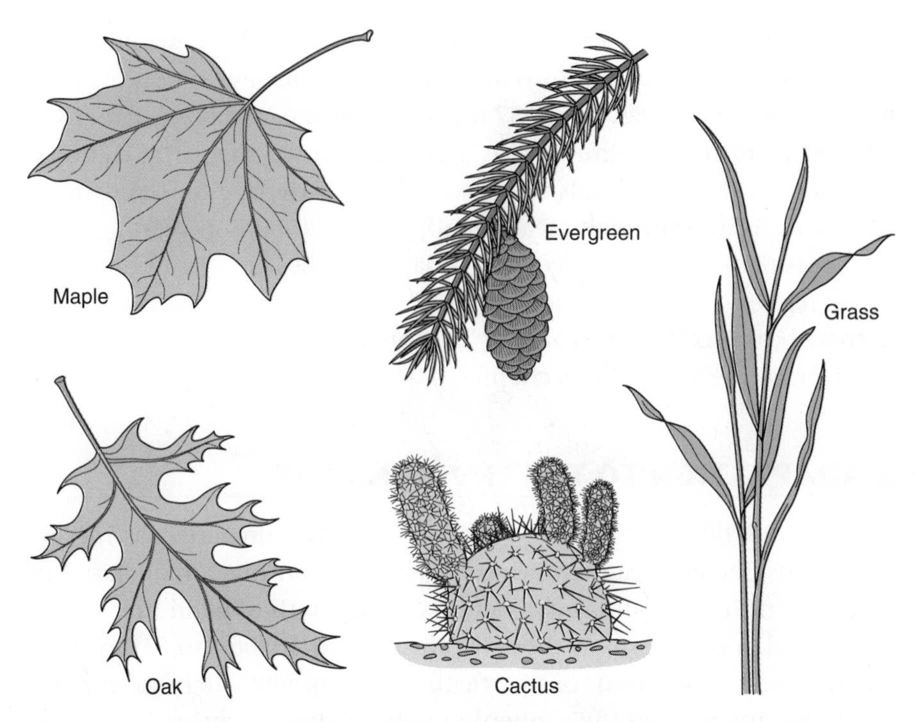

Figure 3-2 Leaves are morphological adaptations of plants.

Maple trees have leaves with a broad surface area. Leaves of this type are adaptations to forests where leaves often block light that strikes other leaves, but where water is abundant. A broad leaf, even though it allows water to escape from its surface, collects abundant light. Very narrow grass blades and the needles of evergreens or cactuses have much less surface area and thus are adapted to drier, sunny climates. In some areas, it is more important to limit water loss than to collect a lot of sunlight. (See Figure 3-2.)

A microscopic view of the cross section of a typical leaf shows more details of how it is a magnificent morphological adaptation. Every living thing is made up of individual cells. If you cut a leaf down the middle and look at it from the edge, that is, as a cross section, you will observe several different types of cells. (See Figure 3-3.) The top of the leaf is covered with a waxy cuticle layer that prevents the evaporation of water. Just beneath the cuticle, the cells are clear, allowing sunlight to shine through to the tall palisade layers where most of the leaf's green pigment, **chlorophyll**, is found. It is the chlorophyll that traps the energy from light. The underside of the leaf has openings called **stomates**. The stomates allow gases to move in and out of the leaf. Under a microscope, you can observe that each stomate is surrounded by two guard cells that can open or close

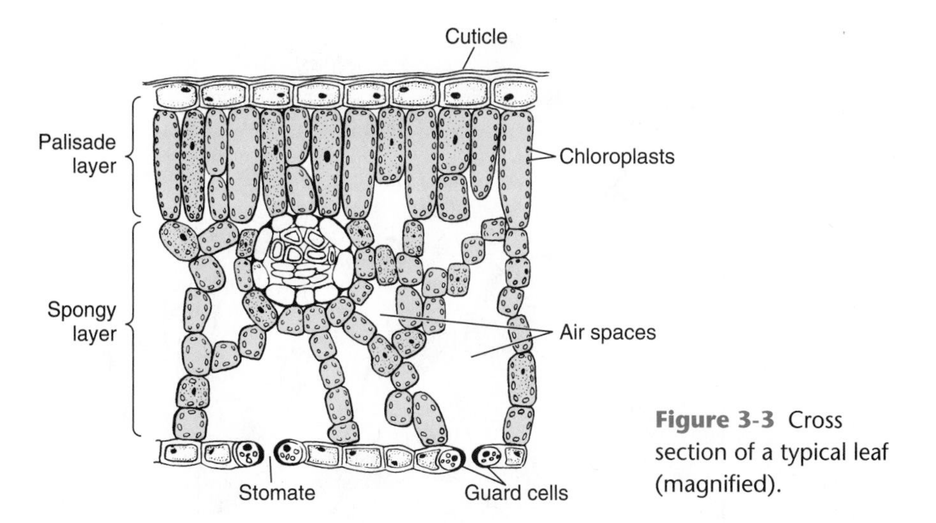

Figure 3-3 Cross section of a typical leaf (magnified).

the stomate to increase or slow water loss and to permit the exchange of gases. Open air spaces in the spongy layer inside the leaf allow carbon dioxide and oxygen to move into and out of the plant cells. This gas exchange must take place *inside* the leaf because of the waxy cuticle that covers the outside of the leaf. So we can see that the leaf is adapted in form and structure to the conditions of the environment, specifically the amount of available light and water. The leaves of each plant species show unique adaptations, such as their shape, size, and placement on the stem. These adaptations have developed over time as a result of the specific environmental conditions the plant species has met.

The adaptation of an organism to its environment may also involve its **physiology**, that is, the way the organism and its internal parts function. For example, bear hibernation is a physiological adaptation. To survive the long, cold, dry winter—a time when food is scarce—the bodily functions of a bear slow down. The bear enters a form of sleep, **hibernation**, which can last for many months. Some bears even give birth during this time and then continue to hibernate while the tiny bear cubs nurse milk from the mother and begin to grow. In our own species, the woman's release of an egg cell once a month, to create the possibility of pregnancy, is also a physiological adaptation.

Behavioral adaptations involve what an organism does. For example, fast-moving rivers are excellent places for fish to live. These rivers teem with a variety of foods that include aquatic plants, insects, worms, and even young fish. However, fish best suited to survive in such rivers must swim all the time. To maintain their position in the river, they use their tails to push back and forth against the water and "swim" at exactly the

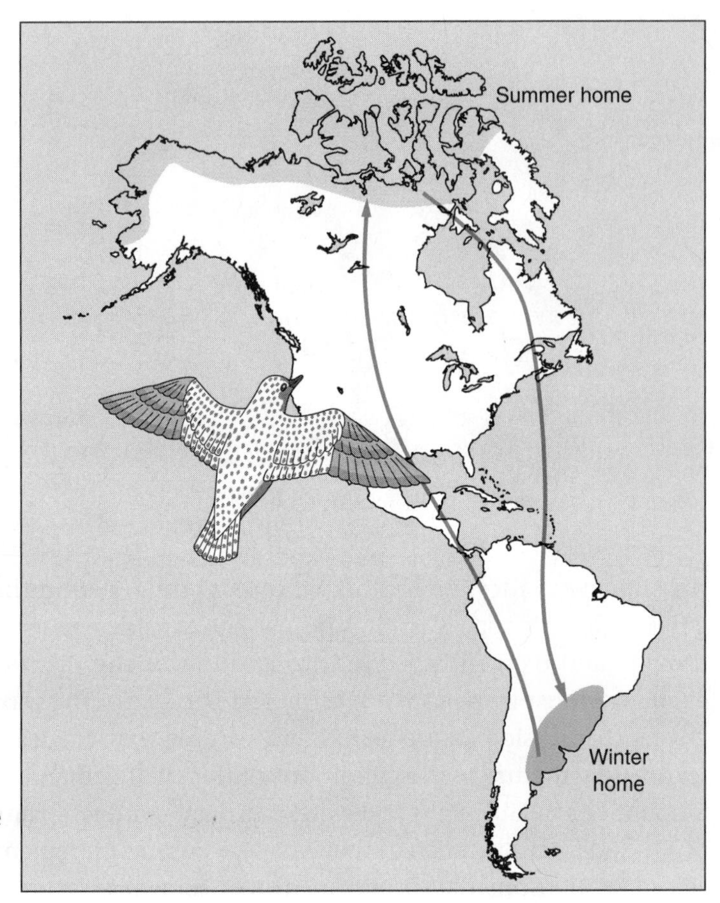

Figure 3-4
Migration route of the Atlantic golden plover.

same speed as the flow of water. If they stopped moving, they would be swept away by the current. The migration of birds, sometimes thousands of kilometers from warm southern regions where they spend the winter to northern destinations where they feed on abundant summer insect and plant life, is another behavioral adaptation. (See Figure 3-4.)

Because environments on Earth are always changing, evolution is an ongoing process. The climate of any particular place is often constantly, although slowly, changing. New or different adaptations in plants and animals occur by chance as a result of the wide genetic variation that exists within any population. Some of these adaptations come to have a positive adaptive value in the changing environment and are "chosen" by natural selection. They are then passed on to future generations as the organisms that have them reproduce. Some adaptations have a negative adaptive value and are eliminated.

SPECIATION

The word *species* comes from the Latin word that means "type" or "kind." After eight years of study in the rain forests of New Guinea, scientists announced in 2002 that they estimate there to be between 4 and 6 million species, or kinds, of living organisms on our planet. Some scientists even estimate there to be 10 million or more species on Earth. All species alive today, and all those that once lived but have become extinct, originated from an ancestral species that lived 3.5 billion years ago. This process of speciation is one of the most important ideas in evolutionary theory.

People have defined species in different ways. Today, the most generally accepted definition for a sexually reproducing species is: a population of organisms with shared characteristics that can breed with other organisms in its group to produce fertile offspring but cannot breed with organisms in other groups. For example, if two groups of similar small, reddish-brown birds lived in the same place but did not interbreed, they would represent two different species. Separation of one group from another group is called isolation, and being unable to breed with that other group is known as **reproductive isolation**. The way in which two groups of organisms become reproductively isolated is part of the process of speciation.

How does a population of organisms change? This does not depend on a particular individual in the population changing. Instead it occurs as the genetic traits common to the group as a whole change. Let's look at an example. Some genetic traits in a population of maple trees in a forest are leaf size, bark color, and tree height. Some maple trees are taller than others. The tall trees have the advantage of getting more sunlight and, so, would be more likely to survive and make seeds. Over time, more tall trees than short trees may grow in the forest. This would be a change in the genetic trait for tree height in the population. Natural selection is the most important cause of this change in genetic traits. Environmental conditions select for particular variations in traits and determine that those traits will be the variations passed on to future generations. Thus, traits present in a population, like our maple trees, can change.

However, the genetic traits of populations can change for other reasons that are not related to environmental conditions. These random changes are called **genetic drift**, and they usually occur in small populations. Suppose a small city park contained 100 pigeons. Some were gray, some dark colored, and 15 were reddish-brown. One day, 13 of the reddish-brown birds were trapped and taken to live in a coop. The next spring, the two remaining reddish-brown pigeons died of old age, without leaving any offspring. The genetic traits of the small population had suddenly changed, or "drifted," because birds with the reddish-brown trait were no longer present.

On the other hand, suppose the first pigeons to arrive in a new park were all, by chance, very dark colored. In the future, the entire population of pigeons descended from those few birds would have the same dark coloration. Scientists call this the **founder effect**. In such a case, a few individuals (which may not represent the original population of organisms as a whole) can give rise to a group with a new or particular set of characteristics.

Bottlenecks, another type of genetic drift, occur when a disastrous event almost wipes out a population. The few survivors, who may not be representative of the original population, begin to repopulate the area. The new population now has a different collection of genetic traits, similar to the traits of the few survivors but different from those of the original population. (See Figure 3-5.)

As long as individuals are able to move freely from one population to another, differences between the populations tend to disappear. In our example, as long as pigeons can fly from one nesting site to another, genetic traits can be shared. A population that is not isolated from other

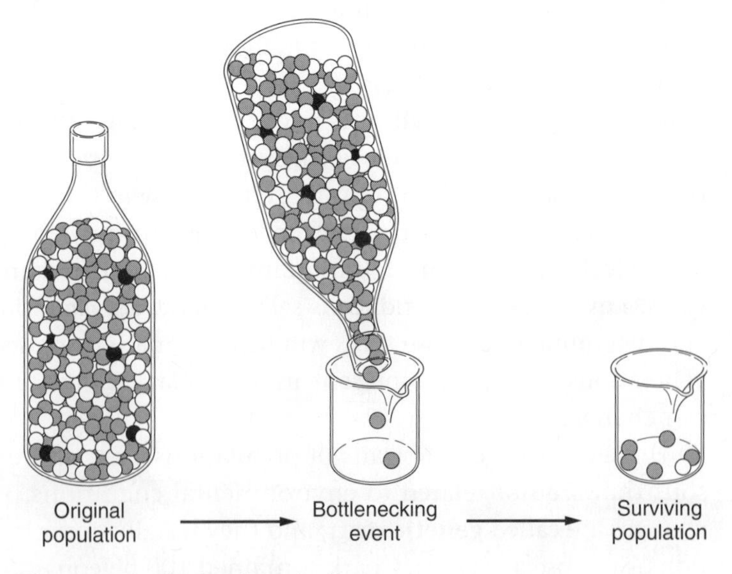

Figure 3-5 The bottleneck effect.

populations will not become increasingly different. The movement of genetic traits into or out of a population as individual organisms come and go is called **gene flow**.

In order for a population to change—through genetic drift and natural selection—and eventually develop into a new species, there must be no gene flow between different populations. If the separation of different populations continues, the populations continue to change genetically until they are no longer able to reproduce together. They are then reproductively isolated, even if they once again live side by side. They are considered to be two separate species.

GEOGRAPHIC ISOLATION

The most common type of isolation that leads to the formation of new species is **geographic isolation**. An actual physical barrier, such as a wide river or a mountain range, prevents organisms from moving from one population to another. Gene flow between the isolated populations ceases; the populations begin to evolve separately from each other. Consider the small group of Galápagos Islands visited by Darwin. Much evidence, including new DNA analyses, supports the idea that ancestral finches arrived from mainland South America, some 950 kilometers away, blown perhaps by a strong storm, sometime after the Galápagos Islands formed from undersea volcanoes several million years ago. (See Figure 3-6.) The arriving ancestral birds colonized all 13 islands. However, each island has different environmental conditions. Due to natural selection, the birds changed in very different ways on the islands. Since the birds were poor long-distance fliers, they tended to stay on their particular

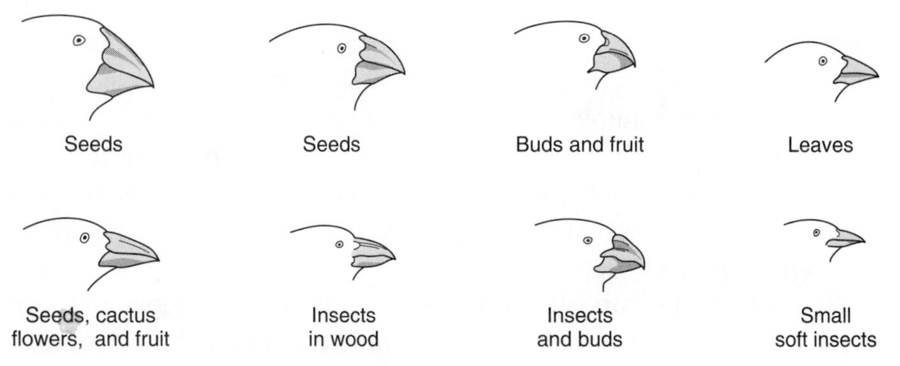

Figure 3-6 Geographic isolation of the ancestral finches led to the formation of several species of Darwin's finches, some of which are shown here.

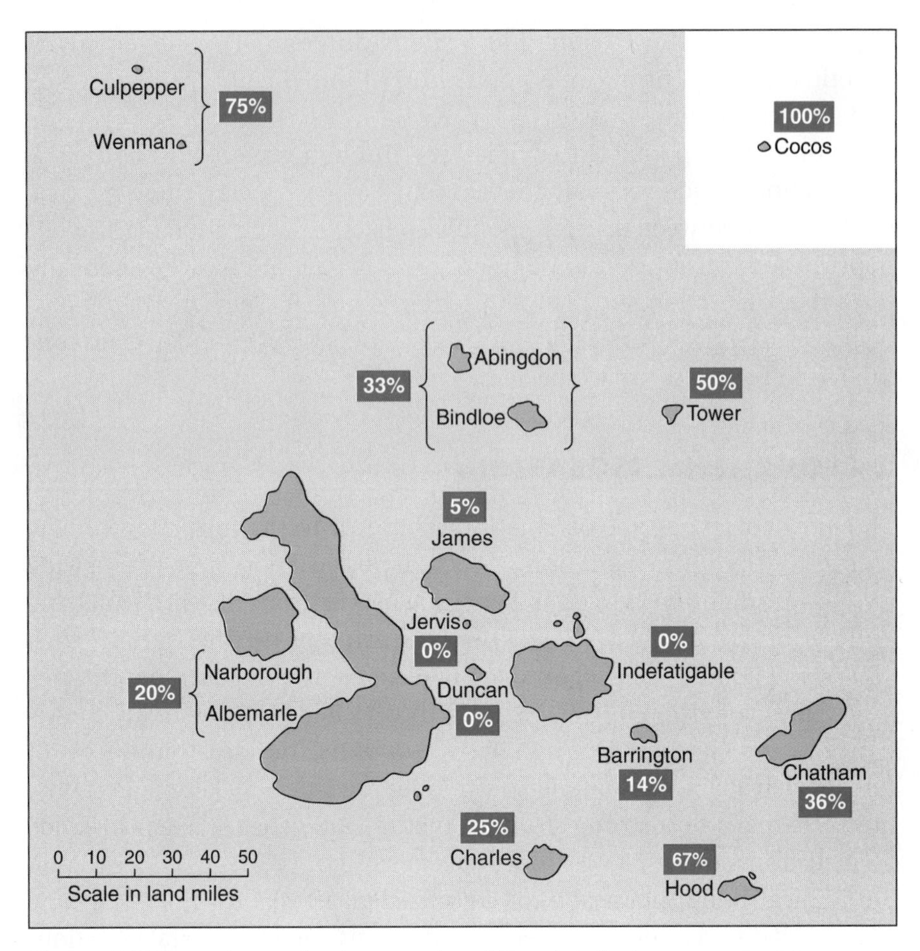

Figure 3-7 Percentages of unique species of Darwin's finches on each of the Galápagos Islands. Those islands that are the farthest away and therefore most isolated have the most unique species.

island, geographically isolated from populations of finches on the other islands. (See Figure 3-7.) With different characteristics being selected on each island—for example, larger beaks on one, the tendency to eat on the ground and not in trees on another—the populations evolved in different directions. Without gene flow between them, the bird populations evolved, each one isolated from the other populations, eventually forming 13 new species.

In addition to ocean islands, there are many types of geographically isolated areas in the world. For example, mountaintops are isolated from each other, lakes are separated by the land that lies between them, even forests are isolated by large open spaces between them. Often, the isola-

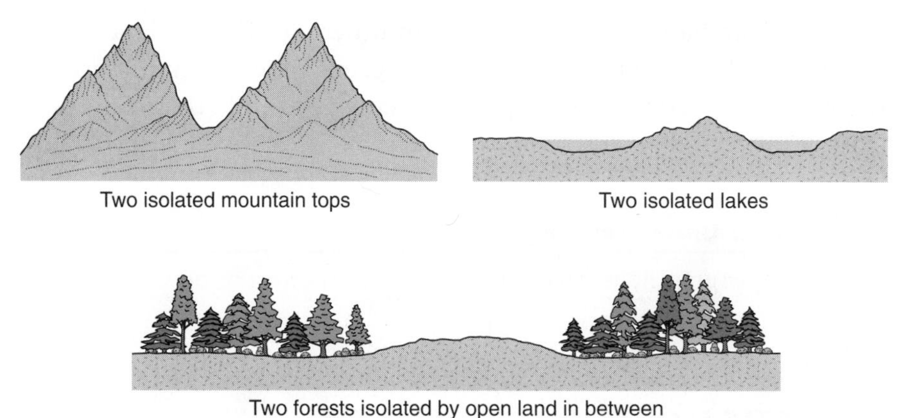

Figure 3-8 Examples of physical barriers that cause geographic isolation.

tion may not seem like an overwhelming obstacle. But any type of geographic barrier may prove an obstacle to a particular species. It is in these geographically isolated areas that speciation is most likely to occur. (See Figure 3-8.)

Scientists have accepted the theory that the world's continents sit on sections, or plates, of Earth's crust. These plates slowly move around, or drift, as massive forces within Earth slide the plates toward or away from each other. As a result, the continents are continually moving in relation to each other. A map of the world 250 million years ago would have shown all the continents locked together, forming one huge "supercontinent." (See Figure 3-9.) Since that time, Earth's plates have moved many times, rearranging the continents. Some of these movements have resulted in the isolation of certain populations. For example, the continent of Australia shows the effects of geographic isolation on the evolution of new

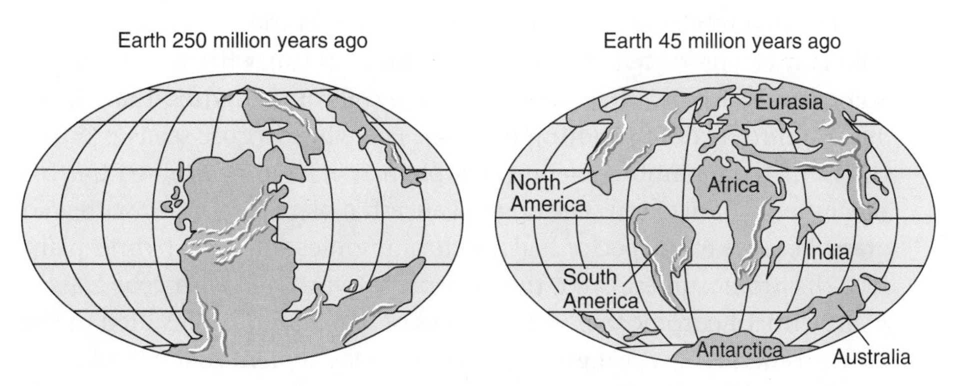

Figure 3-9 Earth's plates have moved many times, rearranging the continents.

species. Australia became isolated from other continental landmasses about 55 million years ago. As a result of geographic isolation, Australia is now home to many unique species of kangaroos and other marsupials that are not found on any other continent.

Check Your Understanding

How can geographic isolation lead to the formation of new species?

GENETIC ISOLATION IN ONE LOCATION

The genetics of plants are as complex as, or even more complex than, those of animals. Two important forms of speciation are common in plants but not in animals. One method of plant speciation involves chromosomes, the bundles of genetic information located in each cell of an organism. Chromosomes are made of DNA. Every species has a precise number of chromosomes in each cell. One important reason why individuals can breed only with other individuals of the same species is that the cells of the interbreeding organisms must have the same number and type of chromosomes. Plants have the ability suddenly to double the number of chromosomes in their cells as they reproduce. The new plants, now with twice as many chromosomes, can no longer interbreed with the original plants. A new species has formed. This form of speciation has produced almost one-half of the species of flowering plants on Earth today.

In addition to this unusual behavior of plants, many plant species can form hybrids with another species. The new hybrid species is often able to reproduce through asexual reproduction, such as by stems called rhizomes that grow along the ground. In asexual reproduction, an individual plant can produce new offspring without mating with another plant. Sometimes a second change occurs in the cells of the hybrid that allows sexual reproduction, resulting in seeds. The wheat whose seeds produce the flour for a hamburger bun is a plant species that evolved by this process of hybridization. The original ancestral plant hybridized with wild grass. Because each species had 14 chromosomes, the new hybrid plant had 28 chromosomes. Later, this new plant hybridized with goat grass, a plant with 14 chromosomes. This series of hybridizations resulted in the modern species of wheat grown on farms today, a plant that has 42 chromosomes. (See Figure 3-10.)

Figure 3-10 The modern wheat species results from a series of hybridizations that have increased the number of chromosomes to 42.

PATTERNS OF EVOLUTION

As speciation proceeds, the direction of change in a species can vary. Most often, as species evolve from a common ancestor, they become increasingly different from each other. They move away, or diverge, from each other. An example of **divergent evolution** is the evolution of monkeys and apes from an ancestral species. Sometimes, an ancestral species arrives in a new uninhabited location, an area with a wide variety of conditions or habitats. Natural selection leads to the evolution of different species in each of these habitats. Referred to as **adaptive radiation**, this process occurred, for example, when amphibians such as frogs, which lay their eggs in water, evolved into an ancestral reptile with a "watertight" egg that could develop on land. Reptiles were free to live on land without needing a source of water for the development of their eggs. Adaptive radiation occurred as a wide variety of reptile species evolved in the many available land habitats. The adaptive radiation of reptiles on land led to the evolution of turtles, lizards, crocodiles, dinosaurs, and eventually birds.

As mentioned in the previous chapter, penguins, sharks, and sea lions show similar adaptations to a life in water, but these three kinds of animals are not closely related. These animals have developed similar adaptations in the process called *convergent evolution*.

Another pattern occurs when two distinct species that have evolved from a common ancestor continue to change in similar ways. Because the changes occur in a similar direction, the process is called **parallel evolution**. One of the most famous examples of parallel evolution is that of pterodactyls and birds. Pterodactyls were a group of flying reptiles that lived at the time of the dinosaurs. In time, unlike birds, pterodactyls became extinct. In the ancient past, pterodactyls came from the same nonflying ancestors as birds did. Both pterodactyls and birds developed the body structures necessary for flying separately from each other, that is, through parallel evolution. (See Figure 3-11.)

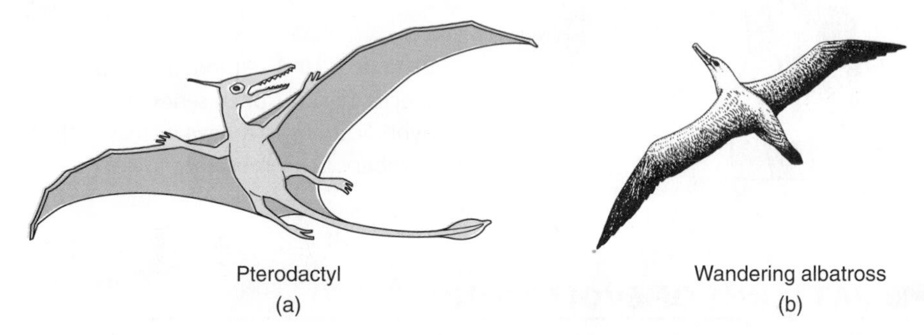

Figure 3-11 Similarities between birds and pterodactyls are the result of parallel evolution.

Species adapt to the conditions of their environments. However, because the environment always includes other living organisms, species frequently change because of, or along with, changes in species around them. This **coevolution** can be seen in many insects and flowers. Gathering food for themselves from juices made by flowers, insects distribute pollen from one flower to another. The pollen joins with an egg; eventually a seed forms. As flowers evolved, so did the insects that pollinated them. For example, sugary nectar evolved in flowers to attract insects. Structures on the heads of insects evolved to collect the nectar as the pollen was being distributed.

PACE OF EVOLUTION

The theory of evolution is supported by an enormous amount of evidence and is accepted by biologists throughout the world. However, disagreements exist about details of the process of evolution. Many scientists are puzzled that few fossils have been found of ancestors that are in-between stages in the evolution of one species and another. If species have been evolving slowly and gradually throughout the long history of life on Earth, shouldn't we find evidence of the intermediate changes in life-forms that occurred at each stage of this process? Many biologists answer simply that we have not yet found enough fossils.

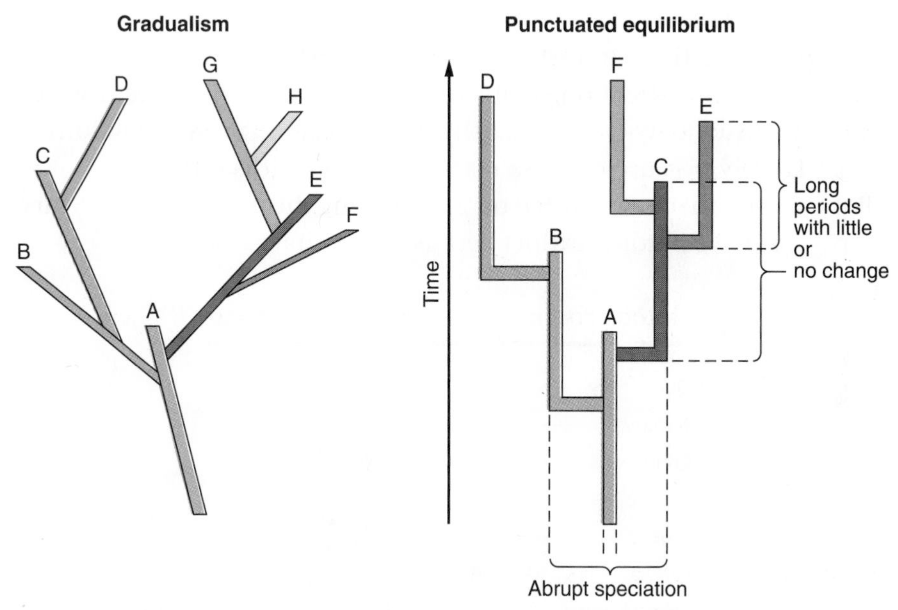

Figure 3-12 Two processes: gradualism and punctuated equilibrium.

Darwin viewed evolution as occurring in this manner, through small, steady steps, a process called **gradualism**. In 1972, two scientists, Niles Eldridge of the American Museum of Natural History and Stephen Jay Gould of Harvard University, suggested that "in-between" or transitional fossils had not been readily found because almost none of them existed. They suggested that species change very suddenly in big jumps, or "spurts," followed by long periods of little or no change at all. Called **punctuated equilibrium**, this process may actually occur in combination with gradualism, with some groups of organisms evolving gradually and others in spurts. (See Figure 3-12.)

EXTINCTION

Another natural part of the process of evolution is **extinction**, the disappearance of a species from Earth. Extinction occurs when a species no longer continues to produce more members of its own type. This inability to reproduce may be due to changes in the environment to which the species cannot adapt. For example, a significant decrease in rainfall or a drop in average temperature are two environmental changes that may affect an organism's ability to survive and reproduce. The extinction of a species may also arise from problems that occur within a population. Harmful genetic traits that become widespread in a population may cause a species' extinction.

At particular times in Earth's history, mass extinctions have occurred. Through a mass extinction, a large number of species disappear forever. Events that suddenly changed Earth's climate caused mass extinctions to occur. Largely because of these mass extinctions, it is estimated that of all the species that have existed on Earth during the past 3.5 billion years, 99 percent have become extinct because of natural causes.

GEOLOGIC PERIOD	MILLIONS OF YEARS AGO
	Present
Quaternary	
	1.6
Tertiary	
	66
Cretaceous	
	138
Jurassic	
	205
Triassic	
	240
Permian	
	290
Pennsylvanian	
	330
Mississippian	
	360
Devonian	
	410
Silurian	
	435
Ordovician	
	500
Cambrian	
	570

The history of Earth is divided into geologic eras and periods, as shown in the table above. Mass extinctions occurred at the end of several of these geologic periods. For example, at the end of the Permian Period, 240 million years ago, nearly all species in the seas and on land disappeared. Another mass extinction occurred at the end of the Cretaceous Period, about 65 million years ago. This mass extinction included the end of the dinosaurs. Fascinating theories and controversies exist today about the causes of these mass extinctions. Massive volcanic eruptions and collisions of Earth with asteroids or comets are often given as causes of mass extinctions. Both volcanic eruptions and collisions with asteroids or comets cause a great deal of dust to be thrown into the air. This airborne dust prevents sunlight from reaching Earth's surface. If this happened in

Black-Footed Ferrets: Back From the Brink of Extinction

Ferrets have become popular as pets in some places in recent years. Their clever antics can be a source of amusement. The unusual appearance of the small, long, slender, furry body of a pet ferret at the end of a leash always draws the attention of onlookers. However, there is a much more serious story about a ferret that is a native of North America—a story of life, death, and near-extinction.

The black-footed ferret is the only ferret species native to North America. These animals lived mostly in western parts of Canada and the United States. The black-footed ferret has a black face, a black tip on its tail, and—as its name suggests—black feet. It is a carnivore and survives mainly on a diet of prairie dogs. The only places where it has ever been found are near prairie dog tunnels. Prairie dogs are not well liked by farmers because their tunnels interfere with the planting of crops. Ranchers don't like them either, because they think their cattle can fall and injure themselves when they step in the openings of prairie dog tunnels. Farmers often put poison in the tunnels to kill prairie dogs. When prairie dogs are eliminated, black-footed ferrets also die. They were last seen in the wild in 1974, and by 1979 they were thought to be extinct.

However, to the great delight of wildlife biologists, a small population of black-footed ferrets was discovered living in a field near Meeteetse, Wyoming, in 1981. A species thought to be extinct was, in fact, still here! However, by the end of 1987, scientists counted only 18 survivors, so these animals were captured. In time, a captive-breeding program produced 400 individuals, which were released in seven different areas. Black-footed ferrets have now been reintroduced into the wild in Wyoming, Montana, South Dakota, and Nebraska. By 2002, they were also being released 100 miles south of the United States border, in Mexico, where many healthy prairie dogs live. The hope of biologists is to reach a goal of 1500 free-living black-footed ferrets by 2010.

The successful reintroduction of black-footed ferrets to the wild—along with the protection of prairie dogs and their grasslands habitat—has brought this interesting animal back from the brink of extinction. Still, we will never know what other species were unintentionally eliminated when the early settlers eradicated 90 percent of the North American prairie dog population.

the past, the temperature of Earth may have lowered dramatically, causing a mass extinction to occur.

Some scientists suggest that the greatest mass extinction of all time is occurring now. While the extinction of species in the past has often been the result of a natural process, the rate at which extinctions are now occurring has increased tremendously due to the effects of the human population on other species. For example, people have hunted animals such as the great auk and the passenger pigeon to extinction. Our destruction of the natural habitat of many species has often resulted in the extinction of a species. In particular, the burning of the tropical rain forests of South America, where naturalists have observed so many diverse species of plants and animals, has led to the extinction of countless species, including many not even discovered by scientists. Industrialized nations in the northern hemisphere, including the United States, have caused the extinction of species through the pollution of land, water, and air. Extinctions have also been produced by profound changes made to the environment, such as the damming of rivers to control flooding and to produce electrical power, the cutting of forests to produce farmland, and the filling in of swamps and other wetlands to build airports and housing. These human activities cause a loss of habitat and may lead to the extinction of certain species.

Before extinction occurs, a species becomes endangered, as has happened to the African elephant, giant panda, peregrine falcon, and black-footed ferret. (See Figure 3-13.) A species becomes endangered when its population reaches critically low levels. Low population numbers may result in too few individuals to maintain a healthy breeding population. The Endangered Species Act is an important federal law that identifies endangered species and protects them—for example, through restrictions on hunting—before they become extinct.

UNITY AND DIVERSITY: TAXONOMY

We have been learning about the constant process of change that occurs on Earth as species adapt to environmental conditions. During this process, some characteristics are lost, while other characteristics are gained. As a result, Earth is populated by many different kinds of organisms. To be able to study the millions of species on Earth today, biologists have for a long time organized them into named groups. **Taxonomy** is the science of naming and classifying organisms according to shared characteristics.

How do biologists decide which organisms to group together? Which

Figure 3-13 The black-footed ferret and the peregrine falcon are endangered species that are protected by law.

characteristics are used to create the groups? Do all flying organisms, such as butterflies, pigeons, and bats, belong in one group? Do all animals with tails, such as salamanders, crocodiles, and cats, belong in one group? These are "shared characteristics," and yet it seems clear that the animals put into these groups are very different from each other.

The system of taxonomy used today groups organisms according to their evolutionary relationships. The smallest group is the *species.* Similar species that have recently evolved from a common ancestor are placed together in a larger group called a *genus.* Next, organisms in similar genera (the plural of *genus*) that evolved from a more distant common ancestor are placed in the same *family.* For example, all felines, from house cats to tigers, belong in the same family. Likewise, all canines, from dogs to wolves, are in the same family, too. (See Figure 3-14 on page 64.) Smaller groupings are organized into larger groups in this manner. The complete sequence of taxonomic groupings is: species → genus → family → order → class → phylum → kingdom → domain. The most widely used system for grouping organisms places all living things in five kingdoms within two domains—prokaryotes and eukaryotes (see Figure 3-15 on page 64). The five kingdoms (with examples) are as follows:

Prokaryotes

Monera: single-celled organisms with no nuclear membrane to surround and contain their DNA: bacteria, blue-green algae.

Taxonomic Group	*Tiger*	*Wolf*
Kingdom:	Animal	Animal
Phylum:	Chordates	Chordates
Class:	Mammals	Mammals
Order:	Carnivores	Carnivores
Family:	Cats	Dogs
Genus:	Panthera	Canis
Species:	tigris	lupus

Figure 3-14 In the system of taxonomy used today, the largest division is the kingdom, and the smallest group is the species.

Eukaryotes

Protists: mostly single-celled; their DNA is surrounded by a nucleus; cells about 10 to 100 times larger than bacteria: amebas, paramecia.

Fungi: mostly multicelled organisms; unable to make their own food; fungi are often decomposers of other organisms: mushrooms, yeast.

Plants: mostly multicelled; able to make their own food by using the energy of sunlight: roses, oaks, cactuses, mosses.

Animals: multicelled; get their energy by eating the energy stored in the tissues of other organisms: humans, fish, worms, sponges.

Recent scientific research has shown that prokaryotes may really consist of two main groups: bacteria and archaea. Archaeans live in some of the most extreme environments on Earth, for example, in the boiling water of geysers or in very salty water. They are not at all like bacteria. Therefore, an even newer grouping of organisms lists three domains: bacteria, archaea, and eukaryotes.

This system of classifying organisms is based on their evolutionary histories, or **phylogeny**. Phylogenetic relationships refer to evolutionary

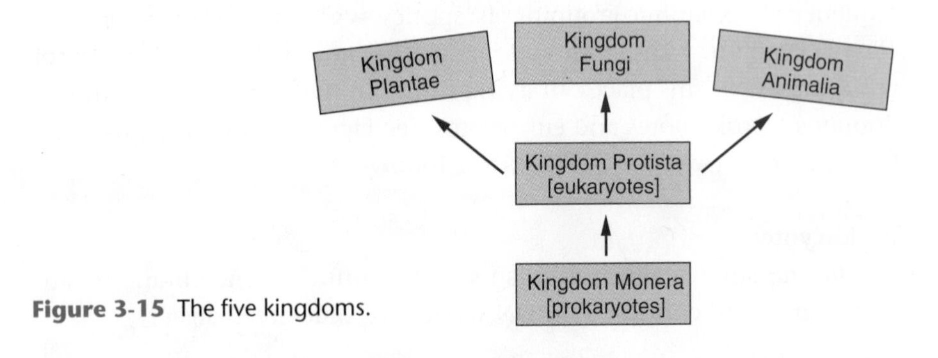

Figure 3-15 The five kingdoms.

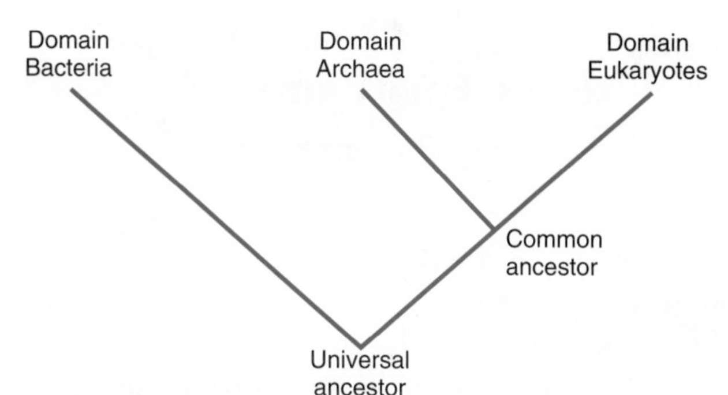

Figure 3-16 The evolutionary relationship of the three domains.

relationships, that is, how recently two types of organisms shared a common ancestor. Scientists use phylogenetic trees to represent the classification of organisms. Figure 2-2 in Chapter 2 is a simple phylogenetic tree. A phylogenetic tree shows the relationships among organisms. In order to figure out how to group organisms, scientists study the types of evidence discussed earlier to determine evolutionary relationships. This evidence includes comparative anatomy, embryology, and biochemistry, as well as the fossil record. For example, because archaean cells are, on the molecular level, in ways more like eukaryotes (than like bacteria), they most likely shared a common ancestor more recently. This relationship is shown in the diagram above. (See Figure 3-16.)

In taxonomy, every type of organism, or species, is given a name. The scientific name consists of two parts. The first part is the genus. The genus name is always capitalized. The second part of an organism's name is the specific name. The specific name is always begun with a lowercase letter. Every organism reading this book belongs to the genus and species *Homo sapiens*. The house cat that may be in your home belongs to the species *Felis catus*. The oak tree outside the window belongs to the genus *Quercus*. If it is a red oak, it is *Quercus rubra*. If it is a white oak, it is *Quercus alba*. The taxonomic naming system we use is called **binomial nomenclature**. The names commonly come from words in Greek and Latin. When scientists speak or write to each other, they use the same scientific names for organisms, no matter what native language they speak. This eliminates the confusion that would otherwise occur if organisms were given different names in every country.

LABORATORY INVESTIGATION 3
How Are Plant Leaves Adapted to Exchange Gases and Water With the Environment?

INTRODUCTION

Like animals, plants have evolved features that help them survive in particular environments. In this investigation, you will use a microscope to examine one of these adaptations. You will examine stomates, the openings in the surface of a plant's leaves. You will observe how the structure of stomates is related to their functions.

MATERIALS

A variety of fresh leaves, forceps, compound microscope, microscope slides, water, coverslips

PROCEDURE

1. Tear a leaf at an angle. If you make the tear correctly, you will see a white or colorless area extending from the leaf edge. This is the epidermis, the thin outer surface of a leaf. By examining the whole leaf, determine if this is the upper or lower epidermis of the leaf.
2. Use the forceps to remove a portion of the epidermis. Place the piece of epidermis on a clean microscope slide. Add a drop of water to the sample on the slide. Place a coverslip on the edge of the water drop. Let the coverslip fall over the piece of epidermis. Your teacher will show you the correct technique to use. You have made a wet mount. Make sure the bottom of the slide is absolutely dry before you place it on the stage of the microscope. (See Figure 3-17.)
3. Place the slide on the microscope stage. If your microscope has clips, move them onto the slide to hold it in place. Make sure the low-power lens is in place. Look through the eyepiece and focus the microscope so that you can see the detail in the leaf clearly. Locate a stomate and its pair of surrounding guard cells. Can you see more than one stomate in your field of vision? If you can, count the number of stomates you can observe. Move the slide to observe several different areas on the leaf. Count the number of stomates in each field. Calculate the average number of stomates you observed in each field for this kind of leaf.

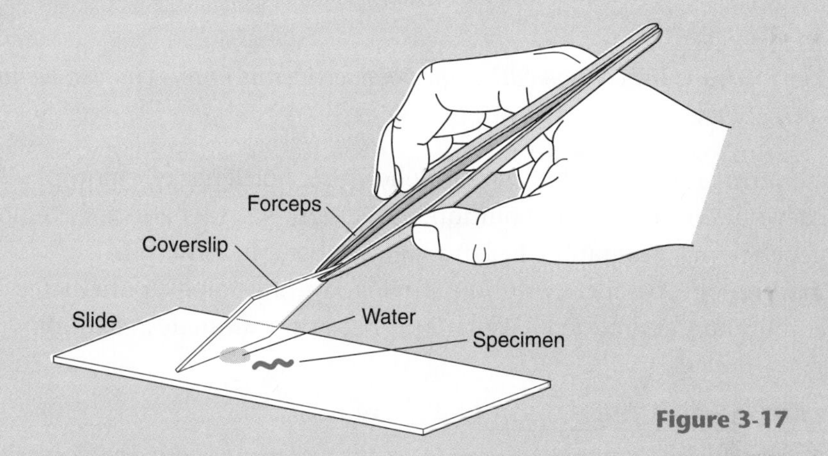

Figure 3-17

4. Repeat steps 1 to 3 for the other surface of the leaf.
5. Repeat steps 1 to 4 for another kind of leaf.

INTERPRETIVE QUESTIONS

1. What happens when the guard cells change shape?
2. Why is it necessary for the guard cells to be able to change shape? Why is this ability important?
3. How does the number of stomates on different types of leaves compare? Why do you think different types of leaves have different numbers of stomates?
4. How does the number of stomates on the upper and lower surfaces of a leaf compare? Explain any differences you observed.

CHAPTER 3 REVIEW

Answer these questions on a separate sheet of paper.

VOCABULARY

The following list contains all of the boldfaced terms in this chapter. Define each of these terms in your own words.

adaptive radiation, binomial nomenclature, bottlenecks, chlorophyll, coevolution, divergent evolution, extinction, founder effect, gene flow, genetic drift, geographic isolation, gradualism, hibernation, morphology, parallel evolution, phylogeny, physiology, punctuated equilibrium, reproductive isolation, speciation, stomates, taxonomy

PART A—MULTIPLE CHOICE

Choose the response that best completes the sentence or answers the question.

1. The science of naming and classifying organisms according to shared characteristics is *a.* phylogeny *b.* binomial nomenclature *c.* taxonomy *d.* speciation.
2. Certain Hawaiian birds have very long, thin, curved beaks, which fit perfectly into the narrow, curved flowers from which the birds obtain nectar and which they pollinate. This best illustrates *a.* parallel evolution *b.* divergent evolution *c.* genetic drift *d.* coevolution.
3. If large stretches of farmland and superhighways prevent the mountain lions in one state from breeding with the mountain lions in a neighboring state, the two populations are
   a. geographically and reproductively isolated
   b. geographically isolated but not reproductively isolated
   c. reproductively isolated but not geographically isolated
   d. neither geographically nor reproductively isolated.
4. Suppose striped cichlid fish and spotted cichlid fish belong to the same species and live in the same lake. If spotted males mate only with spotted females, and striped males mate only with striped females, the two populations are
   a. geographically and reproductively isolated
   b. geographically isolated but not reproductively isolated
   c. reproductively isolated but not geographically isolated
   d. neither geographically nor reproductively isolated.

5. The formation of new species is called *a.* coevolution *b.* speciation *c.* taxonomy *d.* phylogeny.
6. If the orange-and-black coloring of monarch butterflies warns birds that the butterflies are poisonous, but the white-and-black color of mutant monarchs is not recognized as a warning, then the adaptation is *a.* positive *b.* negative *c.* neutral *d.* can't tell from the information given.
7. Suppose birds recognize that monarch butterflies are poisonous because of their shape and the pattern of black on their wings. An adaptation that replaces the orange color on the butterflies' wings with white would be *a.* positive *b.* negative *c.* neutral *d.* can't tell from the information given.
8. The scientific term for the shape and structure of organisms is *a.* taxonomy *b.* phylogeny *c.* physiology *d.* morphology.
9. The way an organism and its internal parts function is its *a.* taxonomy *b.* phylogeny *c.* physiology *d.* morphology.
10. What kind of adaptation is illustrated when birds migrate south for the winter? *a.* behavioral *b.* morphological *c.* physiological *d.* phylogenetic
11. Although cheetahs had genetic variation in the past, they are now genetically almost all identical. Which of the following events was the most likely cause of this loss of genetic variation? *a.* genetic drift *b.* founder effect *c.* bottleneck *d.* gene flow
12. If an organism's scientific name is *Quercus alba,* then it *a.* belongs to the genus *alba* *b.* is closely related to *Quercus rubra* *c.* is closely related to *Rosa alba* *d.* is extinct.
13. In the modern system used to classify organisms, organisms are grouped according to *a.* how they obtain their food *b.* their morphology *c.* whether they are single-celled or many-celled *d.* their phylogenetic relationships.
14. Lionlike saber-toothed "tigers" evolved twice among mammals, in the cat family and in the marsupial family. This best illustrates *a.* parallel evolution *b.* adaptive radiation *c.* coevolution *d.* punctuated equilibrium.
15. The pattern of evolution that consists of long periods of stability interrupted by short periods of rapid change is called *a.* gradualism *b.* punctuated equilibrium *c.* genetic drift *d.* parallel evolution.

PART B—CONSTRUCTED RESPONSE

Use the information in the chapter to respond to these items.

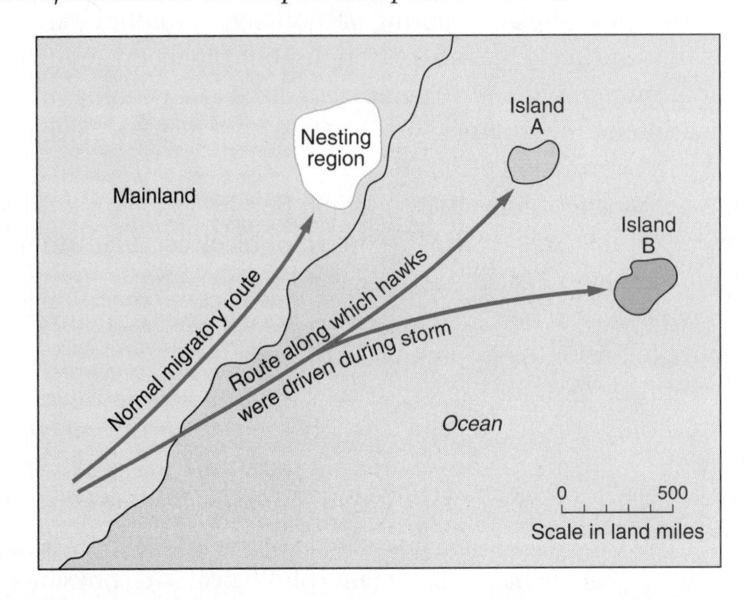

Thousands of years ago, a large flock of hawks was driven from its normal migration path by a severe storm. The hawks found shelter on islands A and B. The environment on island A is similar to that on the mainland, but the environment on island B is very different from both. The hawks have survived on the islands to the present. (See map.)

16. According to the map, how far is island A from the mainland? How far is island B? How far apart are the two islands?
17. Suppose the hawks are not willing to fly more than 400 miles (650 kilometers) over salt water. Describe likely consequences in terms of reproductive isolation, geographic isolation, founder effect, genetic drift, bottlenecks, adaptation, and speciation.
18. Which statement most accurately predicts the present conditions of the island hawk populations? *a.* Island B's hawks have evolved more than island A's. *b.* Island A's hawks have evolved more than island B's. *c.* The hawks on islands A and B have undergone parallel evolution. *d.* The hawks on island A have given rise to many new species.
19. Differentiate among physiological, morphological, and behavioral adaptations. Give a specific example of each type of adaptation.
20. What is a mass extinction? If mass extinctions occur naturally, why is there so much concern about the mass extinction currently in progress?

PART C—READING COMPREHENSION

Base your answers to questions 21 through 23 on the information below and on your knowledge of biology. Source: *Science News* (January 18, 2003): vol. 163, p. 46.

Why Didn't the Beetle Cross the Road?

A road that a person strolls across with barely a thought proves a deadly barrier for many other creatures and can disrupt the usual traffic of their genes throughout a population.

Most of the recent studies of this effect have focused on vertebrates, note Irene Keller and Carlo R. Largiadèr of the University of Bern in Switzerland. These researchers turned their attention to the ground beetle *Carabus violaceus.*

This insect doesn't fly but proves to be a plucky traveler on the ground. Other researchers had observed that another species of ground beetle is extremely wary of crossing roads.

Keller and Largiadèr collected their beetles from eight places in a Swiss forest sliced by three roads. The biggest genetic differences appeared between beetles living on opposite sides of the roads. Also, beetle populations confined to specific forest areas by the roads seemed to have lost some of their genetic diversity, the researchers report in an upcoming *Proceedings of the Royal Society of London B.* The scientists warn that expanding networks of roads could be reducing healthful genetic diversity in invertebrates.

21. State one reason why a road may have an effect on a population of animals.
22. Explain why the researchers from Switzerland collected beetles from a variety of places in a Swiss forest.
23. State one discovery that the researchers described about the beetles in their report.

4 Human Evolution

After you have finished reading this chapter, you should be able to:

Identify important arboreal adaptations of primates and list the distinguishing features of prosimians, monkeys, and apes.

Discuss the development of bipedalism in the early hominids.

Compare and contrast the characteristics of fossil hominid species.

Mankind stood up first and got smarter later.

Stephen Jay Gould

Introduction

The theory of evolution is one of the most important ideas in science. It is also a wonderful example of what science does best: tests ideas against evidence and observations in the real world to determine if the ideas are correct. The study of how the human species evolved is a good example of how science has tested the idea of evolution.

LOOKING FOR HUMAN ORIGINS

It is only natural that we are extremely curious about human origins. "Where did I come from?" is a question that occurs to every person at some point in her or his life. "Where did *we* come from?" is the question we ask now.

Two hundred million years ago, dinosaurs populated Earth. Those great reptiles had come to dominate Earth through the adaptive radiation to life on land that occurred after the evolution of the watertight egg. Living alongside the dinosaurs, but close to the ground and very small indeed—about the size of a mouse—were the first **mammals**. (See Figure 4-1.) Like all mammals, these ancestors had hair, nursed their young, and likely maintained a steady, high body temperature. The early mammals lived between 220 million and 65 million years ago. The most common parts of these animals to be found are their teeth. Because teeth are covered with hard enamel, they are frequently preserved as fossils. Studies of these teeth have shown that early mammals probably ate insects, worms, leaves, and fruits.

Figure 4-1 The first mammals lived alongside the dinosaurs.

For almost 130 million years, reptiles were the dominant life-forms on Earth. Then suddenly, at least by geologic time, dinosaurs became extinct. Faced with fewer competitors, an enormous variety of mammals evolved, again by adaptive radiation. These new groups of mammals included the carnivores (cats, dogs, seals, bears); hoofed mammals (pigs, deer, cattle); rodents (squirrels, porcupines, mice); whales; elephants; bats; insectivores (shrews, moles); and primates (lemurs, monkeys, apes, and humans). (See Figure 4-2.)

Figure 4-2 Representatives of the main groups of mammals.

ADAPTATIONS FOR LIFE IN THE TREES

The fossils of the earliest mammals indicate that they had five separate digits on each of their four feet. As explained in Chapter 2, this was a primitive feature. Fossil remains of various types of later mammals show feet and hands that evolved into hooves for running, feet for digging, wings for flying, or flippers for swimming. Mammals whose feet and hands had fewer than five digits were said to show advanced features. However, one group of mammals—the primates—through natural selection kept five digits on their feet and hands. In fact, these digits became even more fully developed. Eventually the thumb could bend over and easily touch the forefinger. This is called an **opposable** thumb; all primates have this feature. (See Figure 4-3.) What was the great advantage of an opposable thumb? How did this kind of thumb contribute to primate evolution through natural selection? An opposable thumb could hold on to tree branches. Primate evolution began with adaptations suited to an **arboreal** life, that is, a life in the trees. An opposable thumb was a very important adaptation for an arboreal way of life.

Figure 4-3 The opposable thumb is a great advantage to tree-dwelling primates.

The order of primates includes prosimians, known as the "lower primates," and monkeys, apes, and humans, known as the anthropoids or "higher primates." Many characteristics of modern primates are related to their original arboreal way of living. For example, a baseball pitcher uses an amazing shoulder joint, which first evolved to swing from one tree branch to another. Primate hands—with their fingernails, opposable thumbs, and strong, sensitive fingers—helped these animals hang on to branches, hold food, and groom themselves. Primates' eyes are positioned close together on the front of the face. Observe how the eyes of nonprimates, such as horses, are on either side of the head. (See Figure 4-4.) Because their eyes are positioned closer together, primates have stereoscopic (3-D) vision. Knowing how near or far an object is becomes a crit-

Figure 4-4 A horse's eyes are far apart, while a monkey's eyes are close together.

ical piece of information when an animal needs to jump safely from one tree branch to another. You can observe the advantage of 3-D vision if you close one of your eyes. Notice how the arrangement of objects in front of you seems flatter; you lose a sense of "depth" in your field of view.

A life lived in trees poses many hardships and dangers, especially for the young who need time to develop their skills. Because they could easily fall to the ground if left on their own, young primates would have trouble surviving. To ensure the survival of the young, a period of parental care is vital. Primates are great parents, caring for their young for a long time. Although humans no longer live in trees, we share many traits from our arboreal origins with other primates, including prolonged care of our young until they are able to live on their own.

A CLOSER LOOK AT PRIMATES

Prosimians, monkeys, apes, and humans are the main groups of primates alive today. Current research from molecular biology, combined with fossil evidence, indicates that the oldest common ancestor of today's primates lived between 80 and 90 million years ago, long before the dinosaurs disappeared. Weighing less than 2 pounds, this primate ancestor, it is theorized, looked like a very small lemur, lived in tropical forests, and was nocturnal (that is, active at night).

Prosimians include the lemurs of Madagascar and the lorises, bush babies, and tarsiers of tropical Africa and southern Asia. They are relatively small arboreal animals that feed on insects, leaves, fruits, and flowers. Like their ancestors, the prosimians are often nocturnal. These

Figure 4-5 This lemur is a type of prosimian, one of the four main groups of primates alive today.

animals existed in great numbers in the huge forests north and south of the equator 65 to 38 million years ago. Today, because of the destruction of their forest habitats, many prosimians are in grave danger of becoming extinct. (See Figure 4-5.)

Monkeys evolved from prosimian ancestors about 50 million years ago. There are two main groups of monkeys alive today. The New World monkeys, such as capuchins, spider monkeys, squirrel monkeys, and marmosets, are found in Central and South America. (See Figure 4-6.) The Old World monkeys, such as baboons, vervets, langurs, and macaques, live in Africa and Asia.

The higher primates, also known as **hominoids**, include all apes and humans. Ape fossils found in East Africa show that these primates evolved from the monkeys of Africa and Asia. The earliest known ape fossils are of an organism called *Aegyptopithecus,* which means "dawn ape"; they are about 35 million years old. This hominoid lived in trees and was about the size of a cat. *Aegyptopithecus* migrated across Asia around 25 million years ago.

The apes include gibbons, orangutans, chimpanzees, bonobos, and gorillas. Apes are generally larger than monkeys, have larger brains, and lack tails. They can hang upright from tree branches and have relatively long arms and short legs.

Because of the close relationship of apes to humans, we are fascinated by their behavior. Several long-term scientific studies of the apes have

Figure 4-6 Some monkeys have grasping tails, an adaptation to living in trees.

been completed. It is not only our behavior that is so close to that of the apes; most of our DNA is the same as that of the African apes, too. (See Figure 4-7.)

Gorillas live in troops of 8 to 24 individuals with one large male as the leader. Usually peaceful, male gorillas can appear menacing when they threaten an enemy with screams, broken tree branches, and chest pounding. Female gorillas nurse their infants from two to four years.

Chimpanzees are thought to be the primates most closely related to

Figure 4-7 Chimpanzees—intelligent apes closely related to humans—are known to use tools in the wild.

humans. In fact, new research has shown that human and chimpanzee DNA is almost identical. The main difference is that greater quantities of proteins are produced by human genes, especially within brain cells, than by chimpanzee genes. Chimpanzee behavior also shows the evolutionary closeness. They live in social groups, and the males and females form temporary bonds to mate. Friends within the group spend long hours grooming each other. Chimps love to play together and are curious, noisy, and outgoing. However, they can also be quite aggressive and do fight with neighboring groups of chimpanzees. Bonobos (sometimes called pygmy chimps) are also very closely related to humans in their genetic makeup and behavior. They are considered highly intelligent and tend to be more peaceful in their social interactions than the chimpanzees.

HOMINIDS: THE EARLIEST HUMANS

The rising of the Himalayan mountains and drier weather around 20 million years ago caused forest areas to diminish in size. At that time, Asian and African apes became separated from one another. Between 14 and 8 million years ago, from within the African group, the common ancestor of humans and chimpanzees evolved. This does not mean that our ancestors were chimpanzees. We are most closely related to chimpanzees because we shared an ancestor with chimpanzees more recently than with any other animal. Very few details are known about this early stage in the evolution of humans.

One of the most important discoveries in human evolution occurred in 1924 when a small fossilized skull was found in a mine in Taung, South Africa. The "Taung child" skull was sent to a skilled neurologist, Raymond Dart, who recognized that the fossil had humanlike features. (See Figure 4-8.) He concluded that this specimen was the fossilized skull of an early human, a type of **hominid**. The appearance of the skull, the size and shape of the brain case, and the shape of the teeth all showed that this fossil was from an early ancestor on the human family tree. Later fossils confirmed that the Taung child walked on two feet. More than any other feature, walking on two feet is what makes an early human a hominid. Dart named this 3-million-year-old hominid *Australopithecus africanus* (southern ape of Africa). It took 25 years for scientists to accept Dart's conclusions. During that time, many australopithecine fossils were found in different places in Africa. Raymond Dart was indeed correct about the little skull. Discoveries of more fossils provided additional evidence that *A. africanus* walked upright and had hands and teeth similar to ours.

Figure 4-8 Raymond Dart and the Taung skull—the fossilized skull of an early human ancestor.

In 1974, a fossil discovered in Ethiopia became famous worldwide. Significant portions of a 3.18-million-year-old skeleton of a female hominid were unearthed; she was 1 meter tall with a skull about the size of a softball. This fossil also showed evidence of upright walking. She was named *Lucy* after the Beatles song the scientists were listening to at the time of their discovery! Donald Johanson, of the Cleveland Museum of Natural History, led the team of scientists. Further work showed that Lucy's species, *Australopithecus afarensis,* was the ancestor of the *A. africanus* species identified by Raymond Dart. Recently, several other more ancient species have been discovered that bring us ever closer to the dividing point (about 7 million years ago) between humans and apes. (See Figure 4-9.)

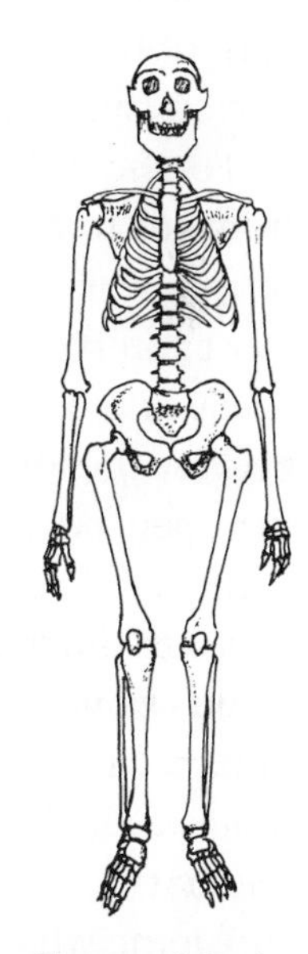

Figure 4-9 A drawing of an *Australopithecus afarensis* skeleton.

The most important point about australopithecine species is that hominids walked on two feet, not four, for at least 2 million years without much enlargement of their brain. **Bipedalism** (walking on two feet) may have helped hominids gather food and care for their young more efficiently by freeing their hands. Tools were not made until much later. That is why evolutionary biologist Stephen Jay Gould of Harvard University said that "Mankind stood up first and got smarter later."

Raymond Dart and the Skull of the Taung Child

In the 1920s, large amounts of limestone were being dug from the ground in Taung, an area of South Africa. Many human fossils were also being dug up in the limestone quarry. Raymond Dart, an Australian doctor teaching at the medical school in Johannesburg, heard about the fossils. Dart was an expert on the anatomy of the human head and was anxious to examine the fossils—a natural curiosity. Dart contacted the owner of the quarry and, in time, two large boxes of fossils arrived at his home.

When he examined the material in the boxes, Dart found a dome-shaped piece of stone and immediately recognized that it was shaped like a brain. In this fossil, Dart saw the folds of tissue that make up the brain and even the blood vessels on the surface. Dart realized what had happened many years before. Long ago, someone had died in the vicinity of this present quarry. Sand and water that contained minerals had entered the skull; eventually these materials hardened into rock in the exact shape of the brain.

WHY DID EARLY HUMANS STAND?

Our earliest ancestors lived in the trees. One of the first big steps on the path of human evolution occurred when early hominids walked on two feet on the ground. Where, when, how, and why did this occur?

For more than 100 years, it has been widely believed that the big step from ape to human occurred the day some apes left the forest. According to this story, the apes had been spending their lives in the manner of most forest animals, enjoying the warm, humid days, with plenty of shade and abundant fruits and berries to eat. For some reason, perhaps because the forest area was decreasing, they now found themselves out on the open, where tall grasses, small shrubs, and occasional trees replaced the forest, in an environment called the savanna. (See Figure 4-10.)

It was drier on the savanna, it took longer to find food, and predators could see you more easily. Life was harder. To survive, you had to walk upright on two feet. By being upright you could see approaching danger more easily. You also had to get smarter. So here are the early hominids out on the dangerous savanna, while back in the forest the other apes are still doing what they always did, going about picking fruits and berries in an environment that was relatively safe.

Interesting story, but does the evidence confirm it? Science is based on proposing ideas, or hypotheses, that can be tested and then seeing if the evidence supports or contradicts the idea. The "savanna hypothesis" is now being thoroughly tested. Evidence against the hypothesis would be

On close examination, Dart felt that the fossil brain looked like it had come from an ape, but he recognized that the fossil also had some similarities to a human brain. The skull, he thought, might provide some clues to the brain's origin. Dart looked again in the box that contained the fossilized brain. Much to his amazement and delight, he found pieces of the lower jaw and the skull. However, the front of the fossil skull—the face—was covered by layers of rock. In a procedure that took several months, Dart chipped away at the rock layers. What he eventually revealed was the face of a young creature, later dubbed the "Taung Child," which Dart believed was an early ancestor of the human species. His find turned out to be one of the most important hominid fossil discoveries ever made, adding crucial details to our understanding of human evolution.

hominid fossils that showed upright walking *and* the ability to climb trees. Lucy had curved fingers that might have been adapted for tree climbing even though she could walk on two feet. To support the savanna hypothesis, fossils of other animals and of plants living at the same time as the hominids would have to show that the climate had become drier, that the forests had disappeared, and that the savannas remained as an exploitable food source. In some places in Africa, such as Tanzania, where 3- to 4-million-year-old footprints of *A. afarensis,* Lucy's species, were found, it was definitely very dry, with no forests. However, in Ethiopia,

Figure 4-10 The savanna has tall grasses, small shrubs, and scattered trees.

where Lucy lived, there were forests as well as open places. Were our bipedal hominid ancestors savanna dwellers or neighbors of other primates that lived in the forests?

The oldest hominid fossils found so far, *Ardipithecus ramidus,* have been dated at 5.8 million years old. The fossils of these individuals, who lived in Ethiopia, show that the skull was balanced at the top of the skeleton for walking erect. Meanwhile, other animal fossils found nearby indicate that *A. ramidus* definitely lived in the forest. If careful studies of the *A. ramidus* bones show that it really did walk upright, the savanna hypothesis will be disproved.

If some apes began walking on two feet not because they left the forest for the open savanna, what other explanations could there be? Why did some apes begin walking upright if not to gain a selective advantage out of the forest and on the open savanna? Perhaps standing up on tree branches, as chimps sometimes do, makes it easier to feed. (See Figure 4-11.) Standing erect to threaten an enemy, as gorillas are known to do, may have provided a survival advantage maintained by natural selection. Another explanation that has been offered is that apes that were born with a slightly greater ability to stand up were better able to gather food, even in the forest. Males with this advantage could bring food back to the females with whom they had mated and to their offspring. In this story, the offspring most likely to survive were those of apes that could walk erect. This would be a tremendous evolutionary advantage and could easily have led to the evolution of walking on two feet.

Figure 4-11 There are many possible advantages to standing and walking upright, as this young chimpanzee is doing.

Once again, science will put this idea to the test, if it becomes a serious hypothesis. Eventually, the question of why bipedal hominids evolved will be answered. By then evolutionary biologists will have new questions to study that have not yet even been asked.

Check Your Understanding

Why is bipedalism such an important characteristic of hominids?

OUR OWN GENUS

None of the fossils that have been discussed so far belong to the genus of modern humans, *Homo*. To be a hominid, as the australopithecines were, you had to walk on two feet. However, to be a hominid in the genus *Homo* (from the Latin word for "man"), you also need to have the enlarged brain that sets you apart from the other primates.

Lucy's species, *A. afarensis,* remained relatively unchanged for almost 1 million years. Then, about 3 million years ago, an adaptive radiation resulted in the Taung child species, *A. africanus,* and several other australopithecine species with heavier bones and much wider faces. (See Figure 4-12.) Known as *A. robustus* and *A. boisei,* these **robust** species were first discovered by Mary Leakey in Tanzania in 1959 and were dated at 1.8 million years ago, using the potassium-argon radioisotope dating technique. Mary Leakey, her husband Louis, their son Richard, and his wife Meave, are among the most important scientists who have studied the fossil evidence of human origins. Much debate has taken place about how *A. robustus* and *A. boisei* fit into the human family tree. Most researchers believe these species to be separate branches on the tree, branches that

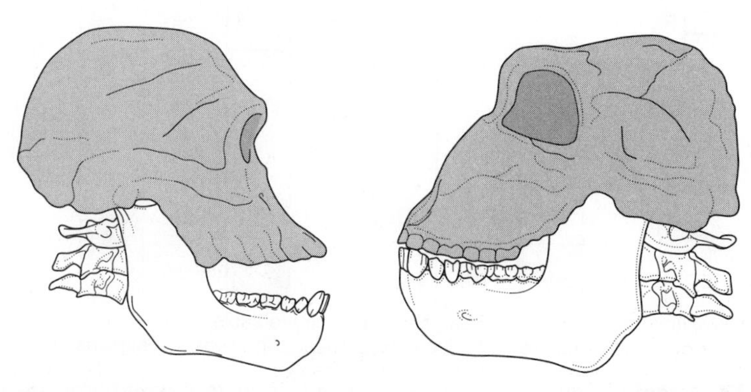

Figure 4-12 Reconstructed fossils of *A. africanus* and *A. robustus* show the differences in their jaws.

ended long ago. These species did not adapt successfully to changing environmental conditions. As a result, they became extinct.

The adaptive radiation that led to the robust, now extinct, australopithecine species also led to hominids with larger brain capacities. The size of the Taung child's brain was about 500 cubic centimeters (cc). These larger hominid skulls, sometimes found along with simple stone tools, were about 650 cc. Great arguments arose when Louis Leakey first stated in 1962 that his 1.75-million-year-old fossils from Kenya with the larger brains belonged to the genus *Homo*. He named the species *Homo habilis,* meaning "handy man." Most scientists now accept Leakey's interpretation. *Homo habilis* is placed on the human family tree. Although there is little agreement on how the *Australopithecus* species are related to *Homo habilis,* it is generally accepted that *H. habilis* led toward modern humans, evolving first into *Homo erectus,* which later evolved into *Homo sapiens* (modern humans).

All of the early hominid fossils discussed so far have been found only in Africa. *Homo erectus* was the first hominid to migrate from Africa to Asia and into Europe. Fossils of this species were found in Java in 1896 (Java Man), in Beijing, China, in 1929 (Peking Man), and in northern Kenya in 1984, where the skeleton of a 12-year-old-boy of this species who died 1.6 million years ago was found in 1984.

Homo erectus had a body skeleton much like that of modern humans. The 12-year-old boy was 1.7 meters tall and walked like modern humans. Differences are found in the size of the skull. Their brains, 700 to 1200 cc, were much larger than those of earlier species and almost as large as those of modern humans. (See Figure 4-13.) However, their jaws and teeth were much larger than those of modern humans. *Homo erectus* skulls had thick, low foreheads and sloping chins.

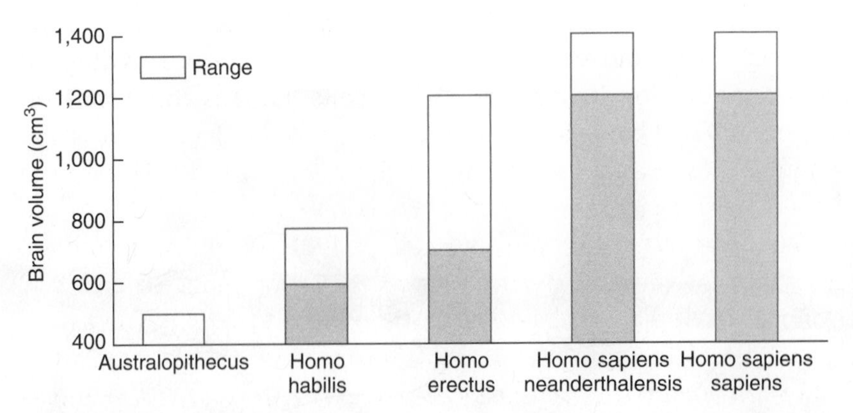

Figure 4-13 A comparison of the brain volume of several important hominids.

By this time, the larger brains of our hominid ancestors showed that they were definitely becoming more intelligent. Much more efficient tools, such as the hand ax (a stone that has a surface for gripping and several cutting edges), are often found with *H. erectus* fossils. (See Figure 4-14.) These are the first hominids known to build fires, live in caves, and clothe themselves. With these skills they were able to migrate to colder northern climates found outside Africa. *H. erectus* existed for a long time on Earth, from 1.8 million years ago to about 300,000 years ago.

Figure 4-14 A hand ax; such tools were made and used by *Homo erectus.*

There is much debate about recent human evolution. In 1997, researchers in northern Spain announced the discovery of yet another ancestor of modern humans, *Homo antecessor.* They think that the 800,000-year-old fossils from Spain belong to the common ancestor of modern humans and other extinct hominids. Today, there are two main views of human evolution. One group of scientists sees it as a ladder, with one species at a time leading to the next species. The other group sees human evolution as a tree, with several branches. One branch leads to modern humans; the other branches lead to extinct hominid species.

THE HUMAN SPECIES

The migration of humans from one place to another on Earth occurred long before travel by ship and plane. These early travelers went over land on foot. Helping them and perhaps encouraging them to move on were the Ice Ages. These were periods when Earth's climate cooled, causing great sheets of ice to move over the land. The levels of the oceans dropped as water remained on land frozen as ice. Thus humans could walk on places once covered by oceans, traveling to the island of Java, to the island continent of Australia, and eventually walking east across the land bridge in the northern Pacific Ocean to North America. The last Ice Age began about 1 million years ago and included several periods of deep cold. Extensive ice sheets covered much of North America and Europe.

Hominids living after *Homo erectus* colonized a variety of places with varying climates, including Africa, Asia, Europe, and Australia. As Earth's climate warmed, and the ice sheets melted and retreated, early humans extended their range. These individuals included the Neanderthals, whose

Figure 4-15 A drawing of how a Neanderthal might have looked.

fossils have been found throughout Europe and the Middle East and who had much larger brain sizes than *H. erectus*—brains about the size of modern human brains. Some scientists consider them to be members of our species, *Homo sapiens*. Others think they made up a separate species, *Homo neanderthalensis*. Their bodies were similar to those of modern humans. However, their faces looked different, with heavier ridges over the eyes, a long, low skull, and small cheekbones. (See Figure 4-15.) Because of these differences, these humans, living from about 400,000 to 35,000 years ago, are usually called early or "archaic" *Homo sapiens*. They may have been descendants of the newly discovered species *Homo antecessor,* representing one branch of the human tree that has ended.

Neanderthals wore animal skins, made better and more varied tools than *H. erectus* did, and buried their dead. We know that they purposely left weapons and flowers with their dead. These were individuals who thought about things, including life after death.

All fossils of hominids that lived during the past 30,000 years are like modern humans both in body and skull size and shape. Modern humans have an average brain size of about 1350 cc. One of the best-known groups of these "modern" *Homo sapiens* is the Cro-Magnons, named for the place in France where their fossils were first found. Other modern *Homo sapiens* fossils, up to 100,000 years old, have been found in Israel and throughout Africa. Cro-Magnons are well known for their advanced tools made of stone, bone, and ivory. These tools included spears, fishing hooks, and needles. In addition, the magnificent cave paintings of these

Figure 4-16 Cro-Magnon drawings show the beautiful forms of many different kinds of animals, and the skill of the early human artists.

humans, which show the beautiful forms of many different kinds of animals, give us a sense that we are seeing humans like ourselves. Cro-Magnon fossils are the remains of people like modern humans who looked and wondered at the world around them, sometimes symbolizing their thoughts and feelings, for whatever reasons, in the form of art. (See Figure 4-16.)

MORE QUESTIONS AND SOME ANSWERS

The study of the history of human evolution is full of controversies, none being debated more intensely than the question of where modern *Homo sapiens* first evolved. This question is considered to be a valid scientific question because it is assumed that it can be put to the test. It is thought that evidence will eventually be found to answer the question. Then it will no longer be just a matter of opinion. This process is an important part of the scientific method. One hypothesis is that the populations of *H. erectus* that had migrated from Africa to a variety of places on Earth each gave rise to archaic and then modern *H. sapiens* independently. In this "multiregional model," human races in each of these areas arose from different populations. Breeding between the various populations would have allowed for gene flow and prevented speciation from occurring. Today, all human races on Earth belong to one species.

The other hypothesis is that modern *H. sapiens* evolved from *H. erectus* in just one place, Africa. According to this "Out of Africa" model, modern *H. sapiens*, moving out from Africa, replaced the archaic *H. sapiens* in the various places where they met. This is a very different proposal. It would mean that the varieties or races in the world's human population arose in just the last 100,000 years since *H. sapiens* left Africa, not more than 1 million years ago when *H. erectus* began migrating. Anthropologists throughout the world strongly support one or the other of the possibilities. Each opposing side claims that the evidence supports its theory. This scientific question continues to be studied.

Another fascinating question about human evolution concerns language. When did humans begin to speak? The answer to this question remains a mystery. Charles Darwin suggested that human speech evolved from animal cries. Critics at the time, who were opposed to Darwin's views, called this the "bow-wow" theory. Noam Chomsky, a famous professor from the Massachusetts Institute of Technology (MIT), has for more than 40 years claimed that the rules of human language are built-in, not learned. How these innate rules could have evolved is difficult to explain. In 1994, another MIT professor, Steven Pinker, defended the idea that language evolved by natural selection, but said he could only guess that it may have begun with primate calls. Other more recent suggestions are that language evolved from primate grooming. Apes and monkeys use physical contact with each other through grooming to establish social connections. Making sounds might have become a more efficient way of doing this. In 2002, a New Zealand psychologist, Michael Corbollis, proposed the idea that human language began with hand and face gestures. He said that the earliest hominid, some 6 million years ago, could not yet have spoken, but would have had the ability to make voluntary hand and face movements. A kind of sign language could have developed, eventually switching from gestures to true speech about 50,000 years ago, after modern humans had evolved.

In spite of these fascinating theories, the question of where language comes from may simply be unanswerable. If that is the case, then this mystery cannot be considered a valid scientific question. Nevertheless, despite the questions that remain unanswered, we have been richly rewarded to date in learning so much about the fascinating story of where we came from.

LABORATORY INVESTIGATION 4

How Can We Determine the Sequence of Hominid Evolution?

INTRODUCTION

In spite of the very incomplete fossil record, scientists who study human evolution have been able to draw some remarkable pictures of what the different hominid species might have looked like. Studying these pictures helps us develop a deeper understanding of human evolution.

One misconception that must be avoided, as the pictures are studied, is the idea that human evolution is like a ladder with a series of steps leading from the most ancient hominid species directly to our own species, *Homo sapiens*. This misconception often has been illustrated as a parade of fossil hominids, with the specimens in the parade becoming more modern as they march across the page.

The more accurate understanding of human evolution is that different hominid species often existed together at the same time and in the same place. Also, many of these species evolved along certain pathways that eventually led to dead ends. Rather than a ladder, a better diagram of hominid evolution would be more like a bush having many branches, with our species being at the end of the only branch that still survives.

MATERIALS

"Hominid Species A–H" and "Hominid Data Sheet" handouts (from the *Teacher's Manual*), scissors, glue or tape, unlined paper

PROCEDURE

1. Examine the drawings A–H. These are artistic impressions based on fossil evidence of different hominid species. Examine them closely. Identify three characteristics that seem to differ and three characteristics that seem to be similar from one figure to another. Share your list of observed characteristics with your group.
2. Determine which figure you think represents the earliest hominid. Determine which figure you think represents the most recent hominid. Give reasons for your choices. Share your choices with the group, discuss all opinions, and then reach a consensus.

3. Cut out the figures and, as a group, arrange them in a sequence from earliest to most recent. Glue or tape the hominid drawings to the unlined paper in the order in which you have arranged them. Make a list of the criteria that guided your choices. Compare your sequence with those of the other groups. Discuss any differences.
4. Compare your time sequence to the one determined by scientists, shown in the Chronology of Hominid Evolution table on the Hominid Data Sheet. Based on this set of data, would you change your sequence? Explain.

INTERPRETIVE QUESTIONS

1. Draw a horizontal timeline that is 15 cm long. Let 3 cm equal 1 million years, going from 5 million years ago (mya) to the present. Place the letter for each hominid species listed in the Chronology of Hominid Evolution table at the correct place on your timeline.
2. Study Diagram A and Diagram B on the Hominid Data Sheet. These diagrams represent alternate ideas of the evolutionary route from ancient hominids to modern humans. Write a comparison of these two different interpretations of human origins.
3. Explain why you think the three different characteristics and three similar characteristics you observed may be important for determining the sequence of hominid evolution.

CHAPTER 4 REVIEW

Answer these questions on a separate sheet of paper.

VOCABULARY

The following list contains all of the boldfaced terms in this chapter. Define each of these terms in your own words.

arboreal, bipedalism, hominid, hominoids, mammals, opposable, robust

PART A—MULTIPLE CHOICE

Choose the response that best completes the sentence or answers the question.

1. Humans belong to the class of animals known as *a.* mammals *b.* carnivores *c.* rodents *d.* invertebrates.
2. Which is *not* considered an adaptation for arboreal life? *a.* opposable thumb *b.* five digits on each foot *c.* stereoscopic vision *d.* prolonged period of parental care
3. The famous fossil known as Lucy belongs to the species *a. Homo antecessor* *b. Homo erectus* *c. Australopithecus afarensis* *d. Australopithecus africanus.*
4. Primates include *a.* porcupines, squirrels, and mice *b.* pigs, sheep, and deer *c.* humans, lemurs, and chimpanzees *d.* shrews, moles, and hedgehogs.
5. The "Taung child" skull is significant because it *a.* was the first fossil of *Homo habilis* to be discovered *b.* belongs to one of the earliest types of hominids *c.* showed that the earliest primates in the human line walked on all fours *d.* indicated that members of its species made tools.
6. The earliest mammals *a.* first appeared 65 million years ago *b.* varied greatly in size, appearance, and lifestyle *c.* had five digits on their front feet and four on their back feet *d.* are known primarily from fossil teeth.
7. The earliest hominoid fossils are of *a. Australopithecus* *b. Aegyptopithecus* *c. Ardipithecus* *d. Homo.*
8. Chimpanzees are classified as *a.* prosimians *b.* monkeys *c.* hominoids *d.* hominids.
9. The oldest hominid fossils found so far are about *a.* 35 million years old *b.* 5.8 million years old *c.* 1.75 million years old *d.* 400,000 years old.

10. Which of these is *not* a general characteristic of mammals? *a.* complex life cycle with alternation of generations *b.* nursing their young with milk *c.* maintaining a high, steady body temperature *d.* body covered in hair
11. An opposable thumb *a.* can bend and easily touch the forefinger *b.* is an adaptation for arboreal life *c.* is a major characteristic of primates *d.* all of these.
12. Animals that live in the trees are *a.* nocturnal *b.* arboreal *c.* diurnal *d.* neanderthal.
13. Which of the following statements is true? *a.* Hominoids include lemurs, lorises, and tarsiers. *b.* Prosimians are higher primates. *c.* Monkeys include orangutans and gorillas. *d.* Prosimians, monkeys, and apes are all primates.
14. *Homo erectus* *a.* could build fires *b.* probably did not make tools *c.* lived in Africa only *d.* is a side branch on the human family tree and not a direct ancestor of modern humans.
15. The "Out of Africa" model of human evolution *a.* is not supported by mitochondrial DNA evidence *b.* states that modern *Homo sapiens* evolved in Africa only *c.* is supported by DNA evidence from Neanderthal fossils *d.* states that breeding among populations of archaic *Homo sapiens* allowed for gene flow and prevented speciation.

PART B—CONSTRUCTED RESPONSE

Use the information in the chapter to respond to these items.

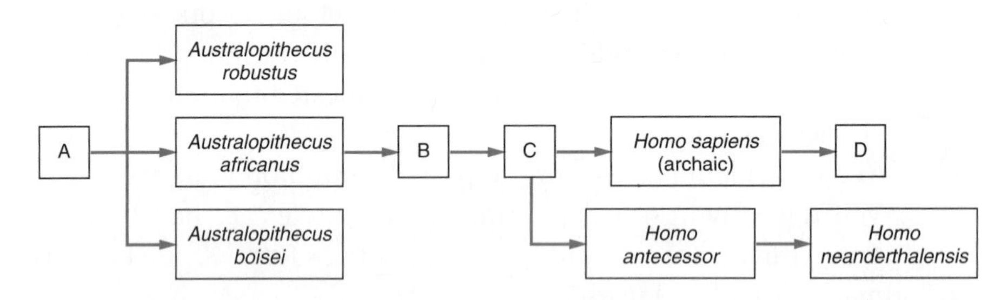

16. The diagram shows one possible pathway of human evolution. What hominid names should appear in boxes A, B, C, and D in the diagram?
17. What do you think the discoverers of *Homo antecessor* might think about the view of human evolution expressed in the diagram?
18. What is the "savanna hypothesis"? What sort of evidence will prove or disprove it?

19. How do australopithecine fossils support the hypothesis that hominids "stood up first and got smarter later"?
20. Why are Neanderthals considered archaic *Homo sapiens* and Cro-Magnons considered modern *Homo sapiens*?

PART C—READING COMPREHENSION

Base your answers to questions 21 through 23 on the information below and on your knowledge of biology. Source: *Science News* (May 10, 2003): vol. 157, p. 302.

New Fossil Weighs in on Primate Origins

Excavations in Wyoming have yielded the partial skeleton of a 55-million-year-old primate that probably was a close relative of the ancestor of modern monkeys, apes, and people. The creature was built for hanging tightly onto tree branches, not for leaping from tree to tree, as some scientists had speculated, based on earlier fragmentary finds. Also, despite expectations, the ancient primate didn't have eyes specialized for spotting insects and other prey.

Jonathan I. Bloch and Doug M. Boyer, both of the University of Michigan in Ann Arbor, unearthed the new specimen. It belonged to a group of small, long-tailed primates that lived just before the evolution of creatures with traits characteristic of modern primates—relatively large brains, grasping hands and feet with nails instead of claws, forward-facing eyes to enhance vision, and limbs capable of prodigious leaping.

The new find, in the genus *Carpolestes*, had long hands and feet with opposable digits, Bloch and Boyer report in the Nov. 22 *Science*. The animal grew nails on its opposable digits, and claws on its other fingers and toes. Unlike later primates, *Carpolestes* had side-facing eyes and lacked hind limbs designed for leaping.

21. State two characteristics of the 55-million-year-old Wyoming primate (fossil) that are different from what scientists had expected.
22. Explain what characteristics are considered to be those of modern primates.
23. State two characteristics of the ancient Wyoming primate that indicate it was *not* a member of the group of modern primates.

Energy, Matter, and Organization

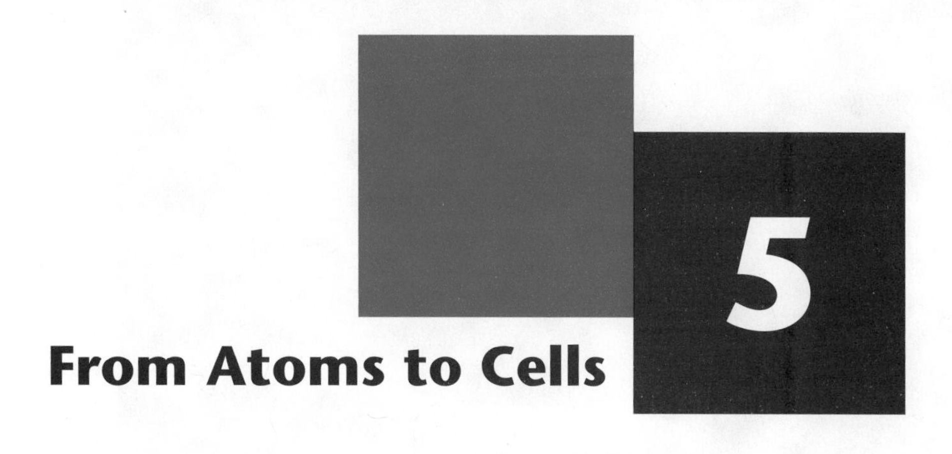

From Atoms to Cells

After you have finished reading this chapter, you should be able to:

Explain that all matter is made of atoms, which combine to form molecules and compounds; and that organic compounds are essential for the chemical reactions that keep all organisms alive.

Describe the four major groups of organic compounds and list their main functions in the human body.

Identify the key points of the cell theory and describe the levels of organization, and defining properties, of living things.

Perhaps the nearest we can come to a definition [of life] is to say: that life is a stage in the organization of matter.

George W. Cray

Introduction

Think about the clothes and shoes in your closet. Every day, as you look for clothes and get dressed, does your closet become more organized? During a thunderstorm, when your little brother plays inside, does his collection of toys in the living room get better organized? Does your family's car ever repair itself and run better as it is used?

Of course we know that the answer to these questions is no. Everything in the world tends to get more disorganized as time passes. Some things wear out, other things fall apart. No one's sneakers ever got newer as they were worn. The tendency for things to become more disordered is described by physicists as the "second law of thermodynamics." According to this law, natural events tend to increase disorder in the universe. Or more simply put, the universe is running down. It becomes more disordered with the passing of time.

Figure 5-1 The muscles of this gymnast work mostly in pairs to move his bones.

All organisms face one important problem. In order to live, it is necessary to remain highly organized. For example, our skeleton consists of 206 different bones, each precisely interconnected with other bones. Six hundred different muscles work mostly in pairs—one contracting while the other relaxes—to move these bones first one way, then the other. (See Figure 5-1.) The digestive system in most animals consists of a series of structures, one after the other, each with a specific function. Food is changed as it moves through this system one step at a time, just like an assembly line in a factory makes changes at each stage of the manufacturing process. However, the digestive system, instead of building something, actually breaks things down step-by-step into its parts.

Living things are highly organized. How do they keep themselves carefully arranged and functioning properly in a universe where things are constantly running down?

To maintain the state of organization necessary for life itself to exist, all living things require energy. We know that energy exists in different forms. Heat, motion, light, and electricity are all forms of energy. And organisms always need more energy—a continuous input of energy—to stay organized and remain alive.

MATTER, ATOMS, AND US

The universe, including everything on Earth, is made of matter. Matter is anything that has mass and takes up space. A hand, a rock, a cup of water, a heart, the root of a tree, a mushroom, and the air in a balloon all have mass and all take up space. (See Figure 5-2.)

Long ago, one of the biggest questions in science—What makes up

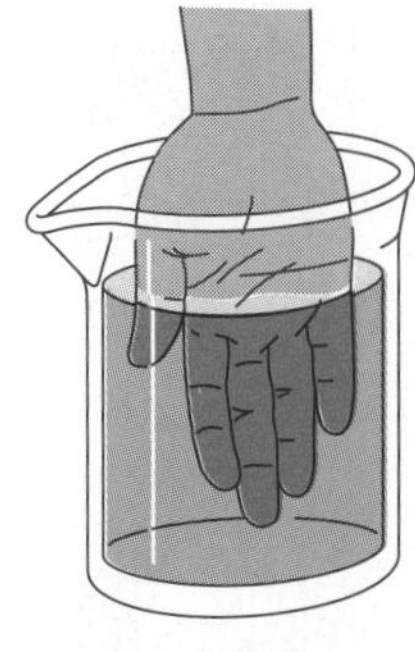

Figure 5-2 The hand and the water are both composed of matter. Placing a hand in the water raises the water level, since the hand takes up space.

matter?—was answered. All matter is made up of **atoms**, particles far too small to be seen with the unaided eye or even through the magnifying lenses of an ordinary microscope. Only recently an extremely powerful microscope, the scanning electron microscope, has been able to photograph atoms. (See Figure 5-3.)

An English schoolteacher, John Dalton, originally stated his theory of the atomic structure of matter in 1810. Dalton described atoms as tiny solid spheres. A century later, the physicist Ernest Rutherford fired tiny particles at thin sheets of gold. Rutherford was amazed to discover that most of the particles passed through the gold sheets, while a very few particles bounced back. Since most of the particles passed through the gold sheets, Rutherford concluded that the atoms that made up the gold sheets were mostly empty space. Rutherford reasoned that each atom has an extremely dense nucleus in its center. The nucleus contains subatomic particles called *protons* and *neutrons*. Distributed in the mostly empty space around the nucleus are the *electrons*. When a particle hits a dense nucleus, it bounced back. Other particles passed through the gold sheets when they moved through the mostly empty space surrounding the nucleus.

While Rutherford studied the nucleus of the atom, another scientist, Niels Bohr, was mainly interested in the electrons. Bohr thought that an

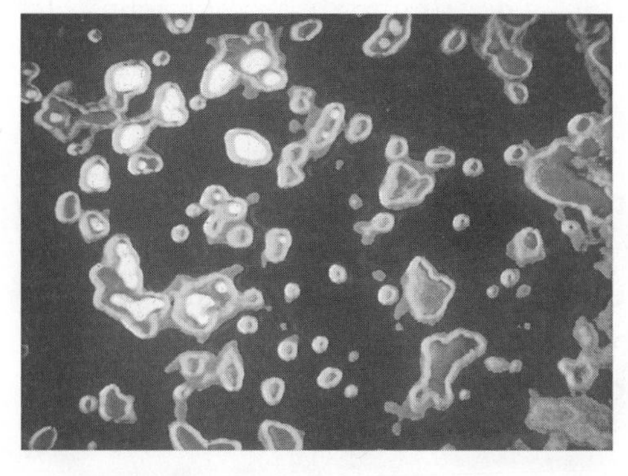

Figure 5-3 A scanning electron micrograph of uranium atoms, magnified more than 3 million times.

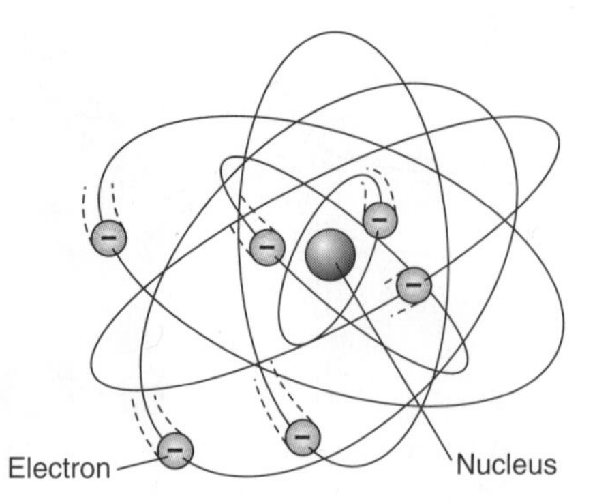

Figure 5-4 Electrons move in "clouds" called orbitals, not in simple, specific orbits around the nucleus.

atom's electrons spin around the nucleus in specific orbits, much like the planets in our solar system move around the sun. Today we know that electrons move in less clearly defined "clouds" called *orbitals*, not in simple, specific "planetary" orbits around a central nucleus. (See Figure 5-4.)

Electrons have energy. The orbitals in which electrons move make up shells. The electron shells closest to the nucleus have the least energy; the electron shells farthest from the nucleus have the most energy. It is possible for electrons to move from one shell to another. As they move, electrons go from one energy level to another. A ball rolling from the top of a hill loses energy as it moves lower. So, too, does an electron as it moves from an outer, higher-energy level to an inner, lower-energy one. The movement of electrons between energy levels in atoms is what produces all changes or transformations of energy.

An organism is really a very complex system for transforming energy. Through natural selection, evolution has produced species of organisms that are efficient energy transformers. Where does the energy come from that keeps organisms as different as grass, ants, and elephants alive? We all know that the light of the sun is the main source of energy for life on Earth. How plants capture this light energy, and how plants, animals, and other organisms use it to live, grow, and reproduce, is the story we can now begin to explore.

ATOMS GET TOGETHER

Atoms, the small particles that make up all matter, are found in only about 100 different types. Of these basic types, called **elements**, 92 are found naturally on Earth. Seventeen more (although this number is subject to change) have been made by scientists in laboratories. Many laboratory elements are not stable and exist for only a fraction of a second.

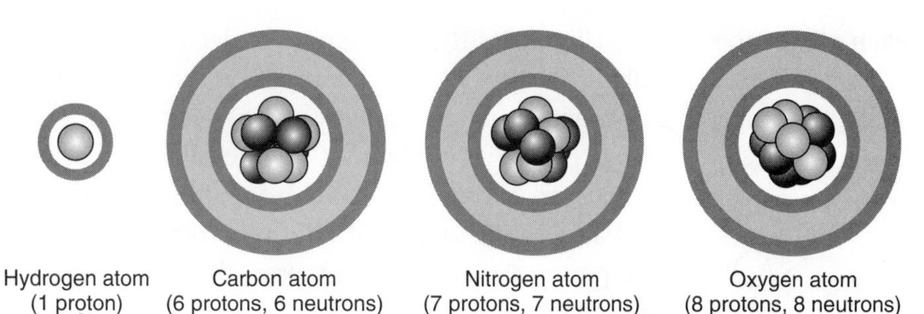

Figure 5-5 Simplified models of hydrogen, carbon, nitrogen, and oxygen. These models show electrons as a cloud of negative charge around the nucleus.

The smallest particles of an element that maintain the characteristics of that element are atoms. Some elements are considered valuable. For example, gold, silver, nickel, and copper, used to make jewelry, art objects, tools, and money, are all elements. Only a small number of the 92 naturally occurring elements are vitally important to living things.

Carbon, hydrogen, nitrogen, oxygen, phosphorus, and sulfur (CHNOPS) are the six most important elements for organisms. (See Figure 5-5.) Also important are sodium, potassium, iron, and a few others. These are the elements that are major components of our bodies and those of all living things. (See Figure 5-6.) However, atoms are almost never found alone. Atoms of most elements combine with other atoms to form larger structures called *molecules*. A substance that consists of molecules made up of two or more different elements is called a *compound*.

What keeps the atoms in a molecule together is a kind of partnership called a *chemical bond*. One type of chemical bond is a covalent bond. In a covalent bond, atoms share electrons with each other, which makes

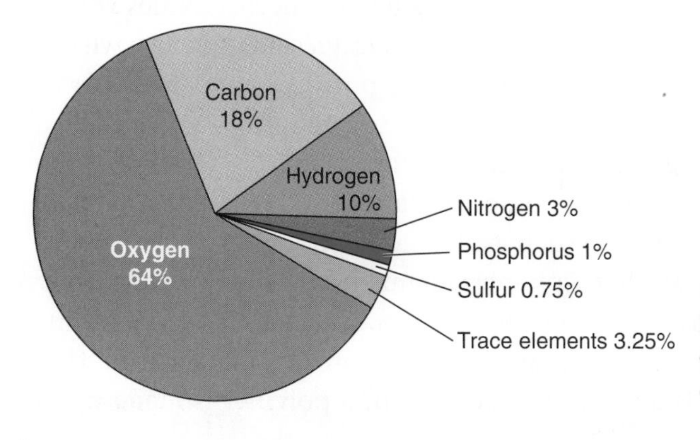

Figure 5-6 Major elemental components of the human body.

each atom more stable. The atoms that make up a molecule of water are held together by covalent bonds. Water is a compound made of the elements hydrogen and oxygen. In a molecule of water, one oxygen atom shares two of its electrons with two hydrogen atoms. The atoms of oxygen and hydrogen remain together as water molecules, and this remarkable chemical compound flows from our taps as the clear liquid we drink. (See Figure 5-7.)

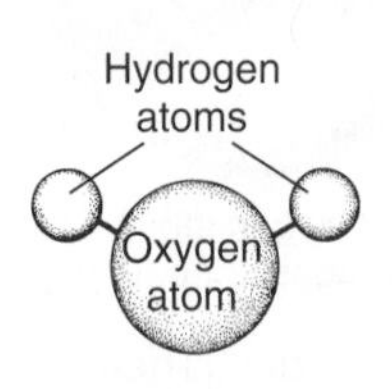

Figure 5-7 A water molecule.

Elements can combine into thousands of different compounds, but only a small number of compounds are important for living things. Many of the compounds found in living things are called **organic** compounds. All organic compounds contain the elements carbon and hydrogen. Of all the elements, carbon has a special importance for life on Earth. Each carbon atom can form bonds with up to four other atoms, including other carbon atoms. Organic compounds usually have a skeleton or backbone of carbon atoms bonded to each other. Atoms, in particular hydrogen and oxygen atoms, as well as atoms of other elements such as nitrogen, phosphorus, and sulfur, can form bonds with the atoms that make up this carbon backbone. (See Figure 5-8.)

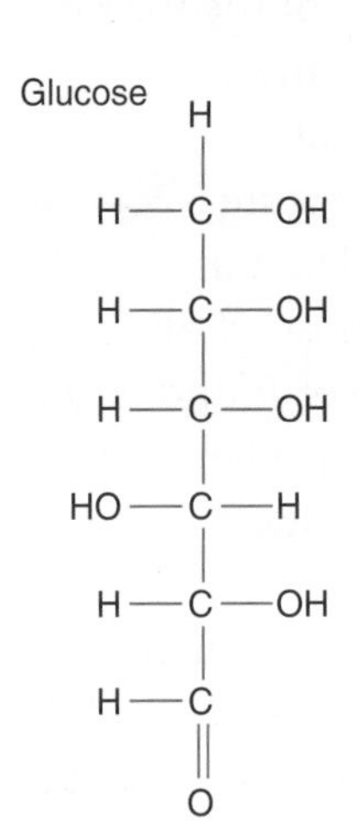

Figure 5-8 Organic compounds, such as glucose, have a skeleton of carbon atoms to which other atoms form bonds.

ORGANIC COMPOUNDS

Organic compounds accomplish many different, complex jobs that keep us and all other organisms alive. These include capturing and transforming energy, building new structures, storing materials, repairing structures, and keeping all chemical activities within an organism working properly. In an organism, chemical reactions are always putting things together or taking things apart. These chemical activities in an organism are called **metabolism**.

Organic molecules are often very large. A large molecule, such as the blood protein hemoglobin, is a **polymer**. Polymers are made by linking smaller molecules together. Imagine a long freight train. The train consists of one engine, many boxcars, and a caboose. Each of the smaller units that makes up the train is a subunit. Like a train, a polymer is made up of many smaller subunits. (*Poly* means "many.") In polymers, the subunits are held together by covalent bonds.

Whenever subunits join, they are linked by a new covalent bond. These covalent bonds form when two hydrogen atoms (H_2) and one oxygen (O) atom are removed. This group of three atoms makes up a molecule of water (H_2O). On a hot day, you lose water when you sweat. If you sweat a lot, you may lose a great deal of water and become dehydrated. So this process of building, or synthesizing, polymers by removing water molecules is called **condensation** (or dehydration) **synthesis**. It occurs constantly in all organisms.

The opposite can happen. In order to be used in cells, large molecules often need to be broken apart. Polymers can be broken into their subunits by splitting the covalent bond links. To do this, the water molecules that had been removed when the polymer formed must be added where the bonds link the subunits. This breaking of polymer bonds is called **hydrolysis**. (*Hydro* is Greek for "water," and *lysis* refers to the process of separating.)

THE FAMILIES OF ORGANIC COMPOUNDS

Even though organisms contain many different molecules, there are relatively few different types of subunits that make up these molecules. The subunits located in the organic compounds found in living organisms from bacteria to butterflies to horses are almost identical. However, the way these subunits are put together creates an enormous variety of polymers. Think of the letters and words in the English language. Our alphabet contains only 26 letters; however, an unabridged dictionary contains 450,000 words. Some words are short, others are long, but all are different combinations of only 26 letters. An English word with 15 letters is very large. Some polymers have more than 100 subunits—a macromolecule of this length is actually quite small. Others have thousands of subunits. Now imagine the variety of different kinds of polymers.

The common subunits and the common way of building polymers from these subunits in all living things are strong evidence that all organisms alive today evolved from a common ancestor long ago. Yet the wide variety of polymers that are constructed from these subunits makes possible the enormous diversity of species and individuals found on Earth.

Organic compounds are grouped into four major families: carbohydrates, lipids, proteins, and nucleic acids. These are the families of compounds found in all living things.

Check Your Understanding

How are organic compounds different from any other compounds? Why are the organic compounds so important to living things?

CARBOHYDRATES

Carbohydrates include the simple sugars, as well as polymers, such as starch, made up of sugar subunits linked together. Carbohydrates are formed from the elements hydrogen, oxygen, and carbon. In all carbohydrates, the number of hydrogen atoms is always twice the number of oxygen atoms. In organisms, the main functions of carbohydrates are energy storage and providing strong building materials for certain types of cells. (See Figure 5-9.)

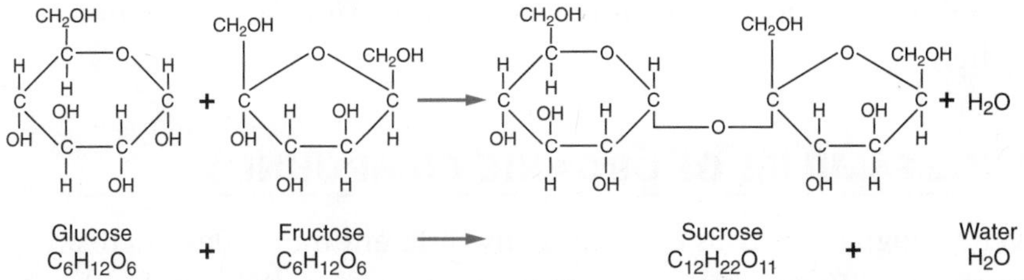

Glucose $C_6H_{12}O_6$ + Fructose $C_6H_{12}O_6$ → Sucrose $C_{12}H_{22}O_{11}$ + Water H_2O

Figure 5-9 Glucose and fructose are monosaccharides. Sucrose is a disaccharide. Two monosaccharides linked together form one disaccharide and water.

Glucose and fructose, sugars in honey, are two of the subunits found in more complex carbohydrates. These **simple sugars** each have six carbon atoms covalently bonded together. Simple sugars that are made up of one subunit are called *monosaccharides*. Two sugar subunits covalently linked together form a *disaccharide* (plus water). Lactose, the sugar in milk, is a disaccharide. Sucrose, or table sugar, is naturally found in plant saps; it also is a disaccharide.

Polysaccharides are made from many sugar subunits joined together. Energy is stored in these large molecules. Our muscles must contain stored energy to allow them to work at a moment's notice. In humans, energy is stored in the polysaccharide glycogen. Our muscles contain enough glycogen to allow us to do strenuous exercise for about 30 minutes. The glycogen stored in muscles and in the liver is made by linking together many glucose subunits.

Plants store energy in the form of **starch**. Like glycogen, starch is a polysaccharide. Potatoes, corn, rice, and wheat all contain starch. Cellulose is another important polysaccharide found in plants. Rather than provide energy, cellulose helps build up the structure of plants. The tough, strong material in grass and wood is made of cellulose.

LIPIDS

Look at a bottle of salad dressing and you will see two layers, one on top of the other. The top layer is oil. It does not dissolve in the other layer (vinegar), which is mostly water. (See Figure 5-10.) Oils, along with fats, make up the second major family of organic compounds, the **lipids**. One type of lipid makes up the basic structure of the cell membrane. However, the main purpose of lipids is energy storage, which they do more efficiently than carbohydrates do. This is because the long chains of carbon atoms in lipid molecules have many more hydrogen atoms bonded to them. It is in each of these C–H bonds that energy is stored.

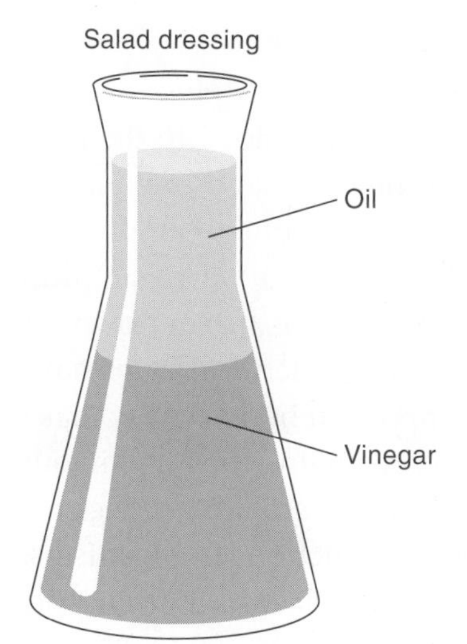

Figure 5-10 Oil is a type of lipid—an organic compound that stores energy.

Energy stored in lipids is for long-term use. During physical activity, the carbohydrate glycogen gets used up quickly. However, anyone who has exercised to lose weight knows that lipids, or fat deposits, do not disappear quickly. The fat deposits in our bodies serve an important purpose, particularly for women, who have a greater percentage of body fat than men do. During the long months of pregnancy, it is essential that a reliable, long-term source of energy be available for the growing baby, in case food becomes scarce. Extra body fat is a trait that has evolved over time, to help ensure reproductive success.

There are many types of lipid molecules. Part of one type of lipid molecule is made up of *fatty acids*, which are long carbon chains. Every food package includes an analysis of the types of fats in the food. In *saturated fat* molecules, almost every position on the carbon chains is saturated, or bonded, to hydrogen atoms. In *unsaturated fats*, some hydrogen atoms are missing. If many hydrogen atoms are missing, the fat is *polyunsaturated*. It is now known that the polyunsaturated fats in vegetable oils are less likely to cause heart disease than the saturated animal fats found in steaks or in dairy products such as cream and butter.

Steroids are a very different group of lipids. Steroids are built with a ring of carbon atoms. Cholesterol is a steroid that is made only in animals. From cholesterol, several important chemical messengers, including the male hormone testosterone and the female hormone estrogen, are made. Cholesterol is made in the body and eaten in foods. High levels of cholesterol in the blood can lead to a condition called *atherosclerosis*, or hardening of the arteries, which can cause heart disease.

PROTEINS

The organic compounds with the most "personality" belong to the family of **proteins**. Their personality comes from the fact that protein molecules are much more diverse than other organic compounds. Hundreds of thousands of different proteins exist. (See Figure 5-11a.) We are all individuals with our own characteristics because each of us has our own set of characteristic proteins. No two people other than identical twins have exactly the same proteins. Proteins are also important because they make up a large portion of all organisms. In fact, after removing the water in our bodies, proteins would make up more than half the remaining weight.

Proteins are responsible for a wide variety of functions in organisms. They are used for building materials (collagen and elastin), to transport

Figure 5-11a An egg white is made of protein molecules.

other substances (hemoglobin), to send signals (insulin), to provide defense (antibodies), and to control chemical and metabolic activities (enzymes).

How can so many different proteins be built? The building plan is the same as for other organic compounds. Proteins are large polymers that are constructed by combining smaller, individual subunits. The subunits of proteins are **amino acid** molecules, of which there are 20 main types. Each amino acid, such as alanine, glycine, cysteine, or tryptophan, has a "mini-personality" of its own. These 20 different amino acids combine in single file in many different orders, one after the other, to form the variety of proteins that exist. (See Figure 5-11b.) This is very much like building words from letters. However, proteins can easily have more than 100 amino acid subunits.

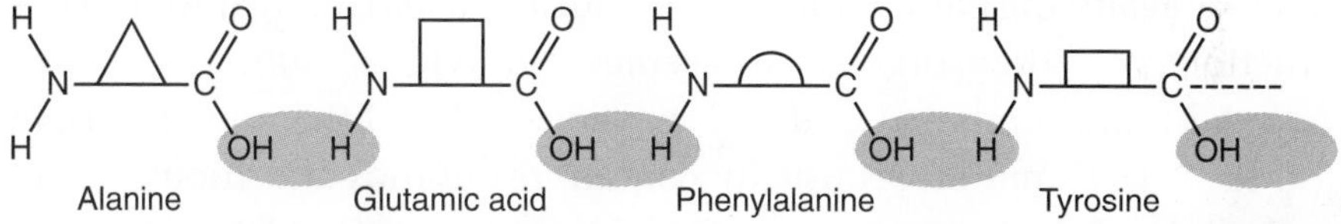

Figure 5-11b Amino acids combine in linear sequences to form a great variety of proteins.

The order in which the amino acids are linked, called the *linear sequence*, determines the characteristics of the protein molecule. Every different linear sequence produces a different protein. However, the sequence does not behave like a string of letters. The long chain of amino acids begins to do strange things as it is assembled. Because of the mini-personalities of the amino acid types—some like to be near each other, while others do all they can to avoid each other—protein chains twist, turn, and bend into a specific three-dimensional shape. This shape of a protein molecule is called its *conformation*. Every protein molecule has a very specific conformation that determines its personality and its function.

NUCLEIC ACIDS

The final family of organic compounds, nucleic acids, is in some sense the master family. Nucleic acids consist of deoxyribonucleic acid (DNA) and ribonucleic acid (RNA). DNA and RNA are responsible for storing the genetic information that contains the directions for building every molecule that makes up an organism. DNA and RNA are also the organic

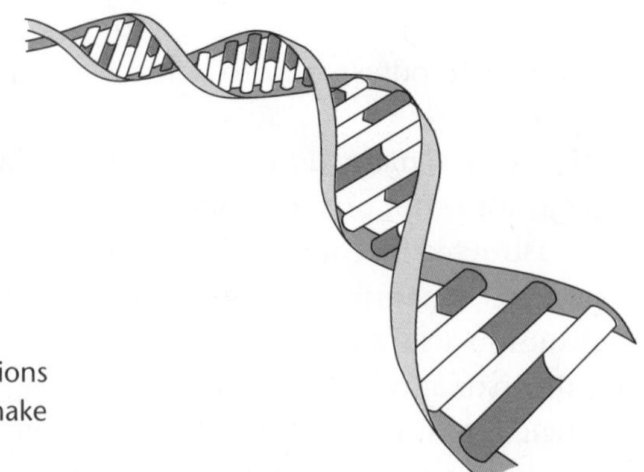

Figure 5-12 DNA—a type of nucleic acid—contains the directions for building the molecules that make up an organism.

compounds most involved in the process of constructing these materials. (See Figure 5-12.)

Once again, the pattern for building nucleic acids is similar to the construction of other organic molecules. Individual subunits, called *nucleotides*, are combined in linear sequence to build the polymers DNA and RNA. DNA molecules usually contain thousands and thousands of nucleotides linked together in a specific sequence. This DNA is copied into a form of RNA called *messenger RNA* with a sequence that follows that of the original DNA. This messenger RNA in turn is moved to a specific location, a "minifactory" in the cell called a *ribosome*. In a ribosome, the messenger RNA molecule is used as a building plan to construct a protein molecule. Specific portions of a DNA molecule are used to make specific proteins. Because proteins make us who we are, and because our DNA makes our proteins, it is really our DNA that makes us who we are.

CELLS ARE US

Some people have unusual hobbies—collecting bottle caps, making dolls from dried apples, or juggling. In the Netherlands in the 1600s, Anton van Leeuwenhoek, who earned his living selling clothes and buttons, had an interesting hobby. He ground pieces of glass into magnifying lenses that he used to make into simple microscopes. (See Figure 5-13.) He was an unusual person. Even after he got a steady job as a janitor, he spent all his free time making his lenses better and better and using them to examine everything he could find—hairs of a sheep, the head of a fly, seeds of plants, the legs of lice. Yet nothing could have prepared him for the surprise he received when he first glanced through his microscope at a small drop of clear rainwater. Calling to his daughter Maria, Leeuwenhoek said, "Come here! Hurry! There are little animals in this rainwater....They

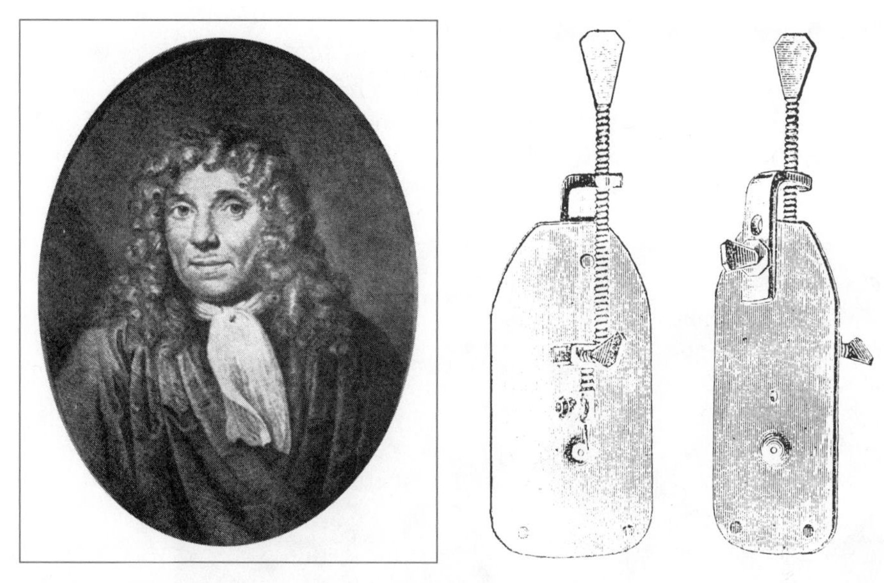

Figure 5-13 Anton van Leeuwenhoek and his microscope.

swim! They play around! They are a thousand times smaller than any creatures we can see with our eyes alone.... Look! See what I have discovered!"

Leeuwenhoek described his discoveries in letters to the Royal Society of England, a group that included the most important scientists and thinkers in the world at that time. They were so surprised that they sent one of their leaders, Robert Hooke, to Holland to investigate.

Hooke also had made a most interesting discovery with a simple microscope of his own design in 1665. Looking at very thin slices of cork (the outer bark of a species of oak that is used to plug bottles), Hooke observed rows of tiny spaces. Under the microscope, the thin slice of cork looked like a honeycomb constructed by bees. Hooke called the spaces he observed "cells."

Robert Hooke examined the microscopic "beasties," as Leeuwenhoek called them. Hooke confirmed the accuracy of Leeuwenhoek's observations when he returned to England. Soon Leeuwenhoek, an untrained amateur scientist with an unusual hobby, became a worldwide celebrity visited by kings, queens, and czars. For more than 50 years, Leeuwenhoek —who had little formal schooling—sent amazing descriptions of his observations to the Royal Society.

In the years that followed, the construction of microscopes improved. More objects were studied under the microscope's magnification, including many different kinds of plants and animals. By 1838, Matthias Schleiden, a German lawyer who had become a botanist, stated that all plants

The Microscopic World on a Pin

As can be seen in these scanning electron micrographs, hundreds of bacteria can live on the point of a pin. The increasing magnification of each picture (more than 50, 300, 1000, and 5000 times enlarged) reveals more and more detail about this microscopic world.

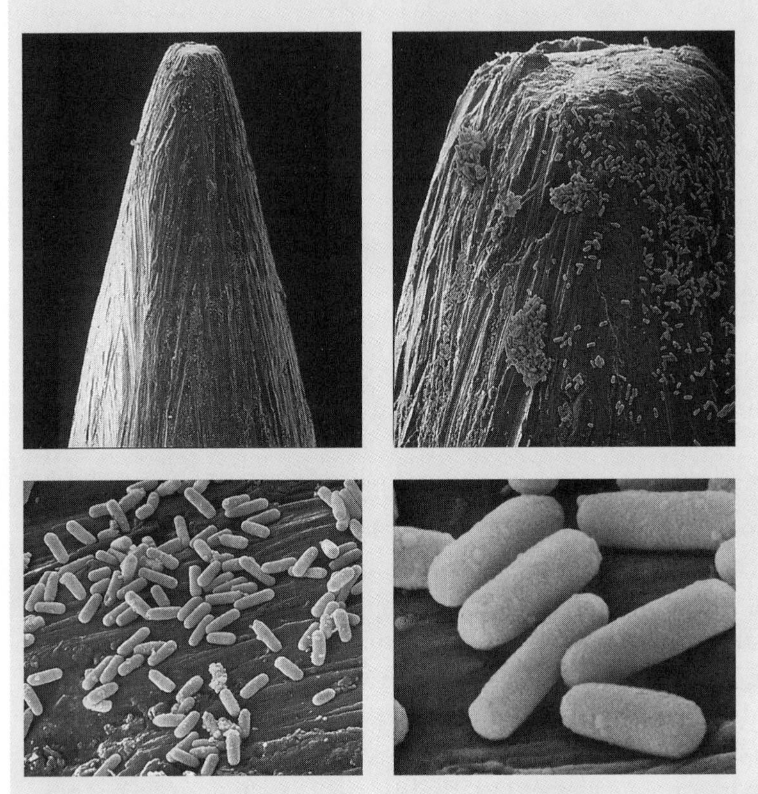

Rod-shaped bacterial cells on the point of a pin as seen by a scanning electron microscope.

were made up of cells. The next year, a German zoologist, Theodor Schwann, said that all animals were made up of cells.

For the first time in the millions of years that humans had been observing the world around them, the world of the very small began to be opened to the microscope's prying eye. People were now able to see the "unseen." The microscope had also opened a window through which people could view the inner structure of living things. What biologists saw was that all living things—from the smallest "invisible" microorganisms in a drop of water to the skin of an elephant to the leaves of a giant redwood tree—are made up of cells. The idea that all organisms are made up of cells is one of the central ideas that forms the foundation of modern biology. It is called the "cell theory." The parts of the cell theory are:

- All organisms are made up of one or more cells.
- The cell is the basic unit of structure and function of all living things.
- All cells arise from previously existing cells.

Although the cell theory describes the astounding unity among all living things, it is not able to address some very difficult questions. Are viruses, which are smaller than cells and cannot reproduce outside of a cell, alive? Also, if cells come only from cells that already exist, where did the first cell come from? And what about muscles, where many cells seem to join together? Do these "supercells" contradict the cell theory? Scientific knowledge is not able to answer every question. It is also not able to make a complicated universe simple.

LEVELS OF ORGANIZATION

We have been studying how living things are put together, beginning with atoms as the starting point. The next **level of organization** above atoms (and the molecules that are made from them) are the families of organic compounds, and beyond that, **cells**. The organization of living things continues. Plants and animals, composed of enormous numbers of cells, have groups of similar cells called **tissues** that are designed to do a specific function. The nervous tissue in your brain consists of billions of nerve cells, or neurons, working together.

A group of tissues, in turn, works together as an **organ**. The brain is an organ made up of different types of tissue, including nervous, blood, and connective tissue. Organs that work together make up **organ systems**. The nervous system, for example, includes the brain, spinal cord, and sensory organs. Finally, different organ systems work together as a functioning organism. (See Figure 5-14 on page 112.)

These levels of organization continue beyond ourselves. Individuals belong to specific populations. All the populations in one place make up a community. The community of living organisms, along with the nonliving environment of water, soil, and air, make up an ecosystem. And finally, all ecosystems on Earth together make up the grandest level of all, the biosphere.

To study life is to learn about all of the levels of life's organization, sometimes one at a time, sometimes all together. Each level adds something to the previous level. At each level there are characteristics that were

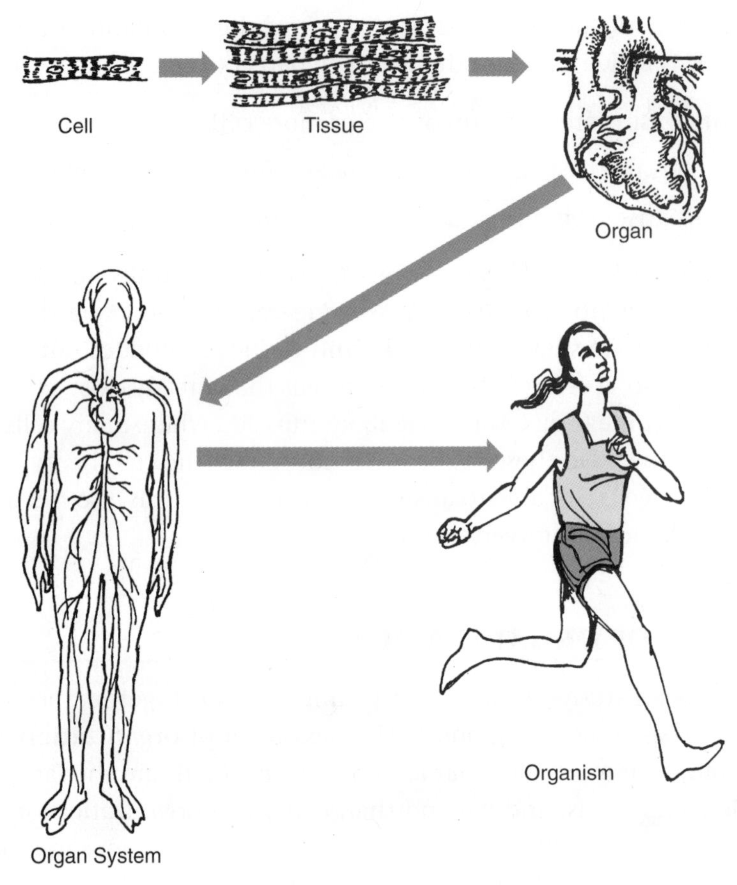

Figure 5-14 Levels of organization in living things.

not present on the previous level. The main characteristic of all, that which we call life, is difficult to define. We know it when we see it; a bug is alive, a rock is not. While we may not be able to define it, we can say what living things do. The properties of living things include:

- Order: organisms have complex organization; even a single cell is organized to perform all functions needed for survival.
- Reproduction: organisms give rise to more of their own kind.
- Growth and development: organisms pass through stages that are the same for all the members of a species; for example, a tadpole becomes a frog.
- The use of energy: organisms take in and change energy for their use; for example, by making, or taking in, food.

- Response to the environment: something happens and an organism responds; for example, sweating on a hot day.
- Homeostasis: the outer environment changes but the inner environment of an organism usually remains the same; for example, the 37°C body temperature of humans remains constant in July and January, even though the outdoor temperatures change dramatically.
- Evolutionary adaptation: life-forms evolve as the organisms interact with their environment; for example, the white fur of a polar bear is an adaptation that helps camouflage the bear, which helps it catch prey in the snowy Arctic.

EXTENDED LABORATORY INVESTIGATION 5
What Does the Microscope Reveal About Changes in Pond Life Over Time?

INTRODUCTION

In 1675, Anton van Leeuwenhoek became the first person to see the "invisible" life in a drop of water from a pond. To see the organisms he termed "wretched beasties," he had to use a microscope.

A drop of pond water contains a world of life. The numbers and types of organisms that live in a pond are probably beyond the limits of most people's imagination. In this investigation, you will study some of the life-forms typically found in a drop of pond water over an extended period of time.

MATERIALS

Pond water, 1000-mL beaker, microscope slides, medicine droppers, coverslips, compound microscope

PROCEDURE

1. Obtain and place 500 mL of pond water into a 1000-mL beaker.
2. Make a Data Table with two columns, one labeled *Date* and the other labeled *Observations*.
3. Make a list of some of the living things you would expect to find in pond water.
4. Place a drop of the pond water on a clean glass microscope slide.
5. Place a coverslip over the drop of water. Place one edge of the coverslip at the edge of the drop. (Make sure there is enough pond water to run along the edge of the coverslip.) Let the coverslip fall onto the drop of water. Try not to trap air bubbles under the coverslip.
6. Place the slide on the microscope. Make sure the low-power lens is in place. Focus your slide.
7. In the Data Table, write a few sentences to describe your first impressions about what you can see. Draw at least three different organisms that you observe on your slide. Include as much detail as you can see. Count or estimate the number of organisms under one field of view of the microscope.

8. Switch to the high-power lens. Use the fine-adjustment knob to focus this lens. (Never use the coarse-adjustment knob to focus when the high-power lens is in place.) Make additional drawings of what you observe.
9. Place your beaker of pond water in a cool location that is exposed to daylight. Repeat steps 4 to 8 each day for two weeks.

INTERPRETIVE QUESTIONS

1. In what ways is the microscopic life you observed in a drop of pond water similar to the macroscopic organisms that live in a pond?
2. How did the numbers and types of organisms change during the two weeks in which you made the observations?
3. Why might certain types of organisms have disappeared after some time has passed?
4. Imagine that you are one of the organisms in the drop of pond water you examined under the microscope. Write a short essay describing a day in your life.

CHAPTER 5 REVIEW

Answer these questions on a separate sheet of paper.

VOCABULARY

The following list contains all of the boldfaced terms in this chapter. Define each of these terms in your own words.

amino acid, atoms, cells, condensation synthesis, elements, hydrolysis, level of organization, lipids, metabolism, organ, organic, organ systems, proteins, simple sugars, starch, tissues

PART A—MULTIPLE CHOICE

Choose the response that best completes the sentence or answers the question.

1. All matter is made up of *a.* molecules *b.* neutrons *c.* tissues *d.* atoms.
2. From smallest to largest, the levels of organization in organisms are *a.* cell, organ, tissue, organ system *b.* cell, tissue, organ, organ system *c.* organ system, organ, tissue, cell *d.* organ, organ system, cell, tissue.
3. Carbohydrates include *a.* glycerol *b.* sugar *c.* cholesterol *d.* fat.
4. Which of the following statements is true? *a.* Matter has mass and takes up space. *b.* All living things require energy. *c.* Life requires a high state of organization. *d.* All of these.
5. Electrons in the shell nearest the nucleus *a.* orbit the nucleus in distinct, predictable paths *b.* have the most energy *c.* have the least energy *d.* lose energy when they move to a different shell.
6. Amino acids are the building blocks of *a.* carbohydrates *b.* lipids *c.* proteins *d.* nucleic acids.
7. A chemical bond in which atoms share electrons is *a.* ionic *b.* covalent *c.* molecular *d.* organic.
8. All organic compounds contain *a.* hydrogen *b.* oxygen *c.* nitrogen *d.* carbon.
9. Polymer bonds are broken by the process of *a.* dehydration synthesis *b.* nuclear fission *c.* hydrolysis *d.* anabolism.
10. The main purpose of lipids is *a.* storing energy *b.* providing strong building materials for certain types of cells *c.* transporting other substances *d.* storing genetic information.
11. Humans store energy for short-term use as *a.* glycogen *b.* glucose *c.* starch *d.* fat.

12. Rutherford contributed to our knowledge of atoms by
    a. inventing a microscope powerful enough to show atoms
    b. discovering that atoms consist mostly of empty space
    c. theorizing that matter is made up of atoms
    d. hypothesizing that atoms spin around the nucleus.
13. According to the second law of thermodynamics, *a.* natural events tend to increase disorder in the universe *b.* matter and energy cannot be created or destroyed *c.* atoms are the building blocks of matter *d.* objects in motions tend to stay in motion.
14. If every position on a fat molecule's carbon chain is bonded to hydrogen atoms, it is said to be *a.* saturated *b.* polyunsaturated *c.* dehydrogenated *d.* polymerized.
15. Which of the following statements is true? *a.* Elements consist of more than one kind of atom. *b.* Elements that do not occur in nature have been created artificially. *c.* The six most important elements for organisms are carbon, oxygen, helium, iron, copper, gold, and nitrogen. *d.* All of these.

PART B—CONSTRUCTED RESPONSE

Use the information in the chapter to respond to these items. Refer to the diagrams below to answer questions 16 and 17.

Reaction I:

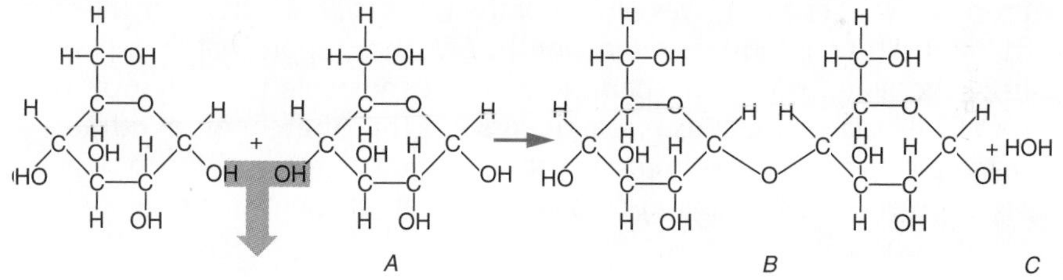

A B C (+ HOH)

Reaction II:

D ($+ H_2O$) E

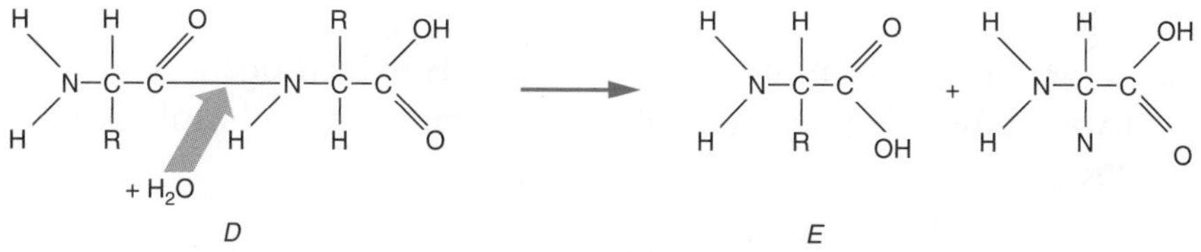

16. (*a*) What process is shown in Reaction I? What process is shown in Reaction II? (*b*) Classify the molecules labeled *A*, *B*, *C*, *D*, and *E*. Use as specific a term as possible. (*c*) To which of the four main groups of organic molecules do molecules *A* and *B* belong?

17. Explain how Reaction I and Reaction II relate to metabolism, polymers, and macromolecules.
18. Why is the element carbon important to life on Earth?
19. Describe, with examples, the four major families of organic compounds.
20. Summarize how Leeuwenhoek, Hooke, Schleiden, and Schwann contributed to the development of the cell theory.

PART C—READING COMPREHENSION

Base you answers to questions 1 through 3 on the information below and on your knowledge of biology. Source: *Science News* (June 28, 2003): vol. 163, p. 408.

Calling Out the Cell Undertakers

Millions of cells die naturally each day in a person. Scientists have now discovered that these dying cells send out a chemical signal to attract other cells that specialize in disposing of cellular corpses.

Over the past few years, biologists have begun to understand how macrophages and other cells recognize dying cells. For example, a cell about to die sprouts what scientists refer to as eat-me signals, which tell a macrophage to consume the cell before it falls apart and triggers inflammation (*SN:* 9/29/02, p. 202).

But what if there is no macrophage close at hand to a dying cell? No problem, say Sebastian Wesselborg of the University of Tübingen in Germany and his colleagues. In the June 13 *Cell,* they report that dying cells from monkeys, mice, and people secrete a molecule called lysophosphatidylcholine. Previous research showed that the chemical attracts macrophages and other immune cells that may be some distance away. This lure ensures that dying cells are removed efficiently, Wesselborg's group concludes.

21. State the purpose of the chemical signal that is sent out by dying cells.
22. Explain why it is important for dying cells to be properly removed.
23. Explain the role that is played by the chemical lysophosphatidylcholine in the removal of dying cells.

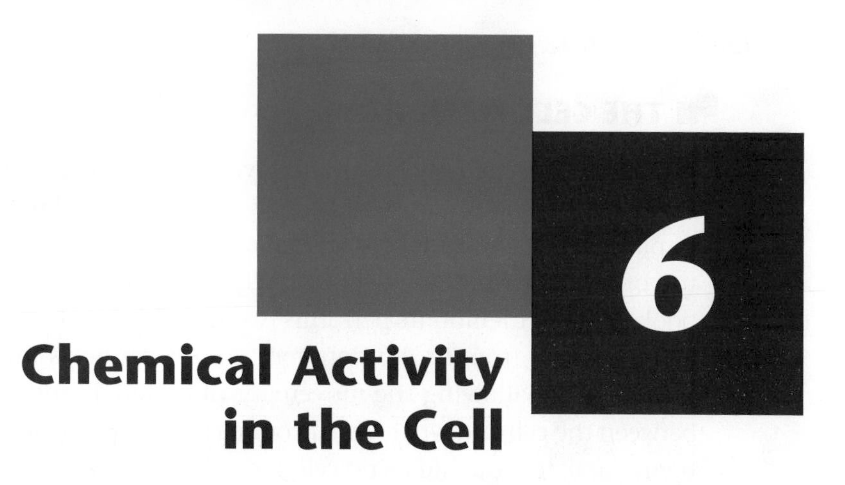

6 Chemical Activity in the Cell

After you have finished reading this chapter, you should be able to:

Describe the structure and function of the cell membrane and of the various cell organelles.

Compare and contrast osmosis and diffusion; passive transport and active transport.

Explain the importance of ATP to energy storage in cells and the role of enzymes and substrates in chemical reactions within cells.

As small as it may be, a cell is a living thing engaged in the risky business of survival.

Cecile Starr

Introduction

Imagine that you are a single-celled organism. No one would ever know of your existence unless someone looked at you under a microscope. Your surroundings would be very important to you—to make you feel comfortable and to supply your needs. Your needs would include food, which you would use to produce energy to stay alive. You would produce waste products that would have to be disposed of. Sooner or later, you would reproduce. Many activities would be going on inside you, and some kind of cell "brain" would have to control and coordinate all this activity. As a cell, you would be very much alive inside. However, the environment outside of you would be the opposite—a physical, nonliving place where changes in temperature occur, where the amount of water and the amount of light also vary. What stands between a cell and the sometimes hostile environment that surrounds it?

THE CELL MEMBRANE

An ultrathin, extremely important layer separates the living world inside a cell from the nonliving world outside. This is the **cell membrane**. A cell, or plasma, membrane is so thin that at least 10,000 of them piled one on top of another would be needed to equal the thickness of a page of this book. The cell membrane performs two functions at the same time. It separates the cell from its environment, and it connects the cell to its surroundings by allowing the movement of materials and communication between the cell and whatever surrounds it. (See Figure 6-1.) Without a cell membrane, there could be no cell. It can be said that the beginning of life had to wait until the formation of the first cell membrane.

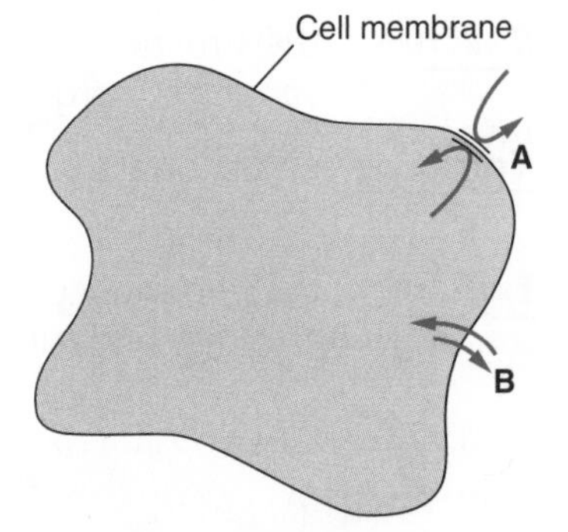

Figure 6-1 The cell membrane acts as a barrier, separating the inside of the cell from the outside (A) and as a bridge, allowing materials to move in and out of the cell (B).

As is true for much of the cell, the cell membrane demonstrates the tremendous unity among living things. The cell membranes of all cells—from bacteria to the cells of mushrooms, roses, lobsters, chimpanzees, and humans—have the same basic structure. Under the powerful electron microscope, the cell membrane appears as a thin double line. What is being observed is a double layer of phospholipids. (See Figure 6-2.)

Phospholipids are a special type of lipids with "split personalities." Unlike in a typical lipid molecule, in a phospholipid molecule, one of its three fatty acids has been replaced by a phosphate group. This is a chemical group that has a special attraction for water molecules. The other two fatty acid chains tend to avoid water. Because of this, the phospholipid molecules arrange themselves in a special way. All the fatty acid "tails" of the molecules move into position to face each other and avoid contact with water. This creates a double layer of molecules, with the fatty acids of both layers facing inward. The phosphate "heads" of the phospholipid molecules face the outside of each layer, where they are in contact with water either inside or outside of the cell. This well-organized arrangement is called a *phospholipid bilayer*. Soap molecules, which behave like phos-

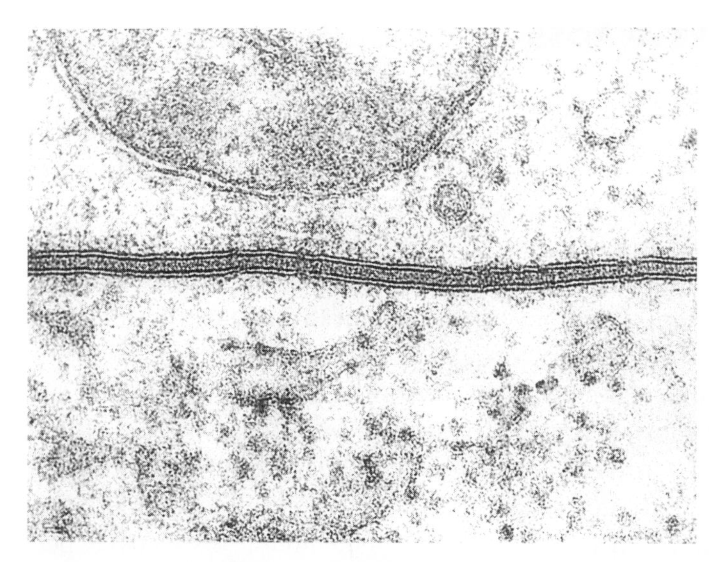

Figure 6-2 Phospholipid bilayer as seen by an electron microscope.

pholipids, arrange themselves into similar layers and then into spherical bubbles. In the same way, phospholipid molecules automatically and spontaneously arrange themselves into double layers that surround tiny spheres, or cells.

The phospholipid bilayer of a cell membrane separates the inside of the cell from the surrounding environment. However, it is absolutely necessary for materials to move into and out of cells. Much of the movement of materials across the cell membrane is done by protein molecules located in the phospholipid bilayer. These protein molecules often extend from one side of the membrane through to the other. The lipid molecules in the membrane are constantly in motion. The protein molecules stuck in them are, therefore, always moving too. This description of the cell membrane is called the **fluid mosaic model**. The protein molecules are thought of as floating in a double-layer sea of phospholipids. (See Figure 6-3.)

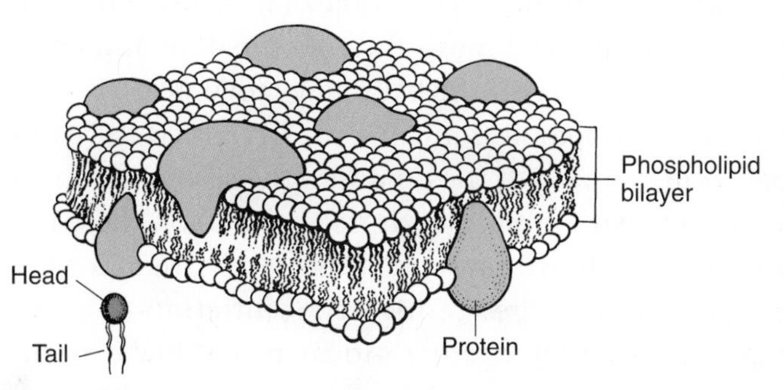

Figure 6-3 The fluid mosaic model of the cell membrane proposes that protein molecules float in a double-layer sea of phospholipids.

TRANSPORT ACROSS THE CELL MEMBRANE

For a cell to remain alive, it must have a very special collection of chemicals inside it. This inside collection is quite different from the chemicals located in the environment outside the cell. Some substances that are abundant outside of the cell are not found inside the cell. Other substances that are scarce in the environment outside the cell are present in significantly larger quantities inside the cell. It is the cell membrane that creates and maintains the special environment inside the cell. How does it do this?

The cell membrane is picky. It allows some substances to pass through but keeps other substances out. This ability to pick and choose which molecules can pass through is called **selective permeability**. (To be permeable is to allow substances to go through. For example, cotton fabric is permeable to water; plastic is not.) The cell membrane is selectively permeable. The cell membrane also makes possible the rapid transport of some molecules across it, while other molecules pass through slowly.

The selective permeability of the cell membrane is due to several factors. Only certain molecules—those that are very small and those that dissolve in lipids but not in water—can move through the phospholipid bilayer. Some other specific molecules, including some larger ones, are allowed through because the protein molecules present in the cell membrane act as carriers. Now you know that the cell membrane acts as a gatekeeper that determines which molecules can pass through. But what determines the direction in which the molecules move? Do molecules go into or out of the cell?

PASSIVE TRANSPORT

Picture a room, perhaps at an amusement park, filled with hundreds of plastic balls being blown in many directions by large fans. Suddenly a door opens connecting this room to another room with fans but only a few balls in it. What will automatically occur? Some of the balls in the first room, bumping into each other, will move through the door and enter the second room. While one or two of the balls may move from the second room back in to the first room, it is much more likely that the overall movement of balls will be through the door into the second room.

A similar process occurs with molecules. Instead of a door, there is a membrane the molecules can move through. There is an overall or net movement of molecules from an area of high concentration—a place where the molecules were crowded together—to an area of low concentration. This kind of movement is called **diffusion**. (See Figure 6-4.) Molecules are constantly in motion, and they naturally move from where they are more concentrated to where they are less concentrated. This movement happens automatically if the membrane is permeable to the substance and if there is a difference in concentration of the substance, a **concentration gradient**, on either side of the membrane. This is called **passive transport**, because no energy is used by the cell and no work is done. For example, one of the basic needs of most cells is oxygen. Many cells use oxygen molecules (O_2). There are few O_2 molecules inside the cell but usually an abundance of them in the water or other liquid that surrounds the cell. Oxygen molecules diffuse across the cell membrane into the cell by passive transport.

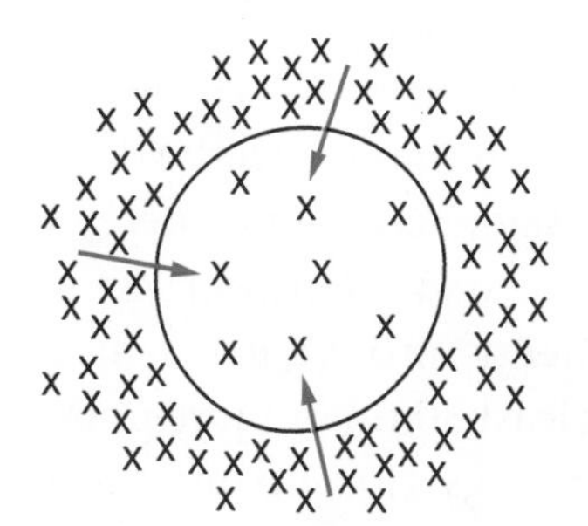

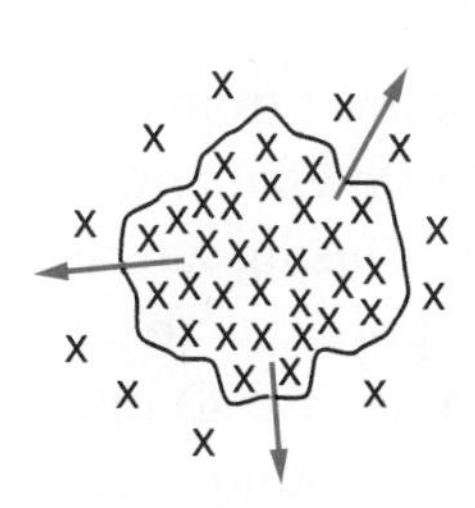

Figure 6-4 Diffusion is the movement of molecules from an area of high concentration to an area of low concentration.

Because they are very small, water molecules can also easily move through the cell membrane. Usually, water moves from a high concentration gradient outside the cell to a region of lower concentration inside the cell. In situations where water is more abundant (high concentration) inside a cell than outside the cell (low concentration), which way will the water molecules move? In this case, they will move outside the cell. The diffusion of water molecules across a cell membrane—so important for living cells—is given a special name, **osmosis**. Many cells, such as plant and fungus cells, have a stiff cell wall that surrounds their cell membranes. If water leaves the cell by osmosis, the contents of the cell shrink. As a result, the cell membrane moves away from the cell wall in a process called **plasmolysis**. Plasmolysis can be seen under a microscope when plant cells are put in a strong salt solution. The abundant fresh water inside the plant cells moves out of the cells, where there is more salt and fewer water molecules. You can actually see the cell membrane pull away from the cell wall. (See Figure 6-5 on page 124.)

The reverse happens when limp celery or lettuce is put in fresh water. The celery stems or lettuce leaves are limp because their cells have too little water in them. When you put the celery or lettuce in water, osmosis

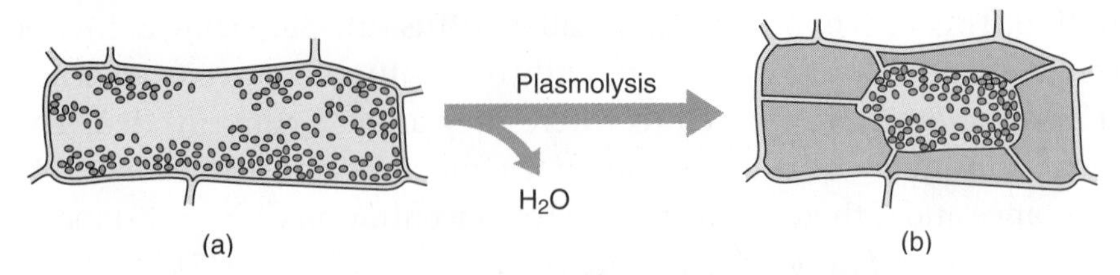

Figure 6-5 Effects of plasmolysis in a plant cell: water moves out of the cell.

occurs and water molecules move into the cells. The cells expand, the cell membranes push against the cell walls, making the cells firm and the celery and lettuce crisp again.

ACTIVE TRANSPORT

Diffusion moves substances from high to low concentration, in the direction of the concentration gradient. As we have said, because no energy is required to move the diffusing substances, this is passive transport. A very different kind of transport, one that requires energy, also occurs. To function properly, human nerve cells must have at least 30 times more potassium inside them than is found in their surroundings. How do nerve cells keep this high concentration of potassium in them? They constantly transport potassium into themselves, even though there are already more atoms of potassium inside the nerve cell than outside. This is done by a protein molecule in the cell membrane that acts like a pump. Like an air pump that fills up a bicycle tire, this pump also requires energy to do work. This movement of a substance against the concentration gradient is known as **active transport**. When substances are moved from a low to a high concentration, energy is used and work is done. While the protein pump is moving potassium into nerve cells, it is also pumping sodium out. Amino acids, sugars, and other nutrients are also moved by active transport. It is one of the most important activities of cells. Other than using energy from your food to keep you warm, the most important use of energy in your body is to help pump substances across the membranes of your cells by active transport—a process going on all the time without your knowledge!

Check Your Understanding

Why is energy required for active transport to take place across a cell membrane? Why is energy not required for passive transport to occur?

ENERGY TRANSFORMATIONS INSIDE THE CELL

You have seen how important the cell membrane is in controlling substances that are kept inside and outside of a cell. You know that life depends on a constant input of energy. You can now begin to consider how a flow of matter and energy through cells is what life is really all about.

The energy used in the chemical reactions that take place inside cells is associated with the electrons of atoms. The greater an electron's distance from its nucleus, the more stored, or potential, energy it has. When some atoms join, the electrons shared between them form the covalent bonds of a new chemical compound. Each type of covalent bond has a specific amount of energy. Whenever covalent bonds are formed or broken, the amount of stored energy changes. (See Figure 6-6.) Chemical reactions are mainly energy transformations in which energy stored in chemical bonds is transferred to other, newly formed chemical bonds or is released as heat or light.

For example, the glucose in a chocolate bar is used by your cells after you eat the chocolate. In a chemical reaction that requires oxygen, the high-energy chemical bonds in the glucose molecules are broken. When glucose molecules are broken apart, carbon dioxide and water form. That is why we breathe in oxygen and breathe out carbon dioxide. The main point, however, is that the energy levels of the chemical bonds in carbon dioxide and water are lower than the energy levels of the chemical bonds in glucose. Energy has been released.

If cells released all this energy at once, a great deal of heat would be produced. This heat would kill the cells. Over long periods of time, cells have developed the ability to control energy-releasing reactions, storing energy in the bonds of special substances and making it available only when and where the organism needs the energy. Much of the metabolism

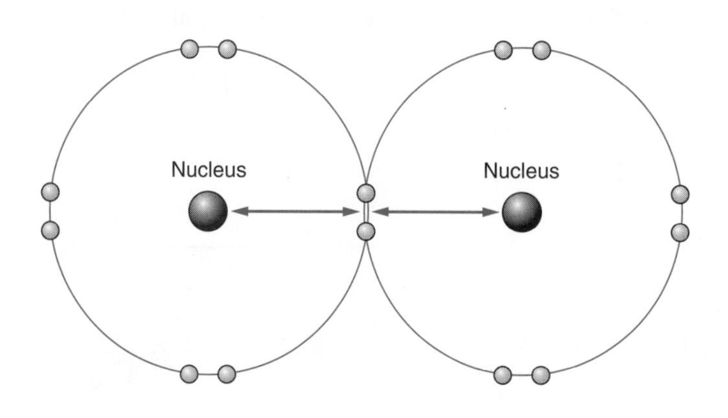

Figure 6-6 The energy in a covalent bond depends on how far the shared electrons are from the nuclei of the atoms.

Who Owns Your Cells?

The spleen is an organ that helps your body fight disease. It also helps break down old red blood cells. For John Moore, removal of his spleen cured his leukemia, a type of cancer of the blood, which he was found to have in 1976. Removing the spleen is a standard treatment for this disease. The leukemia did not return and Mr. Moore was very happy.

But his attitude was soon to change. For what Mr. Moore did not know was that the physicians at the University of California who removed his spleen kept some of the cells from the spleen alive. The physicians put the cells in a special nutrient-rich solution. As the cells continued to reproduce, the physicians studied them. They discovered that cells from Mr. Moore's spleen produced an interesting blood protein. The physicians received a patent on the cells, which made them the legal owners of the cells they removed from Mr. Moore. In turn, the cells and the blood protein they made were being sold to a company that planned to develop a new medicine from the cells—a medicine that would be sold for a large profit. Mr. Moore did not think this was right. He told the physicians, "Don't use my cells—I own them," and then he went to court.

The lawsuit was heard by the Supreme Court of California, which ruled on July 9, 1990 in favor of the physicians. Later, the United States Supreme Court also ruled in the physicians' favor. The courts felt that scientific research would be threatened if researchers did not have the freedom to work with human cells. Besides, the courts said that John Moore never expected to get his cells returned to him when he gave permission for his spleen to be removed in his medical procedure.

As of July 2001, Mr. Moore remains healthy. The drug company stock now owned by Mr. Moore's original doctor—in exchange for the rights to use the Moore cell line—is worth over five million dollars. Were the courts correct? Did Mr. Moore have a legitimate claim to cells from his body? In other words, who owns your cells? What do you think?

of cells involves this interconnecting of energy-releasing and energy-consuming processes.

An important compound in the metabolism of all living organisms is **adenosine triphosphate**, or ATP. ATP is the substance in which cells store this released energy temporarily until it is needed. (See Figure 6-7.) It is

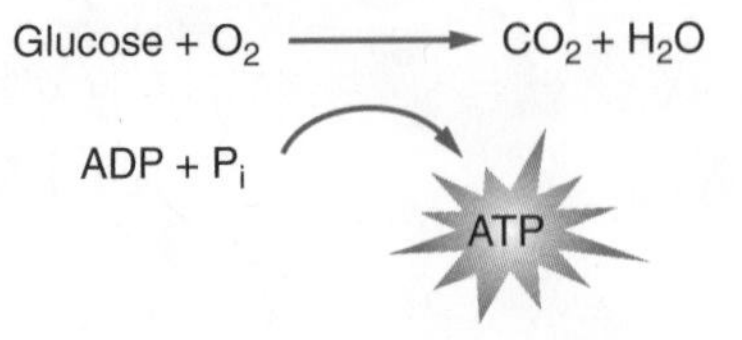

Figure 6-7 ATP is the form in which cells store the energy released by respiration. (ADP is changed to the higher-energy ATP when a phosphate bond is added.)

like change in your pocket. You keep it there until you need to spend it. ATP is the energy "currency" of the cell. ATP molecules are small and contain usable amounts of energy stored in a form that is instantly available when it is needed. Glucose is like the money in your bank account for long-term storage.

GETTING CHEMICAL REACTIONS TO HAPPEN

A chemist wants to cause a chemical reaction. She places two chemical compounds in a test tube, lights a Bunsen burner, and heats the compounds to a temperature of at least several hundred degrees. The molecules heat up, begin to move rapidly, and then begin to collide with each other with enough force to change each other. The reaction happens.

Our cells, thankfully, cannot and do not do this. Our body temperature remains at a steady 37°C and normally does not vary by more than a few degrees higher or lower. In fact, a very high fever of perhaps 42°C may result in death. So how do cells have the amazing ability to perform the many chemical reactions of cellular metabolism at relatively low body temperatures?

ENZYMES: THE CELL'S MIRACLE WORKERS

For chemical reactions in the cell to take place, they must usually occur in a series of small steps rather than in a single large burst of activity, such as in the heated test tube. These steps must also be very precise. They must occur in the correct order, one after the other.

For example, to change substances A and B into substance F, they may first change into C, then D, which may join with E to finally become F. Substances A and B are called the reactants for reaction 1, and C is the product. C then becomes the reactant for reaction 2, changing into product D, and so on. An equation for this series of reactions looks like this:

$$A + B \xrightarrow{(1)} C \xrightarrow{(2)} D + E \xrightarrow{(3)} F$$

In cells, metabolic activity includes hundreds of chemical reactions that occur in steps such as these. Cells must be well organized.

One problem for the cell is that the reactants must get changed into exactly the right products and not something else. Also, as we mentioned, the reaction must occur at a relatively low temperature so that the cell is not harmed. Both problems are solved by **enzymes**. Enzymes are usually proteins, although new research has shown that some may be RNA molecules. Because a particular enzyme is needed for each type of chemical reaction, cells have thousands of different kinds of enzymes. Enzymes are able to do astounding things. For example, a reaction that normally might occur only once every hundred years can occur ten times a second if the correct enzyme is present! And the same enzyme molecule will do its job over and over again, without itself being changed.

A substance responsible for greatly changing the rate at which a chemical reaction occurs without being changed itself is a **catalyst**. Enzymes are types of catalysts. Because they are either proteins or nucleic acids, enzymes are organic catalysts. Each enzyme is important; each has its own name. The name often comes from the name of the substrate, the substance that the enzyme acts on. An enzyme's name usually ends with the letters *ase*. For example, the enzyme that catalyzes the reaction that splits the starch maltose into two glucose molecules is called maltase. Lipases are enzymes that break down lipids; proteases are enzymes that break down proteins.

To get the reactants A and B to come together and change to C is not easy. Molecules behave like individuals who are proud of who they are, and these individuals resist change. You can think of it this way. Their resistance to reacting is similar to climbing a hill. You have to get the reactants "over the hill" in order for them to roll down the other side. That is when the reaction occurs. The energy needed to do this is called the **activation energy**. One of the most important functions of enzymes is lowering the level of activation energy needed to make the reaction occur. (See Figure 6-8.) In a sense, the enzyme lowers the hill. In the presence of

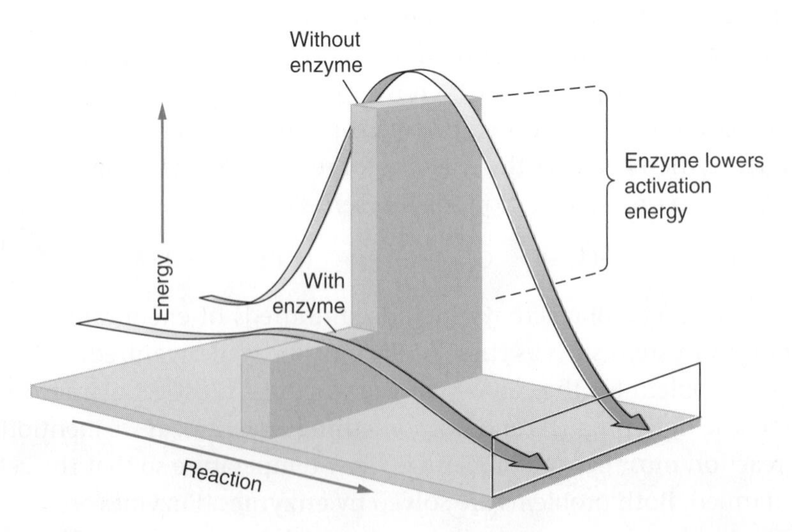

Figure 6-8 Enzymes speed up reactions by lowering the level of activation energy needed to make the reaction occur.

the proper enzyme, the reactants or substrates are much more likely to come together, go over the hill, and change into the new product.

Besides speeding up the rate of a reaction, enzymes are amazingly good at getting reactions done correctly. Enzymes make certain that the product you get after the chemical reaction is the product you want. Consider thousands of parts that make up a car. If all of these parts were thrown into a large container and shaken, would you get a car? Of course not. On a car assembly line, each part is fitted in a very precise manner to another part before going on to the next step. An enzyme is like the assembly-line robot that fits the parts together.

To do this job, enzymes must be very precise in their actions. They are able to be very accurate in their work because they are usually proteins that have a particular shape. Remember that each protein molecule is a long chain of amino acids, which bends and twists into a specific shape. The shape of an enzyme molecule includes a spot somewhere on its surface that is sort of like a pocket. This "pocket" is exactly the right size and shape for a particular substrate molecule. If two different substrates are involved in the chemical reaction, there will be a precise fit in the enzyme's pocket for each substrate molecule. This place on the enzyme molecule is called the **active site**. The fit of the substrate in the active site resembles the fit of a key in a lock, which is why scientists call this the "lock-and-key model" of enzyme action. (See Figure 6-9a.)

An enzyme does its work by joining with the substrate molecules in this close fit. However, this is only a temporary association and is known as the **enzyme-substrate complex**. The chemical reaction occurs while the enzyme and substrate are fitted together. The substrate changes in a specific way, but the enzyme does not change. The product of the reaction is released from the active site and moves to the next step in the process. The unchanged enzyme, with its open active site, gets used again to catalyze the same reaction on another molecule of the same substrate. In fact, the enzymes are recycled.

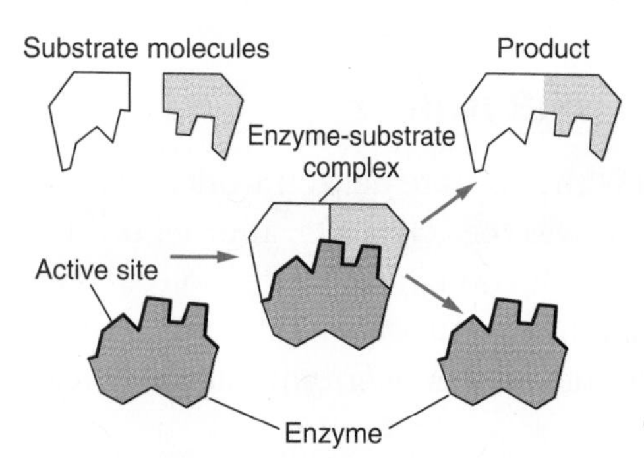

Figure 6-9a Lock-and-key model of enzyme action.

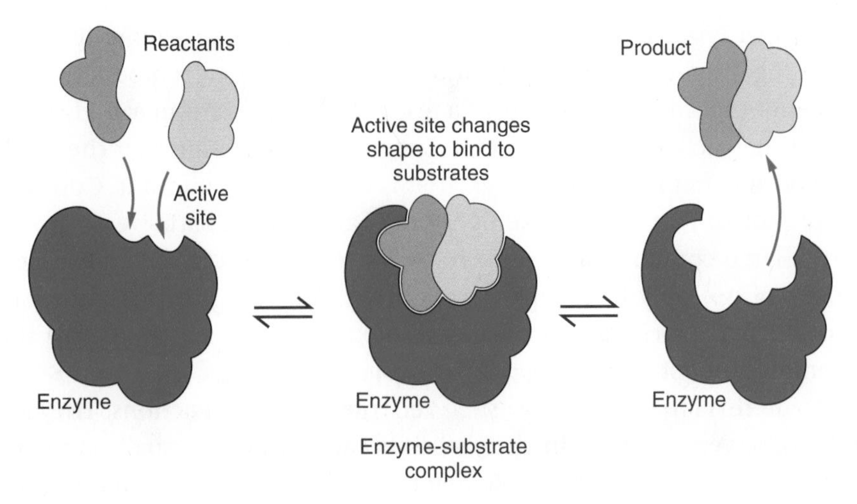

Figure 6-9b Induced-fit model of enzyme action.

Recently, scientists found that when the substrate moves to an active site, a slight temporary change in the shape of the enzyme molecule may occur. The substrate itself is involved in creating or inducing a proper fit with the enzyme. Rather than a key fitting into a lock, it is more like one hand clasping another hand in a handshake. This momentary change in the shape of an enzyme may be involved in getting the substrate to change in a permanent way. Some researchers consider it better to change the name of the enzyme-substrate connection from the lock-and-key model to the "induced-fit model" of enzyme action. (See Figure 6-9b.)

Only small amounts of enzymes are needed to do their work. However, if an enzyme is needed for a particular chemical reaction, the absence of even the small amount needed can be disastrous. A genetic disease called phenylketonuria, which causes severe mental retardation in children, is due to the absence of a single enzyme in the body. Tay-Sachs is another genetic disease caused by the lack of a single enzyme. This disease usually leads to infant death.

HELPING ENZYMES DO THEIR JOBS

Keeping an enzyme well supplied with substrate helps it work at its best. As soon as one substrate molecule leaves the active site, another one can pop into it. However, there are limits. If enough substrate molecules are present to keep the active site full all the time, adding more substrate molecules will not make a difference in the number of enzyme-substrate reactions that occur. The enzyme is said to be saturated and working at its maximum speed and efficiency.

For an enzyme to work properly, the protein molecule must maintain its correct shape. Two conditions in the cell, temperature and pH, are very important for maintaining the shape of an enzyme molecule.

Temperature. Many animals have evolved ways to maintain a relatively constant internal temperature, even as the temperature of their surroundings varies greatly. For example, the huge ears of the desert hare act as radiators. The blood that flows through the ears gives off heat to the air. The heat that is given off prevents the hare from overheating in the hot desert. The main reason animals need to maintain a constant body temperature is to allow cellular enzymes to function properly. The fact that high temperatures can destroy our own enzymes is one reason why a long-term high fever can cause death. (See Figure 6-10.)

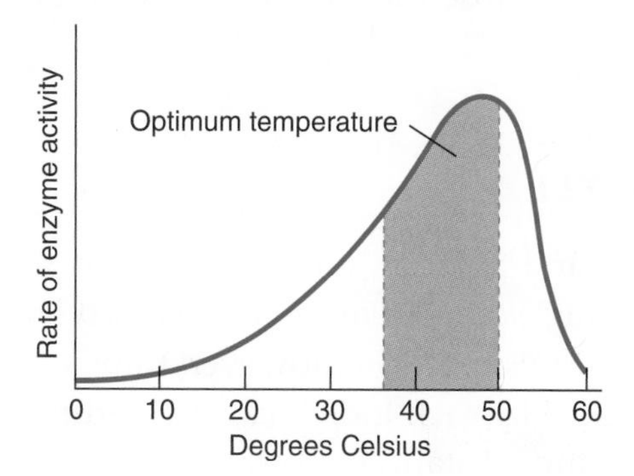

Figure 6-10 A typical enzyme in our body works best at temperatures between 37°C and 50°C.

A look at eggs can help us understand how high temperatures affect proteins, including enzymes. Egg white consists of the protein albumin. When an egg is cooked, a big change occurs in the albumin protein. The heat used to cook the egg causes the protein to change shape. The change is not reversible—the egg cannot be uncooked. If the albumin proteins had been enzyme molecules, they could no longer function.

pH. Scientists use a special scale, the **pH** scale, to measure how acidic a solution is. The opposite of an acid is a base. Both an extremely strong acid and an extremely strong base can burn the skin. The pH scale runs from 0 (highly acidic) to 14 (highly basic). A neutral solution such as pure water has a pH of 7. Vinegar, an acid, has a pH of 2.8; ammonia, a base, has a pH of 11.1. Slight changes in pH quickly change an enzyme's shape

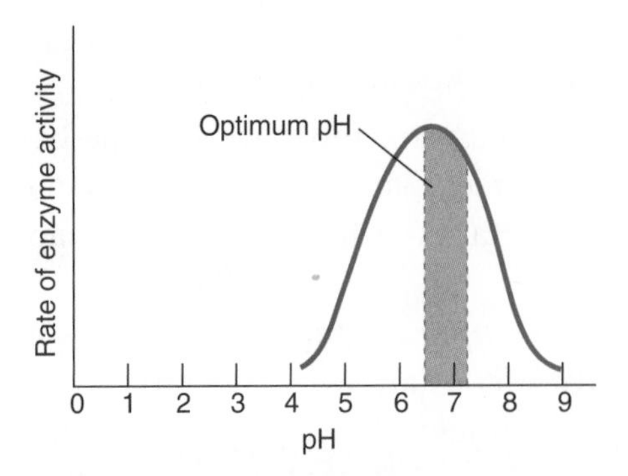

Figure 6-11 Most enzymes function best at about pH 7, which is neutral.

and its ability to affect a substrate. Most enzymes function best near pH 7. Our cells must maintain that pH in order to survive. (See Figure 6-11.)

Many enzymes need nonprotein helpers to do their jobs. These helpers, called **cofactors** or **coenzymes**, may be organic compounds or metal atoms. Most vitamins and some minerals such as zinc, iron, and copper are necessary in our diet because they act as cofactors for enzymes.

TAKING A TOUR OF THE CELL

Cells are the basic units of structure and function for all living things. Some living things are made up of a single cell, but most have many cells that work together on behalf of the organism. Still, almost everything we need to do to stay alive is accomplished by each individual cell: getting food, using food for energy, transporting substances, growing, reproducing, and eliminating wastes. Each of these activities involves a large number of chemical reactions. For all these reactions to take place under the careful, precise control of a great many enzymes involves organization.

Leeuwenhoek and others after him thought that they saw structures inside the single-celled "beasties" in the drops of pond water. Picturing these structures as miniature hearts, lungs, and stomachs, they thought of them as little organs, or **organelles**. We now know that many cells contain internal structures, but they are not miniature organs.

As was mentioned in Chapter 3, all cells can be divided into two main groups: prokaryotic and eukaryotic. Prokaryotic cells lack a nucleus. In structure, they appear "simple." Besides lacking a nucleus, prokaryotes do not have the other membrane-bound organelles found in other cells. All bacteria are prokaryotic. Most scientists marvel at the ability of these so-called simple cells to perform all life functions and survive in a wide variety of different environments on Earth. (See Figure 6-12.)

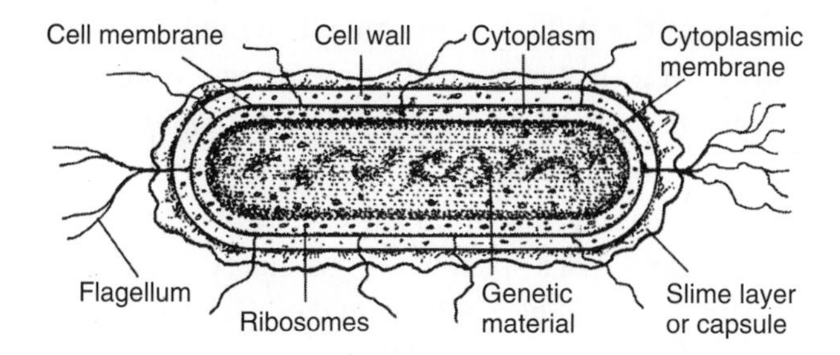

Figure 6-12 Typical prokaryotic cell.

THE EUKARYOTIC CELL

Eukaryotic cells have a separate **nucleus**, that is, a nucleus surrounded by a membrane within the cell. The genetic information of a eukaryotic cell is stored mainly in the nucleus. As you look at a eukaryotic cell through a microscope, you can observe a wide assortment of other complex structures. The cells of all protists, fungi, plants, and animals are eukaryotic. Humans are eukaryotic organisms. A typical eukaryotic cell is about 10 to 100 times larger than a prokaryotic cell.

Let's tour the inside of a eukaryotic cell. To enter a cell, we have to pass through a cell membrane, the phospholipid bilayer. To take our tour, we have to swim, moving through the **cytoplasm** that fills the cell. Cytoplasm is a thin gel, made up mostly of water, with many other chemicals dissolved in it. Around us we observe many *ribosomes*, organelles where proteins are made. We also see *lysosomes*, the structures involved in getting food broken down, scattered throughout the cell. The kidney-bean-shaped *mitochondria* are the "powerhouses" of the cell, the places where energy is released. (See Figure 6-13.)

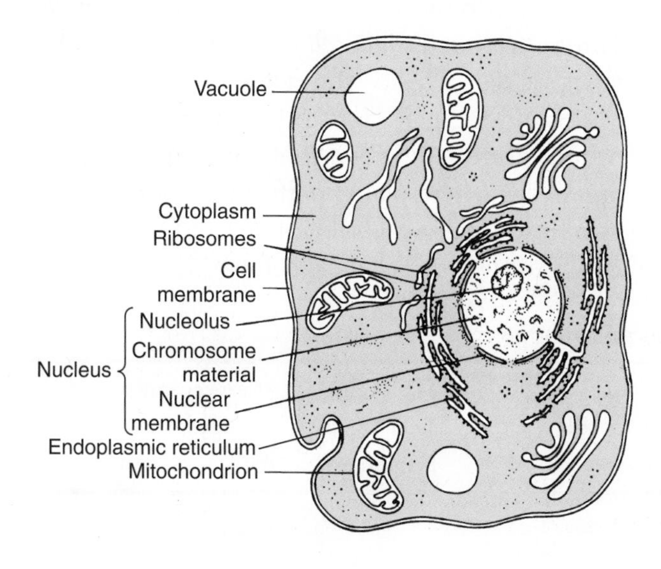

Figure 6-13 Organelles of a typical animal cell.

By far the largest structure in the cell is the nucleus. The nucleus often fills the entire central portion of a eukaryotic cell. We can see a system of canals that spreads out from the nucleus. This system, the *endoplasmic reticulum* (ER), may have ribosomes attached to it. Proteins, made on the ribosomes, and lipids, made in the ER, are then moved through the ER. Near the ER is a structure that looks like a stack of curved plates sitting one on top of another. This is the *Golgi complex*, which makes, packages, and distributes many materials. The organelles mentioned above would be visible no matter what kind of eukaryotic cell we explored.

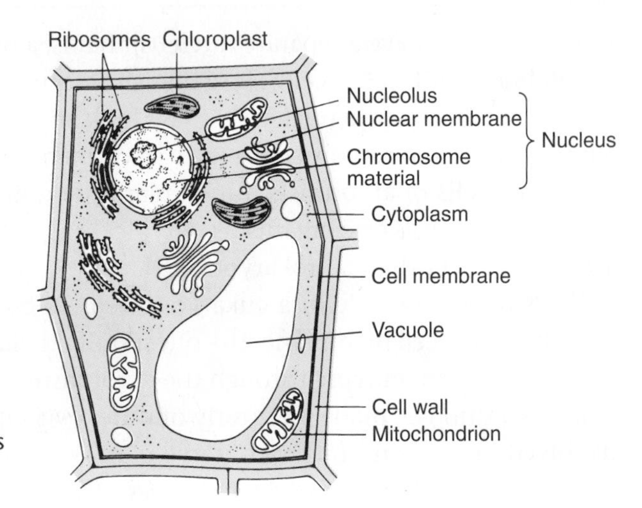

Figure 6-14 Organelles of a typical plant cell.

Other organelles are also present in some cells. If we were touring an animal cell, we would see two sets of rods grouped in rings. These are *centrioles*, structures involved in cell division. If we visited a plant cell, we would first have to pass through a tough, rigid *cell wall* made of the polysaccharide cellulose. (See Figure 6-14.) Inside the cell we would notice a large space occupied by a *vacuole*, the place for storing materials including wastes. Finally, we would notice the green football-shaped objects called *chloroplasts*, the organelles where plants perform the job of converting energy from the sun into food, which the plants and most other organisms—including us—need to live. (See table below.)

SOME CELL ORGANELLES AND THEIR FUNCTIONS

Nucleus	Controls cell activities	**Mitochondria**	Release energy
Vacuole	Stores materials for cell	**Golgi complex**	Packages materials
Ribosomes	Where proteins are made	**Lysosomes**	Break food down

LABORATORY INVESTIGATION 6

What Is the Role of the Cell Membrane in Diffusion?

INTRODUCTION

Diffusion is the movement of molecules from areas of greater concentration to areas of lesser concentration. In organisms, diffusion is limited by the selective permeability of the membrane that surrounds a cell. In this activity, you will study the diffusion of water molecules (osmosis) into and out of red onion cells. You will observe some of the factors that influence the direction of movement of the water molecules.

MATERIALS

Forceps, red onion, glass slides, tap water, coverslips, compound microscope, 5% NaCl solution, paper towels, distilled water

PROCEDURE

1. Use the forceps to remove a very thin piece of the red onion skin and place it flat on a microscope slide. *Note:* The skin is not the brown, papery layer that surrounds an onion. To remove the piece of onion skin that you need, bend a segment of the onion in half and tear one half from the other. If you do this correctly, you will have a very thin piece of onion skin that is transparent and a reddish or light purple in color.
2. Add a drop of tap water to the onion skin. Use your finger to smooth out the onion skin. Add another drop of tap water, if you need to. Place a coverslip over the onion cells. Make sure that the bottom of the glass slide is dry. Place the slide on the microscope stage.
3. Focus the low-power lens on the cells. Make a drawing of what you observe. Switch to high power. Make a drawing of the cells under high power. In your laboratory notebook, use complete sentences to write a description of the cells.
4. Place a small drop of NaCl solution at one edge of the coverslip. Place a small piece of paper toweling at the opposite edge of the coverslip. The paper towel will draw the salt solution over the onion cells.

5. Wait several minutes; then carefully observe the red onion cells. Draw what you observe.
6. Place a drop of distilled water at one edge of the coverslip. Use a piece of paper toweling to draw this water over the cells.
7. Wait several minutes; then carefully observe the red onion cells again and draw them as they now appear.

INTERPRETIVE QUESTIONS

1. How can you explain the changed appearance of the onion cells?
2. Molecules of NaCl do not move through a cell membrane. What explanation can you offer for this?
3. How can the diffusion of water molecules explain what you observed after distilled water was added to your slide?
4. The red onion cells will burst if they remain in distilled water for a long period of time. Offer an explanation for this observation.

CHAPTER 6 REVIEW

Answer these questions on a separate sheet of paper.

VOCABULARY

The following list contains all of the boldfaced terms in this chapter. Define each of these terms in your own words.

activation energy, active site, active transport, adenosine triphosphate, catalyst, cell membrane, coenzymes, cofactors, concentration gradient, cytoplasm, diffusion, enzymes, enzyme-substrate complex, fluid mosaic model, nucleus, organelles, osmosis, passive transport, pH, phospholipids, plasmolysis, selective permeability

PART A—MULTIPLE CHOICE

Choose the response that best completes the sentence or answers the question.

1. The thin layer that separates a cell's contents from the outside environment is the *a.* cell membrane *b.* nucleus *c.* cytoplasm *d.* endoplasmic reticulum.
2. Which structure is missing in prokaryotic cells? *a.* nucleus *b.* cell membrane *c.* cytoplasm *d.* ribosome
3. A nonprotein helper that assists an enzyme is called a *a.* mitochondrion *b.* ribozyme *c.* vacuole *d.* coenzyme.
4. The movement of water molecules across a selectively permeable membrane is called *a.* osmosis *b.* diffusion *c.* passive transport *d.* active transport.
5. The movement of a substance from an area of low concentration to an area of high concentration is called *a.* osmosis *b.* diffusion *c.* passive transport *d.* active transport.
6. Sugars and amino acids move into a cell via *a.* osmosis *b.* diffusion *c.* passive transport *d.* active transport.
7. The system of canals in which proteins and lipids are made and transported in a cell is the *a.* Golgi apparatus *b.* nucleolus *c.* endoplasmic reticulum *d.* mitochondria.
8. Plasmolysis explains why *a.* amebas explode when put into distilled water *b.* plants wilt when they don't get enough water *c.* you can smell the perfume counter from the other side of a store *d.* enzymes lower the activation energy of a reaction.
9. The proteins in the cell membrane *a.* remain in fixed positions *b.* allow small molecules that can dissolve in lipids to pass into

the cell *c.* transport certain molecules into or out of the cell *d.* prevent osmosis from taking place.

10. Adenosine triphosphate (ATP) is important because it *a.* catalyzes important chemical reactions *b.* pumps potassium into and sodium out of nerve cells *c.* holds energy in small amounts so it can be stored or used gradually *d.* helps enzymes do their job.
11. Enzymes *a.* are usually made up of lipids *b.* lower the activation energy of certain reactions *c.* are used up when they catalyze a reaction *d.* operate at any temperature.
12. If a substance has a pH of 6.5, it is *a.* highly acidic *b.* highly basic *c.* mildly acidic *d.* mildly basic.
13. The "powerhouses" of the cell are the *a.* mitochondria *b.* centrioles *c.* chloroplasts *d.* vacuoles.
14. Which organelle would you *not* expect to find in an animal cell? *a.* nucleus *b.* chloroplast *c.* centriole *d.* endoplasmic reticulum
15. A barrier that allows some substances in while keeping other substances out is said to be *a.* subpermeable *b.* selectively permeable *c.* superpermeable *d.* impermeable.

PART B—CONSTRUCTED RESPONSE

Use the information in the chapter to respond to these items.

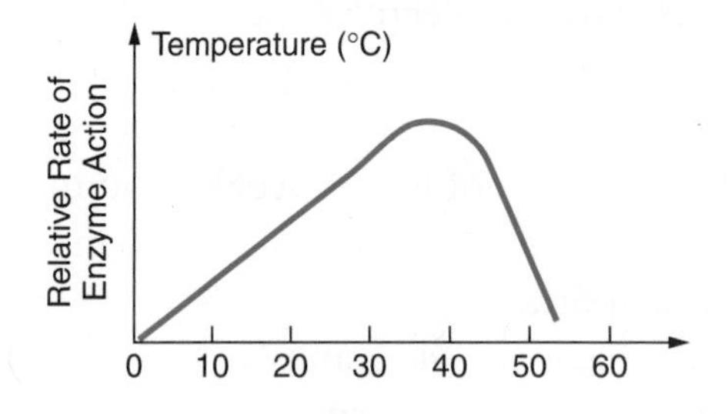

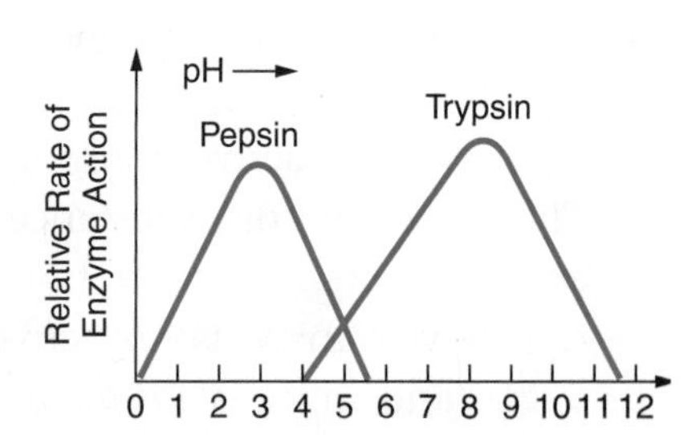

16. According to the graphs, how do temperature and pH affect enzyme action? What might be the reason for these results?
17. Based on the data in the graphs, do you think pepsin and trypsin are found in the same part of the human body? Explain your answer.
18. Using what you know about the movement of materials in and out of cells, explain why traditional methods of preserving food, such as candying fruits and salting meats, worked to some degree.
19. Describe the "lock-and-key model" of enzyme action.
20. Describe the structure of a cell membrane.

PART C—READING COMPREHENSION

Base your answers to questions 21 through 23 on the information below and on your knowledge of biology. Source: *Science News* (December 14, 2002): vol. 162, p. 381.

Bilirubin: Both Villain and Hero?

Bilirubin, the bile pigment that yellows the skin of babies born with jaundice, is generally considered a toxic molecule. According to a new study, however, bilirubin may actually protect cells from dangerous oxygen-containing molecules called free radicals.

Bilirubin forms during the breakdown of hemoglobin, the oxygen-carrying protein in blood cells, and can build up to high concentrations in the blood. Several lines of evidence indicate that bilirubin is toxic, but why then is there a specific enzyme that converts the seemingly harmless molecule known as biliverdin into bilirubin?

Scientists puzzled by this question have unearthed data suggesting that bilirubin, when present at the right concentration, is helpful instead of harmful. A research team headed by Solomon H. Snyder of Johns Hopkins University School of Medicine in Baltimore reports that bilirubin protects brain cells growing in lab dishes from the damage typically caused by hydrogen peroxide, a free radical.

The scientists compared normal cells with ones in which the bilirubin-making enzyme was inhibited. The normal cells were able to survive a dose of hydrogen peroxide 10,000 times greater than the lethal dose for the bilirubin-deprived cells. The investigators report their findings in an upcoming *Proceedings of the National Academy of Sciences.*

Snyder and his colleagues also garnered evidence for a mechanism by which bilirubin, which is altered when it defuses free radicals, is recycled back into its original form. This reuse amplifies its protective powers. A protective role for bilirubin may explain previous findings that have linked low blood concentrations of the molecule to cancer, heart attacks, and other diseases, the scientists note.

21. Describe how bilirubin is produced in the body.
22. Explain why scientists have been confused about bilirubin.
23. State two types of evidence scientists have gathered that point to a protective role for bilirubin.

The Flow of Energy: Photosynthesis and Respiration

After you have finished reading this chapter, you should be able to:

Explain how chlorophyll captures light energy; list five factors that affect the rate of photosynthesis.

Describe how cellular respiration releases the energy stored in food.

Compare the processes of aerobic respiration and fermentation.

Outline the three stages of cellular respiration.

Plant life sustains the living world; more precisely, chlorophyll does. . . . All else obeys the thermodynamic laws that energy forever runs down hill, is lost and degraded. . . . This is the law of diminishing returns, and it is obeyed by the cooling stars as by man and all the animals. Only chlorophyll fights up against the current.

Donald Culross Peattie

Introduction

You are sitting at your desk, next to the window, reading this book. Sunlight is shining through the window and falling on your desk. Next to your book is your afternoon snack, a partially eaten apple. Is there a connection among you, the sunlight, and the apple? There most certainly is. It is a connection that lies at the very center of the theme of life as energy, matter, and organization.

FOOD: MATTER AND ENERGY

The apple is food. It contains complex organic compounds, atoms held together as molecules by chemical bonds that are rich in stored energy.

Figure 7-1 Cows and all other animals are heterotrophs.

You eat the apple, getting both the matter and the energy you need to build your body and to stay alive.

The apple tree that produced the fruit represents one group of living organisms, the group called **autotrophs**, a word that means "self feeding." Unlike humans, the apple tree does not eat. It makes its own food, taking the inorganic substances carbon dioxide and water and changing them into organic compounds such as sugars and starches. Humans are representatives of the other group, organisms that cannot make their own food. Because organisms such as ourselves and all other animals must get complex organic compounds from other organisms, we belong to the group called **heterotrophs**, meaning "other feeding." (See Figure 7-1.)

How does sunlight connect the apple to humans? The connection is energy, of course. For the apple tree to combine inorganic raw materials such as carbon dioxide and water into organic compounds such as sugar and starch, it needs a source of energy. The rays of sunlight, as they fall on the leaves of the apple tree, provide that energy. The process of making this food, by using light as the source of energy, is **photosynthesis**. All green plants are photosynthetic autotrophs.

PHOTOSYNTHESIS

What does a plant need to grow? If you have ever cared for a plant in a flower pot, in a garden, or on a farm, you will probably say: water, soil, and sunlight. Where does a plant get the food it needs to grow? Until 1600, everyone would have said the soil. Animals take in food through their mouths, and people assumed that plants take in food from the soil through their roots.

In one of the first recorded science experiments, Belgian physician

Jean Baptiste van Helmont decided to test this assumption. No one knows for sure why a doctor was interested in this. Maybe it was an overwhelming curiosity that needed to be satisfied. Helmont thought about the question he wanted to study: Do plants get the matter they use to grow from the soil? He took a young willow tree, removed all the soil from its roots, and weighed it. He planted the tree in a tub of soil, which he had also carefully weighed. He then let the tree grow for five years, watering it regularly during that time. After five years, he weighed both the tree and the soil again and discovered that the tree had *increased* in weight by 74 kilograms while the soil had *decreased* in weight by only 57 grams (0.057 kilogram). The willow had grown into a healthy, much taller tree and had increased its weight 1000 times more than the soil's weight had decreased. Helmont had found that plants do not get bigger by simply taking an equivalent amount of matter from the soil. (See Figure 7-2.)

Many other experiments have been conducted since Helmont's time by a wide variety of curious people, including a clergyman, an engineer, and a biochemist. What is now known about photosynthesis, the process by which plants make their own food, is very different from what people once thought was true.

Plants, as autotrophs, are able to make their own energy-rich carbon compounds. In particular, they make the simple sugar glucose, whose chemical formula is $C_6H_{12}O_6$. Plants get the carbon for these glucose molecules from inorganic carbon dioxide, CO_2, in the air. We also know that plants release oxygen, O_2, a gas that can help a candle burn and that animals need to stay alive. The fact that plants give off oxygen was supported by other important experiments done in the 1700s. The final piece of the puzzle was added in the 1940s, when scientists discovered that the oxy-

Figure 7-2 Helmont's plant growth experiment.

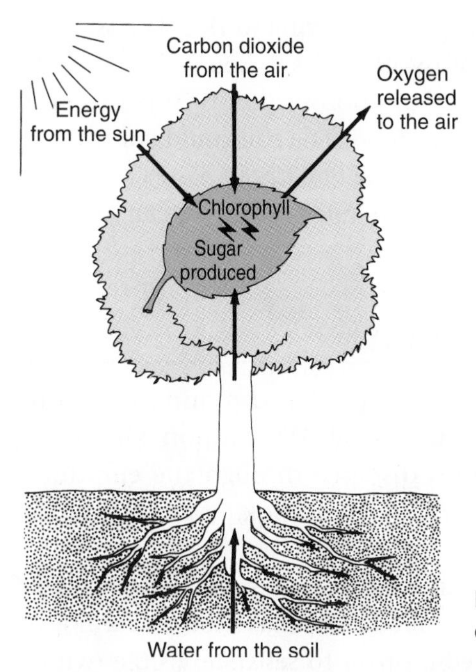

Figure 7-3 The basics of the process of photosynthesis.

gen plants produce comes from water molecules that get split apart, not from carbon dioxide.

Other important information about the process of photosynthesis was learned during the 1800s. Experiments showed that plants must have light to convert inorganic carbon dioxide to glucose and to produce oxygen. Finally, scientists found that photosynthesis also requires the green pigment **chlorophyll**. The chemical reactions of photosynthesis occur within the chlorophyll-containing chloroplasts found in plant leaves and stems. (See Figure 7-3.)

This, then, is the process of photosynthesis. Some consider it the single most important chemical reaction that occurs on Earth. This all-important reaction can be summarized by the following chemical equation:

$$6\,CO_2 + 6\,H_2O \xrightarrow[\text{CHLOROPHYLL}]{\text{LIGHT ENERGY}} C_6H_{12}O_6 + 6\,O_2$$

CARBON DIOXIDE + WATER → GLUCOSE + OXYGEN

Light energy and chlorophyll are needed for this reaction, but the chlorophyll is not used up and the light energy is not a substance. That is why they are written over and under the reaction arrow, rather than with the substances used in the reaction.

Photosynthesis links the nonliving and the living worlds. Almost every organism on Earth depends on photosynthesis as its source of nutrients. Without the marvelous biochemistry of photosynthesis in plants, animals would have no constant source of food. The sun could continue to pour its light energy on Earth without limit. But without plants to capture this light energy and convert it into the chemical forms we call food, we and most other animals would not exist.

CHLOROPHYLL: THE SUN TRAP

The sunlight shining on your desk was produced 8 minutes ago by thermonuclear reactions in the sun, about 150 million kilometers away. After traveling that enormous distance through the emptiness of space, the light energy is now striking your book, or the leaf of the apple tree.

Just as elements consist of fundamental particles called atoms, light energy consists of packets of energy called **photons**. Photons cannot be further divided. You may have used a prism to separate visible (white) light into the spectrum of colors that make it up. Each color in the spectrum has a different wavelength. The visible light that comes from the sun has wavelengths that vary from the 700 nanometers found in red light to the 400 nanometers found in violet light. The amount of energy in light depends on its wavelength. The shorter the light's wavelength, the higher its energy level. For example, a photon of violet light has more energy than a photon of red light. The range of colors in the spectrum of visible light includes red, orange, yellow, green, blue, and violet (ROYGBV). Our eyes are able to detect these colors. The pigments in plants are sensitive to the same visible colors of light. (See Figure 7-4.)

Objects either absorb or reflect a particular color of light. A red apple absorbs all colors of light except red. It reflects red light, sending it away from its surface. That is why the apple appears red. The photosynthetic pigment chlorophyll absorbs all colors of light except green. Plant leaves appear green because they reflect green light. Both the red lower-energy light and the blue higher-energy light are absorbed by chlorophyll. A graph can show the amount of light energy absorbed by chlorophyll at different wavelengths. This is called the **absorption spectrum** for chlorophyll. (See Figure 7-5.)

Through evolution, plants have developed other pigments that use some of the wavelengths of light that chlorophyll does not. These pigments absorb green and blue light but reflect yellow, orange, and red. We

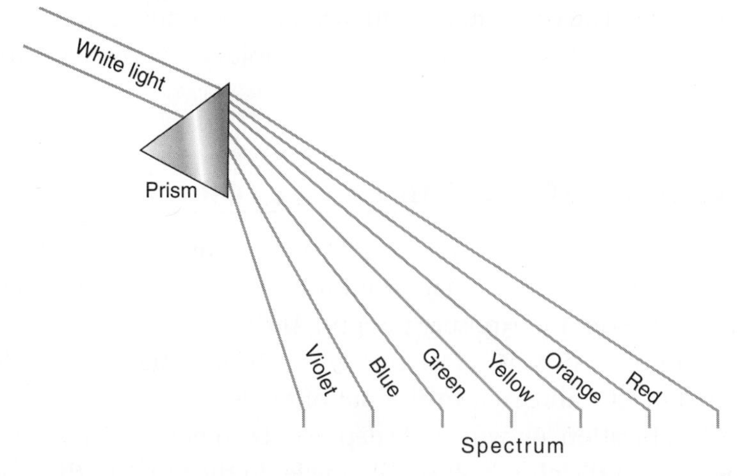

Figure 7-4 A prism separates white light into its spectrum of colors.

do not usually see these pigments because the chlorophyll in the leaves hides the other colors. However, plants need warm weather to make chlorophyll. When summer passes into autumn and cool temperatures arrive, plants stop making chlorophyll. The chlorophyll begins to break down and green light is no longer reflected. The other pigments last a little longer, though, and the glorious colors of yellow, orange, and red

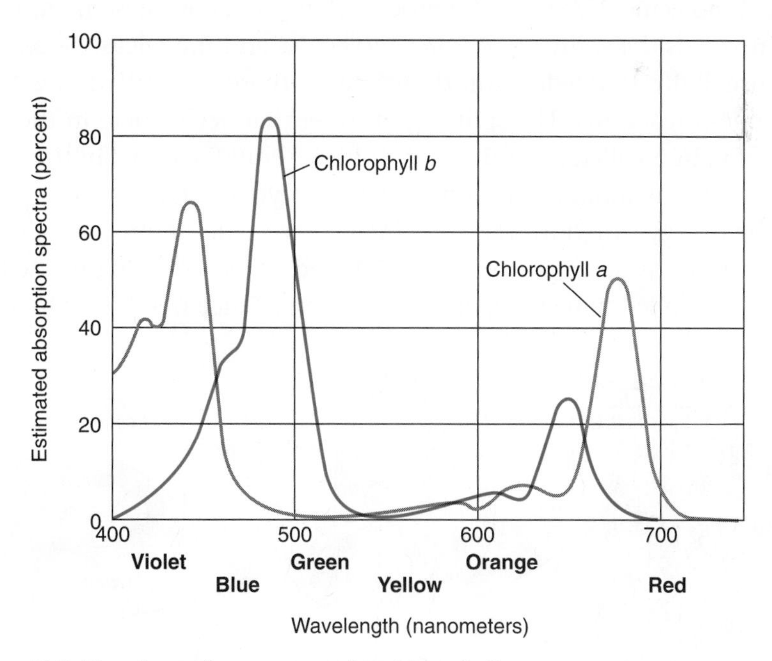

Figure 7-5 The absorption spectrum for chlorophyll.

become visible. The colors of autumn leaves are a result of the chemistry of pigments, the physics of light, and the biology of leaves all working together.

LEAVES: PHOTOSYNTHETIC FACTORIES

You have already seen in Chapter 3 how the structures present inside the leaf are well organized. Such organization allows cells that contain chlorophyll to get maximum exposure to light. At the same time, the leaf controls the amount of water lost to the air. It also makes possible the movement of CO_2 and O_2 into and out of the leaf.

This organization, on which life depends, becomes even more obvious when we look at the **chloroplasts**, organelles in the leaf cells that contain chlorophyll. A typical leaf cell may contain as many as 60 chloroplasts. (See Figure 7-6.) A cross section of an individual chloroplast shows many stacks of membranes. One of the two main steps of photosynthesis occurs in these membranes; the other step occurs in the liquid material in the spaces between the membranes.

Photons of light energy are captured by chlorophyll molecules embedded in the membranes. In the molecules, the sun's energy is converted from light energy to chemical energy, in the form of ATP and another related molecule, NADPH. The process that occurs on these membranes also involves the splitting of water molecules and the release of oxygen. Because light is needed for these steps, these are called the light-dependent reactions. The splitting of water that occurs here in the presence of light is called **photolysis**. Within the liquid material in the spaces between the membranes, a complex set of enzymes catalyze reactions that use the energy stored in ATP and NADPH molecules to produce glucose. These reactions make glucose from CO_2 and water. They do not use light and so are called light-independent reactions. Since free-floating carbon

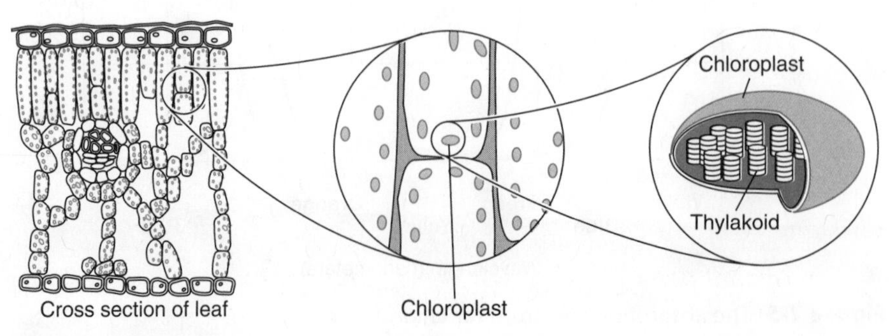

Figure 7-6 Light energy causes chemical reactions to occur inside chloroplasts.

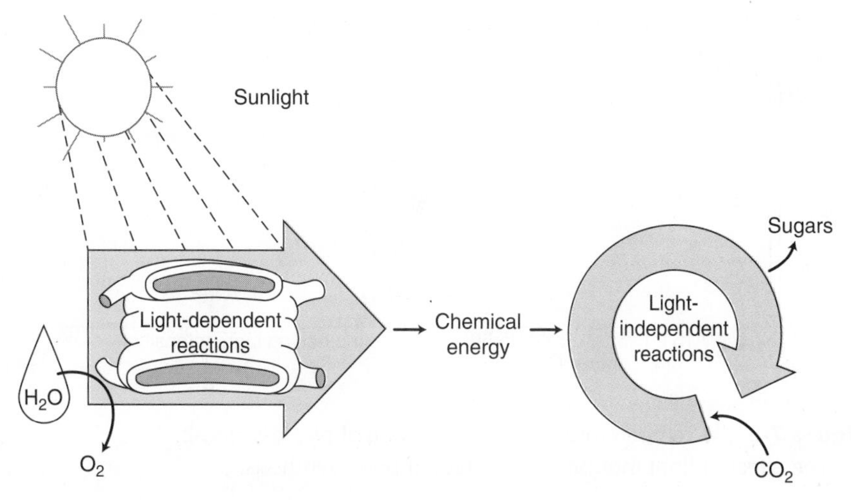

Figure 7-7 Carbon fixation turns inorganic matter into organic compounds.

from CO_2 gets combined into the carbon backbone of glucose molecules in these steps, this process is also called **carbon fixation**.

Carbon fixation is the single most important way in which inorganic matter from the world around us, mainly CO_2 in the air, gets turned into organic carbon compounds that make up the bodies of all living things. (See Figure 7-7.)

THE RATE OF PHOTOSYNTHESIS

As with any chemical reaction, the reactions of photosynthesis can occur at different rates. What factors affect the rate at which photosynthesis occurs?

- *Temperature.* When temperature increases, molecules move more quickly. The rate at which a reaction occurs also increases. However, at temperatures higher than 35°C, the rate of photosynthesis reactions decreases. Higher temperatures begin to destroy the enzymes needed for the reactions. (See Figure 7-8, graph A, on page 148.)
- *Light intensity.* An increase in light intensity increases the rate of photosynthesis in a leaf until a maximum rate of photosynthesis is reached. Beyond that level, additional light has no further effect. (See Figure 7-8, graph B, on page 148.)
- *CO_2 concentration.* Because CO_2 is used by plants in photosynthesis reactions, increasing the concentration of CO_2 in the air around a plant usually increases the rate of photosynthesis.

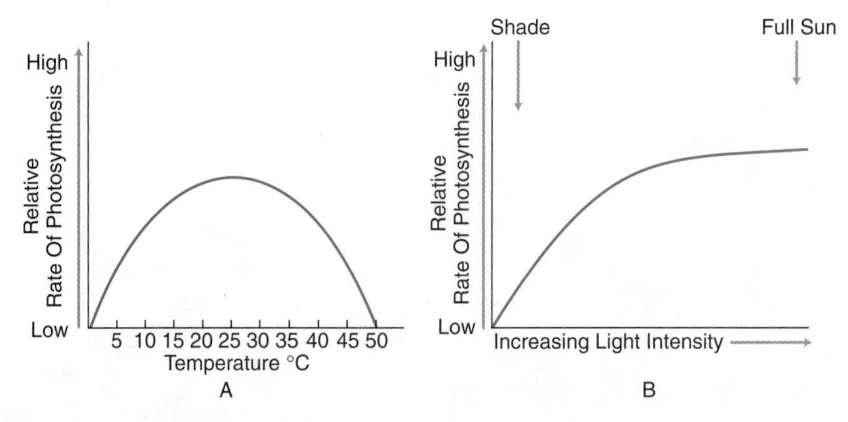

Figure 7-8 (A) Temperature's effect on the rate of photosynthesis. (B) The effect of light intensity on the rate of photosynthesis.

- *Water.* Even though only a small amount of water is needed for photosynthesis, a shortage of water slows the process. Also, when water is scarce, the stomates close. The rate at which the gases CO_2 and O_2 can be exchanged is therefore reduced, and photosynthesis is slowed.
- *Minerals.* Certain minerals are important for photosynthesis reactions. For example, magnesium and nitrogen are needed to make chlorophyll molecules. Zinc, manganese, iron, and copper are needed for some of the reactions. Actually, it was the removal of these minerals from the soil that caused the slight decrease in soil weight in Helmont's experiment. The total absence of any one of these minerals would have a negative effect on the entire process of photosynthesis.

DISCOVERIES OF LIFE ON THE OCEAN BOTTOM

In 1977, researchers from the Woods Hole Oceanographic Institution in Cape Cod, Massachusetts, led by scientist Robert Ballard, used a camera on a small remote-controlled submarine to record observations made on the seafloor, deep in the Pacific Ocean near the Galápagos Islands. The researchers knew that there was a crack in Earth's crust at this place. Hot water, rich in chemicals, was escaping from vents in the crack into the surrounding ocean water. To their great surprise, the researchers found a very busy scene of life clustered around the vents. Viewed through a television camera, the large numbers of organisms looked like they had escaped from a science-fiction movie. Giant red tube worms, huge clams, and blind white crabs were found in the water near the vents. The scientists took water samples near the vents and found great concentrations of bacteria.

Drifters in the Sea

We walk on land. Even the very name Earth is used to mean land. But look at a world map and you will see a lot of blue space. In fact, more than 70 percent of Earth's surface is covered by water, mostly oceans. Unseen in these waters—drifting along with waves and currents—are countless numbers of tiny organisms. Photosynthetic bacteria, protists, and plants are included in these drifters. Some of these unicellular species are so small that if 12 million cells were lined up in a row, the line would be only about 1 centimeter long. In some places in the oceans, these microscopic organisms are so numerous that a cup of seawater may hold 24 million individuals of a single species, and that cup would contain other species as well!

These species are very small, but their importance to the overall life on the planet is huge. Tiny sea-dwelling organisms are the beginning food source for almost all living things in the oceans. It is easy for us land dwellers to understand that many animals eat plants to get food. We have seen cattle and sheep grazing on grasses in a pasture. The drifting cells in the ocean could be called the grass fields or pastures of the sea. Just like grass on land, the sea drifters capture energy from the sun and convert inorganic CO_2 and water into organic molecules, which become important foods for other organisms. On land, plants bloom with wild displays of colorful flowers in spring. The photosynthetic drifters in the pastures of the seas are said to "bloom" in the spring, too, as the water warms and nutrients from ocean depths are brought to the surface by currents. A great deal has been learned recently about the seasonal explosive growth of these photosynthetic cells in the ocean from photographs taken by orbiting satellites.

These organisms perform another vital function for life on Earth. They are our planet's most important absorbers of CO_2 from the air as they photosynthesize. People have been changing Earth's atmosphere by burning enormous quantities of coal, oil, and gas as fuels. Burning these fossil fuels increases CO_2 levels in the air. This change is thought to be increasing the average temperature of Earth. The tiny photosynthetic cells drifting in the oceans are absolutely necessary to help prevent global warming, since they absorb the CO_2. So the drifters of the seas are vitally important for all life on Earth, not only life in the oceans.

How does this fantastic array of life survive in the total darkness thousands of meters below the ocean's surface? The bacteria were the key. They were the food source for the other organisms. The bacteria made their own food from the chemicals present in the hot water escaping from the vents. Because chemical reactions involving inorganic substances, and *not* light, are used as the source of energy, the process is called **chemosynthesis**. Chemosynthetic bacteria are also found in other places, such as the mud in marshes at Earth's surface.

Much research has continued on hot-water vents over the past 25 years. In 2002, scientists were investigating how the unique organisms found at the many ocean vents, which are widely separated from each other on the ocean floor, got to these various locations in the first place.

CELLULAR RESPIRATION: RELEASING THE STORED ENERGY IN FOOD

We have spent some time considering the relationship between the sunlight that falls on the leaves of an apple tree and the chemical process of photosynthesis. During photosynthesis, the light energy of the sun is converted into the stored chemical energy of glucose in the apple. After we eat the apple, our cells are ready to use that stored chemical energy. How does this happen?

We know that the release of energy cannot occur all at once. For example, if you hold a lighted match to a marshmallow, the carbohydrates in the marshmallow will burst into flames. Energy is being released, and quickly, too. Obviously this quick energy release is not what happens inside our cells. Instead, the release of energy occurs in a series of enzyme-controlled small steps. The energy stored in organic compounds is eventually converted to a usable form, the energy currency of all cells, ATP. This process is known as **cellular respiration**. (See Figure 7-9.)

Cellular respiration is basically the opposite of photosynthesis. Instead of building energy-rich glucose molecules, glucose molecules are taken apart to release their stored energy. We have said that we can consider the glucose molecules like money stored in the bank. The energy the cell needs to do the many different tasks involved in living is like the pocket money you need to buy groceries and clothes. Cellular respiration converts the energy stored in glucose into the easily spent energy currency called ATP.

Cellular respiration occurs in several stages. The first stage releases only a small amount of the stored energy from glucose. This partial breakdown of glucose occurs in the absence of oxygen. To completely take apart glu-

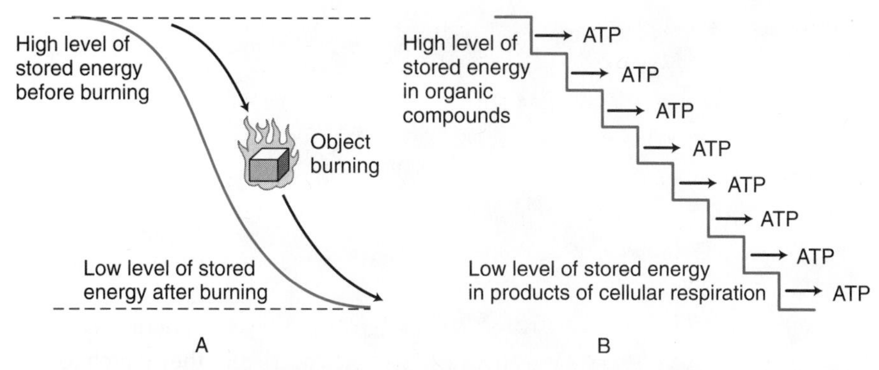

Figure 7-9 (A) The burning of an object is due to the sudden release in one step of the energy stored in the object. (B) In cellular respiration, energy is released from organic compounds as ATP in a series of enzyme-controlled steps.

cose to produce carbon dioxide and water, and to release a large amount of energy, oxygen must be present. Respiration that occurs in the presence of oxygen is called **aerobic respiration.**

Glucose, with a large amount of energy stored in it, can be compared to an object on top of a high hill. Allowing the object to move all the way down the hill releases all of its stored energy. This is what happens after aerobic respiration. Moving the object just a little way down the hill releases a small amount of energy. The object now has less stored energy in it. This is the case after fermentation.

Check Your Understanding

In what ways are cellular respiration and photosynthesis opposite reactions?

FERMENTATION

Every baker knows that living yeast cells can turn a lump of dough into a loaf of bread that is then baked. Every winemaker knows that growing yeast cells turn grape juice into wine. These are **fermentation** reactions. Microorganisms such as yeast and bacteria are capable of fermentation.

Fermentation begins with the splitting of a glucose molecule that has six carbon atoms into two molecules, each with three carbon atoms. This glucose-splitting process is called **glycolysis.** During glycolysis, ten individual chemical reactions take place, each reaction controlled by a different enzyme. The net result is the release of energy in the form of two ATP molecules for every glucose molecule that is split in half. No oxygen

Figure 7-10 When muscles do not get enough oxygen, they switch to fermentation to produce energy.

is used in these reactions. In yeast, the anaerobic process known as alcoholic fermentation produces alcohol and carbon dioxide. The alcohol produced turns grape juice into wine. The bubbles of CO_2 make bread dough rise and make the inside of a loaf of bread look like a sponge.

Our own cells also carry out fermentation under certain conditions. How many push-ups can you do? You could try right now if you don't know. Can you keep on doing more and more push-ups until *you* decide to stop? Try it. You will find that no matter how determined or how strong you are, your muscles will continue to work only until a certain point. (Some people will reach this point sooner than others.) Then your muscles will stop working for you. You will reach a point where you can do no more push-ups. Why? (See Figure 7-10.)

In order for your muscles to do work, you need oxygen to release the energy from glucose through aerobic respiration. If you work your muscles strenuously enough, glycolysis occurs in the muscle cells faster than oxygen can be delivered to them. In the absence of oxygen, the muscles switch to fermentation to produce energy. In muscles, the final product of this fermentation is not alcohol but a compound called lactic acid. When lactic acid builds up in muscle cells, the pH becomes more acidic, the muscles get tired, and cramps develop.

AEROBIC RESPIRATION: GETTING THE MOST OUT OF IT

Under normal conditions, when the supply of oxygen is sufficient, cells can carry out aerobic respiration. This process begins with the same ten steps of glycolysis. The fact that the same ten chemical reactions, in exactly the same order, occur in each of our cells and in microorganisms such as bacteria and yeast is strong evidence that these chemical processes developed long ago before single-celled organisms evolved into multicellular organisms.

What is so special about being able to go beyond the steps of glycolysis and limited energy release to the process of aerobic respiration? Remember the object on a hill, which moved only a short way down during fermentation? In that example, the glucose was split into only two pieces. Now, in the presence of oxygen, the molecule of glucose can roll all the way down the hill. The glucose molecule gets completely taken apart. The result is the release of a large amount of energy. Rather than only the two molecules of ATP produced during glycolysis and fermentation, we now get an average of 36 ATP molecules for every glucose molecule. For large, active, multicellular organisms like ourselves, aerobic respiration is a necessity. Fermentation, with its low level of energy production, may be enough to provide for a single yeast cell living in grape juice or bread dough. However, only the high-energy production of aerobic respiration is sufficient to meet the energy needs of active animals and plants.

THE THREE STAGES OF CELLULAR RESPIRATION

To understand cellular respiration, we need to resume our tour of the eukaryotic cell. Let's imagine that we can see glucose molecules being transported into the cell. These glucose molecules might have come from your apple. They eventually arrived in one of the muscle cells in your arm. To get into the cell, they were transported across the cell membrane by carrier proteins.

Inside the cell, the first stage of respiration occurs. Enzymes responsible for the ten chemical reactions of glycolysis are present in the cytoplasm. To proceed to the second stage of cellular respiration, the products of glycolysis are transported to numerous membrane-filled organelles called **mitochondria**. These are the powerhouses of the cell. In the mitochondria, the second stage of aerobic respiration will begin only if oxygen is present.

The products that result from glycolysis, although smaller than molecules of glucose, still contain most of the original energy stored in the glucose. The energy associated with the electrons that form the chemical bonds in these molecules is still there. Once these products are inside the mitochondria, an entirely new process begins. This second stage, called the **Krebs cycle**, involves eight separate enzyme-catalyzed reactions

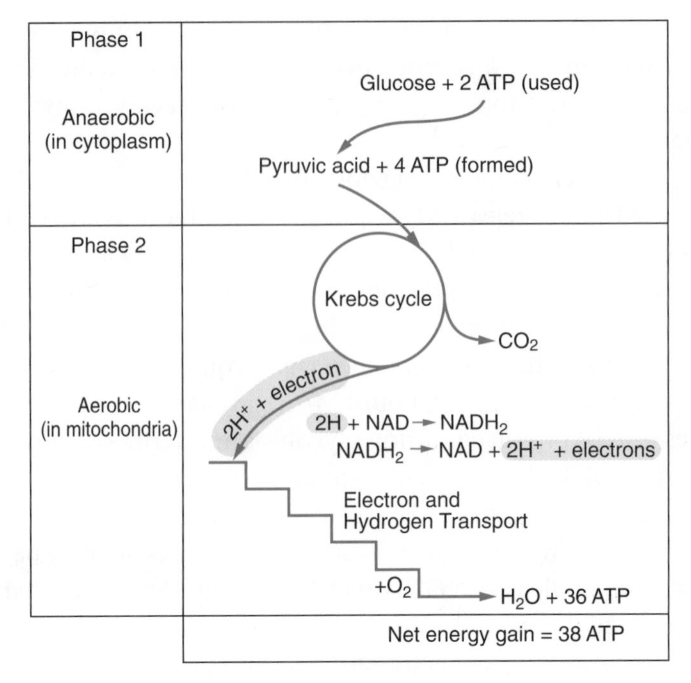

Figure 7-11 Summary of energy release in cells.

linked to one another in a circle. (See Figure 7-11.) The enzymes for the Krebs cycle are all found in the liquid interior of the mitochondria. Oxygen is required for the Krebs cycle; some ATP is produced, and CO_2 is released as the energy-containing molecules get broken down, with one carbon atom at a time being removed from them. The most important result of the Krebs cycle is the stripping away of hydrogen atoms from the energy-rich molecules. These hydrogen atoms are taken away by carrier molecules called NAD^+ and FAD, which become NADH and $FADH_2$, to the third and final stage of cellular respiration.

To find where this final stage occurs, we must take a closer look at the mitochondria. (See Figure 7-12.) You will see an inner membrane that is folded back and forth many times inside a mitochondrion's outer membrane. These outer and inner membranes have the same phospholipid bilayer structure as the cell membrane. A series of carrier molecules is attached to the inner membrane. While some of the energy released during the Krebs cycle is used to make molecules of ATP, most of it is released as high-energy hydrogen atoms, which become attached to either NAD^+ or FAD. Each of these hydrogen atoms consists of one electron and one proton. These particles are high up on the "energy hill." To get them to release their energy, other particles must be willing and able to accept

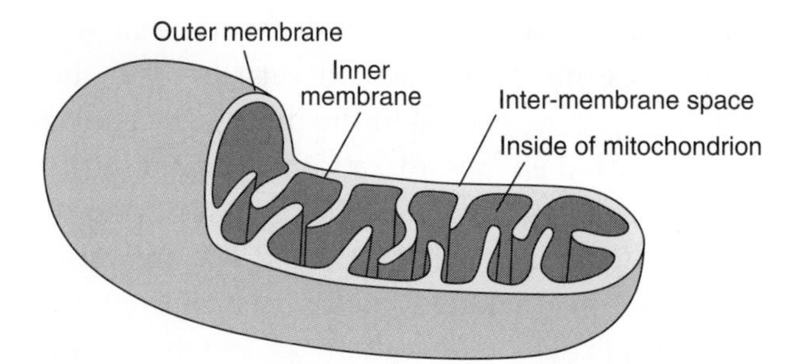

Figure 7-12 The final stage of cellular respiration occurs in the inner membrane of the mitochondria.

them at a much lower energy level. The high-energy hydrogen atoms begin to get passed from NADH and $FADH_2$ to the electron carriers attached to the inner membrane, moving from one carrier molecule to another along a chain. In terms of energy, this chain, called the **electron transport chain**, is moving downhill. The process works only if, at the end, something is waiting to accept the electrons. What do you suppose are the final electron-acceptor particles at the bottom of the hill? Take a deep breath. You just took them into your body.

Oxygen molecules are the final hydrogen and electron acceptors that make possible the efficient release of large amounts of stored energy from glucose molecules. What is the result of this final stage? As the oxygen accepts the electron/proton combinations (a hydrogen atom), the result is H_2O, or water. If you breathe out on a mirror, you will observe moisture on the glass. Some of this is the water produced during the final stage of cellular respiration. Most important, about 36 molecules of ATP are released from every molecule of glucose that is completely disassembled by aerobic cellular respiration.

We can now summarize cellular respiration in the following chemical equation:

$$C_6H_{12}O_6 + 6\,O_2 \xrightarrow{\text{ENZYMES}} 6\,CO_2 + 6\,H_2O + 36\text{ ATP}$$

GLUCOSE + OXYGEN → CARBON DIOXIDE + WATER + ENERGY

The ATP is the energy source in your cells, and every cell in your body uses it to do the work that keeps you in that highly organized state we call life.

A FINAL VISIT BACK TO PLANTS

Remember that apple and the tree it came from? The leaves of the tree are necessary in this entire flow of energy from the nonliving to the living world. Photosynthesis occurs in them, converting the sun's energy into energy stored in chemicals. In leaves, the carbon dioxide from the air gets transformed into organic glucose molecules, which get stored in the apple. We eat the apple, and the glucose is used in the process of cellular respiration that transforms the stored energy into a form we can use.

But does the apple tree specifically go through the process of photosynthesis in order to make food for us and other animals? The answer is a definite no! The apple tree is simply making food for itself to live long enough to reproduce successfully. The apples contain seeds, which may get carried away to new places by animals that eat the apples. This makes it possible for the apple tree to produce more apple trees in other places. However, most of the glucose made in the leaves of the tree does not go into storage in the form of apples. Rather, it gets taken to all the different parts of the tree and is used by the tree to stay alive. And what process does the tree use to get the energy it needs from the glucose it has made? The same process we just finished studying, cellular respiration.

To summarize: All plants are autotrophs and are able to make their own food by photosynthesis and use it for energy through cellular respiration. All animals are heterotrophs and must obtain food energy from other organisms. They use cellular respiration, just as plants do, to obtain energy from the food they eat.

LABORATORY INVESTIGATION 7

How Is Chromatography Used to Separate Pigments in Chlorophyll?

INTRODUCTION

Chromatography is an important technique used by scientists to separate the components in a mixture. Some components dissolve more easily in one solvent than in another. If you place a mixture on a piece of paper and place the end of the paper in a solvent, the components of the mixture that are more soluble will be carried faster and farther up the paper. This technique is called paper chromatography.

Chlorophyll, the pigment that gives green plants their color, has been analyzed in this way. As a result of these studies, scientists have learned that chlorophyll is made up of several pigments: chlorophyll A (yellow-green), chlorophyll B (blue-green), carotenes (orange), and xanthophylls (yellow). In this investigation, you will use paper chromatography to separate the pigments in an artificial chlorophyll mixture.

MATERIALS

Filter paper, scissors, ruler, pencil, test tube, toothpicks, mixture of dyes (food colorings in water), solvent solution (2% concentrated ammonia, 3% isobutyl alcohol, and 95% water)

PROCEDURE

1. Cut a piece of filter paper about 2 centimeters wide and as long as the test tube. The paper strip should be just a little wider than the diameter of the test tube. Cut one end of the strip into a V.
2. Draw a pencil line across the width of the paper, 2 centimeters from the pointed end of the paper. DO NOT USE INK. Ink contains dyes that will interfere with your results.
3. Use a toothpick to place some of the dye mixture along the pencil line. Let the material dry for a few minutes. Repeat this process 10 times.
4. Carefully pour a small amount of the solvent mixture into the test tube, to a depth of 1 to 2 centimeters.
5. Slide the filter paper strip down the test tube until the point of the V touches the solvent. The line of dye on the strip must not touch the solvent.

6. Place the test tube in a rack. DO NOT shake the tube.
7. Record your observations immediately. Then record your observations after 2, 5, and 10 minutes.
8. Remove the paper strip after the solvent line has reached the top of the filter paper. If this happens before 10 minutes have passed, remove the paper immediately.

INTERPRETIVE QUESTIONS

1. What does the strip of filter paper look like at the end of this investigation?
2. How many different bands of colors can you see? List them, in order, from the bottom of the paper strip to the top.
3. The leaves of many varieties of Japanese maples are red all year long. How could you find out if chlorophyll is present in these leaves?
4. Why are the leaves of most deciduous trees green? Why do the leaves of these trees change color in the fall?

CHAPTER 7 REVIEW

Answer these questions on a separate sheet of paper.

VOCABULARY

The following list contains all of the boldfaced terms in this chapter. Define each of these terms in your own words.

absorption spectrum, aerobic respiration, autotrophs, carbon fixation, cellular respiration, chemosynthesis, chlorophyll, chloroplasts, electron transport chain, fermentation, glycolysis, heterotrophs, Krebs cycle, mitochondria, photolysis, photons, photosynthesis

PART A—MULTIPLE CHOICE

Choose the response that best completes the sentence or answers the question.

1. Which organism is an autotroph? *a.* mushroom *b.* oak tree *c.* parrot fish *d.* *E. coli* bacterium
2. Which of these processes takes place in the mitochondria? *a.* Calvin cycle *b.* Krebs cycle *c.* glycolysis *d.* photolysis
3. Carbon fixation and photolysis usually take place in a plant's *a.* roots *b.* nuclei *c.* seeds *d.* chloroplasts.
4. Photosynthesis requires *a.* light energy, chlorophyll, carbon dioxide, and water *b.* light energy, soil, oxygen, and water *c.* light energy, chlorophyll, oxygen, and water *d.* light energy, chlorophyll, glucose, and oxygen.
5. The products of cellular respiration are *a.* glucose and oxygen *b.* carbon dioxide and alcohol *c.* carbon dioxide, water, and energy *d.* ATP, NADH, and $FADH_2$.
6. When water is scarce, *a.* photosynthesis speeds up *b.* the stomates close *c.* CO_2 concentration inside the leaf increases *d.* the mineral copper is used up at a faster rate.
7. The "energy currency of all cells" is *a.* ATP *b.* glucose *c.* NADP *d.* light.
8. For carbon fixation, deep-sea vent organisms depend on the process of *a.* photolysis *b.* photosynthesis *c.* fermentation *d.* chemosynthesis.
9. The absorption spectrum for chlorophyll reveals that it *a.* reflects ultraviolet light *b.* absorbs green light *c.* absorbs red and blue light *d.* reflects red and yellow light.

10. Glycolysis *a.* produces a net gain of 2 ATP molecules *b.* produces a net gain of 4 ATP molecules and 2 NADH molecules *c.* uses 1 O_2 molecule *d.* uses 2 CO_2 molecules.
11. The end products of fermentation in yeast are *a.* carbon dioxide and water *b.* glucose and oxygen *c.* carbon dioxide and alcohol *d.* alcohol and oxygen.
12. Which of the following would speed up photosynthesis?
    a. an increase in light intensity *b.* an increase in temperature above 35°C *c.* the absence of zinc, manganese, iron, or copper *d.* a decrease in CO_2 concentration
13. Which of the following statements is correct?
    a. Glycolysis produces 2 ATP, and respiration produces 34 ATP.
    b. Glycolysis produces 2 ATP, and respiration produces 26 ATP.
    c. Glycolysis produces 10 ATP, and respiration produces 20 ATP.
    d. Glycolysis produces 4 ATP, and respiration produces 32 ATP.
14. Most of the energy released during the Krebs cycle is *a.* used to make ATP *b.* used to build an electron transport chain *c.* in the form of $FADH_2$ and NADH *d.* in the form of high-energy hydrogen atoms.
15. Which of these events occurs during the light-dependent reaction?
    a. ATP loses a phosphate group to form ADP *b.* NADPH becomes NADP *c.* photolysis *d.* carbon fixation

PART B—CONSTRUCTED RESPONSE

Use the information in the chapter to respond to these items.

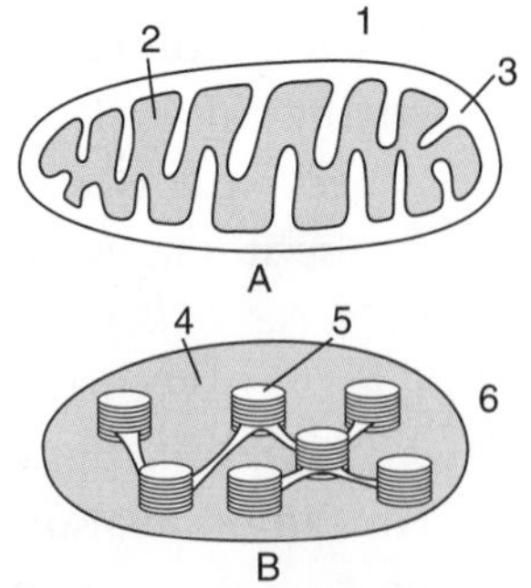

16. What are the names of the structures labeled *A* and *B* in the diagram? In which structure does photosynthesis take place?
17. Identify the numbered areas in which glycolysis, photolysis, and the Krebs cycle occur.
18. Why is photosynthesis considered a bridge between the nonliving and living worlds?
19. Explain how it is that muscle cramps and champagne originate from the same process.
20. Imagine that you are a carbon atom. Describe your journey from the air into a dandelion, then into a rabbit that eats the dandelion, and finally back into the air.

PART C—READING COMPREHENSION

Base your answers to questions 21 through 23 on the information below and on your knowledge of biology. Source: *Science News* (July 12, 2003): vol. 164, p. 29.

Flight Burns Less Fuel Than Stopovers

The first measurements of energy use in migrating songbirds have confirmed a paradox predicated by some computer models of bird migration: Birds burn more energy during stopovers along the way than during their total flying time.

Martin Wikelski of Princeton University and his colleagues monitored 38 Swainson's and hermit thrushes during the nights of their spring migration through the northern United States. The researchers injected the radio-tagged birds with chemical-isotope tracers that enabled the scientists to measure the birds' metabolism. The team members spent their nights driving a car, trying to keep up with a tagged bird. "We got stopped by a cop just about every night, not because we were speeding, but because they wanted to know what somebody was doing in a little town in Wisconsin at 4 A.M. with a giant antenna on the roof of a car," says Wikelski.

A dozen birds took night flights covering up to 600 kilometers. The rest stayed put. The scientists determined that the birds that flew burned 71 kilojoules [kJ] of energy on an average night's flight of 4.6 hours. The birds that didn't fly burned energy at 88 kJ per day.

Since the birds spent about 24 days and nights on stopovers during a typical 42-day journey from Panama to Canada, actual flying consumed only 29 percent of the total energy budget for the migration, Wikelski and his coworkers report in the June 12 *Nature*.

21. How did researchers from Princeton University measure the metabolism of thrushes while they were migrating?
22. Explain what comparison the researchers were interested in making.
23. State the conclusion that was made by the scientists after they finished their investigation of the migratory birds.

8 Getting Food to Cells: Nutrition

After you have finished reading this chapter, you should be able to:

Understand the importance of proper nutrition for good health.

Outline the process of digestion in different types of organisms.

Describe the structures and functions of the human digestive system.

For some animals, obtaining food is a relatively simple task. Tapeworms, for example, have no mouth or digestive tract; they simply attach to the wall of an animal's intestine and absorb digested nutrients across their outer body surface. For these animals, eating is not necessary. Unfortunately, as humans, we can't enjoy the same advantage. We can't recline in a bathtub of oatmeal for breakfast and chicken soup for lunch and simply soak up the nutrients.

Gil Brum

Introduction

A nutritionist once said, "You are what you eat." This seems a simple phrase. In this chapter, you will examine this statement to learn how it is true. In addition, you will examine ways your body uses the food you eat.

An adolescent's body is growing rapidly. The food you eat provides the matter your body uses to build cells, tissues, and organs. The atoms in a carrot you eat will take their place in the new cells you are building. The carrot's atoms may become part of skin cells, muscle cells, or blood cells. However, is it only growing adolescents who need to eat? Or does a person who is no longer growing also need to eat?

The answer is pretty obvious. You need to eat for as long as you live; our bodies are in constant need of repair even if we are no longer growing. Every day of our lives, some old cells and tissues are replaced by new

Figure 8-1 Food provides us with the matter and energy that keep us alive.

ones. Atoms in the carrot may be used in the process of replacing cells and tissues. In addition to using food to build cells and tissues, food provides the energy an organism needs to remain alive. Food provides our matter (what we are made of) and our energy (what keeps us alive). In effect, we really are what we eat. (See Figure 8-1.)

KNOW YOUR FOOD: PROPER NUTRITION

In the 1880s, a nerve disease called *beriberi* affected many soldiers being cared for at a military hospital in the Dutch East Indies. The patients who developed beriberi were extremely tired. In time, their muscles became weak, and some even became paralyzed. Scientists studying this disease spent years searching for the microorganisms they assumed must be the cause of beriberi. They were unsuccessful.

Then in 1896, one scientist, Christian Eijkman, noticed that some of the chickens that lived near the hospital also showed symptoms of beriberi. Within four months, many but not all of the sick chickens died. Eijkman studied the chickens that survived. He learned that a new worker in the hospital had stopped feeding the surviving chickens food that was left over from feeding the hospital patients. Most of the hospital leftovers consisted of white polished rice, in which the outer shells (hulls) of the rice seeds (kernels) have been removed. Eijkman learned that the new hospital worker was feeding the chickens whole rice that still had outer hulls. Later, Eijkman found that whole rice reversed beriberi in both chickens

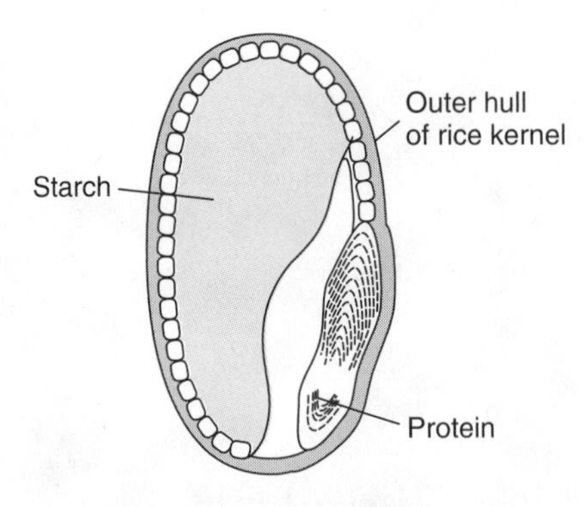

Figure 8-2 Cross section of the inside of a rice kernel.

and humans. Something present in the outer hulls of rice kernels had cured the disease. (See Figure 8-2.)

This discovery, made over a century ago, provided the first clue that something missing from a diet can produce a specific disease. Years later, researchers discovered that rice hulls contain vitamin B_1, also called *thiamin*. Thiamin is a coenzyme needed by an enzyme in one of the chemical reactions that occurs in cellular respiration. Without vitamin B_1, the enzyme does not work and a basic function of cell metabolism is disrupted.

Useful, often essential, molecules in food are known as nutrients. The six main groups of nutrients are carbohydrates, lipids, proteins, vitamins, minerals, and water. The first three types—carbohydrates, lipids, and proteins—can be used by the body as sources of energy. These organic compounds can be broken down and used in chemical reactions that eventually produce ATP. The breaking down of more complex substances into simpler substances is called *catabolism*. The chemical pathways of cellular respiration that produce energy from nutrients are *catabolic pathways*. Chemical reactions that break down substances are a major part of metabolism.

Organisms also combine simpler molecules to make more complex substances. This building-up part of metabolism is called *anabolism*. For example, proteins are broken down into amino acids. These amino acids are then used in anabolic pathways to build other proteins an organism needs. Lipids are also broken down and reassembled into new lipids. It is through *anabolic pathways* that nutrients provide the matter that living organisms need.

Amino acids are the building blocks of proteins. Humans can make only 12 of the 20 types of amino acids the body needs. The eight amino acids that cannot be made by humans, called the *essential amino acids*, must be obtained from food. Rice and beans are a popular combination of foods. You might be surprised to learn that there is a sound nutritional reason for eating rice and beans together. Neither rice nor beans alone provide all eight of the essential amino acids. To get all eight amino acids, you must eat both rice and beans together!

What nutrient makes up most of what we eat? Most of our food consists of water. For example, almost 95 percent of a tomato is water. (See Figure

8-3.) While we need about 50 grams of protein a day, we need 2 liters or 2000 grams of water a day. We also need vitamins and minerals—simple chemical elements such as calcium, magnesium, iron, sodium, potassium, and chlorine—in our food to allow for the proper functioning of a wide variety of chemical reactions that occur in cellular metabolism.

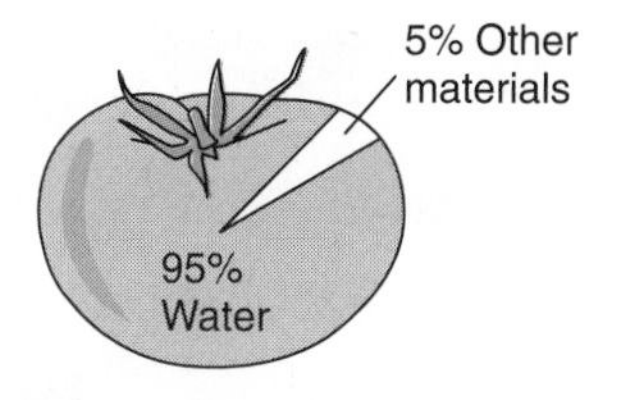

Figure 8-3 A tomato is about 95 percent water.

It is also recommended that a person's diet include foods high in fiber, also called *roughage*. Fiber is the part of plants that cannot be broken down by our digestive system. Fiber consists mostly of the complex carbohydrate cellulose. Cellulose is what makes up the tough wall that surrounds plant cells. If fiber cannot be digested, why do we need to include it in our diet? In fact, it is now believed that a high-fiber diet reduces the risk of developing certain kinds of cancer. Dietary fiber works in this way: Because it cannot be digested, fiber must be eliminated by the body. Having significant amounts of dietary fiber to eliminate keeps food and other ingested materials moving through the body. Organs that make up the digestive system have many muscles. Like the muscles in our arms and legs, muscles in the organs of our digestive system must be kept in good shape. Fiber makes the muscles work by forcing them to move it through the body. As a result, fiber helps prevent constipation, a condition that occurs when food moves too slowly through the digestive system. In addition, fiber helps remove dangerous substances, or *toxins*, that may be present in food. For all of these reasons, eating food with sufficient amounts of fiber will help maintain good health.

LET'S EAT RIGHT, BUT HOW?

In 1977, the United States government published a set of dietary guidelines, recommending what people should eat. There was great concern that the incidence of heart disease was increasing, and something had to be done. The best-known form of these guidelines is found in the Food Guide Pyramid that was developed by nutritionists at the United States Department of Agriculture in 1991. It groups together foods of certain types and shows how many servings of each type of food the average person should eat each day. The pyramid is designed to be a clear and simple way to help people select the correct foods to eat, that is, foods that contain the proper nutrients. (See Figure 8-4 on page 166.)

There are six types of food in the food pyramid. The food types you should eat the most of according to the USDA are at the broad base of the

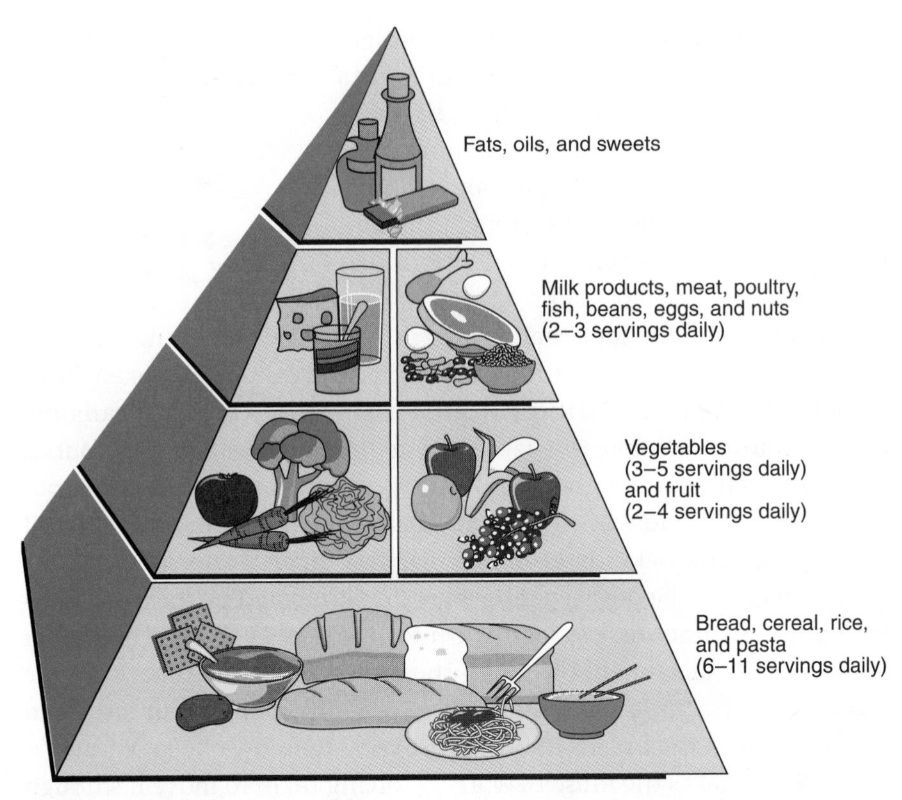

Figure 8-4 The food pyramid shows how many servings of each type of food a person should eat each day.

pyramid. The foods that you should eat the least of are at the narrow top of the pyramid. Notice that the grain group is in the bottom level of the pyramid. The next layer up contains the fruit and vegetable groups. In the next higher layer are the foods in the milk group and the meat, poultry, fish, eggs, beans, and nuts group. Finally, at the top of the pyramid are fats, oils, and sweets. According to these guidelines, these foods should be consumed in limited amounts. The food pyramid is meant to be a visual representation of a healthful diet. Foods in a healthful diet should be selected in the same way the food pyramid is constructed, with more servings of foods at the bottom and fewer servings of foods at the top.

However, a great deal of controversy surrounds these guidelines. What is troubling to many prominent scientists and nutritionists is that, in spite of all the attention given to healthy eating (especially for low-fat, high-carbohydrate diets, as recommended by the Food Guide Pyramid and encouraged by the food industry), the levels of obesity and diabetes have increased significantly in the United States in the last 25 years. In addition, heart disease has not decreased. The head of the department of nutrition at the Harvard School of Public Health declared in 2001 that the food pyramid is not at all accurate. The food pyramid says that all fats are bad and all complex carbohydrates are good. In a major article in *The New York Times* in July 2002, this nutritionist and a variety of other researchers stated that such thinking may, in fact, be all wrong. They state that it is necessary to understand the hormone insulin and how it works in relation to carbohydrates, levels of sugar in the blood, and the conversion of stored sugar into fat. Also, we must learn what determines whether the body uses stored fat or carbohydrates for energy, and what determines the feelings of hunger and fullness in a person. More scientists and nutritionists are beginning to suggest that fat in our diet may be good and that too many carbohydrates, like white bread and pasta, may be bad for your health. Intensive scientific research is just beginning to get answers to the question: How do we eat right? One thing is certain—it will be careful scientific investigation and not just government policy makers that will give us reliable answers to such important questions.

THE NEED FOR DIGESTION

Food comes in many different forms. For a single-celled ameba, food may be a bacterial cell one-tenth the ameba's size. For an earthworm, food may be a piece of a leaf that fell to the ground. A fish may eat a small crayfish; a bird may eat an earthworm. For us, food may be a hamburger or a piece of celery. For all organisms, food must be digested. Why?

Every organism is a single cell or a collection of cells. The cell theory states that cells are the basic units of structure and function of all organisms. In other words, what an organism needs is what its cells need. An organism needs food because its cells need food. You may be 1.5 meters tall and weigh 60 kilograms, but, from a biological point of view, a human —or a cat, a tree, or any other organism—is simply a collection of a very large number of cells. Each and every one of the 40 trillion (12 zeros!) or so cells in your body needs food to survive. So in order to nourish you, the food you eat must get into each of your cells.

That is why digestion is necessary. The foods mentioned—a piece of leaf, a crayfish, an earthworm, a hamburger, or a piece of celery—are all too large, their molecules too complex, to get inside of a single cell. For substances to get inside the cells, they must be broken down into relatively simple molecules. **Digestion** is the process of breaking down food particles into molecules small enough to be absorbed by cells.

Check Your Understanding

Why do all organisms need food? How is that need related to the processes of catabolism and anabolism?

DIFFERENT ORGANISMS: DIFFERENT METHODS OF DIGESTING FOOD

Usually, before food can be digested, it must be taken into an organism. The process of taking in food is appropriately called **ingestion**. For most organisms, the food starts to be broken down right after ingestion. Often this begins with food being made into smaller pieces by **mechanical digestion**. In humans and many other animals, this is accomplished by chewing. Teeth break the food into smaller pieces. Mechanical breakdown is followed by chemical digestion. During chemical digestion, enzymes break large molecules into small, simpler molecules. These small molecules can then get into the cells by **absorption**. Finally, **elimination** occurs as indigestible material, such as fiber, passes out of the organism.

The ameba ingests its food by surrounding it. In a flowing motion, called **phagocytosis**, an ameba engulfs a food particle such as a bacterium. Once inside the ameba, the bacterium is contained within a food vacuole. **Lysosomes**, organelles that contain enzymes, join with the food vacuole to deliver the enzymes that digest the food. This process that occurs in the cell is **intracellular** (chemical) **digestion**. (See Figure 8-5*a*.)

Have you ever seen a stump of a dead tree covered with mushrooms? The mushrooms survive by feeding on the nutrients contained in the dead wood cells. To take the complex molecules of wood into its cells, the mushroom secretes enzymes into the tree stump. Wood molecules are broken down outside the fungus and then absorbed. The work done by the digestive enzymes outside the mushroom's cells is called **extracellular digestion**. (See Figure 8-5 *b*.) Organisms in the fungus kingdom are an important exception to the usual order of things. Chemical digestion occurs outside these organisms before food gets taken in. The matter that once made up the bodies of most organisms gets broken down by fungi after the organism's cells have died. This process, which we call rotting, or decomposition, may happen in your refrigerator if food is left there too long; but it is happening in the natural world all the time. By the process of decomposition, fungi, along with many types of bacteria, make it possible for the organic compounds in dead organisms to get recycled and reused by living organisms.

Remains of food vacuole

Nucleus

Large food particle

Lysosome

1. Large food particle is ingested
2. Food vacuole is formed
3. Enzymes of lysosome enter food vacuole where digestion occurs
4. Small food molecules diffuse into cytoplasm

a. Intracellular digestion

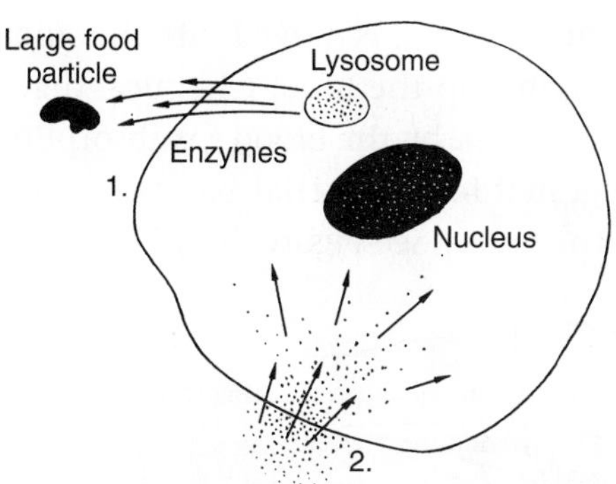

1. Enzymes leave cell and break down large food particle
2. Small food molecules diffuse into cell and then through cytoplasm

b. Extracellular digestion

Figure 8-5 Types of digestion: (*a*) intracellular and (*b*) extracellular.

Another very different digestive system is found in the *Planaria,* a type of flatworm often found living under rocks in streams and ponds. A planarian's mouth is on its underside, or ventral surface. Extending from the mouth is a muscular tube, the pharynx, which the planaria can push into its prey to release digestive enzymes there. Partially digested food then gets sucked into a pouch, or sac, inside the planaria, called the **digestive cavity.** Further digestion and absorption into cells occur here; undigested materials leave the flatworm through its mouth. (See Figure 8-6.)

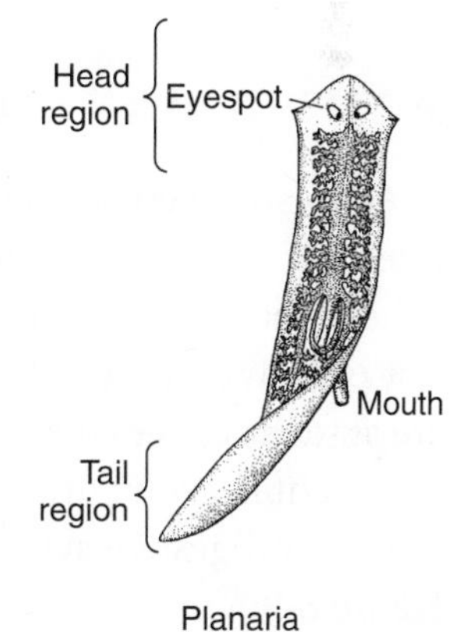

Figure 8-6 The planarian takes in food and gets rid of undigested material through its mouth.

Crayfish eat plants, dead organisms, and even other crayfish. They grab food with their claws and tear it into smaller pieces. Next, mandibles crush the food before it gets passed into the crayfish's mouth. Then the food passes into the **esophagus**, the tube located at the beginning of the animal's digestive system. The crayfish's digestive system consists of a pathway beginning with one opening, a mouth, and ending with another, the anus. As food travels through the pathway, the process of digestion occurs in a series of steps. The pathway of digestion is similar to an assembly line in a factory, except that the reverse process is occurring. Instead of putting parts together in a specific series of steps, we are now taking particles of food apart, one step at a time. This *one-way digestive system* is found in many animal species, including humans.

So, in crayfish, the digestive pathway or tube begins with the mouth and esophagus. Leaving the esophagus, the food is moved into the stomach. Here, hard teethlike structures grind up the food. Chemical digestion then occurs, and the food is transported by the blood for absorption into the cells of the crayfish. Undigested food material passes into the intestine to be eliminated through the anus. (See Figure 8-7.)

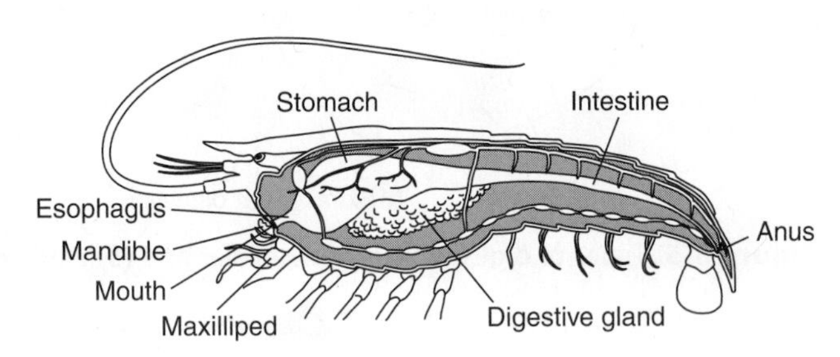

Figure 8-7 The crayfish has a one-way digestive system: food enters the mouth and waste exits the anus.

THE HUMAN DIGESTIVE SYSTEM

The digestive system in humans has exactly the same purpose as the digestive system in any other organism: to get nutrients from food into cells. When does food enter your body? Not when you put it in your mouth; not even when it enters your stomach or your intestine. These locations are actually spaces directly connected to the outside world. Food does not really enter "you" until it is absorbed across the membranes of cells that line your digestive system. Only then is the food truly inside of you. (See Figure 8-8.)

Think about what happens to a hamburger as it travels through the human digestive system. To begin, of course, the hamburger must enter the mouth. Teeth in the mouth provide mechanical digestion. The meat and bun get broken up into smaller and smaller pieces. This chewing process increases the surface area of the food. Smaller pieces of food provide more surface area for the body's enzymes to work on. When in the

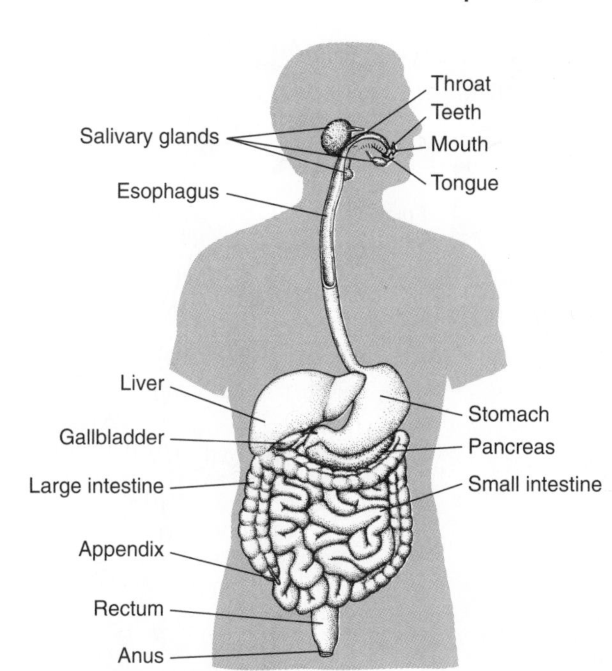

Figure 8-8 The organs of the human digestive system.

mouth, the hamburger is mixed with **saliva**, a fluid produced by salivary glands in the mouth. Chemical digestion begins as the enzyme amylase in the saliva starts the digestion of starch in the hamburger bun. The large polysaccharide molecules begin to be broken down into simpler molecules, eventually to the monosaccharide sugars. (See Figure 8-9.)

Swallowing occurs in two steps. The tongue pushes the food to the back of the mouth where it enters the esophagus, a muscular tube that carries the mouthfuls of hamburger down to the stomach. As the food starts its 10-second journey through the esophagus, an involuntary action begins. Waves of muscle contractions, called **peristalsis**, occur. As the process of peristalsis is initiated, the pathways up to the nose and down to the lungs are closed off. The epiglottis is a small flap of tissue that is pushed down by the food being swallowed to close off the trachea.

Figure 8-9 Food does not really enter you until it is absorbed across the membranes of the cells that line the digestive system.

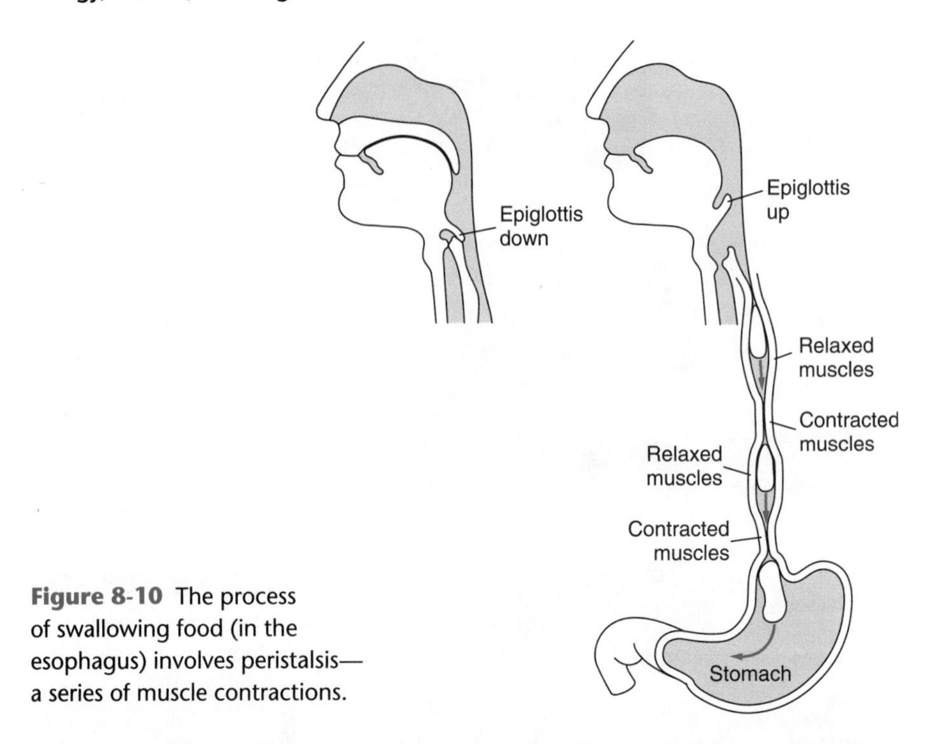

Figure 8-10 The process of swallowing food (in the esophagus) involves peristalsis—a series of muscle contractions.

The trachea is the pathway for air to the lungs; if food enters it by mistake, uncontrollable coughing and difficult breathing occur. (See Figure 8-10.)

When the waves of contraction along the esophagus arrive at the stomach, a muscular valve quickly opens and closes. The valve allows the food to enter the stomach, without letting out the acidic contents of the stomach. Occasionally, acid backs up through the valve, causing a pain commonly called "heartburn." This pain has nothing to do with the heart. Heartburn occurs when stomach acid reaches the lining of the esophagus, causing a painful feeling in the chest.

The stomach is a large muscular organ on the left side of the abdomen, just below a muscular sheet called the *diaphragm*. Having many folds in its wall, the stomach can stretch from its normal 1-liter size to hold 4 liters of food. The stomach evolved long ago in animals because it was an advantage to store food. With this ability, an animal could eat less frequently. Eating is often a dangerous time for many animals. A grazing animal, for example, has to lower its head to eat grasses. In that position, it is more difficult for the animal to notice a predator. Therefore the ability to eat quickly, and store food in the stomach for digestion later, was an adaptation that contributed to an animal's survival. It is interesting to note that some people survive quite well without a stomach. In some cases the stomach, or part of the stomach, is surgically removed in order to fight stomach cancer. A person without a stomach has a digestive system that is basically one long tube. These people eat small amounts of food many times during the day.

The muscular movement of the stomach churns the food and continues the process of mechanical digestion. As muscles in the stomach wall contract and relax, food gets mixed with **gastric juice**. Gastric juice, made and released by cells in the stomach wall, contains hydrochloric acid, pepsinogen, and water. The hydrochloric acid kills many bacteria that may be present in the food, some of which could cause illness. The acid also turns the pepsinogen into the protein-digesting enzyme pepsin. Therefore, a second step in the chemical digestion of the hamburger begins in the stomach with the breaking down of the protein in the meat by pepsin.

Muscular contractions of the stomach walls occur in waves of peristalsis, pushing the contents of the stomach into the small intestine about 3 or 4 hours after your meal. The small intestine, which is a muscular tube like the esophagus, is about 6 to 7 meters in length. (The small intestine takes its name from the small size of its diameter, only about 2.5 centimeters. It is actually longer than the large intestine, which is about 1.5 meters long.) Again, peristalsis moves food material along the length of the small intestine. Most of the chemical digestion of food occurs in the small intestine, not in the stomach. Large quantities of different enzymes are used to accomplish this task. These enzymes come from two major sources—the lining of the small intestine itself and the pancreas. Enzymes that break down all four major types of organic compounds come from the pancreas. These include the enzyme trypsin that breaks down proteins, nucleases for nucleic acid, carbohydrases for carbohydrates, and lipases for lipids.

Of the nutrients in the hamburger, the fat—usually a substantial amount—causes the most problems. Fat particles remain stuck together, making it difficult for the lipase enzymes to break them down. (Remember that enzymes work only on the surface of a food particle.) The liver, the largest organ in the abdomen, helps the body digest fat by producing **bile**. Bile is stored in the gallbladder until it is needed. Then bile is emptied into the small intestine through a small duct. When it comes in contact with fat, bile acts like the detergent in dish soap and breaks up the fat into smaller droplets. This process is called *emulsification*. Bile emulsifies fats into tiny droplets, allowing them to be more effectively digested by enzymes.

By now, the hamburger is very much changed. The starch that was in the bun has been changed into monosaccharide sugar molecules. The

When Dieting Becomes Dangerous

There is a common idea held by many people that fat is dangerous to the body. In fact, fat is important for the body's health. Fat surrounds and protects important tissues and organs. And the fat-storing cells of the body are sources of energy, especially when there is a shortage of food. Some of our body's fat is an adaptation for survival, a result of millions of years of evolution. The normal amount of fat, expressed as a percentage of body mass, is about 18 to 24 percent for females under thirty and 12 to 18 percent for males in the same age group.

As we all know, the percentage of body fat for many people is not within the normal range. Having too high a percentage of body fat is known to increase a person's risk of heart disease and certain cancers.

People try to lose weight by dieting. Researchers now know that dieting may reduce the amount of fat in the body, but it also changes how the body functions. When people diet, they usually limit the amount of food calories they take in. A dieter's body reacts to protect itself. Because of our evolutionary history, the body thinks that there is an actual shortage of food —that starvation is imminent. The body doesn't know that the dieting person is intentionally limiting the amount of food taken in, and it reacts by slowing down to survive the food shortage. The result? Fewer calories are burned during normal activities, and not much weight loss occurs. The fat cells that normally store fat are being emptied, but they still remain in the body. The person feels hungry and may end the diet. The "starved" fat cells

meat protein exists as amino acids, and the fat as fatty acid molecules and glycerol. Remember, all of this food is still outside of you! Now, these small molecules are able to pass through the wall of the small intestine into your blood vessels. In the blood, the food is carried to your cells. Finally, your food is really inside of you!

In order to make this absorption of food molecules more efficient, the inside of the small intestine has many folds covered by large numbers of fingerlike projections called **villi**. These projections, and even smaller projections on each of them, greatly increase the surface area inside the small intestine. The surface area of all the villi is equal to the size of an entire tennis court. This huge surface area is all wrapped up inside your small intestine! (See Figure 8-11.)

Almost all of the useful nutrients get absorbed in the small intestine. Anything that passes on from there into the large intestine is primarily indigestible material, such as the cellulose in the lettuce that was on the hamburger. Compared to the small intestine, the large intestine is large quickly refill their reserves. On again, off again, or "yo-yo," dieting may result. The fluctuating weight loss and gain can be dangerous.

Most researchers now realize that the best way to avoid becoming overweight is to reduce the amount you eat somewhat and to increase physical activity. Exercise increases the amount of energy used by the body. It also increases the amount of muscle tissue, which even when resting burns more calories than other types of body tissues.

There are other serious health risks involved in severe weight loss that is caused by a refusal to eat. The disorder called *anorexia nervosa* is most common in young women. Abnormal fears of being overweight, as well as other fears, may lead to anorexia nervosa. An anorexic person appears unhealthy. This disorder can be fatal.

Bulimia is another eating disorder. Unlike most anorexics, a person with bulimia might appear healthy. However, this person swings between overeating and getting rid of the food, often by taking laxatives or inducing vomiting. Some studies show that as many as 20 percent of college-age women suffer from some form of bulimia. This disorder can be dangerous. It can damage the heart, kidneys, or digestive system. Counseling to help a person understand the reasons behind these eating disorders is important. It is also important to learn how to make wise choices about what one eats. In some severe cases, hospital treatment may be necessary.

in diameter, about 6 centimeters, but it is short in length. In the large intestine, water from the remaining material is reabsorbed into the body. Feces, the solids that remain, are pushed along by peristalsis and forced by muscles of the rectum out through the anus.

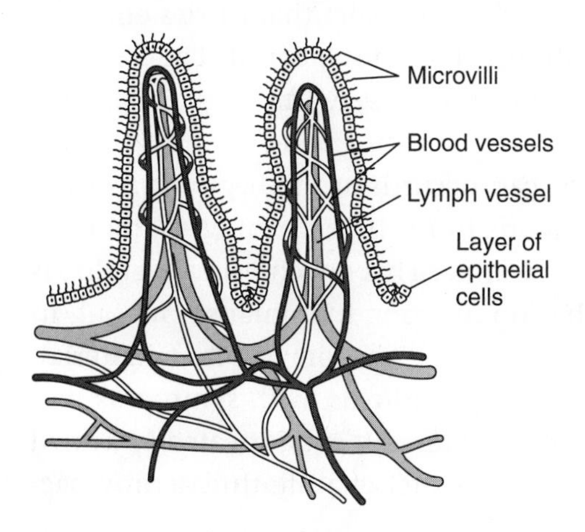

Figure 8-11 Villi increase the surface area of the small intestine, which makes absorption of food molecules more efficient.

Before concluding this tour of the digestive system, credit must be given to some helpers. Throughout the intestines, huge numbers of bacteria help break down food, producing certain gases as by-products. In fact, almost half the mass of feces, other than the water in them, is made up of bacterial cells. In addition to helping the process of digestion, these intestinal bacteria make several important vitamins. They also help rid the body of harmful bacteria. Fortunately for us, useful "friendly" bacteria keep us company and give important assistance to us in our intestines.

WHEN THINGS GO WRONG: DISEASES OF THE DIGESTIVE SYSTEM

In the stomach, gastric juice begins to digest protein in the foods we eat. However, the stomach itself is made of protein. To prevent the stomach from digesting itself, cells in the wall of the stomach produce a thick layer of mucus. This mucous lining protects stomach tissue from the acid and the pepsin it produces in the gastric juice. However, sometimes the layer of mucus protection fails. Gastric juice reaches the wall of the stomach and begins to break it down. The results are painful and serious. The eating away of tissue produces an ulcer. When this happens in the stomach, it is called a *peptic ulcer*.

For many years, it was thought that ulcers should be treated by reducing the amount of acid in the stomach. People who had ulcers were given antacids, chemicals to neutralize the acidity. Although the people often felt better, the ulcers almost always returned. Recently, researchers have shown that an infection by the acid-resistant bacteria *Helicobacter pylori* is really the main factor in causing ulcers. Treatment with antibiotics to kill the bacteria is much more effective for eliminating ulcers than the use of antacids.

A related but rare disease occurs when the duct that carries enzymes from the pancreas to the small intestine becomes blocked. The enzymes build up in the pancreas, and it rapidly digests itself. This condition, called *pancreatitis*, is very serious.

Another place where problems can occur is at the beginning of the large intestine. A small pouch extends from the beginning of the large intestine where it joins the small intestine. This pouch, the appendix, is a vestigial organ. The appendix has no function in humans now, but, in our ancestors, it probably helped in the digestion of plant material. Inflammation of the appendix causes appendicitis, with symptoms that include pain in the abdomen, nausea, and fever. If left untreated or treated incorrectly, the appendix may burst, producing a life-threatening bacterial infection in the abdomen. (See Figure 8-12.)

The main job of the large intestine is allowing water to be reabsorbed from the feces. If too much water is reabsorbed, the feces cannot move easily through the large intestine, and constipation occurs. If too little water is reabsorbed, the feces are too liquid, and diarrhea is the result. This can happen if a viral or bacterial infection irritates the lining of the large intestine. Diarrhea can cause the body to lose a great deal of water. Dehydration due to diarrhea is the main cause of infant death in many countries of the world.

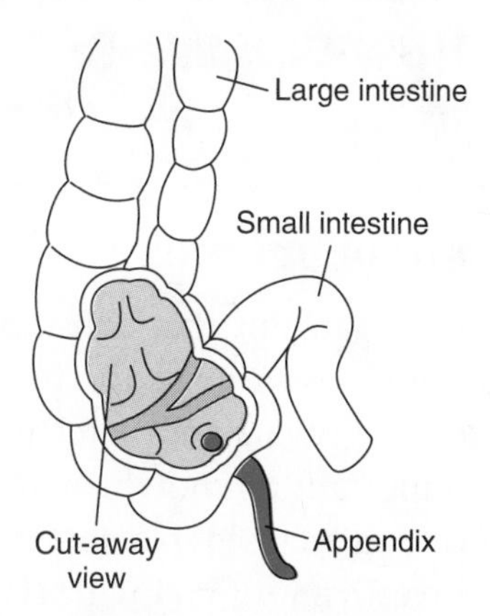

Figure 8-12 The appendix, a vestigial organ, is a small pouch that extends from the large intestine.

One of the most common types of cancer in North America is colon cancer. The large intestine is made up of the colon and the very end of the intestine, the rectum. The typical North American diet contains low levels of fiber. As a result, the feces move too slowly through the colon. This is directly related to colon cancer. Physicians also suggest that there may be a strong hereditary predisposition to colon cancer.

Refer to the table below to review the main types and functions of digestive organs, glands, and enzymes in humans.

DIGESTIVE ORGANS: THEIR GLANDS, ENZYMES, AND FUNCTIONS

Organ	Gland	Enzyme	Digestive Function
Mouth	Salivary	Ptyalin (amylase)	Starches into sugars
Stomach	Gastric	Pepsin (protease)	Large proteins into small proteins
Small intestine	Pancreas	Lipase, trypsin, carbohydrase, nuclease	Fats, proteins, starches, nucleic acids
Small intestine	Liver	Bile	Digest and absorb fats (fatty acids)
Small intestine	Intestinal	Intestinal	Sugars; proteins to amino acids

LABORATORY INVESTIGATION 8
How Can We Test for the Presence of Certain Nutrients in Foods?

INTRODUCTION

All organisms that consume food need to obtain certain nutrients from the foods. Among the most important of these nutrients are carbohydrates (sugars and starches) and fats. In order to eat a balanced diet, it is important to know which nutrients are found in which foods. A variety of tests are used to detect the presence of certain nutrients in foods. In this investigation, you will test for the presence of glucose (a sugar), starch, fats, water, and protein in a variety of foods.

MATERIALS

Food samples, test tubes, test-tube rack, graduated cylinder, scapula, Lugol iodine solution, Benedict solution, alcohol burner or other source of heat, test-tube clamp, biuret solution, brown paper or a brown-paper supermarket bag, safety goggles

PROCEDURE

Try to predict which nutrients will be found in the food samples before you actually test them. After each test, record your results in a data table. **Note:** *Wear safety goggles and follow all other safety precautions.*

1. *To test a food sample for the presence of starch:*
   In a test tube, add a drop of Lugol iodine solution to the food sample. A blue-black color indicates the presence of starch.
2. *To test for the presence of a sugar:*
   In a test tube, cover a food sample with Benedict solution. Place the test tube in a test-tube clamp and heat the test tube until the contents boil. An orange color indicates that sugar is present.
3. *To test for the presence of protein:*
   In a test tube, cover a food sample with colorless biuret solution. Observe. A pink or purple color in the biuret solution indicates the presence of protein.

4. *To test for the presence of fat:*
   Gently rub the food sample on a piece of brown paper or a brown paper bag from a supermarket. A translucent grease spot shows that fat is present.
5. *To test for the presence of water:*
   Gently heat a food sample in a test tube. Place the test tube in a rack and observe the top of the test tube when it cools. Droplets from steam show that water is present.

INTERPRETIVE QUESTIONS

1. How did your predictions compare with the results of your tests on the food samples?
2. List some foods that are good sources of the nutrients you tested for.
3. Why is water considered an important nutrient?
4. Research: What is the importance of, and some uses for, the following nutrients?
   a. carbohydrates
   b. fats
   c. proteins
   d. water
   e. minerals
   f. vitamins

CHAPTER 8 REVIEW

Answer these questions on a separate sheet of paper.

VOCABULARY

The following list contains all of the boldfaced terms in this chapter. Define each of these terms in your own words.

absorption, bile, digestion, digestive cavity, elimination, esophagus, extracellular digestion, gastric juice, ingestion, intracellular digestion, lysosomes, mechanical digestion, peristalsis, phagocytosis, saliva, villi

PART A—MULTIPLE CHOICE

Choose the response that best completes the sentence or answers the question.

1. Which of the following are groups of nutrients? *a.* vitamins, minerals, water, and carbon dioxide *b.* carbohydrates, lipids, proteins, and nucleic acids *c.* carbohydrates, proteins, water, and minerals *d.* carbohydrates, vitamins, water, and oxygen
2. Starch is broken down by enzymes in saliva. This is an example of *a.* catabolism *b.* peristalsis *c.* anabolism *d.* mechanical digestion.
3. Amino acids are assembled together to build protein molecules. This is an example of *a.* catabolism *b.* peristalsis *c.* anabolism *d.* mechanical digestion.
4. Most of the food you eat consists of *a.* proteins *b.* carbohydrates *c.* lipids *d.* water.
5. Plant-eating dinosaurs had a muscular organ that contained stones that helped to grind up food. This organ was primarily involved in *a.* chemical digestion *b.* mechanical digestion *c.* intracellular digestion *d.* nutrient absorption.
6. Beriberi is a type of *a.* bacteria that live in the human digestive system *b.* disease caused by a vitamin deficiency *c.* enzyme found in gastric juice *d.* microorganism involved in decomposition.
7. Amino acids from the proteins you eat are used to make the proteins that make up your body. This is an example of *a.* absorption *b.* catabolism *c.* anabolism *d.* phagocytosis.
8. Fiber is important to your body because it *a.* helps keep food moving through the body *b.* is rich in vitamins *c.* is an important source of protein *d.* contains essential amino acids.

9. A spider injects enzymes into its prey, then sucks up the liquefied tissues. This is an example of *a.* phagocytosis *b.* absorption *c.* extracellular digestion *d.* intracellular digestion.
10. Which of these organisms has a one-way digestive system with two openings? *a.* planarian *b.* jellyfish *c.* mushroom *d.* crayfish
11. Swallowed food is pushed to the stomach through the process of *a.* phagocytosis *b.* peristalsis *c.* ingestion *d.* mechanical digestion.
12. The food types you should eat most frequently are in the *a.* top of the food pyramid *b.* fruit and vegetable group *c.* meat, eggs, beans, and nuts group *d.* grains group.
13. The process of breaking down food particles into molecules small enough to be absorbed by cells is called *a.* digestion *b.* ingestion *c.* phagocytosis *d.* exocytosis.
14. Most of the chemical digestion of food takes place in the *a.* esophagus *b.* pancreas *c.* stomach *d.* small intestine.
15. Villi *a.* emulsify fats *b.* increase the surface area for absorption *c.* play an important role in mechanical digestion *d.* increase the chances of colon cancer.

PART B—CONSTRUCTED RESPONSE

Use the information in the chapter to respond to these items.

16. Identify structures A through I in the diagram on the next page.
17. At what time does:
    a. most of the digestive action of bile and pancreatic juice occur?
    b. the digestion of the starch in a whole-wheat bun begin?
    c. the digestion of the meat protein in a hamburger patty occur?
18. Explain why a two-opening digestive system is more effective than a one-opening digestive system. Name an animal that has each type of digestive system.
19. Describe how products from the salivary glands, stomach, pancreas, and liver aid in the process of digestion in humans.
20. When people are treated with antibiotics for long periods of time, they may begin to have digestive problems. Explain why this is so.

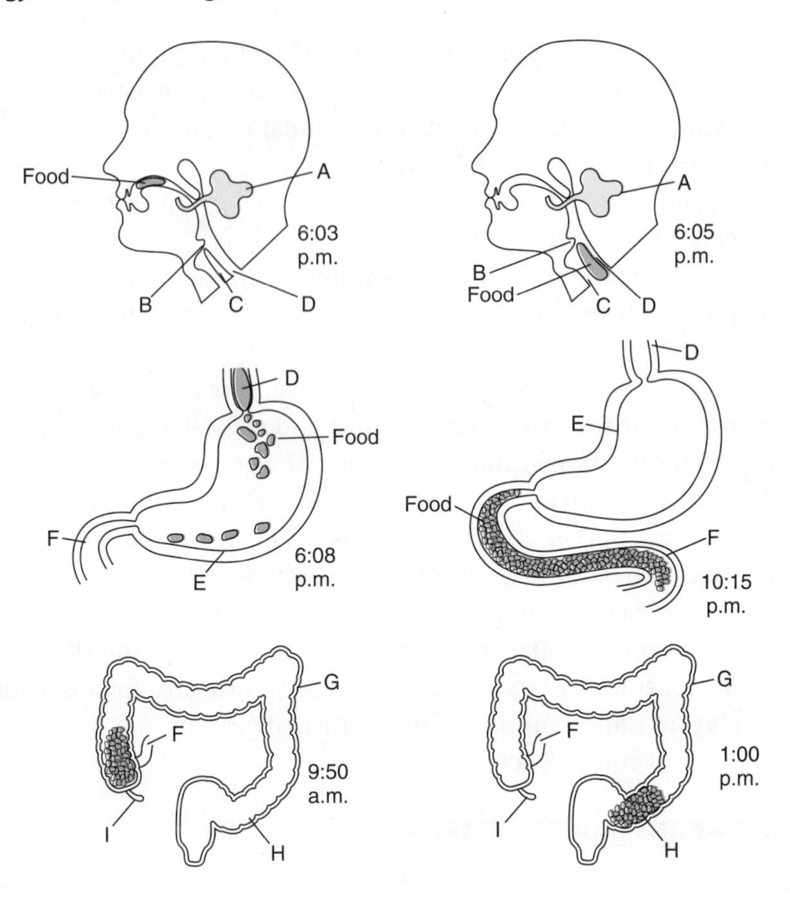
Food
A
6:03
p.m.
B
C
D
A
6:05
p.m.
B
Food
C
D
D
Food
F
E
6:08
p.m.
D
E
Food
F
10:15
p.m.
G
F
9:50
a.m.
I
H
G
F
1:00
p.m.
I
H

PART C—READING COMPREHENSION

Base your answers to questions 21 through 23 on the information below and on your knowledge of biology. Source: *Science News* (May 24, 2003): vol. 163, p. 333.

Eating Right Early Might Reduce Premature Births

A new study of sheep suggests that malnutrition around the time of conception may promote early delivery of offspring.

In the April 25 *Science*, Frank H. Bloomfield of the University of Auckland in New Zealand and his colleagues followed the pregnancies of 8 ewes that were consistently well fed and 10 ewes that were undernourished from 60 days before conception to 30 days after. Sheep in the latter group, whose weights fell to about 15 percent below normal, had an average pregnancy of 139 days, while the well-fed ewes were pregnant an average of 146 days.

The investigators found that modest undernutrition altered a crucial surge of the hormone cortisol that normally occurs in a mammalian fetus as birth approaches. This surge triggers maturation of organ systems and also seems to provide a signal to the mother that it's time to give birth. In half of the undernourished ewes, this cortisol spike came early.

"If these findings are applicable to human pregnancy, then a focus on events around the time of conception may hold the key to prevention of one of the major causes of preterm birth," Bloomfield and his colleagues conclude.

21. Explain the difference in how the two groups of female sheep in the study were treated.
22. State the finding that resulted from this study of the two groups of sheep.
23. Explain why this research could be important to knowledge about human birth.

Matter on the Move: Gas Exchange and Transport

After you have finished reading this chapter, you should be able to:

Describe the structures and functions of the respiratory system of humans and other organisms.

Describe the structures and functions of the transport (circulatory) system of humans and other organisms.

Discuss diseases of the respiratory and transport systems of humans.

I celebrate myself, and sing myself,
And what I assume you shall assume,
For every atom belonging to me as good belongs to you. . . .

My respiration and inspiration, the beating of my heart,
the passing of blood and air through my lungs . . .

Walt Whitman, *Song of Myself*

Introduction

A woman collapses on the sidewalk. The packages she is carrying fall in a jumble around her. "Does anyone know CPR?" a bystander yells. A policeman rushes over. Kneeling beside her, he asks the woman if she is all right and receives no response. Through his years of training, he knows that in a medical emergency, getting oxygen to the brain is the highest priority. Without breathing, brain cells lacking oxygen begin to be damaged within 4 minutes. After 8 minutes, damage to the brain occurs that cannot be corrected. After 12 minutes without oxygen to the brain, death may occur. The policeman carefully assesses the stricken woman's condition and determines that she is not breathing. After attempting unsuccessfully to restore her breathing, he fills her lungs with his own air by

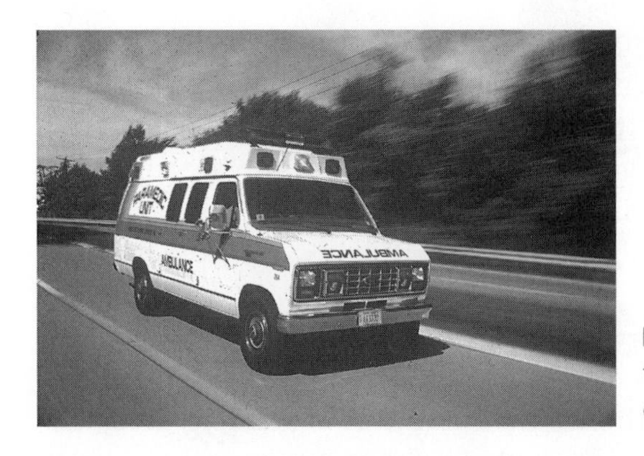

Figure 9-1 Emergency medical technicians are trained to use defibrillators to restart hearts.

breathing into her mouth. However, again through careful training, he knows that just getting air into the lungs is not enough. Getting oxygen from the lungs to the brain is also necessary. Is this happening? Checking for her pulse, he finds that her heart is not pumping. He remembered the words of his CPR trainer: "Replace what's missing." So he begins the chest compressions, acting as a temporary heart pump for the woman. Within minutes, an ambulance arrives. (See Figure 9-1.) The woman's heart begins to beat again after several electric shocks are delivered with a defibrillator. She is rushed to the hospital and later makes a full recovery.

Remember from Chapter 8 that the basic structure in the organization of life is the cell. All matter and energy used by an organism are used within its cells. The process of digestion makes food small enough to get it into cells. To make certain that food can be used for energy, gases must be delivered to and removed from cells. The heart attack victim had to get oxygen to her brain cells. A system must be able to move food molecules and oxygen to cells, and to remove carbon dioxide and waste molecules from cells. This chapter examines the means of gas exchange—the respiratory system—and the means of delivery—the circulatory system.

THE RESPIRATORY SYSTEM, BREATHING, AND CELLULAR RESPIRATION

In Chapter 7, you learned that the chemical energy stored in food molecules is released through the process of cellular respiration. A dog chases a Frisbee using the energy it got from its food through aerobic cellular respiration. The dog uses oxygen to release energy from food and gives off carbon dioxide as a waste product.

Animals move oxygen into—and carbon dioxide out of—the body with a respiratory system. Although cellular respiration refers to an energy-releasing series of chemical reactions that occurs in cells, the word *respiration* also means the process of exchanging gases. Finally, in many but not all animals, air is physically pumped into and out of the body by the process of breathing. Putting it all together, breathing or respiration —the process of gas exchange—is necessary to allow the life-sustaining metabolic activities of cellular respiration to occur.

GAS EXCHANGE SURFACES

All aerobic organisms—both plants and animals—exchange oxygen (O_2) and carbon dioxide (CO_2) with their environment. For example, the single-celled ameba exchanges these gases with the watery environment in which it lives. Oxygen in the water and carbon dioxide in the ameba move by diffusion across the cell membrane. As O_2 gets metabolized inside the ameba, the concentration of O_2 molecules in the cytoplasm becomes less than in the water outside the cell. Then, O_2 diffuses into the cell through the cell membrane. At the same time, CO_2 diffuses out of the cell as its concentration increases as a by-product of cellular respiration. (See Figure 9-2.)

Respiration in all organisms involves the diffusion of gases across cell membranes. You read in Chapter 3 that this occurs for plants in the spongy layer of cells within the plants' leaves. Although gas exchange in animals may involve a complex respiratory system, the actual process of taking in and getting rid of gases is identical to that of the ameba. Gases must be moved to and from the environment. They must cross a barrier to be moved in or out of the animal. This barrier, a part of the animal's body, is known as the **respiratory surface**. Respiratory surfaces in different animal species vary in shape and size. However, all share certain requirements:

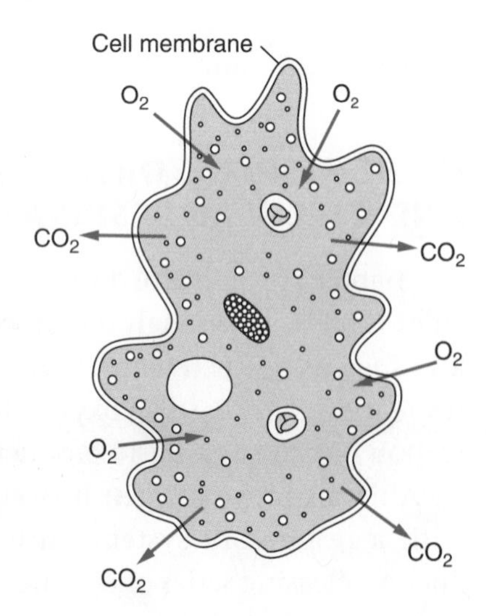

Figure 9-2 The single-celled ameba exchanges oxygen and carbon dioxide with the watery environment in which it lives.

- The respiratory surface has to remain moist at all times. Oxygen and carbon dioxide must be dissolved in water in order to diffuse across a cell membrane. A dry earthworm is a dead earthworm because its skin is its respiratory surface. When its skin dries, the earthworm suffocates.
- The respiratory surface must be very thin so that gases are able to pass through it. Not surprisingly, the cells that make up the respiratory surface in many animals are of the same type as skin cells—very thin and flat in shape.
- There must be a source of oxygen. The source may be dissolved oxygen in the water, or it may be oxygen mixed with other gases in the air. In either case, the respiratory surface must be near the water or air that is supplying the oxygen.
- The respiratory surface must be closely connected to the transport system that delivers gases to and from cells. For example, next to the respiratory surfaces in our lungs are the blood vessels that speedily move the oxygen to all parts of the body.

ADAPTATIONS FOR GAS EXCHANGE

Animals have evolved a variety of structures for exchanging gases. As has been mentioned, some small animals, such as the earthworm, use the entire outer surface of their bodies as their respiratory surface.

Some aquatic animals have portions of their body surface that project out into the water around them. These specialized respiratory organs are called **gills**. Water continually passes over these gills. Oxygen passes through the gills into the animal's blood, which transports the oxygen throughout the animal's body. At the same time, carbon dioxide is released from the blood in the gills into the water. (See Figure 9-3.)

Over time, animals such as spiders, ticks, and insects have evolved thickened skins to help them survive in dry environments. How does gas

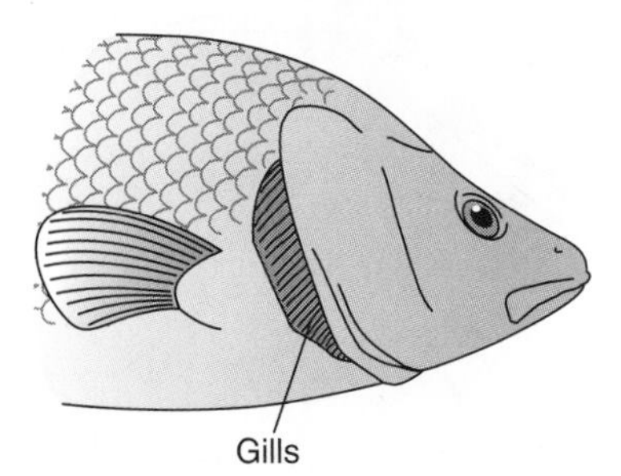

Figure 9-3 When water flows over the gills of a fish, oxygen enters the animal's blood and carbon dioxide leaves.

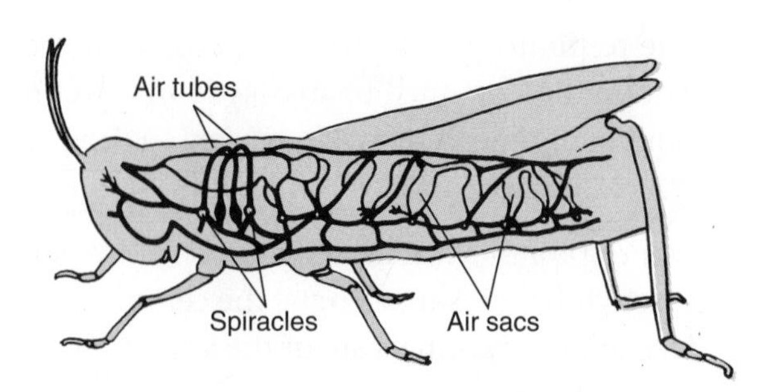

Figure 9-4 Respiratory system of the grasshopper.

transport occur in these animals? Look closely along the side of a cockroach or a grasshopper. You will notice a row of tiny openings in the insect's hard covering, which lead inside the animal to a series of branching tubes. These tubes, called **tracheae**, provide pathways that travel close to all the cells of the insect's body. Air is delivered directly to these cells through the tracheae. (See Figure 9-4.)

In their 300-million-year history on Earth, the success of insects has been in their numbers, not in their size. There will never be a cockroach the size of an elephant. One reason for this is that having tracheal tubes deliver air directly to all cells in the body works only for small animals. Large animals need a different way to exchange gases. In these animals, gases are delivered to blood vessels that transport the gases throughout their body. All terrestrial vertebrates—animals with backbones that live on land, such as frogs, snakes, birds, and humans—have two lungs (although some snake species have a single lung). In shape, lungs look somewhat like balloons. Located inside an animal, lungs are directly connected by tubes to the air outside. Blood vessels next to the lungs transport gases to and from body cells. (See Figure 9-5.)

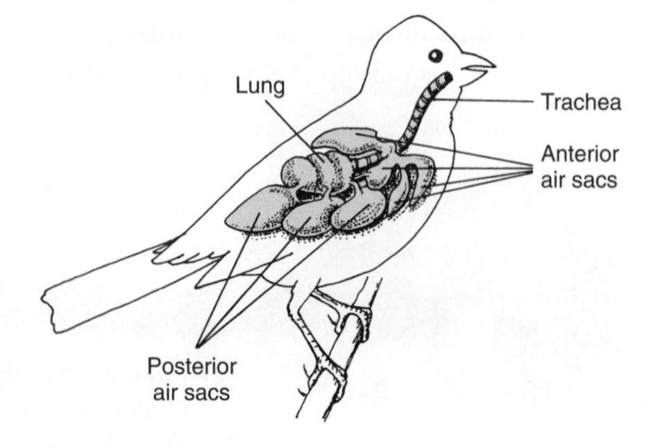

Figure 9-5 The respiratory system of birds and all other vertebrates includes two lungs.

A TYPICAL MAMMAL: THE HUMAN RESPIRATORY SYSTEM

The human respiratory system is similar to the respiratory system of other mammals. (See Figure 9-6.) Air moves through the nostrils into the **nasal cavity**, where dirt and other particles in the air are trapped by hairs and mucus. The air is also warmed, humidified, and tested for odors. The nasal cavity leads to the pharynx, where it meets air and food arriving from the mouth. Air continues flowing down, passing by the **larynx**. The vocal cords are in the larynx. When air moves out of the body, it passes over the vocal cords. The air causes the vocal cords to vibrate, producing the sounds of our voice. The hollow voice box formed by the flexible cartilage that surrounds the larynx helps amplify the sounds. In males, hormones increase the size of the voice box. This enlarged voice box is called the Adam's apple. Continuing down the tube, the **trachea** is surrounded by rings of stiff cartilage that act like the rings on a vacuum-cleaner hose. The rings help maintain the tubelike shape of the trachea, thus keeping it open.

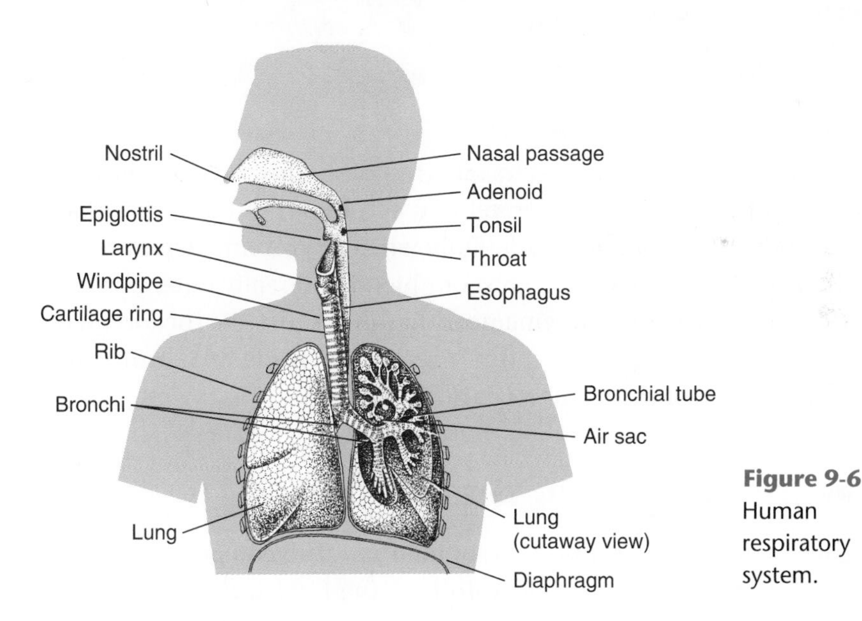

Figure 9-6 Human respiratory system.

The trachea branches into two **bronchi**. Each bronchus leads to a lung. In the lungs, the bronchi continue branching into smaller and smaller tubes called **bronchioles**. Most of these tubes are covered on the inside by mucus and by tiny hairlike extensions called **cilia**. The cilia move back and forth in a wavelike motion. As dust and other particles in the air get trapped in the mucus, the beating cilia help move them up the trachea.

The mucus then gets swallowed down the esophagus or expelled by the body. The cilia help keep clean the delicate tissues that line the lungs. One of the many harmful effects of cigarette smoking is that the cilia are paralyzed by inhaled smoke. When cilia are paralyzed, the lungs lose much of their ability to keep themselves clean.

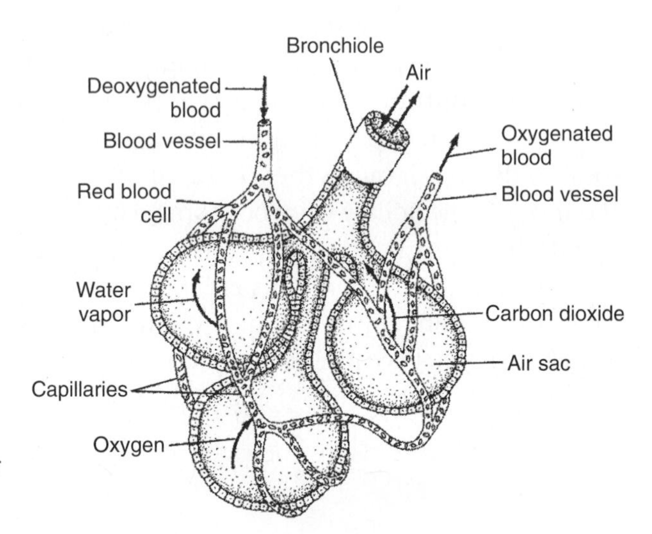

Figure 9-7 The lining of the alveoli acts as our respiratory surface.

The tiniest bronchioles end in bunches of microscopic air sacs, the **alveoli**. It is the lining of the alveoli that acts as our respiratory surface. Blood vessels surround the alveoli. Only when oxygen molecules diffuse across this lining do they enter into our blood and really enter our bodies. (See Figure 9-7.) You will remember that this is just like the food molecules of the hamburger that did not really get inside of us until they crossed the lining of the small intestine.

BREATHING: MAKING THE AIR MOVE

In the lungs of mammals, hundreds of millions of alveoli greatly increase the size of the respiratory surfaces. In fact, the total surface area of our alveoli is 60 square meters, about one-sixth the area of a basketball court. However, the lungs on their own are completely helpless in moving air into or out of our body. Lungs have no muscles. They cannot move and they cannot change size on their own.

All mammals, including humans, move air into their lungs by lowering the air pressure in the lungs. How do we lower the air pressure in our lungs? Two sets of structures are involved. You can easily feel your ribs with your fingers. Between your ribs, and harder to feel, are muscles.

These muscles move your ribs. When the muscles move the ribs upward and outward, the rib cage expands. At the same time, the diaphragm, a large flat muscle that lies across the bottom of the chest cavity, contracts and moves down. This movement also increases the size of the chest cavity. The air pressure lowers in the lungs because the same amount of air suddenly has more space to fill. **Inhalation** occurs when air rushes into the lungs through the respiratory tubes. The lungs fill with air. In other words, air gets sucked into the lungs.

It takes work to move air into our bodies. To move air out, the diaphragm and the muscles between the ribs relax. When they return to their original positions, the volume of the chest cavity decreases. With less space to fill, the pressure of the air in the lungs increases and air moves out. **Exhalation** occurs. You have taken one breath. Breathing is the physical process of inhalation—an active process that occurs when the muscles contract—and exhalation—a passive process that occurs when the muscles relax. About 12 times a minute, 720 times an hour, 24 hours a day, when we are asleep and when we are awake, we inhale and exhale. Each breath is the breath of life. (See Figure 9-8.)

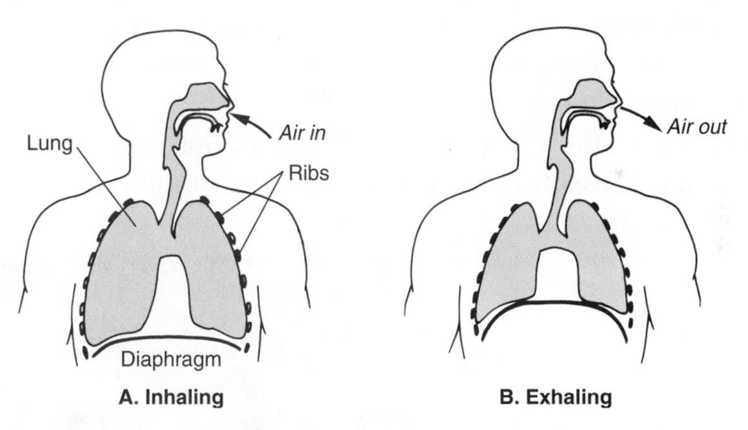

Figure 9-8 Breathing is controlled by movements of the diaphragm.

WHEN THINGS GO WRONG: DISEASES OF THE RESPIRATORY SYSTEM

Inhaling and filling the lungs with tobacco smoke is one of the most harmful activities in our society today. In the United States, smoking cigarettes causes more deaths that could be prevented than any other activity. It has been estimated that each cigarette reduces a smoker's life by about five minutes. The average smoker cuts six to eight years off his or her life by smoking cigarettes.

Why is smoking so harmful to your health and the health of people around you? How does smoking affect a fetus in a pregnant woman? Tobacco smoke contains many irritating, cancer-causing chemicals. Cancer results from uncontrollable cell division. Evidence has shown that the chemicals in tobacco smoke cause cancer in the lungs, esophagus, larynx, mouth, and even in the pancreas. Also, as mentioned before, the hairlike cilia that sweep away dirt particles in the respiratory system are paralyzed by cigarette smoke. This built-in cleaning system stops working. The enormous quantities of smoke particles from the cigarette go directly into the lungs. Without the action of the cilia, these smoke particles remain there. (See Figure 9-9.) In addition, smoking cigarettes increases the risk of other diseases of the respiratory system. These include:

- *Bronchitis.* This is an inflammation and swelling of the inside of the bronchial tubes. Mucus accumulates, plugging up the air passages. Breathing is difficult and painful.
- *Asthma.* The walls of the bronchi contract. These narrowed passages restrict the flow of air. As a result, the lungs do not fill or empty normally. People with asthma feel like they are suffocating. Although rare in 1900, by the year 2000 asthma, for a wide variety of reasons, had grown into an epidemic. It is now the most common chronic childhood disease in the United States.
- *Emphysema.* This is a very serious chronic disease in which the alveoli break down, greatly reducing the total area of the respiratory surface.

Figure 9-9 This micrograph shows the tiny hairlike cilia (magnified) that line the surface of the trachea and bronchioles. The cilia's wavelike motion helps keep the lungs clean. However, smoking paralyzes the cilia that sweep away dirt particles in the respiratory system.

In time, the lungs are unable to inflate. A person with emphysema suffers from shortness of breath. Eventually, even the smallest physical effort becomes difficult.

- *Pneumonia.* This disease is caused by a bacterial or viral infection that causes the alveoli to fill with fluid. Breathing becomes difficult. There is a vaccine available that prevents certain types of bacterial pneumonia.

Check Your Understanding

Why is it important that the respiratory surface in an animal is closely connected to its transport (circulatory) system?

WANTED: A TRANSPORT SYSTEM

It has become very clear that for an aerobic organism to live and grow, matter in the form of food molecules must be delivered to every cell. To release the energy stored in those molecules, oxygen must make that same trip to all cells. Finally, to prevent harmful buildups, carbon dioxide gas and other wastes produced by cells must be taken away.

Imagine that your house is a cell. Think of everything that exists to bring things to your house—water pipes, fuel trucks, mail and grocery deliveries, and so on. Then there are things that have to be taken away from your house. There are sewer pipes, garbage collectors, and recycling bins. All of these efforts to deliver needed materials and remove wastes from your house depend on transport systems. In the same way, an organism needs a transport system to service its cells.

Transport involves two processes. We have already met the first process, absorption. To be transported to cells, materials must cross cell membranes from the outside environment into the body. Absorption of food molecules occurs through the lining of the small intestine. Absorption of gas molecules occurs through the lining of the lungs' alveoli. Then materials must be circulated throughout the body to where they are needed. Transport involves absorption and **circulation**.

All organisms transport materials inside of them within a liquid, and the transport fluid must reach every cell in the body. There have to be pathways, vessels that are able to deliver materials around the body. These vessels are similar to the water pipes in a house. Finally, there must be something that forces the fluid through the transport vessels; the transport system must have a pump.

How to Help a Person Who Is Choking

Have you ever eaten a meal in a restaurant and noticed a poster about what to do when a person chokes on a piece of food? While it may seem out of place in a restaurant, these posters are required by law in many places, such as New York City. On a typical poster, the pictures show how to help a person who is unable to breathe when an object blocks the air passage to the lungs. People in a restaurant need to know the procedure—called the *Heimlich maneuver*—because a piece of food may cause a person to choke. However, food is not the only cause of choking. It is just as important for people at home to know the Heimlich maneuver. Chewing gum or, in the case of a child, small toys or a piece of balloon can block the air passage.

The Heimlich maneuver uses the air that is already present in the lungs to push an object up and out of the throat. The Heimlich maneuver is actually quite simple. First, you must determine if a person has an object blocking the throat or is, in fact, experiencing another condition, such as a heart attack. Ask the person, "Are you choking?" If so, he or she will not be able to speak to answer you. For a person who is standing, you administer the Heimlich maneuver by standing behind the choking person and grasping him or her with both arms around the waist. Place one hand, now closed as a fist, just below the bottom of the rib cage and above the navel. Grasp the closed hand with your other hand and make an upward, not inward, thrust. This thrusting motion puts pressure on the victim's diaphragm. The increased pressure forces air up and out of the lungs in order to force the blockage out of the throat. If the choking person is lying on the floor, turn the person so that he or she is facing upward. Straddle the choking person's hips and place one hand over the other on the abdomen below the rib cage and above the navel. Again, thrust in an upward direction with the heel of the hand in order to force air out of the lungs to expel the blockage.

ADAPTATIONS FOR TRANSPORT

Why does a single-celled organism such as the ameba not need a transport system? The moment materials are absorbed by the ameba across its cell membrane from the aquatic environment, they are right where they need to be, inside the cell.

However, all multicelled organisms, including animals and plants, must transport materials around their bodies. One of the largest plants on Earth is the redwood tree. Materials have to travel over 100 meters, the height of a 30-story building, between the redwood's roots in the soil and the leaves at the top of the plant. The redwood tree absorbs water through the cells of its roots. The plant's root cells have extensions, **root hairs**, which increase their surface area, similar to the villi that increase the surface area of the small intestine. The water and minerals dissolved in the water travel upward along pathways formed from **xylem** cells, which are lined up one after another to form long, hollow tubes. Water leaves a plant by evaporation from the stomates in its leaves, a process called **transpiration**. At the same time, food produced in leaves high above the ground is moved downward through tubes made of **phloem** cells. Plants have two sets of transport vessels: xylem for water and minerals, phloem for food. However, they do not have pumps. Instead, fluids are moved by the interactions of molecules. In xylem tubes, the interactions involve water molecules; in phloem tubes, the interactions involve sugar molecules. (See Figures 9-10a and 9-10b.)

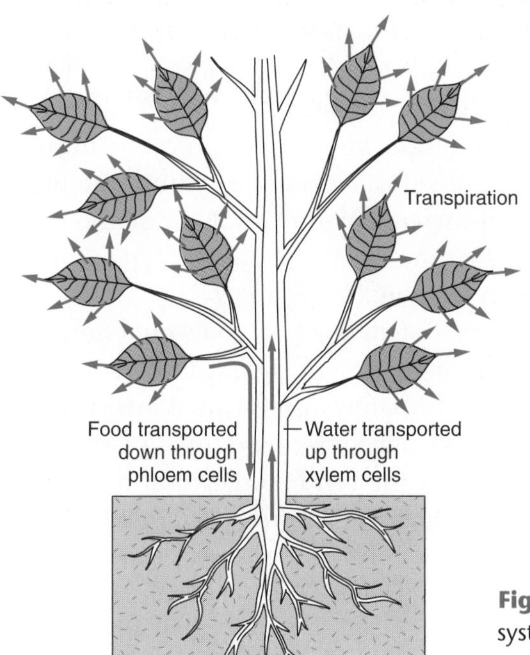

Figure 9-10a The transport system of a plant is made up of the roots, xylem, and phloem.

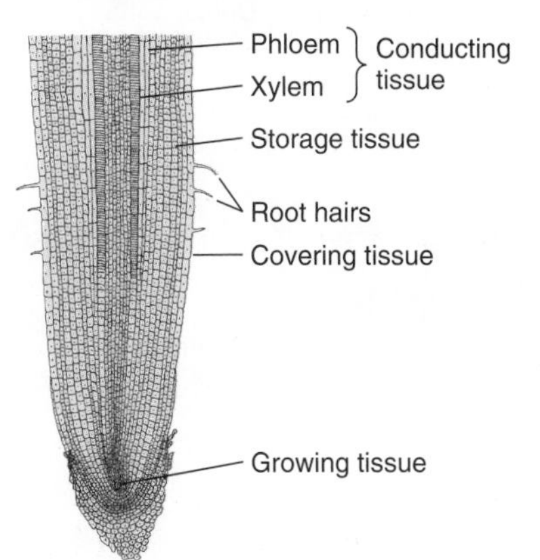

Figure 9-10b Root hairs increase the surface area of the roots.

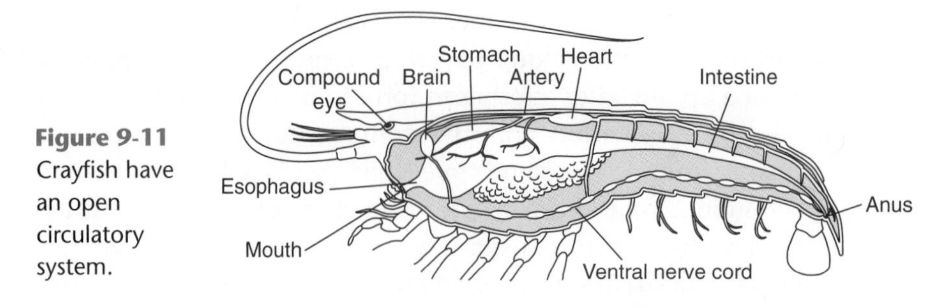

Figure 9-11 Crayfish have an open circulatory system.

Now let's examine the transport system in an animal—the crayfish. The crayfish's heart is near the animal's dorsal surface. The heart pumps blood into seven large blood vessels called *arteries*. The arteries release the blood into the spaces that surround the organs and cells of the crayfish. The blood bathes the cells by flowing through these spaces. After a while, the blood gathers again in a space near the heart. The blood passes into the heart through three small openings. The pumping process continues. Note that much of the time the blood is *not* confined within the blood vessels; instead it moves around the animal's organs. Because the blood is not contained in vessels all the time, scientists call this type of transport system an **open circulatory system**. (See Figure 9-11.)

If you examined the circulatory system of a fish, you would notice an important difference. A fish's transport system is typical of vertebrates. Blood leaves the fish's heart through large vessels called **arteries**. The arteries repeatedly branch into smaller vessels, eventually becoming microscopic **capillaries**. Every cell of the fish's body is near a capillary. The capillaries in turn begin to join together to form **veins** that lead back to the heart. Blood is always contained inside the vessels. Scientists call this a **closed circulatory system**. (See Figure 9-12.)

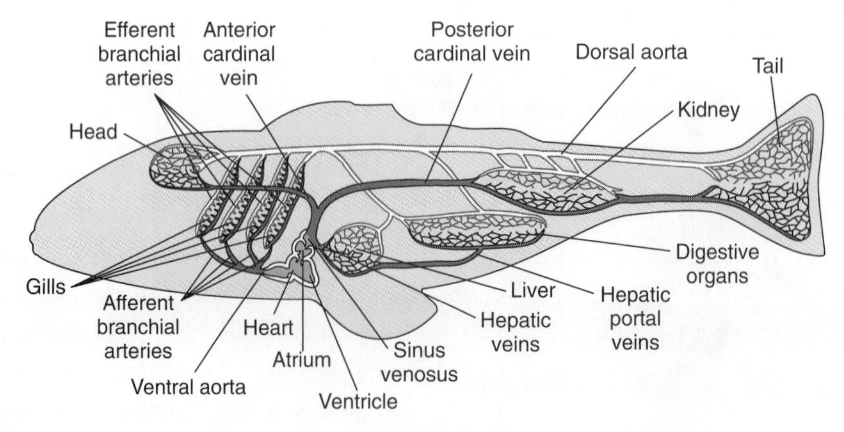

Figure 9-12 Fish have a closed circulatory system.

Where does the blood go when it leaves the heart? First it travels to the fish's gills to receive a fresh supply of oxygen and to give up the waste gas carbon dioxide. Blood then travels through more arteries to all parts of the body before returning to the heart. The blood travels in a circular path around the body of the fish. Upon returning to the heart, the blood enters the **atrium**, the first part of the heart. The blood is then pushed to the **ventricle**, a thick-walled, muscular pumping chamber, before being pushed out into the arteries. The heart of the fish has only two chambers: a single atrium and a single ventricle.

Fish first evolved about 500 million years ago. It took another 125 million years for the first amphibians to evolve. If you examine the heart of a frog, you can see one of the big advances amphibians achieved in this enormously long period of evolution. Unlike a fish's, the frog's heart has three chambers, not two. Blood arrives in the left atrium from the lungs, having been given a fresh supply of oxygen in exchange for carbon dioxide exhaled. The blood passes into the ventricle and gets sent throughout the body. Returning to the heart after delivering its oxygen, the blood enters the right atrium and again goes into the ventricle. While the frog is underwater, the blood that enters the right atrium may also have gotten some oxygen that diffused from the water directly into blood vessels in the skin. Now the blood gets sent to the lungs, getting a fresh supply of oxygen—if the frog has been out in the air, not underwater—and then it returns once again to the left atrium. You will notice that in the frog, the blood travels twice along two different pathways. The pathway to and from the lungs is the **pulmonary** circulation. The pathway to and from the rest of the body is the **systemic** circulation. (See Figure 9-13 on page 198.) Amphibians were the first vertebrates to evolve this more efficient system of transporting materials around the body. The blood really gets pumped twice as it circulates—once after coming back from the lungs and once after coming back from the body. Blood pressure is kept higher. In addition, oxygen-rich and oxygen-poor blood are, for the most part, kept separated by structures inside the ventricle.

THE HUMAN CIRCULATORY SYSTEM

Every cell in our body needs food and oxygen, and a way to get rid of the wastes it produces through normal life processes. In order for the entire organism to survive, every cell in the body must be taken care of. Our circulatory system accomplishes this vital task through the many thousands of kilometers of blood vessels in our body. Blood from the lungs enters the left atrium of the heart, then passes to the left ventricle. The blood then

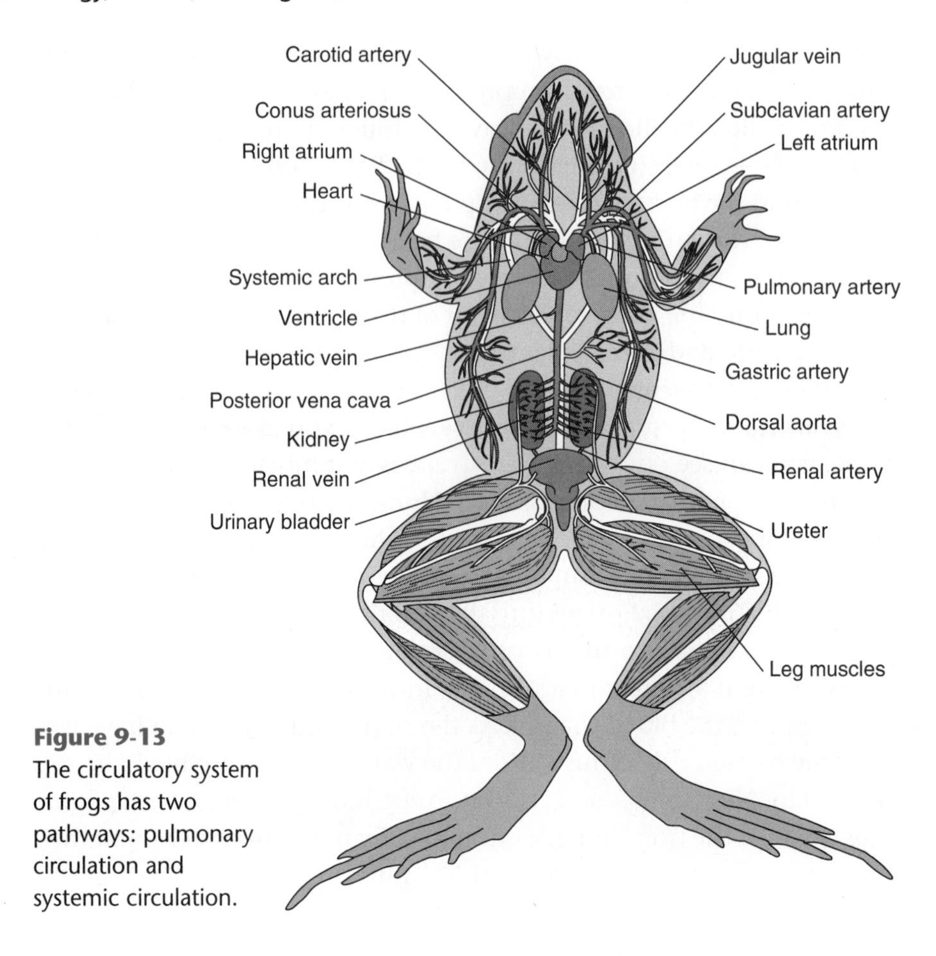

Figure 9-13
The circulatory system of frogs has two pathways: pulmonary circulation and systemic circulation.

begins its journey throughout the body through the aorta (the body's largest artery) to the other arteries.

Arteries have thick, muscular walls. They are very elastic, expanding and contracting as blood from the heart is pumped through them in pulses—one pulse for every heartbeat. The elasticity of the arteries' walls exerts pressure against the blood inside. This pressure is measured as your blood pressure. Every time the heart contracts, a pulse of high blood pressure occurs. This is called the **systolic pressure** and it is 120, on average, in a healthy young adult. This number represents the height in millimeters to which the pressure of blood in an artery forces a column of the heavy liquid element mercury. The pressure drops in the moments between contractions. This is the **diastolic pressure**, normally about 80 millimeters of mercury. A person's blood pressure is given using the two values, for example, 120/80. If a person's systolic and diastolic values are above 140/90, the person is usually said to have high blood pressure.

The arteries become smaller and smaller as the blood moves away from the heart. Eventually the blood enters capillaries, the smallest vessels. Capillaries are close to every body cell. And it is in the capillaries that the exchange of nutrients and gases to and from the blood takes place. After moving through capillaries, blood returns through the thin-walled veins, which get larger and larger closer to the heart. Unlike the walls of arteries, the walls of veins have little elasticity; blood is under low pressure in them. In fact, it would be very easy for blood to flow backward. For example, in the veins of the leg, as the blood returns to the heart it flows upward, against the pull of gravity. To prevent the blood in veins from moving backward, one-way valves work to trap the blood. The blood remains stationary in the veins until the beating of the heart and other muscular activities, such as the contracting of leg muscles, force the blood back up toward the heart. (See Figure 9-14.)

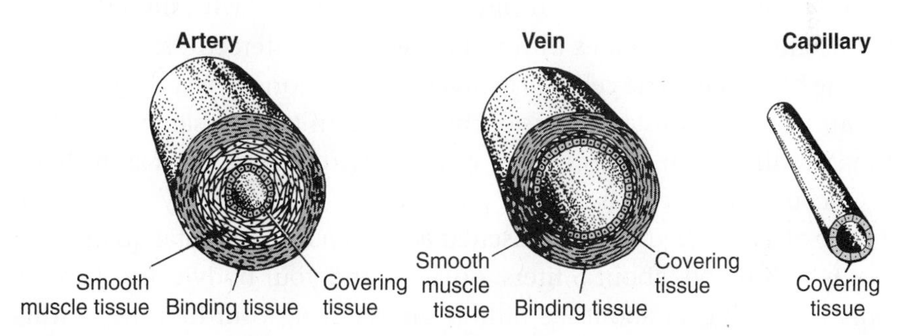

Figure 9-14 Cross sections of the three kinds of blood vessels.

The blood from the body enters the right atrium of the heart. The systemic circuit has been completed. After entering the right ventricle, the blood then enters pulmonary arteries that take it to the lungs. After the exchange of gases that occurs in the lungs, the blood returns by pulmonary veins to the left atrium.

The heart of mammals has four separate chambers. (See Figure 9-15 on page 200.) Our circulatory system, which has the same layout as that of all other mammals, works more efficiently than the circulatory system of any other animal group.

The entire purpose of the transport system is to move materials to and from cells. A capillary is so small that red blood cells must travel through it in single file. Laid end-to-end, the capillaries in your body would circle the globe at the equator. (That's more than 40,000 kilometers!) Molecules, including water, diffuse through the capillary walls and enter the spaces

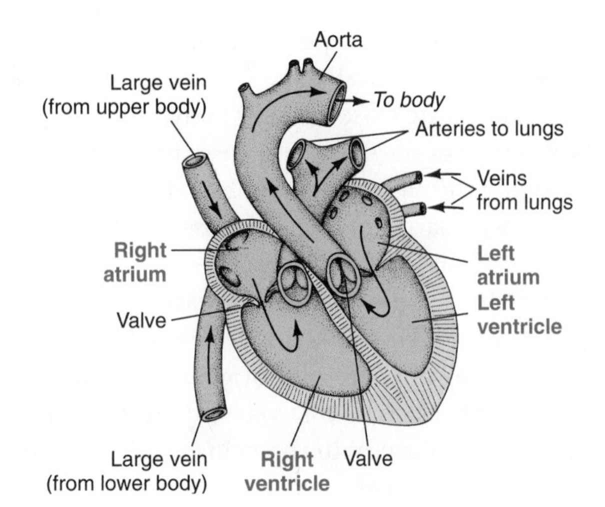

Figure 9-15 The heart of mammals, including humans, has four chambers.

around body cells. These intercellular spaces are filled with fluid that surrounds all cells. Molecules diffuse between this intercellular fluid (ICF) and the body cells. The cells we evolved from billions of years ago existed in warm, chemical-filled waters of the oceans. In just the same way, all of our cells still exist in a warm, salty, watery environment. (See Figure 9-16.)

Of course, the heart is the star player in our circulatory system. About the size of a clenched fist, its muscular actions pump the total quantity of your body's blood, about 5 liters, once around your body every minute. Your heart will beat about 2.5 billion times during your lifetime, resting between each beat. Each beat begins with an electrical signal that spreads outward from a spot in the wall of the right atrium to the rest of the heart. This spot is called the *pacemaker*. An electrocardiogram is a measurement of electrical activity in the heart during each heartbeat. Abnormal activity in the heart can be detected by looking at a person's electrocardio-

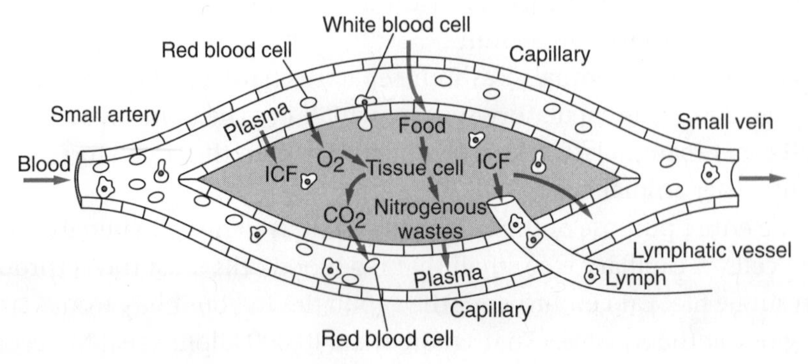

Figure 9-16 Molecules diffuse between capillaries, ICF, and body cells.

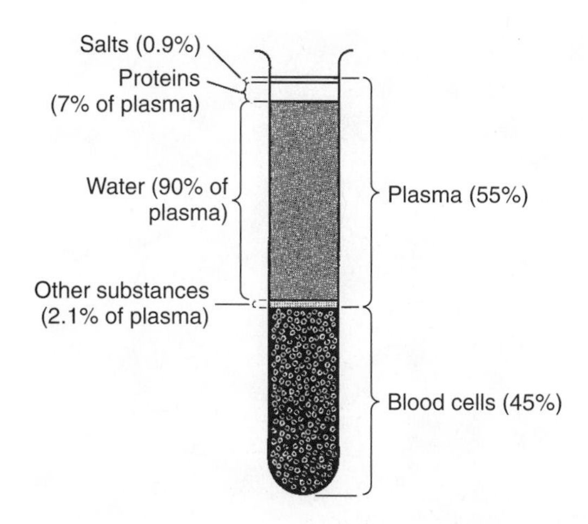

Figure 9-17 Blood is a tissue made up of plasma and blood cells.

gram. If a person's heart is not functioning normally, an artificial pacemaker that provides the correct electrical signal can be implanted in the heart by a surgeon.

As the blood moves from the atria to the ventricles, valves between the two chambers close. The closing of a valve creates the first part of the "lub-dup" sounds of the heartbeat. The second sound, the "dup," comes as the blood leaves the ventricles and valves close behind it.

Blood itself is a tissue. But unlike any other tissue in the body, blood is a liquid. Blood is made up of cells, cell parts, and a clear, light yellow-colored liquid called **plasma**. Plasma is 90 percent water, plus many important proteins, salts, vitamins, hormones, gases, sugars, and other nutrients. One of the proteins, fibrinogen, helps in the clotting process that stops bleeding caused by an injury. (See Figure 9-17.)

The cells in the blood include red blood cells, which contain the oxygen-carrying protein hemoglobin, and white blood cells. There are five types of white blood cells, all of which are involved in protecting the body from disease-causing foreign substances. Blood also contains **platelets**—fragments of cells that plug "leaks" where an injury occurs. Platelets also begin the complex chemical process that results in production of a blood clot. A blood clot stops the flow of blood out of a damaged blood vessel. (See Figure 9-18 on page 202.)

By 2002, after decades of research, at least six blood substitutes were close to being approved for use in patients. The most common of these substitutes contains a synthetic chemical called *perfluorocarbon*, or PFC, which can deliver oxygen where needed in the body just like hemoglobin does.

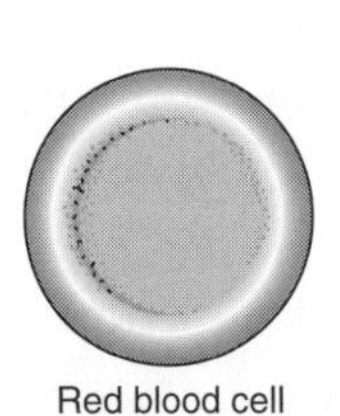

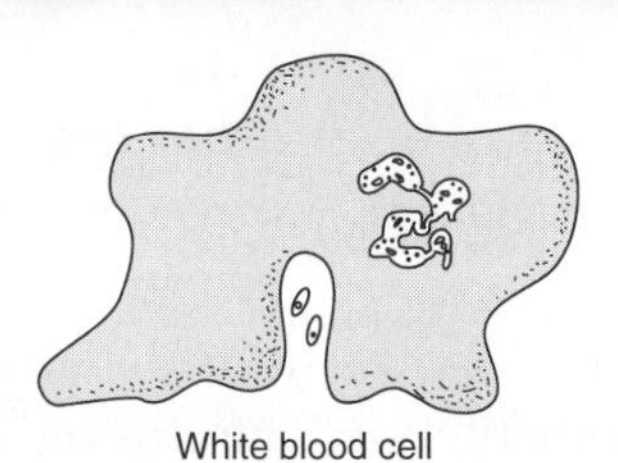

Platelets

Figure 9-18 Scanning electron micrograph of red blood cells, which carry oxygen. White blood cells protect us from disease, and platelets begin the clotting process.

WHEN THINGS GO WRONG: CARDIOVASCULAR DISEASE

The human transport system is a cardio- (heart) vascular (vessels) system. Cardiovascular disease includes several important diseases and is the cause of more than one-half of all deaths in the United States. A heart attack occurs when muscle tissue in the heart dies. The heart is made of muscle tissue that contracts and relaxes many times a minute. Like any other muscle, heart muscle must be supplied with blood, which brings nutrients and oxygen and removes wastes produced by the heart. When the coronary arteries that bring blood to the walls of the heart get blocked, the heart tissue beyond the clot is not supplied with blood. When this happens, a heart attack occurs. If the clot affects only a small portion of the heart muscle, the effects are relatively minor. If the clot affects a large part of the muscle tissue of the heart, the effects might be devastating.

Unlike a heart attack, which often occurs suddenly, some forms of cardiovascular disease develop slowly over a long period of time. Atherosclerosis is the gradual buildup of layers of fatty deposits on the insides of arteries. Included in these deposits is the lipid cholesterol. (See Figure 9-19.) Because of these deposits, the inside diameter of the blood vessel becomes

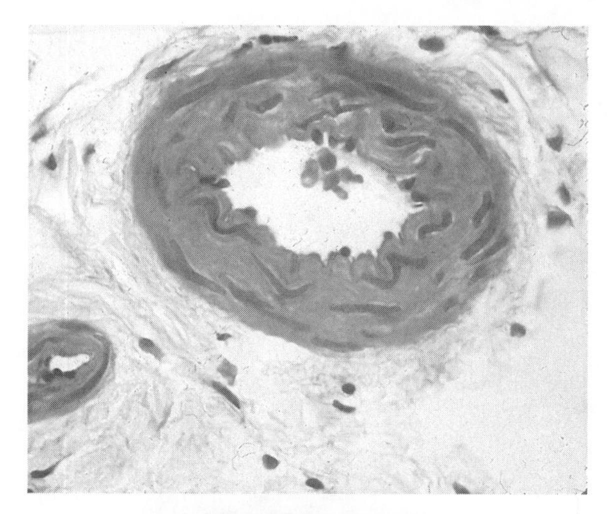

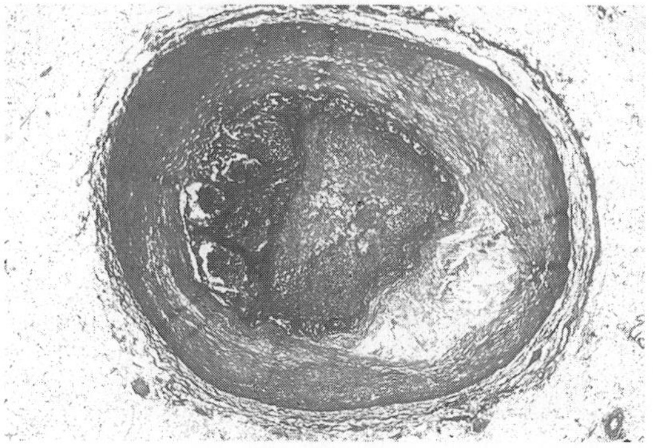

Figure 9-19 A normal artery (top). An artery clogged as the result of atherosclerosis (bottom).

smaller and the blood cannot flow so easily. In time, the arteries become "hardened"; the less flexible walls do not expand and contract as easily with each beat of the heart. Blockage of the blood vessel becomes much more likely. Cardiovascular disease also includes strokes that may block an artery in the brain. Depending on what area of the brain is affected, strokes damage a person's ability to feel things or to speak and move.

Atherosclerosis, heart attacks, and strokes are more likely to occur if a person has hypertension, or high blood pressure. Sometimes called the "silent killer," abnormally high blood pressure often shows no symptoms. Dietary and hereditary factors may contribute to high blood pressure. It is important to have your blood pressure checked regularly. How can hypertension be prevented or reduced? The answers are no surprise: Do not smoke, exercise more, and eat properly. Sometimes prescription medicines are used to treat high blood pressure. Your physician is your best source of information about proper methods of treating high blood pressure.

LABORATORY INVESTIGATION 9
How Does Exercise Affect Our Heart and Respiratory Rates?

INTRODUCTION

Exercise increases the demand by the cells of your body for oxygen, resulting in an increase in the rates of respiration and circulation. This can be measured by comparing the output of carbon dioxide with the pulse rate before and after exercise. In this investigation, you will participate in a study to determine the effects of jogging on your respiratory and circulatory systems.

MATERIALS

Test tube rack with three test tubes, phenolphthalein solution (pink), drinking straws, watch or clock with a second hand

Note: *Caution should be used when handling chemicals such as phenolphthalein. Protective eyewear should be worn while performing the CO_2 test.*

PROCEDURE

1. Prepare a Data Table to show the heart-rate data and carbon dioxide (CO_2) test data (1) at rest, (2) after 1 minute of jogging, and (3) after 5 minutes of jogging, for four trials (one per student) and for the average of the four trials for each of the three test conditions.
2. For this laboratory investigation, you will work in groups of four. Each member of the group will take turns being the jogger. While one person is jogging, the other three group members will keep time, take the pulse of the jogger, and record the results.
3. For Trial 1, select one member of the group as the jogger. Pour an equal amount of pink phenolphthalein (about 2.5 centimeters) into each of three test tubes. Be sure that the level in all three tubes is the same. The rate of respiration is determined by the CO_2 test, which measures the amount of carbon dioxide released. Due to CO_2 in your breath, blowing gently through a straw into phenolphthalein in a test tube causes the pink color to disappear and the solution to become clear. The time required (in seconds) to change the color of a measured amount of phenolphthalein by exhaling into it is the CO_2 test result, and it can

be used as an indicator of respiratory rate. Have the jogger at rest perform the CO_2 test. Take his/her pulse rate to measure heart rate. Record your results in the Data Table.

4. The jogger now jogs in place for 1 minute. As soon as the exercising is done, repeat the measurements in step 3 for the CO_2 test using the second test tube and the heart rate. Record your results in the Data Table.
5. The jogger now jogs in place for 5 minutes. As soon as the exercising is done, repeat the measurements in step 3 for the CO_2 test, using the third test tube and the heart rate. Record your results in the Data Table.
6. Rinse out the test tubes and refill with pink phenolphthalein, again about 2.5 centimeters per tube. Select another group member as the jogger. Repeat steps 3 through 5 with the second jogger for Trial 2. Record your results in the Data Table.
7. For Trials 3 and 4, repeat as above with the other two members of the group as the joggers.
8. Compute the averages for the CO_2 test and the heart-rate data for the four trials of the three test conditions: at rest; after 1 minute of jogging; and after 5 minutes of jogging.
9. Using the averages of the results, prepare three graphs as follows: Graph 1—Heart Rate *vs.* Minutes of Exercise; Graph 2—CO_2 Test/Color Change (in seconds) *vs.* Minutes of Exercise; and Graph 3—Heart Rate *vs.* CO_2 Test/Color Change (in seconds).

INTERPRETIVE QUESTIONS

1. Explain why carbon dioxide exhalation can be used as a measure of respiratory rate in humans.
2. How might the concentration of carbon dioxide in the blood act as a signal for an increase in heart rate?
3. Explain the correlation between the heart rate and the respiratory rate shown in Graph 3.

CHAPTER 9 REVIEW

Answer these questions on a separate sheet of paper.

VOCABULARY

The following list contains all of the boldfaced terms in this chapter. Define each of these terms in your own words.

alveoli, arteries, atrium, bronchi, bronchioles, capillaries, cilia, circulation, closed circulatory system, diastolic pressure, exhalation, gills, inhalation, nasal cavity, open circulatory system, phloem, plasma, platelets, pulmonary, respiratory surface, root hairs, systemic, systolic pressure, trachea, tracheae, transpiration, veins, ventricle, xylem

PART A—MULTIPLE CHOICE

Choose the response that best completes the sentence or answers the question.

1. Gas exchange is also known as *a.* respiration *b.* circulation *c.* exhalation *d.* transpiration.
2. The vessels that carry blood to the heart are called *a.* arteries *b.* veins *c.* bronchioles *d.* xylem.
3. In humans, the respiratory surfaces are found in the *a.* larynx *b.* trachea *c.* capillaries *d.* alveoli.
4. Which of these animals has an open circulatory system? *a.* crayfish *b.* ameba *c.* chicken *d.* mouse
5. Which of these organisms uses gills for gas exchange? *a.* butterfly *b.* seal *c.* frog *d.* tadpole
6. A grasshopper has respiratory organs known as *a.* lungs *b.* gills *c.* tracheae *d.* bronchioles.
7. A thin, moist membrane that is near a source of oxygen and an organism's transport system is necessary for *a.* transpiration *b.* respiration *c.* circulation *d.* exhalation.
8. The structures that line the air tubes in humans and help to keep the lungs clean are *a.* bronchioles *b.* alveoli *c.* cilia *d.* villi.
9. The part of the blood closely associated with clotting is the *a.* plasma *b.* platelets *c.* hemoglobin *d.* red blood cells.
10. The diaphragm contracts and moves down, lowering the air pressure in the chest cavity, during *a.* exhalation *b.* inhalation *c.* emphysema *d.* transpiration.

11. The process by which materials are moved throughout the body is best described as *a.* exhalation *b.* transportation *c.* absorption *d.* circulation.
12. Valves are found in the *a.* capillaries *b.* arteries *c.* veins *d.* bronchioles.
13. In plants, water is carried in the cells of the *a.* xylem *b.* phloem *c.* bronchi *d.* ventricles.
14. If your blood pressure is 120/80, the number 120 refers to your *a.* systolic pressure *b.* diastolic pressure *c.* cholesterol level *d.* pacemaker activity.
15. The liquid part of blood is the *a.* platelets *b.* leukocytes *c.* intercellular fluid *d.* plasma.

PART B—CONSTRUCTED RESPONSE

Use the information in the chapter to respond to these items.

16. Identify the structures labeled A through L on the diagram. Which structure is associated with the sound of a heartbeat? Which structures carry oxygen-rich blood?

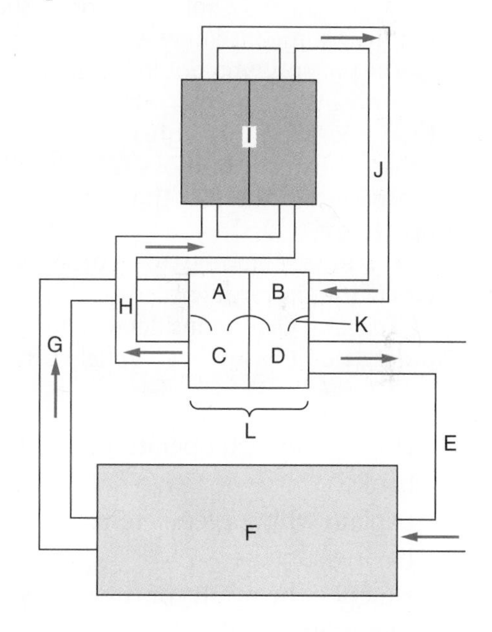

17. Which structures in the diagram are involved with pulmonary circulation? Why is pulmonary circulation important?
18. Explain how the respiratory, circulatory, and digestive systems each contribute to providing energy to cells.
19. Compare and contrast the circulatory systems in a crayfish, a fish, a frog, and a human.
20. Describe the journey of an oxygen atom from the air into a muscle cell in your leg and back into the air. Be creative!

PART C—READING COMPREHENSION

Base your answers to questions 21 through 23 on the information below and on your knowledge of biology. Source: *Science News* (February 1, 2003): vol. 163, p. 77.

Clot Promoter Cuts Surgical Bleeding

Blood banks face a perpetual supply shortage, but a clot-promoting agent known as recombinant activated factor VII (FVIIa) might offer a new means to staunch the demand for blood. When administered during surgery, the lab-generated enzyme can reduce a patient's bleeding and need for transfusions, a new study indicates.

Dutch surgeons tested the drug or a placebo in 36 men undergoing removal of a cancerous or seriously enlarged prostate glands. The surgery often causes substantial bleeding. Early in their observations, 24 of the patients received injections loaded with either of two amounts of FVIIa. The other 12 volunteers got a sham injection. Neither the patients nor their surgeons knew which treatment went to whom.

The patients who got FVIIa lost less blood during surgery and needed fewer transfusions than their placebo-treated counterparts. None of those receiving the higher dose of FVIIa needed a blood transfusion, while 38 percent of those receiving the lower dose and 58 percent of those getting the placebo required extra blood to get through their operations.

There were no negative consequences from the treatment, Marcel Levi of the University of Amsterdam and his colleagues report in the Jan. 18 *Lancet*. FVIIa is considered safe for patients with blood clotting disorders, but it hasn't been widely tested in other people.

21. How do surgical operations add to the difficulties faced by blood banks?
22. Explain why surgeons tested the enzyme FVIIa on a group of men having surgery.
23. Describe the results of the research on FVIIa and explain the importance of these results.

Maintaining a Dynamic Equilibrium

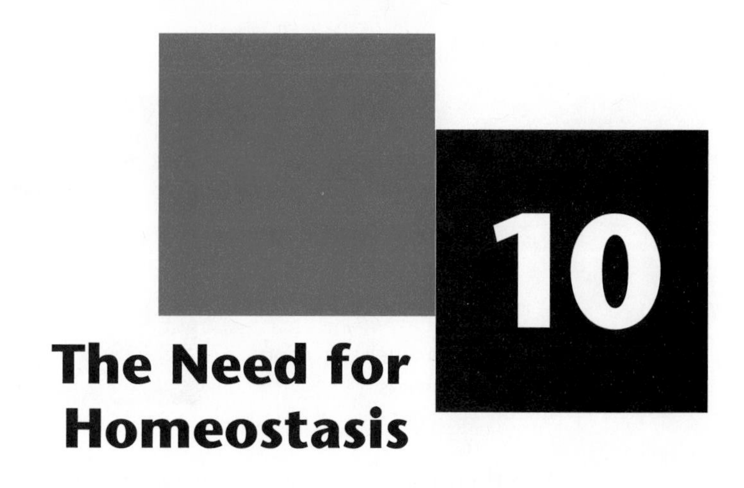

The Need for Homeostasis

After you have finished reading this chapter, you should be able to:

Describe how animals maintain homeostasis through thermoregulation.

Explain how plants adjust their water balance to maintain homeostasis.

Identify the steps by which the body automatically adjusts its breathing rate.

It is the fixity of the internal environment which is the condition of free and independent life. All the vital mechanisms, however varied they may be, have only one object, that of preserving constant the condition of life in the internal environment.

Claude Bernard

Introduction

It is a difficult life for Terry the ameba, who lives in a thin layer of water in the soil. This layer of water, Terry's home, surrounds the particles of minerals and decaying parts of plants that make up much of the soil. For nutrition, Terry ingests pieces of food that include plant-decaying bacteria. An ameba carries on all the normal functions of staying alive, including digestion, respiration, and getting rid of wastes it produces. An ameba even has the ability to divide into two cells to reproduce. (See Figure 10-1 on page 212.) But right now the ameba is having a difficult time. Terry is only a single cell, and that is the problem.

For Terry, a long period without rain could be disastrous. The soil will become dry, the thin layer of water will disappear, and Terry, without a moist environment, will probably die. Or perhaps the temperature of the soil will become hot or cold. When it is too hot or cold, Terry's metabolism will be disrupted. If the temperature changes too much, Terry might even die.

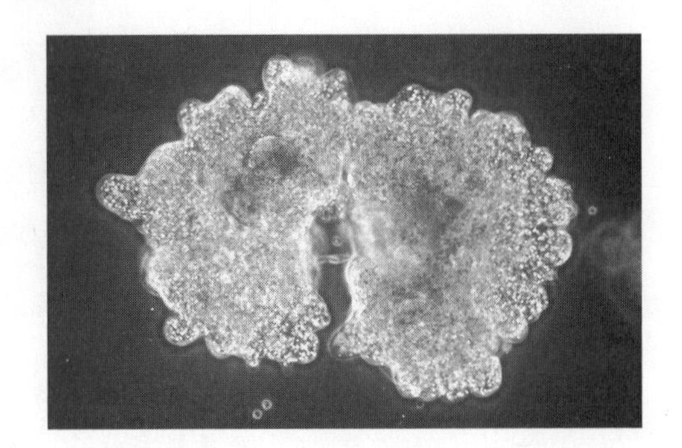

Figure 10-1 An ameba dividing into two new amebas.

And then there are the chemicals in the soil to deal with! Terry lives in the soil under a person's lawn. In order to improve the appearance of his lawn, the homeowner spreads fertilizer. The fertilizer he uses is a mixture of chemical nutrients for the grass. Suddenly, the chemicals change the ameba's watery home. Some chemicals may cause a change in the pH of Terry's layer of water. Other chemicals, being much more concentrated outside of Terry's cell membrane than inside, may cause the rapid osmosis of water out of its single-celled body. These changes too may result in Terry's death. Changes in temperature, pH, and chemical concentration are all problems that organisms must deal with. However, they are actually smaller pieces of one larger problem—maintaining a constant state.

HOMEOSTASIS

All organisms live in a world of changing conditions. But, to remain alive, the conditions inside of every organism need to remain fairly constant. An organism must have ways to keep the conditions inside of itself from changing as its external environment changes. One of the most important characteristics of all living things is the ability to maintain a constant internal environment. This ability is known as **homeostasis**.

The word *homeostasis* was first used by Harvard University physiologist Walter Cannon in 1926. During World War I, Cannon had seen many European soldiers recover from severe wounds. It amazed him that wounded bodies could return to the normal, orderly conditions necessary for long-term survival. *Homeo* means "sameness," and *stasis* means "standing still" or "balance." Cannon used the word *homeostasis* to express the idea that the body constantly makes adjustments and changes. The purpose of this is to maintain a kind of balance. Internal conditions remain within certain acceptable ranges. This ability is the key to staying alive. Through the process of evolution, organisms have developed structures and systems to maintain homeostasis in their environments and niches. It is this need for homeostasis in all living things that you will study in this and the following four chapters.

IS BIGGER BETTER?

The first forms of life on Earth 3.5 billion years ago were single cells. All organisms alive today are descendants of these ancestral cells. The tiny nematode worm *Caenorhabditis elegans,* or *C. elegans,* lives in the soil. When it is fully grown, it consists of 959 cells. *C. elegans* is the first organism made up of many cells that has had every one of its cells counted and studied by scientists. It is a multicellular organism. Every organism in the plant and animal kingdoms is multicellular. A human body is not made up of 959 cells, but perhaps 40,000,000,000,000—40 trillion—cells. It is unlikely that anyone will ever be able to count all of the cells in a human. (See Figure 10-2.)

One of the most obvious changes that has occurred through the long history of evolution on Earth is the development of larger, multicellular organisms from microscopic, single-celled ones. Is there an advantage to being multicellular? Is being "bigger" better? The problem that Terry the ameba encountered provides one obvious and important answer. Being microscopic and single-celled makes it difficult for an organism to

Figure 10-2 The human body is made up of about 40 trillion cells.

maintain homeostasis. Having a multicellular body makes possible many types of protection against changes in the environment. As a biologist would say, multicellularity and increased body size have evolved because of the need for homeostasis. An organism with many cells is able to have structures and systems that protect its individual cells from external changes. Being able to maintain a stable internal environment means being able to stay alive.

THE CELL AND ITS ENVIRONMENT

One of the most fascinating facts about our bodies is that each of our many, many cells is surrounded by liquid. We have seen that life on Earth began with single cells living in the oceans. Today, humans live as complicated multicellular land organisms, and yet each of our cells exists in a fluid environment surprisingly similar to that of the ancient ocean. How could this happen?

It is necessary to look again at the smallest blood vessels in our bodies, the capillaries. Capillaries are close to every cell in our body. However, as mentioned in the previous chapter, there is a small amount of space between the capillaries and the body cells. This space is filled with fluid. In a sense, each of our cells is still living in its own miniature ocean. (See Figure 10-3.)

In 1843, a young Frenchman named Claude Bernard was trying to decide if he should become a pharmacist or spend his life writing plays. After he moved to Paris and decided to go to medical school, his future was set. He became fascinated by how the body functions, and he went on to become one of the most important physiologists of the nineteenth century.

What fascinated Bernard most were the fluids that surrounded all of the cells in an animal's body. Bernard made many careful measurements

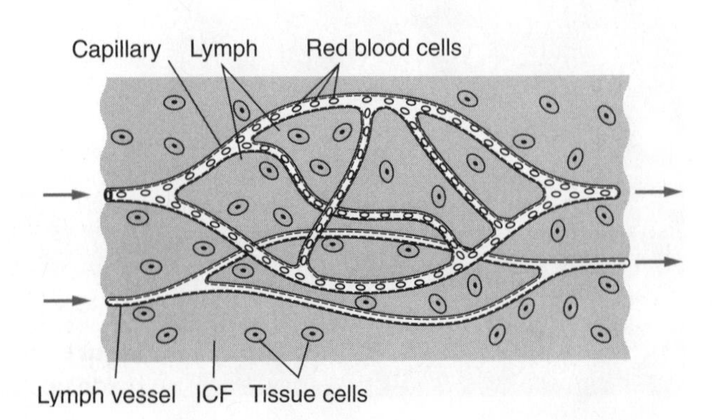

Figure 10-3 Each cell in our body is surrounded by its own miniature ocean.

of temperature, acidity, and the concentrations of salt and sugar. He was amazed to learn that intercellular fluids remained constant despite changing conditions.

We saw how helpless an ameba is when conditions in the soil or in a pond change. Animals, in contrast, have this wonderful ability to control the conditions of their own internal "ponds" in which their cells live.

The intercellular fluid (ICF) that surrounds cells is made up mostly of water, with many substances dissolved in it. Intercellular fluid is important in helping to maintain stable conditions inside each of our cells. Many materials are exchanged between cells and ICF. In turn, materials may be exchanged between the ICF and the blood in the capillaries. All of this is done to make sure that each and every body cell is able to maintain homeostasis and remain healthy.

MAINTAINING HOMEOSTASIS—EXAMPLE 1: THERMOREGULATION

One of the obvious problems the ameba had was dealing with the changes in soil temperature. A hot or a cold day could stress an ameba's ability to survive. On the other hand, look at where animals are found on Earth today. Polar bears live on sheets of floating ice near the North Pole, where the temperature can reach –4°C. Parrots fly through the steamy jungles of South America, where temperatures may reach 48°C. (See Figure 10-4.) Look at where a human can go on a cold winter's day—perhaps from a hot shower out into bitter-cold winter air, then into a warm classroom and back to the frozen outside again. Two groups of animals, birds and

Figure 10-4 Mammals and birds, such as parrots, have evolved the ability to maintain a fairly constant internal body temperature.

mammals, have evolved the ability to maintain a fairly constant internal body temperature. This ability is called **thermoregulation**.

How does thermoregulation occur? It is easy to see how the fur of a mammal and the feathers of a bird keep these animals warm when the environment around them is very cold. However, humans are mammals with very little fur, and no feathers. How do we maintain a steady temperature? When you go outside on a frigid winter's day, the cold air quickly cools your body. The temperature of your blood falls a little. In all cases, in order to maintain homeostasis, the body must be able to detect the change that occurs in order to help prevent too much change. The slightly cooler blood travels around the body. Eventually, the cooler blood reaches the brain. A structure in the brain, the **hypothalamus** (see Figure 10-5), has thermoreceptors that detect the temperature change. This was demonstrated by Yale University scientist Henry Barbour in 1912.

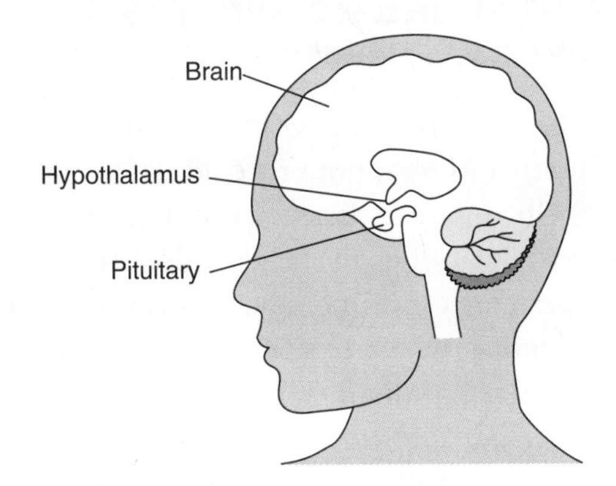

Figure 10-5 The hypothalamus in the brain has receptors that detect temperature change.

Barbour implanted fine silver tubes into the hypothalamus of laboratory animals. He could deliver water of various temperatures through the tubes. When he sent warm water to the hypothalamus, the animal acted as if it were in a hot room. Sending cool water to the hypothalamus made the animal act as if it were outside on a cold night.

When body temperature drops, the hypothalamus immediately sends signals to blood vessels that lie just under the surface of the skin. Since the skin is next to the cold air, this is where the body is losing the most heat. In this situation, it would be better to keep blood away from the skin. Therefore, the hypothalamus signals the blood vessels near the surface of the skin to become narrower. When the blood vessels narrow, the amount of blood moving through them is reduced, and heat is conserved. (See Figure 10-6.)

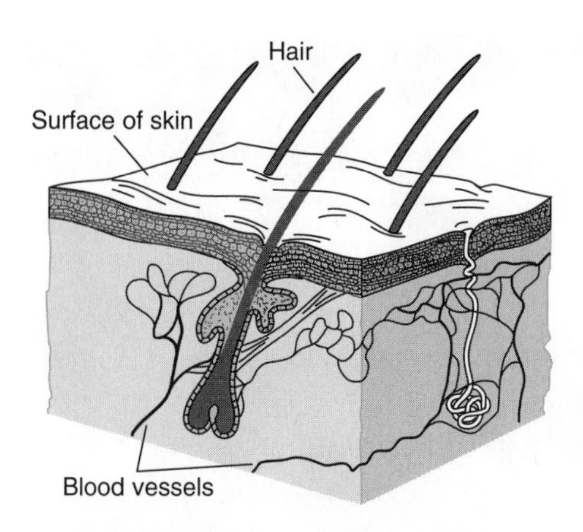

Figure 10-6 When body temperature drops, the hypothalamus signals the blood vessels near the skin's surface to become narrow.

Everyone has experienced another way the body keeps warm. If narrowing the blood vessels is not enough to limit heat loss on a really cold day, the brain signals the muscles to start shaking. You start to shiver. Shivering forces the muscles to burn food for energy. This increased energy production results in heat to warm the body.

What happens if the temperature of the air becomes extremely high or if the body temperature gets too high from exercise or a fever? What signal will the hypothalamus send to the blood vessels near the skin when it detects a warming of the blood? The blood vessels are signaled to expand, allowing a greater amount of blood to flow through. The blood then releases heat through the skin to the outside air. Our skin feels and looks flushed because of the increase of blood in the blood vessels near the skin's surface.

Animals have other ways to lower body temperature. For example, your dog pants when it gets too hot. (See Figure 10-7 on page 218.) We perspire when we get too hot. In both cases, the moist surfaces of the tongue or the skin are exposed to the air. Water evaporates, causing cooling. The body temperature lowers.

Reducing or increasing blood flow, panting, and sweating are examples of how animals regulate the temperature of their body, maintaining it within narrow ranges for health.

Your body's temperature is regulated somewhat as a thermostat regulates the temperature of your house. A thermostat measures the temperature of the air in a room. Electrical wires connect the thermostat to the furnace. When the air temperature in the house falls below a preset figure, the thermostat turns the furnace on. The furnace produces heat, and the temperature of the air in the house increases. In time, the temperature of the air rises above the preset temperature of the thermostat. Another

Figure 10-7 Dogs lower their body temperature by panting.

message from the thermostat tells the furnace to shut down. When the furnace stops producing heat, the temperature in the house stops rising; soon the air begins to cool, and the thermostat continues the cycle of telling the furnace to produce heat or to shut down. (See Figure 10-8.)

This is a type of self-regulation called **negative feedback**. In negative feedback, a change occurs that produces another change, which in turn reverses the initial change. Negative feedback is an important process in maintaining homeostasis. We will see how it occurs in other examples later.

The following parts of a home thermostat and a human body are necessary in homeostasis:

- *Sensor.* Something must be able to *detect* a change. A thermometer attached to a thermostat is a sensor. In the body, thermoreceptors in the skin and inside the body detect temperature changes.
- *Control unit.* Something must know what the correct level should be. The thermostat in the house has been set to a particular comfort level. An area in the hypothalamus in the brain is preset at the correct temperature, usually about 37°C. Other animals have different temperatures preset in their hypothalamus. For example, camels, adapted to a life in the hot desert, have their thermostats set higher, at about 40°C.
- *Effector.* Something must take instructions from the control unit and make the necessary changes. The effector would be a furnace or an air conditioner in a house. In the body, the effector may be the blood vessels in the skin as they narrow or widen, or the muscles as they begin to shiver.

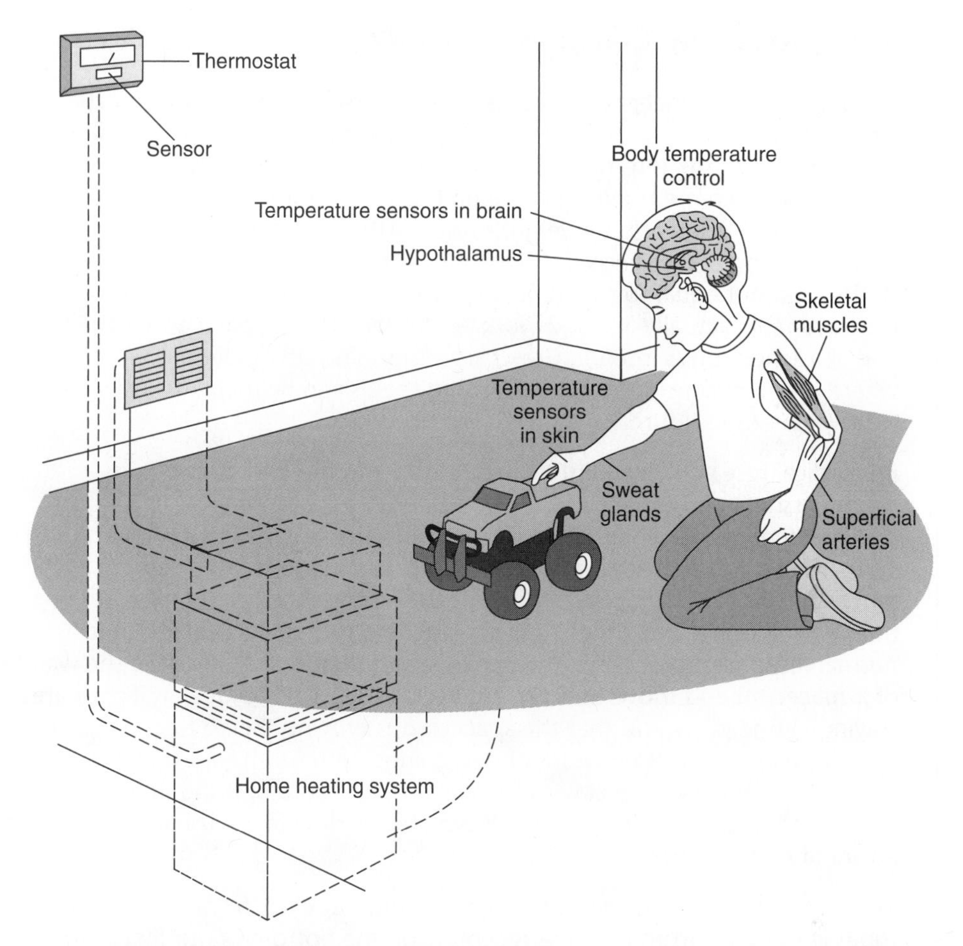

Figure 10-8 Body temperature is regulated somewhat as a thermostat controls the temperature in a room, through negative feedback.

While negative feedback works to decrease or reverse a change, another type of process—**positive feedback**—protects the body by causing a particular change to occur. For example, blood clotting is a kind of positive feedback. An injury to a blood vessel causes a small amount of proteins in the blood to clot. This in turn affects more and more proteins in the blood, which continue to increase the size of the clot until the loss of blood has stopped. In this example, by not losing any more blood, the body maintains homeostasis.

Check Your Understanding

What would happen to the body if it could not thermoregulate? Why would that be a threat to survival?

Homeostasis in Space

In 2002, flight engineers Carl Walz and Dan Bursch set the record for the longest United States space flight, with 196 days in space as members of Expedition 4 on the International Space Station (ISS). Typically, ISS crews have six or seven members who live on the station for three to six months. The crews live in a world of weightlessness—the station has no up or down, so there are no real ceilings or floors. While the total inside space of the station is about equal to that of a jumbo jet, the individual spaces in which the astronauts actually live and work are relatively small, each about the size of a school bus's interior. Crews sleep standing up or camping out where they feel comfortable by attaching their sleep restraints to the wall with Velcro. (See Figure 10-9.)

Biomedical researchers are interested in studying the effects of weightlessness on humans. Being "weightless" is a brand-new challenge never experienced before in the millions of years humans have lived on Earth. And yet, time and again, space travel has demonstrated the marvelous, and often subtle, abilities of the human body to adapt. The body's reactions to weightlessness are teaching us a great deal about its normal responses to gravity. Astronauts report that when they grab the wall of a spacecraft and move their bodies back and forth, they feel as if they are staying in one place and that the spacecraft is moving. Being free of gravity's effects makes us aware of new things. Humans have evolved many automatic reactions to deal with the constant pressure of living in a downward-pulling world. Until we leave that world, we are usually not aware of such reactions.

These reactions include the use of signals from our eyes, from the fluid-filled tubes in our ears, from pressure receptors on the bottom of our feet, and from the distribution of liquids in our blood vessels. A sophisticated control system has evolved to keep gravity from pulling all the liquid in our body to our legs. Within minutes of being in a weightless environment, the veins in an astronaut's neck begin to bulge. The astronaut's face begins to fill out and become puffy. In this situation, the fluids in an astronaut's body are not being pulled down by gravity. The fluids spread throughout the body.

MAINTAINING HOMEOSTASIS—EXAMPLE 2: WATER BALANCE

Maintaining water balance is a major concern for all living things. An organism with too much water in it will have problems, as will one with too little water in it. Just as you would have to constantly shift your weight from one side to the other in order to walk along a balance beam, living organisms have to constantly make adjustments that maintain correct water balance.

Figure 10-9 Astronauts live in the weightlessness of space for months at a time aboard the International Space Station.

Because the body seeks to maintain homeostasis, this new distribution of fluid causes other changes in the body in order to control fluid movement. Included in these are changes in hormone levels, kidney function, and red blood cell production. Keeping things stable even when conditions change —that is, dynamic equilibrium—is as necessary for life in space as it is on Earth. The unexpected result of "living" in space is a better understanding of how the body works here on Earth.

It may seem strange, but to maintain homeostasis, organisms must constantly make changes. This is why homeostasis is often referred to as maintaining a **dynamic equilibrium**. *Dynamic* means "active," and *equilibrium* means "balanced." Homeostasis requires active balancing.

It is not only animals that must maintain homeostasis. This is a requirement of all living things. In Chapter 3, you learned how plant leaves have excellent structural adaptations. For example, stomates—openings in the surfaces of a leaf—are adapted to control the loss of water.

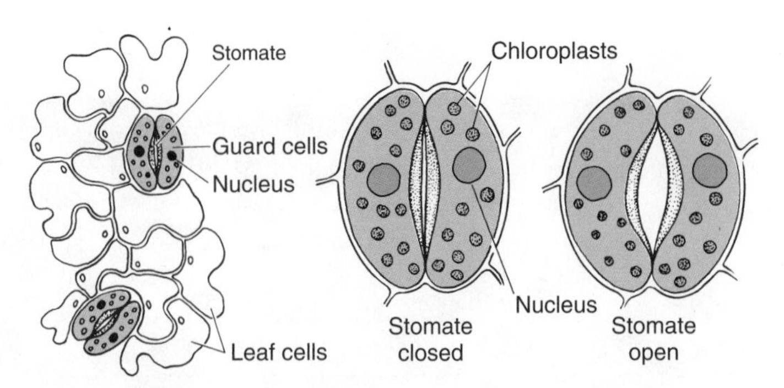

Figure 10-10 In plants, the stomates maintain water balance.

How a plant uses its stomates is a form of regulation. Plants, just as animals, have evolved structural and regulatory adaptations that maintain water balance. Let's see how. (See Figure 10-10.)

If you examine a stomate under a microscope, you will see that the opening of each stomate is surrounded by two guard cells. These guard cells, like any cells, allow water to diffuse through their cell membrane by osmosis. When water is abundant, it moves into the guard cells. The increased quantity of water increases the pressure within the cells. What is unique about guard cells is that they are somewhat curved in shape. When they are filled with water, they become even more curved. The space between them expands, the stomate opens, and excess water is allowed to evaporate out of the air spaces inside the leaf into the air that surrounds the plant.

How do guard cells behave when water becomes scarce? Again, watching the stomates through a microscope, we see the guard cells becoming less curved in shape. This is happening because osmosis is moving water out of the guard cells. Having less pressure in them, the cells become less curved, and the stomate closes. And this is exactly what the plant needs to happen. When the stomates close, water loss is reduced; once again the plant is able to maintain its water balance.

Some plants have evolved unique adaptations for maintaining water balance. Most plants tend to have their stomates wide open during the day when photosynthesis is occurring, in order to take in CO_2 and release O_2. Every evening the stomates close. However, plants living in the hot, dry desert have a particular problem. They need to save water, and if their stomates were open during the heat of the day, saving water would be impossible. What adaptations do these plants show? Dry-climate plants close their stomates during the day and open them at night instead. Cactuses, pineapples, and stonecrops are plants that close their stomates during the day. They have also evolved a special kind of chemistry that stores the CO_2 taken into the leaves at night for use in photosynthesis during the day.

The abilities that plants have evolved, in order to maintain water balance, demonstrate once again that nothing is more important for staying alive than the maintenance of homeostasis.

MAINTAINING HOMEOSTASIS—EXAMPLE 3: BREATHING RATES

How often do you breathe each minute? Do you constantly think about and decide how fast to breathe? The answer, of course, is no. Unless you stop and count, you may not even be aware that humans breathe about 12 times each minute. Even though we do not usually think about breathing, however, our bodies are constantly making adjustments in our respiration rate. What and where are the sensors and the control center for the regulation of our breathing rate?

When the level of carbon dioxide increases in the blood, a chemical reaction causes the blood to become more acidic. The pH of the blood becomes lower. Receptors in several places in the body detect this change in blood chemistry. In the aorta (the large artery that leaves the left ventricle of the heart) and in the carotid arteries in the neck, chemoreceptors detect this chemical change in the acidity of the blood. In the brain, very sensitive chemoreceptors also detect changes in the pH of the blood.

Once detected, this information is sent to the breathing control center in a portion of the brain called the *medulla*. (See Figure 10-11a.) The medulla controls activities that occur automatically without conscious

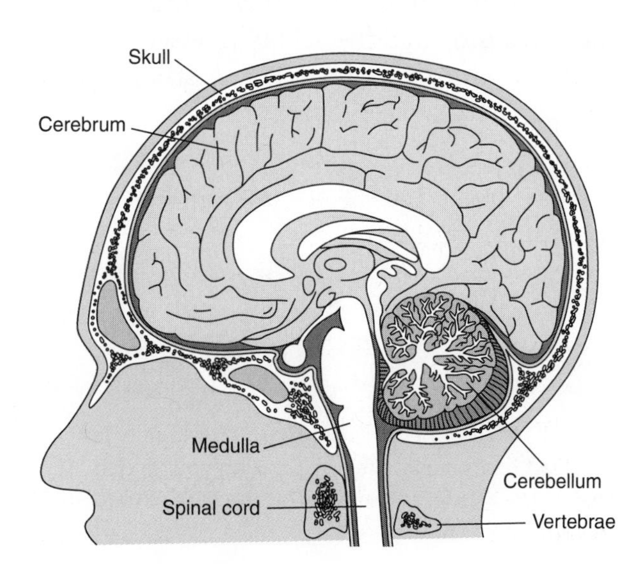

Figure 10-11a The breathing center in the medulla monitors the amount of carbon dioxide in the blood.

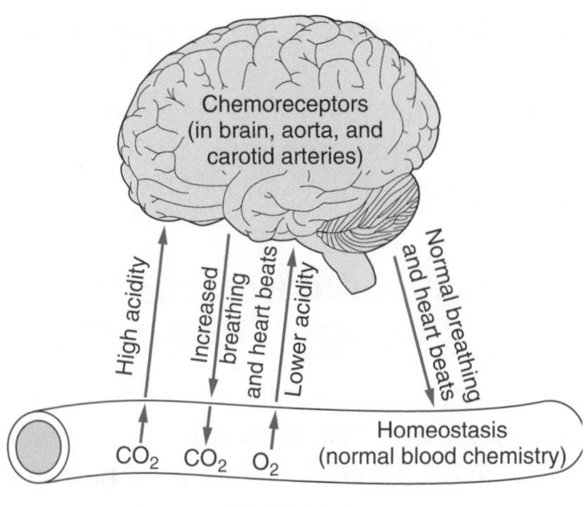

Figure 10-11b When chemoreceptors in the brain and arteries detect increased acidity in the bloodstream, the breathing rate and heart beats are increased to lower the acidity and maintain homeostasis of the blood's oxygen and carbon dioxide levels.

control, such as the beating of the heart and peristalsis (the rhythmic contractions of the digestive tract that move food along). The breathing control center sends instructions to the body to change our breathing rate. (See Figure 10-11b.) The effectors that act on these instructions involve three types of adjustments: (1) how rapidly we breathe; (2) how deeply we inhale, that is, whether we are taking small breaths or large breaths; and (3) how quickly our heart beats in order to deliver the oxygen that is being breathed in. (See Figure 10-12.) These are the steps, or effectors, which make the necessary changes to maintain the homeostasis of oxygen and carbon dioxide levels in our blood.

As in any negative feedback system, the effectors reverse the initial change, which in turn ends their own action. As carbon dioxide levels in the blood drop and the acidity is lessened, breathing and the beating of the heart return to normal. Homeostasis is attained once again!

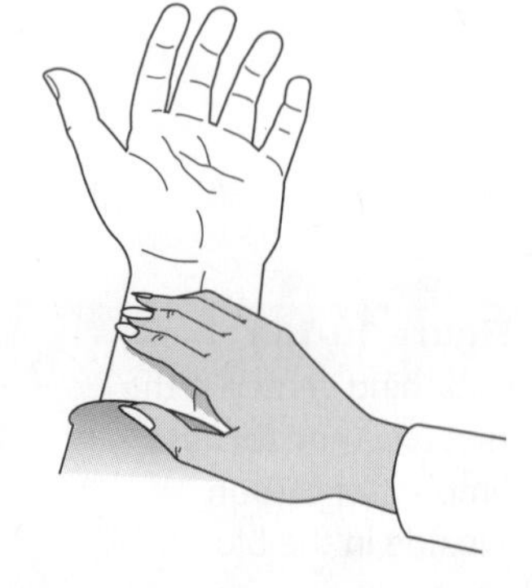

Figure 10-12 A person's pulse, which tells how fast the heart beats, can be felt by pressing your fingers against the arteries in the wrist (at the base of the thumb).

SYSTEMS FOR MAINTAINING HOMEOSTASIS

While Terry the ameba does its best to maintain homeostasis by using its selectively permeable cell membrane, or by moving away from harmful conditions, its ability to control its own internal environment is limited. Large, multicellular animals, on the other hand, have evolved highly organized, complex organ systems especially suited to maintaining a relatively constant internal environment. These organ systems will be the focus of other chapters. They include the excretory system, which regulates the chemistry of the body's fluids while removing harmful wastes; the nervous system, which uses electrochemical impulses to regulate body functions; and the endocrine system, which produces hormones, chemical messengers essential in regulating the functions and behavior of the body. Finally, the immune system stands guard, with a wonderful set of defenses that protects the body from a huge variety of dangerous substances and microorganisms that could upset the internal balance on which life itself depends.

EXTENDED LABORATORY INVESTIGATION 10

Promoting Wellness Through Community Resources

INTRODUCTION

Wellness is made up of the habits, decisions, and patterns of behavior that are part of one's lifestyle. More information about wellness can be found in Chapter 14. Wellness includes emotional, physical, social, and intellectual dimensions. The community also plays an important role in individual wellness. In this investigation, you will work in a group to explore the resources that promote wellness in your community. Your group will then design and produce a pamphlet that will help other people become aware of these resources. (No laboratory materials required.)

PROCEDURE

1. Form a group that will cooperate together on this project.
2. Identify the dimension of wellness that interests your group. For example, you may be interested in social wellness. How does the community provide opportunities for people to socialize? Or, for intellectual wellness, what are the educational resources?
3. Develop a plan to conduct research about your community's available resources. The purpose of the research is to identify health resources that relate to your group's area of interest.
4. Strategies for conducting your community research may include library research, telephone calls, personal interviews, and newspaper, magazine, and Internet research.
5. Collect and edit the data from your research. Determine what information you will use in your pamphlet.
6. A good size for a pamphlet is a three-fold document with six panels. The six panels are printed on the front and back of an 8½″ × 11″ sheet of paper. Design the layout for the pamphlet.
7. Produce your pamphlet. Prepare a presentation for your class that introduces your findings. Distribute your pamphlet at that time.

CHAPTER 10 REVIEW

Answer these questions on a separate sheet of paper.

VOCABULARY

The following list contains all of the boldfaced terms in this chapter. Define each of these terms in your own words.

dynamic equilibrium, homeostasis, hypothalamus, negative feedback, positive feedback, thermoregulation

PART A—MULTIPLE CHOICE

Choose the response that best completes the sentence or answers the question.

1. The internal balance within an organism is called *a.* physiology *b.* homeostasis *c.* thermoregulation *d.* hypothalamus.
2. Which organism maintains a constant body temperature despite temperature changes in the environment? *a.* frog *b.* maple tree *c.* goldfish *d.* human
3. When a leaf has plenty of water, *a.* the guard cells swell and the stomates open *b.* the guard cells swell and the stomates close *c.* the guard cells lose pressure and the stomates open *d.* the guard cells lose pressure and the stomates close.
4. When the level of carbon dioxide in the blood increases, the *a.* pH of the blood goes up *b.* receptors in the blood vessels sense the pH change *c.* medulla sends a signal to slow the breathing rate *d.* all of these.
5. The first forms of life on Earth *a.* appeared about 10,000 years ago *b.* could easily maintain a stable internal environment *c.* were single-celled *d.* depended on the hypothalamus for thermoregulation.
6. In an embryo, the tissues that receive a nerve produce sprouting factor, which encourages nerves to grow and branch, while the nerve produces anti-sprouting factor, which inhibits nerve growth. This is an example of *a.* a positive feedback system *b.* a negative feedback system *c.* thermoregulation *d.* homeostasis.
7. Intercellular fluids *a.* tend to remain constant despite changing conditions *b.* were found by Claude Bernard to have widely varying pH *c.* do not play a role in homeostasis *d.* are contained in the capillaries in vertebrates.

8. You observe that your cat sprawls on the cool tile floor in the kitchen on hot summer days and curls up in front of a heating vent in the winter. This suggests that *a.* cats have trouble thermoregulating *b.* something is wrong with the cat's hypothalamus *c.* behavior can play a role in homeostasis *d.* cats know how to find the most comfortable place in a house.
9. Which structure detects and helps to maintain body temperature in humans? *a.* spine *b.* hypothalamus *c.* pituitary gland *d.* medulla
10. A child having a tantrum in a department store screams louder the more upset his parent gets. This is an example of *a.* negative feedback *b.* positive feedback *c.* survival of the fittest *d.* dynamic equilibrium.
11. Which of the following events occurs first? *a.* The brain sends a signal to the muscles. *b.* The blood vessels just under your skin become narrow. *c.* The hypothalamus detects a drop in body temperature. *d.* You shiver.
12. A dog pants because *a.* evaporation causes cooling *b.* evaporation gets rid of excess water in the body *c.* it is tired *d.* panting helps to raise body temperature.
13. Which is necessary for a feedback mechanism? *a.* control unit *b.* effector *c.* sensor *d.* all of these.
14. In moist soil, Terry the ameba absorbs as much water from its surroundings as diffuses out of its cell. This is an example of *a.* homeostasis *b.* dynamic equilibrium *c.* thermoregulation *d.* feedback.
15. Which structure regulates breathing rate in humans? *a.* diaphragm *b.* lungs *c.* hypothalamus *d.* medulla.

PART B—CONSTRUCTED RESPONSE

Use the information in the chapter to respond to these items.

[The diagram below shows how the body regulates the amount of sugar (glucose) in the blood.]

16. According to the diagram, what happens when blood glucose levels go up?
17. Is the regulation of glucose an example of positive feedback or of negative feedback? Explain.

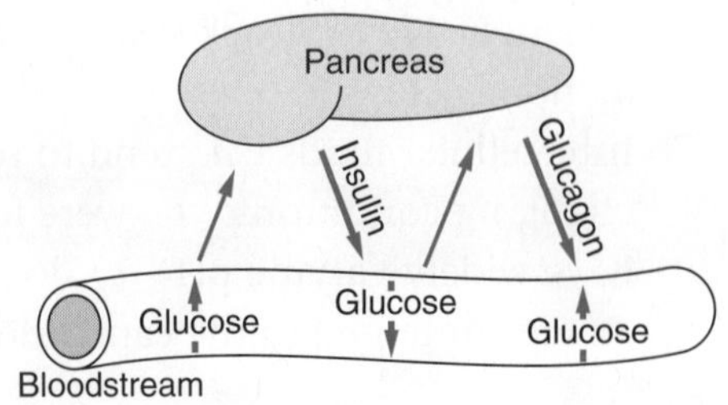

18. Using the diagram, explain why quickly eating a lot of candy can make you feel ill from a temporary *drop* in blood-sugar levels.
19. Describe how the work of Claude Bernard and Henry Barbour contributed to the understanding of homeostasis.
20. Link the regulation of breathing in humans to the three components of any homeostatic process.

PART C—READING COMPREHENSION

Base your answers to questions 21 through 23 on the information below and on your knowledge of biology. Source: *Science News* (May 10, 2003): vol. 163, p. 301.

Second Cold-Sensing Protein Found

They're on a hot streak. Researchers who last year discovered a mammalian cell-surface protein that senses coolness—and the presence of menthol—have now found a protein that enables nerve cells to recognize much colder temperatures (*SN:* 2/16/02, p.101).

Whereas the cool-menthol receptor kicks in around 25°C, the newly identified receptor doesn't trigger nerve cells until the thermometer falls below 15°C, Ardem Patapoutian of the Scripps Research Institute in La Jolla, Calif., and his colleagues report in the March 21 *Cell.* They had suspected that an additional temperature sensor exists because other scientists had recently documented nerve cells that respond to cold temperatures but not to menthol.

The two cold-activated receptors are related in structure, but their amino acid makeup is very different, the researchers report. Curiously, the new cold receptor is found on nerve cells that also sport a receptor for hot temperatures and capsaicin, the chemical that gives chilies and other foods their fiery kick.

21. State the difference between the cold-temperature sensor that has been recently discovered and the one that scientists already knew about.
22. Explain the evidence scientists knew of which suggested that a second temperature sensor existed.
23. Explain what is remarkable about the newly discovered cold receptor.

Integration and Control: Nervous and Hormonal Regulation

After you have finished reading this chapter, you should be able to:

Describe how the nervous system works to transmit messages and coordinate responses.

Explain how the endocrine system works to regulate body growth and maintain homeostasis.

Discuss some diseases of the nervous and endocrine systems of humans.

Insects have compound eyes. Some compound eyes have many thousands of units. The units must be very good at detecting movement; have you ever tried to sneak up on a fly?

Cecile Starr

Introduction

Did you ever wonder what a frog sitting on a lily pad sees as it looks around? Scientists at the Massachusetts Institute of Technology wondered, and they took careful measurements of cells inside a frog's eye as the frog observed life around it. The scientists found that the frog's eye responded to lights that were turned on and off, and to large moving shadows. Most important, the scientists found that a frog's eye responds to small moving objects. In the world of a frog, a small moving object is most likely a bug, and bugs are dinner. (See Figure 11-1.) The scientists learned that the frog sees, or detects, moving bugs very well. Frogs did not notice objects bigger than bugs or bug-sized objects that did not move. In reality, a frog's eyes are efficient bug detectors. When it sees a moving bug, a frog's response is immediate—a lightning-quick leap and a flick of its tongue to capture a meal. (See Figure 11-2.)

Figure 11-1 Frogs' eyes are good bug detectors.

We have seen that all living things interact with their environment in many ways. Conditions outside and inside the organism are constantly being checked. When needed, adjustments are made to maintain homeostasis. Life goes on. Whatever the interaction is, whether it is finding food, maintaining the correct temperature, or protecting oneself from disease, communication is required. Information must be received from the environment, processed, and responded to. Organisms, particularly complex, multicellular ones, must organize the information they receive and respond to it. This makes it necessary for all parts of an organism to work in a coordinated fashion. Therefore, to maintain homeostasis, an organism must have a means for integration (making all of its body parts work together) and a means for control (acting in an organized and appropriate fashion).

Every function of an organism must, as the cell theory states, involve cells. This includes the communication of an organism with its environment and the communication within an organism among all of its parts.

Figure 11-2 When a frog sees a bug, it responds with a quick leap and a flick of the tongue to capture its meal.

Most important, the only way in which cells communicate is chemically. Communication for a cell means having chemicals moving into and out of it. We will now see how the work of the two organ systems responsible for integration and control, the nervous system and the endocrine system, is based on the chemical communication between cells.

THE NEURON: A CELL FOR RAPID COMMUNICATION

When a frog sees a bug, it opens its mouth wide and flips its tongue over and outward. The insect is caught and instantly drawn into the frog's mouth, which quickly snaps shut. For all this to happen, messages must travel through the frog's body very rapidly. It is through the frog's nervous system that these rapid messages travel.

How does a message travel through the nervous system? The cell theory tells us that the messages must travel along pathways composed of cells. The very specialized cells that make up these pathways are nerve cells, or **neurons**. The message itself is a **nerve impulse**. Nerve impulses travel through neurons very rapidly, at speeds of up to 100 m/sec.

Every neuron does three things: It receives, conducts, and sends impulses. In most instances, the structure of a part of an organism is closely related to its function. This is especially true with neurons. The receiving end of a neuron is made up of a series of branching extensions called **dendrites**. Incoming impulses are gathered by the dendrites, which are attached to the **cell body** (cyton) of the neuron. The cell body contains the nerve cell's nucleus. Usually extending out of the cell body is a single long **axon**, which carries impulses away from the cell body. An axon typically makes up most of the length of a neuron. Neurons range from several millimeters in length to the 3-meter-long neurons that reach from a giraffe's legs to its spinal cord. These are among the longest cells in the animal world. The axon ends in a small series of *terminal branches* (end brushes), which send the impulse on its way. (See Figure 11-3.)

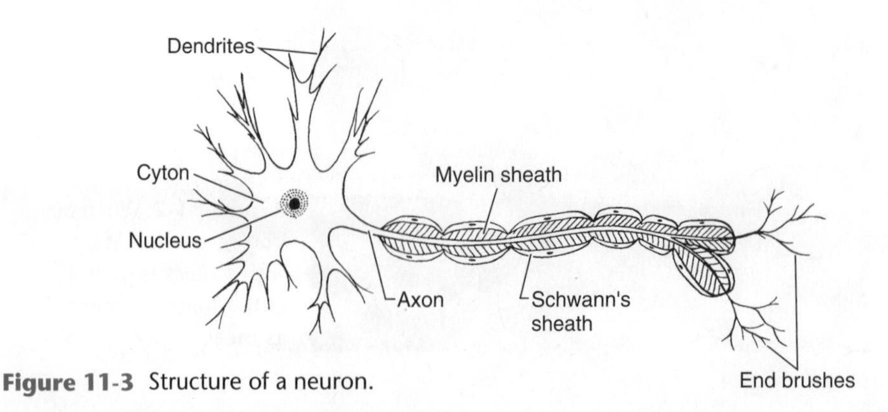

Figure 11-3 Structure of a neuron.

All communication within organisms is chemical. In what way is a nerve impulse a form of chemical communication? The most important part of a neuron involved in transmitting an impulse is the cell membrane. Through the rapid movement of positive sodium ions (Na^+) [and potassium (K^+) ions] across the cell membrane, an electrical voltage is created. Electrical voltage is a form of energy. It is the same type of energy that is stored in a battery and converted to light energy when a flashlight is turned on. In a neuron, the voltage changes that occur at one place on the membrane trigger the same kind of changes at the next spot on the membrane. The movement of these cell membrane voltage changes along the length of the axon is the nerve impulse. (See Figure 11-4.)

Before a dentist begins to drill your tooth, she usually injects an anesthetic such as Novocain into your gum. This anesthetic works by preventing the movement of sodium and potassium ions across the cell

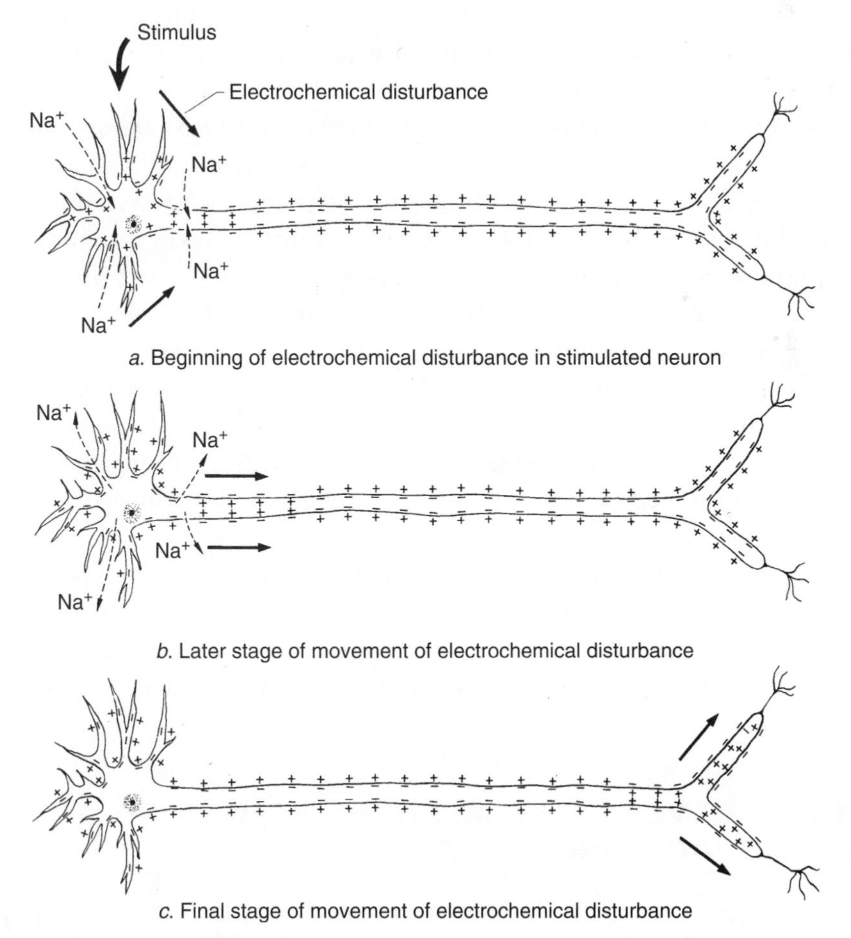

Figure 11-4 A nerve impulse is the movement of cell membrane voltage changes along an axon.

membranes in the region around the tooth she is working on. With the interruption of ion movement, impulses cannot travel through the neurons from your tooth to your brain. The messages of pain do not get sent to your brain and, as a result, you cannot feel what the dentist is doing to your tooth.

CROSSING THE GAP: NEUROTRANSMITTERS

If you accidentally touch a hot pot on a stove, you immediately pull your hand away. Almost instantly, messages have traveled through your body, resulting in this reaction. The nerve pathway that carries messages from your hand to your spinal cord and brain, and then back to your hand, consists of many neurons. An impulse travels along a neuron as a wave of chemical and electrical changes in the cell membrane. Close examination shows that neurons do not touch each other. They are separated by a gap called a **synapse**. How does the impulse get from one neuron to another? How does the nerve impulse cross the synapse? Extremely important chemicals known as **neurotransmitters** are released by the terminal branches of one neuron. The neurotransmitter is released as the impulse arrives at the terminal branches. These chemicals diffuse across the synapse to the dendrites of the next neuron. Once received by the dendrites of the next neuron, a neurotransmitter makes a new nerve impulse possible. In this way, the message continues along the entire nerve pathway, moving from one neuron to another. (See Figure 11-5.)

So far, at least 30 different neurotransmitters have been identified, including **acetylcholine**, the best known. Acetylcholine is the neurotransmitter involved in getting the muscles of a frog to contract so that it can jump to catch a bug. At the same time, in the frog's heart, the nerves that cause the heart muscles to contract release a different neurotransmitter, norepinephrine.

Two words show us how important the synapse is: nerve poisons! Nerve poisons are substances that block neurotransmitters from crossing the synapse. They include the jungle plant–derived drug curare and the bacterial toxins that cause botulism and tetanus. Each of these can cause death!

THE NERVOUS SYSTEM: AN INTERCELLULAR COMMUNICATION NETWORK

Any event, change, or condition in the environment that causes an organism to react is a **stimulus**. The resulting reaction of the organism is a

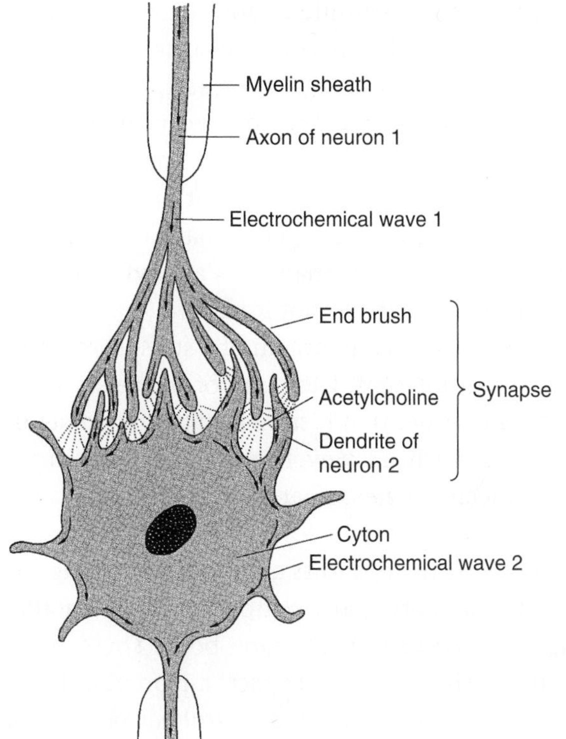

Figure 11-5 Nerve impulses travel from neuron to neuron by way of neurotransmitters.

response. For the frog, the moving bug is a stimulus; the frog's leap to catch the bug is a response. The nervous system of an organism makes both the detection of a stimulus and the response to a stimulus possible. The nervous system is a complex organization of cells and organs in vertebrates such as frogs and humans.

The pattern of evolution that has led to the vertebrate nervous system can be traced back through invertebrate organisms. If you have ever walked along the beach near the ocean's edge, you may have noticed transparent, glistening jellyfish that have washed ashore. The name jellyfish is somewhat misleading: Jellyfish are not fish, and they are not made of jelly. Jellyfish are invertebrates and are much less complex than fish, which are a class of vertebrates. Jellyfish can look very beautiful as they float through the water, with their mouths pointing downward. The tentacles that hang from their bodies are used to capture food. Jellyfish have only two layers of cells. They lack the middle layer of cells that makes up the bones and muscles of more complex organisms. (See Figure 11-6 on page 236.)

A jellyfish has a nerve net. Its nerves are distributed equally in all directions throughout its body's cells. If the nerves at one point of a jellyfish

Figure 11-6 Jellyfish have a nerve net; they have no brain and cannot learn.

detect a stimulus, for example being touched, the message spreads throughout the animal's body. As a result, the jellyfish may change its shape or direction of movement. A jellyfish cannot send messages for a response to a particular location—a jellyfish has no brain. Researchers have been unable to change the behavior of a jellyfish in any way. Jellyfish cannot learn.

The nervous system of a sea star is somewhat different. Each arm of the sea star has a nerve net that is similar to the nerve net in a jellyfish. However, one main nerve from each of the sea star's arms connects the nerve net to a central nerve ring. (See Figure 11-7.)

The small planaria, a type of flatworm, shows an important change in its nervous system. In this animal, nerves are grouped into two parallel nerve cords that run along the length of a planarian's body. These nerve cords meet in a cluster of nerves in the head. In fact, a planarian has a brain. The nerve cords and brain can be called a *central nervous system* (CNS). Between the two nerve cords, an entire series of nerves makes up the *peripheral nervous system* (PNS). For example, if a planarian's brain receives information about a stimulus such as a bright light, it then can send out messages to the muscles of its body to respond. The planaria can crawl to a safe dark place under a rock. (See Figure 11-8.)

Unlike the two nerve cords in a planaria, an earthworm has one dou-

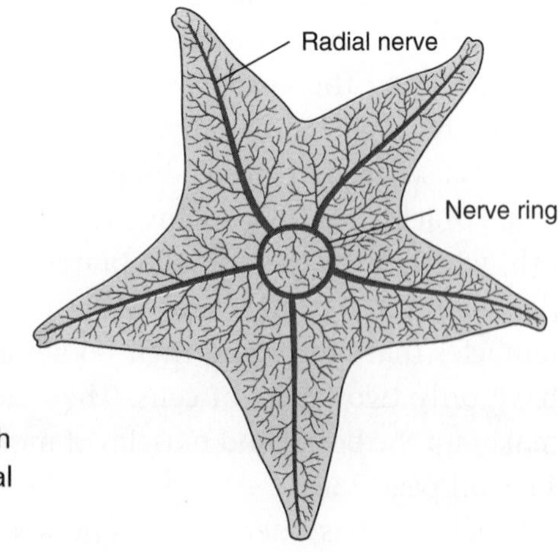

Figure 11-7 In a sea star's nervous system, one main (radial) nerve in each arm connects the nerve net to a central nerve ring.

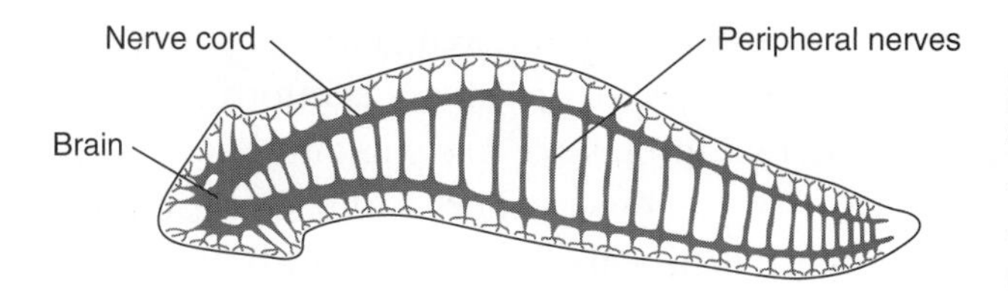

Figure 11-8 Planaria have a central nervous system (the brain and nerve cords), and a peripheral nervous system.

ble nerve cord that runs along the ventral side of its body. In each segment, a cluster of neurons on the nerve cord coordinates information for that segment. A larger brain in the earthworm makes up the rest of its CNS. Nerves that connect to each of the nerve cord clusters make up its PNS. (See Figure 11-9.)

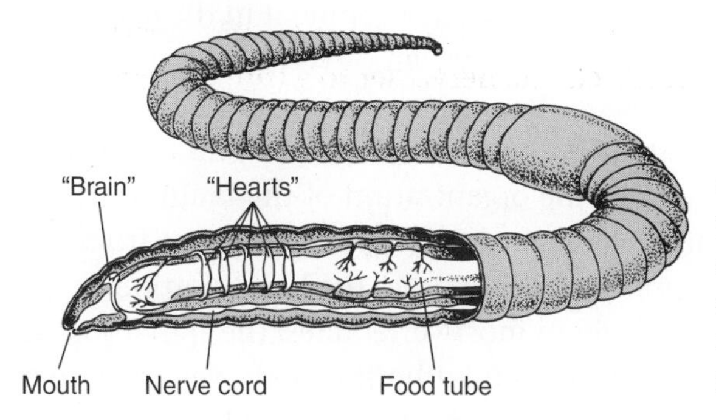

Figure 11-9 The central nervous system of the earthworm is made up of a double ventral nerve cord, a cluster of neurons in each segment, and a brain.

This same tendency toward centralization occurs in the nervous system of the crayfish, an animal that looks like a miniature lobster. In a crayfish, there is a double ventral nerve cord. While large clusters of nerve cells in the head make up a brain, other clusters of nerves along the ventral nerve cord control the movements of the many appendages of a crayfish. (See Figure 11-10.)

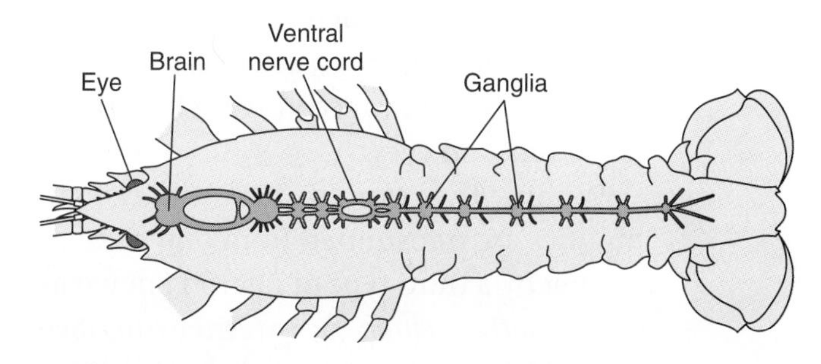

Figure 11-10 The central nervous system of a crayfish has a double ventral nerve cord, clusters of nerves along the cord (the ganglia), and a brain.

The nervous system of the crayfish is well designed for receiving sensory information. Specialized structures that receive this information are called **sensory receptors**. The two impressive-looking antennae of the crayfish are sensory receptors that can detect odors and tastes. Each of a crayfish's eyes moves around on a little stalk, independently of the other. Each eye has 2000 lenses. These lenses are able to detect the slightest movements within the crayfish's field of vision. Sensory bristles on the antennae and on other parts of the body are sensitive to touch.

THE VERTEBRATE NERVOUS SYSTEM

The nervous system of vertebrates—fish, reptiles, amphibians, birds, and mammals—can be seen as further stages of development in the pattern of evolution we are studying. The nervous system of vertebrates includes a central and a peripheral nervous system.

The **central nervous system** (CNS) consists of the brain and spinal cord. (You will learn more about the organization of the brain in Chapter 12 when you study animal behavior.) The spinal cord runs from the base of the brain to the lower portion of the back. In most vertebrates, the spinal cord is surrounded by hollow bony vertebrae that make up the backbone. (See Figure 11-11.)

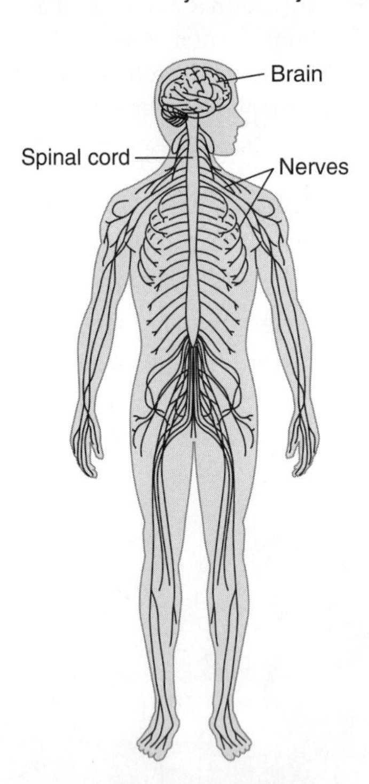

Figure 11-11 The central nervous system of a human is made up of the brain and the spinal cord.

The **peripheral nervous system** (PNS) consists of neurons with axons that travel out of the CNS to all parts of the body. Those neurons that carry signals out of the CNS are **motor neurons**. The signals carried by motor neurons are delivered to effectors, such as muscles or glands. They, in turn, put into effect the instructions carried in the message. The **sensory neurons** carry signals into the CNS from receptors such as those in the ears and eyes.

Within the central nervous system, impulses are transmitted from one place to another by a third type of neuron known as an *associative neuron*, or **interneuron**. (See Figure 11-12.) All vertebrate behavior, from a simple reflex action to a complex learned behavior, involves interactions of these various types of neurons.

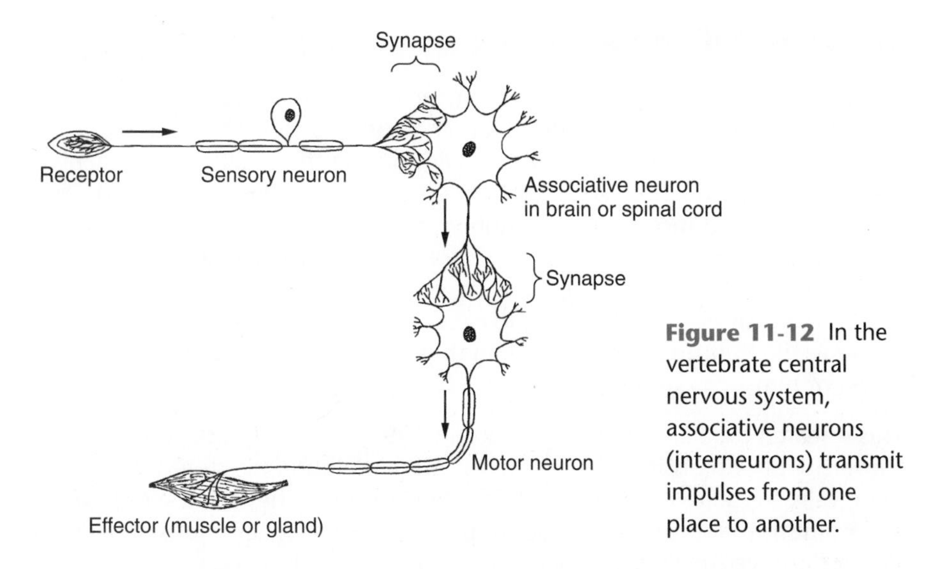

Figure 11-12 In the vertebrate central nervous system, associative neurons (interneurons) transmit impulses from one place to another.

The PNS includes the *autonomic system,* which controls the involuntary activities of the body. This system is made up of the sympathetic nervous system and the parasympathetic nervous system. (See Figure 11-13.) Instructions from each of these two systems are generally opposite each other. The careful balance, so important for homeostasis, is often maintained by the relationship of instructions from these two systems. For example, heart rate is precisely controlled by the balance of impulses from the sympathetic neurons that stimulate or excite the heart and impulses from the parasympathetic neurons that inhibit or slow down the heart.

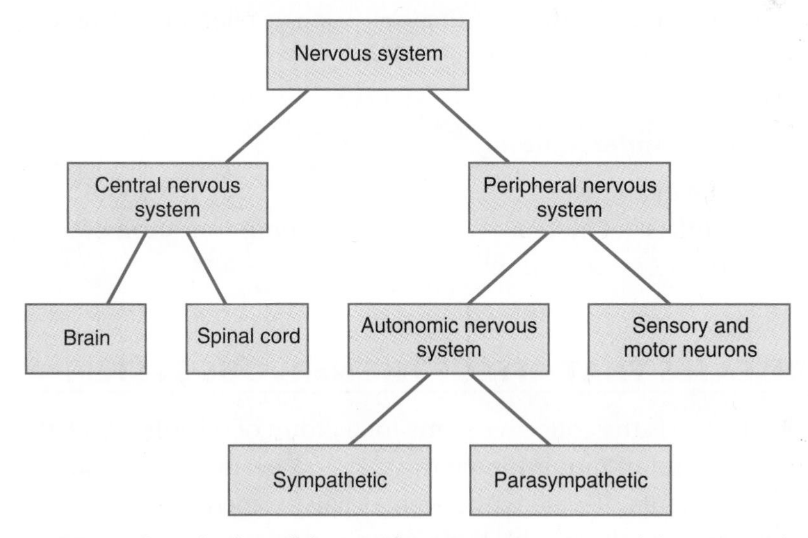

Figure 11-13 Organization of the vertebrate nervous system—it is composed of a central nervous system and a peripheral nervous system.

The Stem Cell Debate

On Memorial Day weekend in 1995, the actor Christopher Reeve was thrown from his horse Buck during a riding competition. The actor who had played the part of *Superman* in movies during the 1980s had suddenly become paralyzed with a spinal-cord injury. Years later, Reeve regained some very limited movement through intensive exercise, but his paralysis largely continued. Unfortunately, in October 2004, Reeve died of an infection due to complications from his paralysis. However, just as he had fought hard for his own recovery, Reeve had also become an important leader in the fight to support research that could lead to a cure for others, if not for him.

This research took a large step forward in 1998 when scientists at the University of Wisconsin isolated the first human embryonic stem cells. Stem cells are quite unique. Found in tiny quantities, they are the cells in an organism that have not yet developed to do their specific jobs. In fact, stem cells have the ability to become almost any kind of tissue in the body. Researchers hope that they can use stem cells to produce specific tissues such as heart, lung, kidney, or nerve tissue. There is some evidence now that tissues grown from stem cells may offer cures to millions of people

Impulses from the sympathetic nervous system speed up the heart, add sugar to the blood, and increase the level of oxygen in the blood. These changes prepare the body to deal with difficult, dangerous, or stressful situations. The body is ready for action. The parasympathetic system is in charge of more ordinary functions, such as releasing saliva to chew food and emptying the bladder of urine.

Check Your Understanding

What is the difference between the central nervous system and the peripheral nervous system? Why is speedy communication between these two systems important?

DISEASES THAT AFFECT THE NERVOUS SYSTEM

Cerebral palsy is the collective name for a group of disorders that affect a person's control of motor function—that is, a person's ability to control body movements. People with cerebral palsy experience brain damage just before or after birth. This brain damage does not get worse over time. Persons with cerebral palsy are taught to be as independent as possible while living with the effects of this disease.

who suffer from conditions such as diabetes, Alzheimer's disease, Parkinson's disease, and spinal-cord injuries.

However, there is much controversy about stem-cell research. Since the 1998 discovery, scientists have had the means to collect stem cells that are very easy to grow from human embryonic tissue. But is it right to use cells from human embryos? In 2001, a committee of scientists was formed by the National Academy of Sciences and the National Research Council to study this problem. It concluded that public policy should keep as many methods of research open as possible, including the use of adult and embryonic human stem cells, to speed the way toward finding cures. In the same year, because of the ethical questions, the United States government placed strict limits on stem-cell research. It is because of the controversy that, by the year 2002, Christopher Reeve began to be very vocal in support of stem-cell research. Though not in time to help Reeve, such research may someday provide a cure for the paralysis of others with spinal-cord injuries. Regardless of one's opinion on stem-cell research, you can be sure that much more will be heard about it in the years ahead, from scientists and politicians, here and abroad.

Multiple sclerosis occurs when myelin, the fatty substance that covers axons in the brain and spinal cord, is destroyed gradually. A wide variety of symptoms—including shaking of the hands, blurred vision, and slurred speech—occur in people with multiple sclerosis. Symptoms may appear and disappear for many years.

Alzheimer's disease is a progressive, degenerative disease. Eventually, memory loss and the inability to think, speak, or care for oneself occur. This disease is usually fatal. The exact cause of Alzheimer's disease is currently unknown.

Parkinson's disease also involves the brain; however, its cause is known. A group of neurons in the brain use dopamine as their neurotransmitter. Loss of function in these neurons produces the typical shaking motion, poor balance, lack of coordination, and stiffening of the muscles that occur with this disease.

THE ENDOCRINE SYSTEM: ANOTHER COMMUNICATION NETWORK

The mineral calcium is necessary for our body. For example, if calcium is not present in a person's diet, it disappears from the blood. The effects of a calcium deficiency are profound. Neurons are not able to release

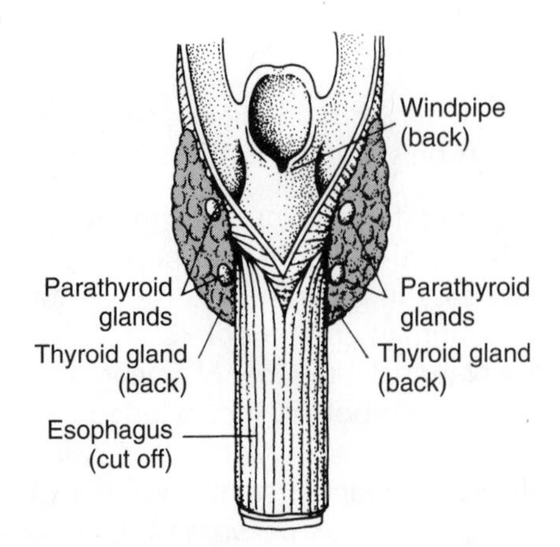

Figure 11-14 The parathyroid glands, part of the endocrine system, help regulate the amount of calcium in the blood.

neurotransmitters and so cannot function properly. Muscles become stiff, and movement and breathing become difficult. Blood cannot clot properly, so even a small wound can become life threatening. Fortunately, none of these ailments occur, because four small glands in the neck can detect a drop in blood calcium levels and release a hormone that acts throughout the body to solve the problem. (See Figure 11-14.) This hormone causes a small amount of calcium to be removed from the bones and less calcium to be excreted in the urine. However, these steps could put too much calcium in the blood. So another gland releases a hormone that reverses the effects of the first hormone, thus stabilizing the amount of calcium. Two sets of glands with two opposing hormones work to maintain the careful balance of calcium that life requires. This system of glands and hormones, including the glands that regulate calcium levels, comprises the **endocrine system**. The key function of this important system is maintaining homeostasis.

Endocrine glands produce hormones, chemical messengers that are released into the blood and carried throughout the body by the circulatory system. At some place or places in the body—often far away from the endocrine gland that made it—the hormone arrives at its special target cells. It is at its target cells that the hormone puts into effect whatever changes it has been designed to produce.

How do hormones do their work? Some hormones bind to specific receptor proteins found in the cell membranes. The binding of a hormone with a receptor protein then causes a change inside the cell, usually involving the cell's enzymes. Other hormones pass right through the cell membrane and bind to receptor proteins in the cytoplasm. The hormone-receptor complex may then move to the nucleus and interact with the cell's DNA, affecting gene activity.

Both the nervous system and the endocrine system are communication networks. In fact, many scientists are beginning to see that these two systems are really parts of one big system, now called the **neuroendocrine system**. However, there are important differences in the two systems. Impulses sent by the nervous system usually produce rapid responses. Frequently, these responses are produced by the actions of muscles. Hormones generally produce slower, more long-lasting changes that often involve metabolic activity within the target cells.

HORMONES: IN ANIMALS AND PLANTS

Vertebrate animals are not the only organisms that rely on chemical messengers. Hormones are also extremely important in the life of plants. Hormones regulate the growth, development, and metabolism of tissues throughout a plant's body. Plant hormones called **auxins** cause the part of a plant where the hormone is concentrated to grow faster than other parts. Auxins are responsible for the more rapid growth of plants at the tips of stems and roots; for the development of the large, fleshy fruits on tomato, apple, and strawberry plants; for the growth of stems upward and roots downward (called *geotropism*); and for the bending of stems and leaves toward light (called *phototropism*). (See Figure 11-15.)

Arthropods are a group of invertebrate animals that have a hard external skeleton, appendages with joints, and a body divided into parts or segments. Because growth is a major problem for animals with a hard

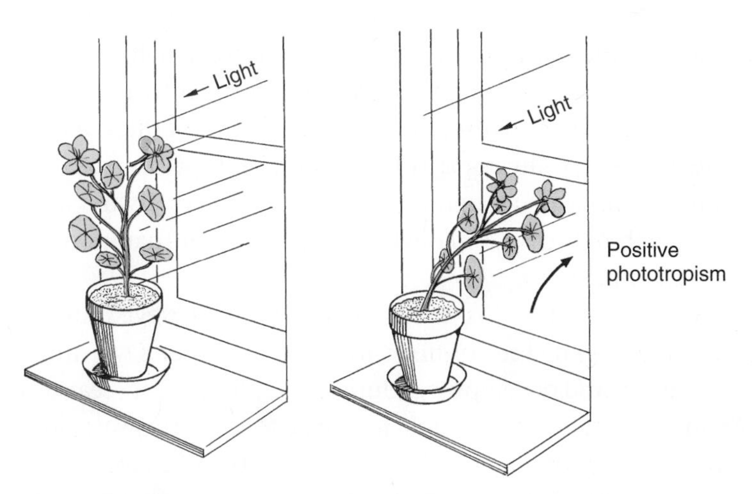

Figure 11-15 Auxins are the plant hormones responsible for the bending of plants toward light.

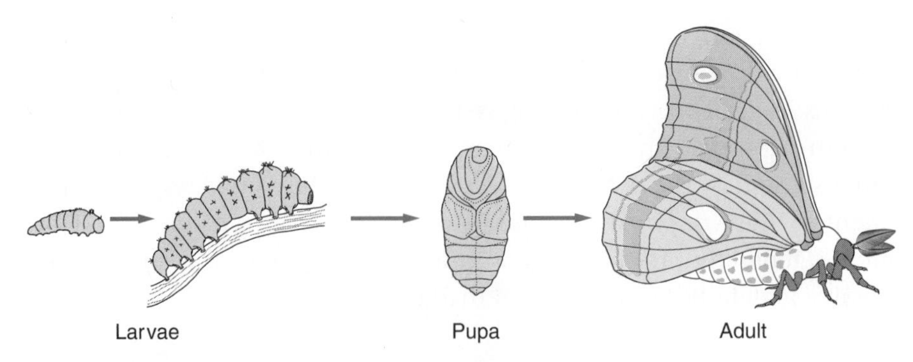

Figure 11-16 The molting hormone causes an insect to molt, or shed its hard outer layer. The juvenile hormone keeps an insect in its larval form.

exoskeleton, hormones are important in an arthropod's life. Arthropods, such as insects and crustaceans, can grow only by shedding their exoskeleton and replacing it with a new, larger one that accommodates their larger body size. This shedding is called molting. The young of some insect species resemble the adult, only smaller. Eventually, after a series of moltings, the insect reaches adult size. In other insect species, the young forms do not resemble the adults. For example, young moths and butterflies—which are actually caterpillars—look rather like worms with legs. However, as adults, they are winged insects that are able to fly. (See Figure 11-16.)

Two types of hormones work in the growth of arthropods. The molting hormone causes an insect to shed its outer layer. An insect molts every time the molting hormone is released from its brain. The juvenile hormone keeps an insect in its young, or larval, form. In butterflies and moths, while the juvenile hormone is present, the insect remains a larva. Once the level of this hormone decreases, the larva turns into a pupa, or cocoon. After its next molt, the insect becomes an adult, perhaps a butterfly. And more amazingly, during this last molt, the pupa undergoes a dramatic metamorphosis, or change. The insect replaces tissues and organs in its body with new ones and develops into an adult that is unlike its juvenile form in appearance.

An amphibian's development is just as amazing. A frog's egg hatches into a swimming tadpole, its life confined to a pool of water. Over time, the tadpole loses its tail and develops legs, lungs, and all the other features of an adult. As an adult, it is even possible for a frog to leave the water and move about on land. It is, of course, no surprise that the dramatic metamorphosis of a frog is also controlled by the actions of hormones. (See Figures 11-17a and 11-17b.)

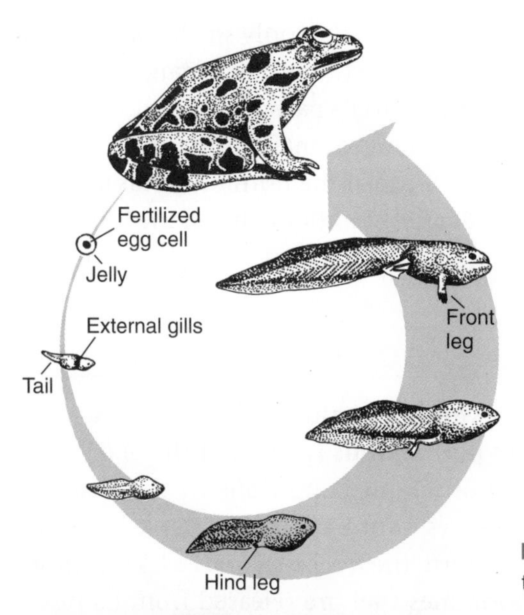

Figure 11-17a Hormones control the metamorphosis of frogs.

Figure 11-17b Three stages in the life cycle of a frog.

TYPES OF HORMONES AND HOW THEY ACT

By 1970, only about 30 hormones had been studied by scientists. Today, almost 200 hormones have been investigated. Almost all hormones can be placed into two main groups, based on their chemical makeup and structure. **Steroid hormones** are formed from the ring-shaped lipid molecule called *cholesterol*. The sex hormones in vertebrates are steroid hormones. The other main group of hormones is **protein hormones**. They are either protein molecules or modified amino acids, the subunits of proteins. Insulin, the hormone that controls the level of sugar in the blood, is a protein hormone.

A fascinating characteristic of hormones is that only small amounts are usually needed to produce the required effect. A group of target cells for a hormone is usually exquisitely sensitive to its particular hormone. Negative feedback works to control the amounts of many hormones that are released. As we discussed earlier, the release of the hormone ends up having the effect of stopping any further release until the hormone is needed once again.

A TOUR OF THE HUMAN ENDOCRINE SYSTEM

The close link between the nervous and endocrine systems can be seen at the beginning of our tour. The **hypothalamus** is a part of the brain. The hypothalamus receives information about conditions in the body as blood passes through it. It also receives information from nerve impulses that are carried to it by neurons. In turn, the hypothalamus uses the information it receives to control hormones that are released from its next-door neighbor in the brain, the pituitary gland.

The **pituitary gland**, only about the size of a pea, is sometimes called the "master gland" because it controls the activities of so many other glands of the endocrine system. The pituitary gland consists of a forward, or anterior, part and a rear, or posterior, part. The anterior pituitary produces at least six hormones. When the hypothalamus detects a need for one of these hormones in the body, it sends a tiny amount of a releasing factor to the anterior pituitary to secrete the correct hormone into the blood. The posterior part of the pituitary produces two hormones. (See Figure 11-18.)

The **adrenal gland** attached to the top of each kidney consists of two glands, the adrenal cortex and the adrenal medulla. The most important hormone from the adrenal cortex, cortisol, is released only after that

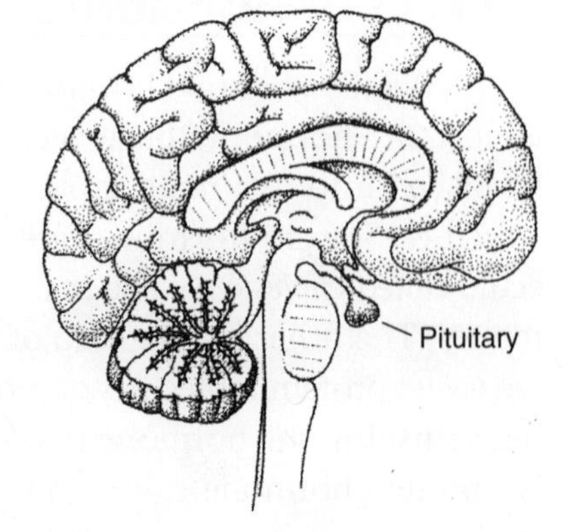

Figure 11-18 The pituitary gland controls the activities of other glands in the endocrine system.

gland is instructed to do so by adrenocorticotropic hormone (ACTH) made in the pituitary. This process of a pituitary hormone triggering another gland to release its hormone occurs throughout the body.

Other major glands of the endocrine system are the thyroid gland in the neck, the four small parathyroids connected to the thyroid, the pancreas, and the ovaries and testes. The locations of these glands, as well as the hormones they control and secrete, are shown in Figure 11-19.

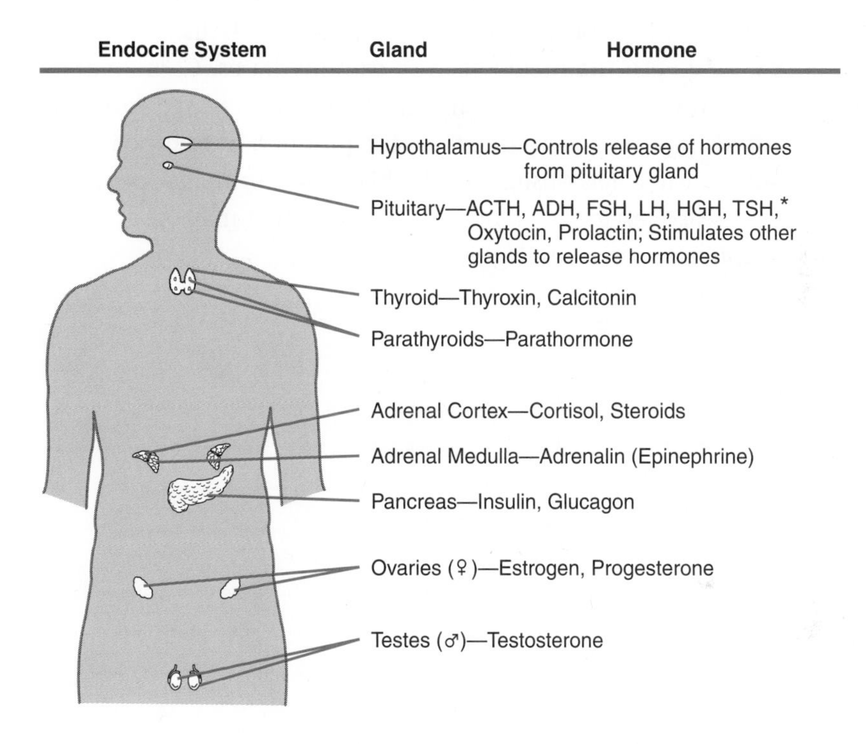

*Key: ACTH—Adrenocorticotropic hormone; ADH—Antidiuretic hormone (vasopressin); FSH—Follicle stimulating hormone; LH—Luteinizing hormone; HGH—Human growth hormone; TSH—Thyroid stimulating hormone.

Figure 11-19 The human endocrine system—the major glands and their hormones.

WHEN THINGS GO WRONG: DISEASES OF THE ENDOCRINE SYSTEM

Goiter is a disease of the thyroid gland. Too little iodine in the diet causes a goiter to form. The chemical element iodine is needed for the manufacture of the hormone thyroxin. Too little thyroxin slows down a person's metabolism, the rate at which chemical reactions occur in the body's cells. Iodine is now added to table salt so that people get enough in their diet to prevent goiter. Grave's disease, in which the thyroid releases too much hormone, makes a person's metabolism overactive.

Dwarfism, in which a person's body is much smaller than is normal, can result from too little growth hormone being produced by the anterior pituitary gland. A kind of dwarfism can also occur when the thyroid gland becomes diseased. This kind of dwarfism is called cretinism. Mental retardation is associated with cretinism but not with pituitary dwarfism.

Diabetes is a disease that affects about one in 20 Americans. Diabetes occurs for a variety of reasons. However, in all people with diabetes, something goes wrong with the metabolism of carbohydrates. Carbohydrate metabolism involves the hormone insulin, which is released from the pancreas. Anything that goes wrong with the production or functioning of insulin affects the levels of sugar in the blood and urine. Proper treatment of this disease can greatly reduce the damage that may occur to the eyes, kidneys, heart, arms, and legs.

LABORATORY INVESTIGATION 11

How Is a Hydra Able to Respond to Its Environment?

INTRODUCTION

A tissue is a collection of similar cells that have a common purpose. The hydra is a freshwater animal that shows some of the organization of body tissues commonly seen in most multicellular organisms. The hydra is about 5 millimeters long. It consists of a two-layer hollow cylinder of cells. At one end of the hydra, a group of tentacles surrounds the single opening of the digestive cavity, or gut. The tentacles are armed with stinging cells that help the hydra capture prey organisms, which are passed into the gut to be digested. The other end of the hydra has a base that can attach the animal to one place. In this investigation, you will observe how a hydra acts, reacts, and interacts.

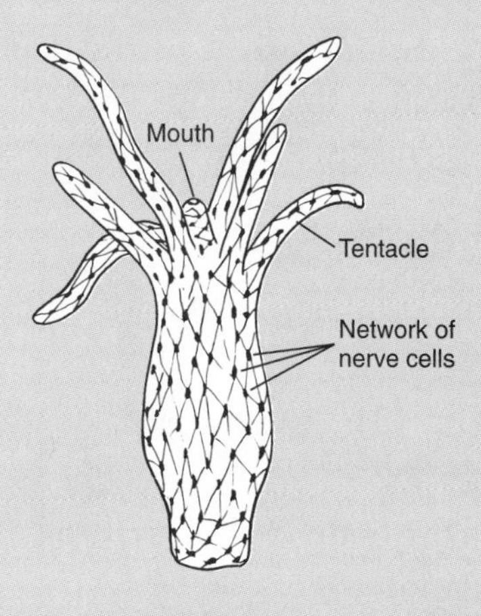

MATERIALS

Hydra, depression slide, medicine dropper, fresh water, compound microscope, dissection microscope, dissection probe, daphnia (water fleas), culture dish

PROCEDURE

1. Use the medicine dropper to place a single hydra and a drop of water on the depression slide.
2. Look at the hydra under low power and make a drawing of what you observe.
3. Use complete sentences to describe the appearance of the hydra and any movement you observe.
4. Use the medicine dropper to gently move the water around the hydra. Observe and record the hydra's reactions to the water movements.
5. After several minutes, gently touch the hydra with the dissection probe. Observe and record the hydra's reactions.
6. Put the hydra in a culture dish. Add the daphnia. Place the culture dish under the dissection microscope. Focus and observe the interactions of the two organisms. Record your observations.

INTERPRETIVE QUESTIONS

1. Explain the evidence you observed that indicates the hydra is a multicellular organism.
2. How does the hydra take in food? How does it get rid of undigested wastes?
3. In what ways does the hydra demonstrate an ability to respond to the environment?
4. How does the feeding process demonstrate the ability of this organism to coordinate its movements?
5. From your research and observations, explain how the hydra is able to respond to environmental stimuli even though it lacks an organizing center such as a brain or a nerve cord.

CHAPTER 11 REVIEW

Answer these questions on a separate sheet of paper.

VOCABULARY

The following list contains all of the boldfaced terms in this chapter. Define each of these terms in your own words.

acetylcholine, adrenal gland, auxins, axon, cell body, central nervous system, dendrites, endocrine system, hypothalamus, interneuron, motor neurons, nerve impulse, neuroendocrine system, neurons, neurotransmitters, peripheral nervous system, pituitary gland, protein hormones, response, sensory neurons, sensory receptors, steroid hormones, stimulus, synapse

PART A—MULTIPLE CHOICE

Choose the response that best completes the sentence or answers the question.

1. The system of the body that rapidly receives, relays, and responds to internal and external stimuli is the *a.* nervous system *b.* autonomic system *c.* endocrine system *d.* hormone system.
2. Which organism has a distinct central nervous system? *a.* sea star *b.* jellyfish *c.* crayfish *d.* clam
3. Hormones control *a.* the amount of sugar in the blood *b.* the development of a tadpole into a frog *c.* human growth *d.* all of these.
4. The anesthetic Novocain works by *a.* blocking the reception of acetylcholine *b.* preventing the breakdown of norepinephrine *c.* preventing the movement of sodium and potassium ions *d.* preventing the release of neurotransmitters across the synapse.
5. Another term that means *nerve cell* is *a.* steroid *b.* axon *c.* dendrite *d.* neuron.
6. The best-known neurotransmitter is *a.* acetylcholine *b.* calcitonin *c.* ecdysone *d.* insulin.
7. Eyes, ears, and antennae are *a.* parts of the sympathetic nervous system *b.* hormone-producing structures *c.* parts of the central nervous system *d.* sensory receptors.
8. If you walk past a bakery, and the delicious scent of cookies causes your mouth to water, the stimulus is *a.* walking *b.* the bakery *c.* the smell *d.* your mouth watering.

9. In the stem of a plant that is bending toward the light, auxins are most concentrated in *a.* the top surface of the leaves *b.* the bottom surface of the leaves *c.* the side of the stem facing the light *d.* the side of the stem away from the light.
10. The process in which insects and crustaceans shed their exoskeleton is *a.* metamorphosis *b.* molting *c.* neurotransmission *d.* cephalization.
11. Which organism has a nerve net? *a.* jellyfish *b.* grasshopper *c.* flatworm *d.* frog
12. The endocrine disease caused by insufficient insulin is *a.* goiter *b.* cretinism *c.* Grave's disease *d.* diabetes.
13. An animal's "fright, fight, flight" responses are controlled by its *a.* sympathetic nervous system *b.* parasympathetic nervous system *c.* somatic nervous system *d.* autonomic nervous system.
14. Which of these is *not* associated with the endocrine system? *a.* regulation of the amount of calcium in the bloodstream *b.* goiter *c.* metamorphosis *d.* reflex reactions
15. Steroid hormones are based on *a.* cholesterol *b.* hemoglobin *c.* tryptophan *d.* thyroxin.

PART B—CONSTRUCTED RESPONSE

Use the information in the chapter to respond to these items.

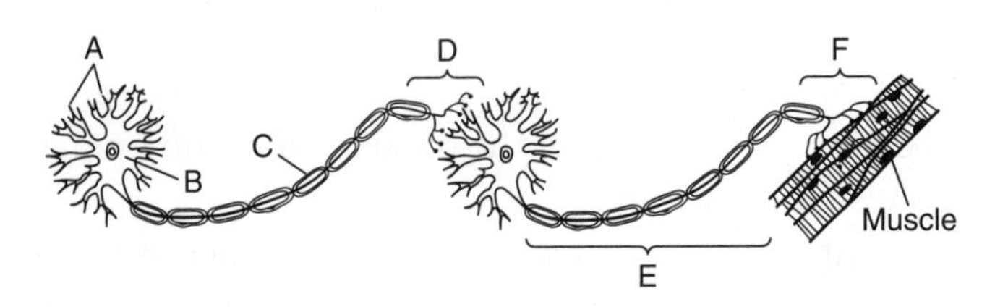

16. What structures are shown in the diagram? Identify the parts labeled *A*, *B*, and *C*.
17. What happens in the part of the diagram labeled *E*? In the parts labeled *D* and *F*?
18. Differentiate and explain the relationships between these terms: *central nervous system* and *peripheral nervous system*; *motor neurons* and *sensory neurons*; *axon* and *dendrite*.
19. Compare and contrast the functioning of the nervous and endocrine systems.
20. Do you think it is appropriate that the pituitary is called the master gland? Justify your answer.

PART C—READING COMPREHENSION

Base your answers to questions 21 through 23 on the information below and on your knowledge of biology. Source: *Science News* (March 29, 2003): vol. 163, p. 206.

Protein Protects Rat Brain from Strokes

A protein related to oxygen-carrying hemoglobin in blood cells may protect the brain during strokes.

Scientists discovered the hemoglobin cousin several years ago and dubbed it neuroglobin because only nerve cells in the brain of vertebrates make it.

Seeking to uncover neuroglobin's role, David A. Greenberg of the Buck Institute for Age Research in Novato, Calif., and his colleagues recently induced strokes in rats whose brains had been injected with viruses genetically engineered to churn out the protein. The amount of brain tissue damaged by the strokes was significantly less in those animals than in rats not given the virus, or in rats whose brains had less-than-normal amounts of neuroglobin, the investigators report in the March 18 *Proceedings of the National Academy of Sciences.*

Greenberg and his colleagues conclude that neuroglobin naturally protects brain cells faced with too little oxygen. They speculate that drugs that increase the production of neuroglobin could become a new stroke therapy.

21. State two facts about the brain-protecting protein that led to its being given the name *neuroglobin.*
22. Describe the basic design of the experiment that was conducted to study the role of neuroglobin.
23. Explain how this understanding of the role of neuroglobin could be used to help stroke victims.

12 Animal Behavior

After you have finished reading this chapter, you should be able to:

Discuss how behavioral adaptations aid an organism's survival.

Distinguish between inborn and learned patterns of behavior.

Identify and explain the importance of social behavior in animals.

as a representative of the insect world, i have often wondered on what man bases his claims to superiority. everything he knows he has had to learn whereas insects are born knowing everything we need to know.

Don Marquis, *the lives and times of archy and mehitabel* (Archy the cockroach)

Introduction

Many years ago, Japanese researchers began throwing sweet potatoes on the beach for the macaque monkeys that lived on the island of Koshima. Up until then, the monkeys were never observed on the beach. However, they came out of the forest, walked to the beach, picked up the potatoes, brushed off the sand, and ate them. Then, in 1953, one year after the feeding began, one female—named Imo by the scientists—was seen carrying a piece of potato to the water. She dipped the potato in the water and washed the sand off of it. Soon other monkeys who lived with Imo were seen washing their potatoes, too. Adult male monkeys, who were strangers to Imo's troop, never learned to wash their potatoes. However, younger monkeys who learned to wash their potatoes continued the habit as they grew older. Today, all the monkeys wash their potatoes. Later, Imo and other monkeys learned to use seawater to rinse other foods, to add the flavor of salt to their food. They even learned to swim. All these behaviors were learned by animals that originally had never been seen on the beach or anywhere near the water. (See Figure 12-1.)

Figure 12-1 The macaque monkeys of Koshima Island in Japan have learned to wash their food in the salty water at the beach. Here a mother macaque is washing her baby.

What is behavior? How do animals learn behaviors? Are there some behaviors that do not need to be learned? These and other questions will be explored in this chapter.

BEHAVIOR: HOMEOSTASIS AND EVOLUTION

Everything an organism does is its **behavior**. This includes finding a place to live, finding food, avoiding predators, and mating and reproducing. It even involves dying.

The purpose of animal behavior is to allow the organism to maintain homeostasis, survive, and reproduce. Birds build nests and keep their eggs warm until they hatch. (See Figure 12-2.) In Africa, huge herds of wildebeests migrate first northward and then southward each year, following seasonal rains. When they are very young, discus fish swim around their parents' bodies and feed off a nutritious slime produced on the surfaces

Figure 12-2 Animal behaviors, such as nest building, allow the organism to maintain homeostasis and survive.

of their scales. Maintaining constant temperature, water balance, and level of nutrition are all behaviors that control aspects of homeostasis. Homeostasis requires that the conditions within an organism remain relatively constant. This is a requirement for life. Organisms behave in a wide variety of ways to maintain these conditions.

Honeybees lay eggs in individual wax cells that make up the structure of the beehive. The eggs hatch and develop through larval and pupal stages in the cells before they develop into adults. Sometimes the developing bees die in their individual cells. Some types of honeybees remove young pupae that die in their cells. Other types of honeybees do not. How has this particular behavior developed? It has evolved over time, like any other adaptation that has survival value. Although requiring more effort, removing dead pupae may have the survival value of keeping the beehive free from disease. How have all other organisms developed the behaviors they now have?

In Chapter 3, you learned that adaptations may be morphological, physiological, or behavioral. The process of natural selection acts on behavior just as it does on any physical or physiological characteristic. Within a population of organisms, there are variations among individuals in appearance. There are also variations in how they show a particular behavior. The behavior may actually be determined by genetic traits already present in the organism since birth. Those individuals whose behavior makes them more likely to survive will pass on their genetic traits to offspring. Others, with behavior that is not as well adapted to survival, will be less likely to reproduce. In this way, certain behaviors—like physical traits—are naturally selected. The behavior of a species of organism will evolve just like other characteristics such as fur color and body size do.

SENSORY PERCEPTION

Behaving in an appropriate manner that maintains homeostasis and increases one's chances of survival depends on receiving accurate information about the environment. Sense organs provide this information. Receiving this information is called **sensory perception**.

Our eyes, ears, nose, mouth, and the surface of our skin act as sense organs. You receive information about your environment through these organs. Imagine what life would be like without these sense organs. You would have no awareness of the world around you. Experiments have been done on people who have been put in a completely dark room with no sound and no temperature changes. These individuals were unable to remain in the room for very many hours. They quickly became agitated and upset. Clearly, we need to receive information about the world around us.

We also need to know about the conditions inside us. We depend on internal sensors to keep track of body temperature, blood pressure, water balance, and many metabolic activities. Without these internal sensors, homeostasis could not be maintained.

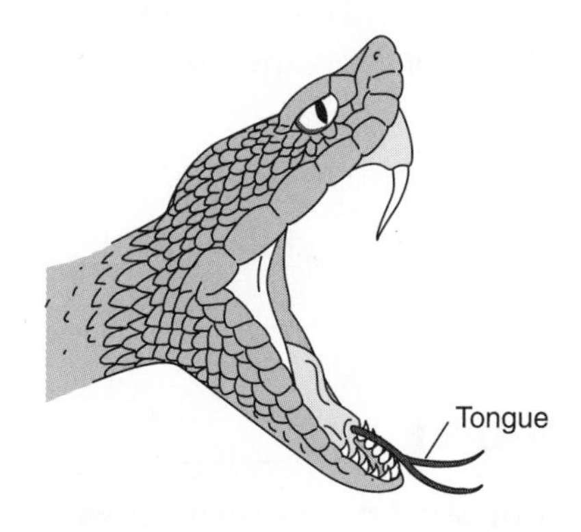

Figure 12-3 Many snakes can "smell" molecules in the air through special organs in their mouth, which helps them locate their prey.

Organisms show an amazing diversity of sense organs and sensory abilities. Many snakes have organs in their mouth that "smell" molecules in the air that are gathered with their tongue. (See Figure 12-3.) Dogs are able to hear sounds at a frequency much higher than the human ear can hear. Elephants are able to hear sounds at a frequency much too low for humans to hear. These low-frequency sounds can be heard by elephants over extremely long distances. Fish have a series of tiny sense organs in a row along the side of their body. These organs, called the *lateral line,* are able to detect chemicals in the water, wave movements and vibrations, and water temperature. (See Figure 12-4.) Sharks are able to detect electrical signals from prey. Beluga whales are able to sense Earth's magnetic field. Despite the great variety of types of sensory perception in animals,

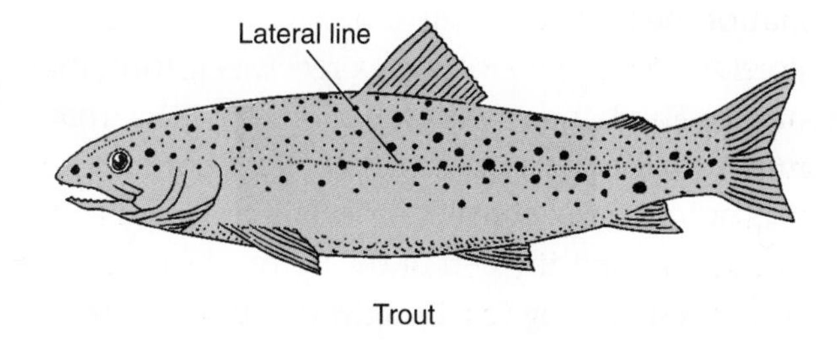

Figure 12-4 The lateral line in fish detects chemicals in the water, wave movements and vibrations, and water temperature.

each type of stimulus from the environment excites only one type of sensory receptor. The types of sensory receptors include:

- Mechanical receptors respond to mechanical pressure and can detect motion, body position, touch, pressure, and sound. Mechanical receptors are found in the skin, within muscles, and in the ears.
- Temperature receptors detect heat and cold. They are found both in the skin and in the hypothalamus of the brain.
- Chemical receptors respond to specific chemicals. These are the receptors of taste and smell; they also detect levels of gases and nutrients in the blood.
- Photoreceptors, such as those in the eyes, respond to light.
- Pain receptors are stimulated by heat, pressure, and the chemicals released by damaged tissues.

THE BRAIN: THE ORGAN IN CHARGE

All behavior depends on the integration of sensory information with appropriate behavioral responses. In vertebrates, the brain is the organ responsible for this integration and control. The human brain is without doubt the most complex, highly organized structure on Earth. Will any person ever learn enough about the brain to be able to understand how the brain itself works? This is a question that remains unanswered, so far.

The brain, made up of billions of interconnected neurons, has a mass of only about 1.5 kilograms. However, it uses 25 percent of the body's oxygen supply. The brain can be divided into three major parts: the cerebrum; the cerebellum; and the brain stem, or medulla oblongata. (See Figure 12-5.)

The **cerebrum** is the most highly developed part of the human brain. The left side, or hemisphere, of the brain controls sensory and movement functions of the right side of the body, and vice versa. Scientists have mapped the outer surface of the cerebrum, the **cerebral cortex**, for sensory and movement, or motor, functions. Sensory information from each part of the body comes to a specific location on the cerebrum. The largest portion of the sensory cortex receives information from the face and the hands. The hands are much more sensitive than the toes, which have a much smaller portion of the cerebral cortex assigned to them. Similarly, the motor cortex consists of several regions, each of which controls movements for a specific part of the body. The cerebrum is the area of the brain that is responsible for thought processes and for the creativity we associate with humans.

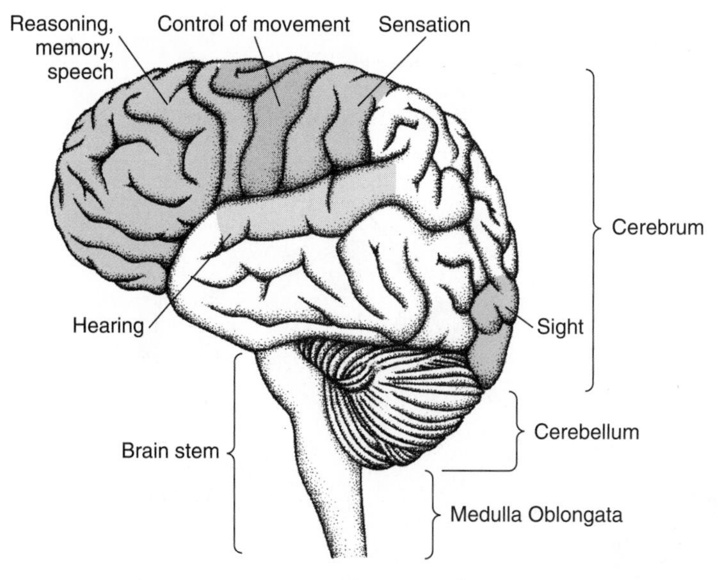

Figure 12-5 The brain can be divided into three major parts: cerebrum, cerebellum, and medulla oblongata.

The **cerebellum**, below and toward the rear of the cerebrum, also processes a vast amount of sensory information and coordinates body movements. It processes such complex movements as bicycling, skating, or dribbling a basketball. In the brain, information is processed by interconnected neurons. The greater the complexity of processing needed, the greater the number of neurons that are interconnected. It is therefore logical, although amazing, that one neuron in the cerebellum may receive impulses from up to 80,000 other neurons!

The part of the brain that evolved most recently is the cerebrum. The part that is the oldest and most similar to the brains of vertebrates such as fish and amphibians is the brain stem. The lowest part of the brain stem, or **medulla oblongata**, controls and coordinates involuntary activities such as breathing, heart rate, swallowing, peristalsis, and blood pressure.

BEHAVIORS THAT DO NOT NEED TO BE LEARNED

You step off the curb while talking with a friend. An enormous truck blasts its horn, and you both leap back instantly, safely out of harm's way. As soon as you are back on the sidewalk, you become weak as you realize how close you came to being an accident victim.

The behavior that just occurred is a **reflex**. You did not have to learn it; you did not even have to think about it. A reflex happens so quickly

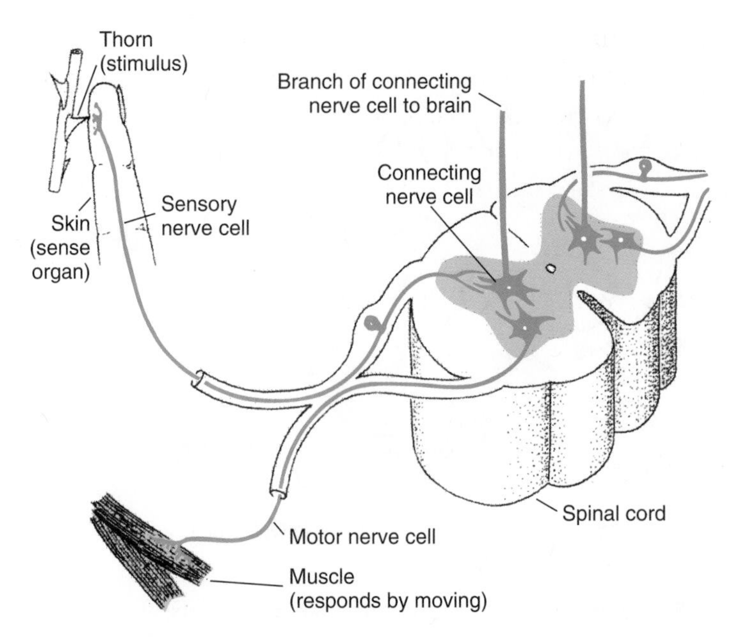

Figure 12-6 In a reflex, nerve impulses travel from the sensory receptors to the central nervous system and directly back to an effector.

because the nerve impulses travel the shortest distance possible, from the sensory receptors that detect the stimulus to the central nervous system, and directly back to an effector that produces the response. (See Figure 12-6.) For example, the automatic knee-jerk response tested by a physician begins with a receptor, which detects when the knee is tapped. An impulse travels from the knee to the spinal cord along a sensory neuron. Immediately an impulse returns along a motor neuron to the leg muscle, the effector. The leg moves, kicking upward. In the meantime, information travels to the brain, telling it the knee has been tapped. Just as in the example of jumping out of the way of a speeding truck, the awareness of the knee-tap event occurs after the reflex response has already happened.

Reflex behaviors do not have to be learned. In terms of evolution, reflex actions have been selected for since they help to ensure the survival of an organism. Scientists have also observed a wide variety of more complex behaviors in animals that are not learned. Animals are born with these behaviors, also called instincts, which are a series of reflexes. All animals' instincts are inherited from their parents, just as their physical characteristics are inherited. Because these behaviors do not change or improve with repetition, they are called **fixed action patterns**. The sucking of a newborn baby on a mother's breast and the flick of a frog's tongue to catch a fly are both examples of fixed action patterns.

Konrad Lorenz and Nikolaas Tinbergen were two European zoologists who studied animal behavior and shared the Nobel Prize for their efforts. The most famous example of a fixed action pattern described by Lorenz and Tinbergen involved the greylag goose. The greylag goose makes its nest on land. Whenever an egg rolls out of the goose's nest, the goose returns the egg with exactly the same series of movements. These movements involve stretching its neck and twisting its head to the side in order to roll the egg along the ground. (See Figure 12-7.) Sometimes the egg

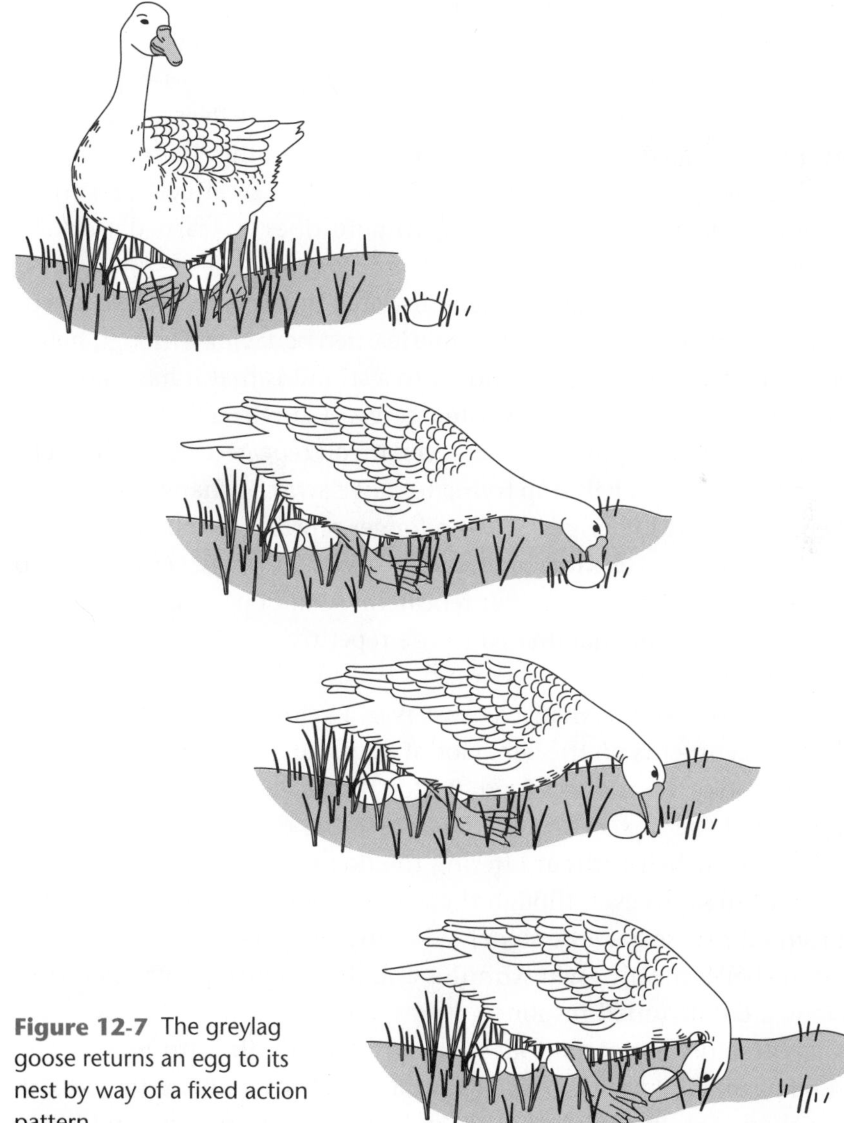

Figure 12-7 The greylag goose returns an egg to its nest by way of a fixed action pattern.

slips away, but the goose continues the same series of movements all the way back to the nest, without the egg. The behavior is built in, fixed, and cannot change.

Check Your Understanding

Why are inborn behaviors important to the survival of organisms? How are they related to sensory perceptions?

LEARNED BEHAVIOR

Much of animal behavior, including most of our own, is not fixed. It is learned. Learned behavior changes over time as a result of experience. Organisms such as honeybees and cockroaches, whose lives are relatively short, seem to have little ability to learn. On the other hand, primates with long life spans and complex brains develop a great deal of learned behavior. In fact, much of what a primate does is learned through experience.

Learned behavior can also be demonstrated in relatively simple organisms. **Habituation** is one type of simple learned behavior. Through habituation, an organism stops responding to a stimulus that it has learned is unimportant, in order to not waste energy. A single-celled ameba will move away from a bright light. However, if you repeatedly shine a bright light on the ameba, it will stop trying to move away. It changes its behavior because the steady bright light no longer represents a threat. Habituation has occurred. The same process happens in us whenever we learn to ignore annoying noises that occur repeatedly. Out of necessity, many city dwellers have become habituated to the repetitive, meaningless wailing of car security alarms.

In the 1920s, the Russian scientist Ivan Pavlov performed a famous series of experiments on the behavior of dogs. He noticed that a dog salivates, or drools, when it sees food. Pavlov rang a bell at the same time as he presented the dog with food. Of course the dog salivated. In time, Pavlov rang the bell without offering the dog food. Amazingly, the dog still produced saliva even though there was no food for it to eat. The dog had learned to associate the signal—the ringing bell—with food. Pavlov called the bell a **conditioned stimulus** and the production of saliva without food a **conditioned response**.

Pets often display conditioned responses. Goldfish learn to associate your hand moving near the water's surface with your giving them food. When they see your hand, the fish quickly swim to the top and begin

Figure 12-8 Fish can develop conditioned responses. Here carp await food from visitors at a botanical garden.

feeding. After only a few feedings, the fish learn to associate your hand with food. Simply putting your hand near the tank brings the fish to the surface, even without any food being present. (See Figure 12-8.)

How much of our behavior is learned through conditioning? Some examples may offer clues to answer this question. Many people eat at about the same times each day. As the days pass, they realize that they feel most hungry at these "mealtimes." The hunger has become a conditioned response that is now associated with the time of day rather than with the need, thought, or sight of food.

Fears can also develop through conditioning. For example, a child may be repeatedly frightened by a neighbor's dog jumping against their fence. Years later, in another house, the child may again experience fear at the sound of a fence shaking, even though this time it is only the wind.

Included among other types of learning is **imprinting**, which occurs when an animal, such as a baby goose, follows the first moving object it sees after birth. In nature, the first moving object a baby goose sees is usually its mother; but it may be you, if you are hatching the eggs. Lorenz, in a series of famous photographs, leads a trail of baby geese that imprinted on him soon after they hatched. (See Figure 12-9 on page 264.) **Imitative learning** occurs when animals observe and learn from each other. The macaque monkeys that learned to wash their food showed imitative learning.

Figure 12-9 Konrad Lorenz researched imprinting in geese.

Habit formation is another type of learning. A **habit** is a common type of behavior that results from repeating an action over and over again until the action is done automatically. Try writing an entire paragraph without dotting an *i* or crossing a *t*. You will find that it is hard not to make these marks. Writing letters correctly is a habit that is hard to alter. To change or break a habit takes a great deal of concentration. This is because you are trying to learn a new behavior to replace an old behavior you have learned so well that it occurs automatically. Learning to break a habit such as smoking requires a great deal of determination, a strong desire to learn the new habit of not smoking, and much praise and reward for doing so.

SOCIAL BEHAVIOR

The vast majority of animals, from crayfish to frogs to tigers, live solitary lives. Other than coming together for the purpose of reproduction and perhaps caring for their offspring, each individual is on its own. Each animal is in a competition for survival, competing not only with other types of organisms but also with other members of its own species.

However, some animals live in groups called **societies**. These animals often share responsibilities for finding food, building a home, providing protection for other individuals, and even reproduction. In these groups, the animals depend on each other.

Certain insects, along with humans, have developed the most complex societies in the animal kingdom. Termites, ants, and some species of wasps and bees have highly structured societies. The honeybee society is one of the best-known and probably the most-studied example.

A honeybee society, or colony, consists of 30,000 to 40,000 female

Figure 12-10 The members of a honeybee colony communicate by means of pheromones and visual signals.

worker bees and one queen. The colony may also include some males, or drones. The worker bees live for about six weeks, first acting as nurses for young bees. Next, they become houseworkers, taking care of the hive. Finally, they act as food gatherers, flying out to collect nectar and pollen from flowers. The queen bee's only responsibility is to mate with male drones, often from another colony, and lay eggs. (See Figure 12-10.)

As in any society, communication is important in a bee colony. One way honeybees communicate with each other is through chemicals they produce, known as **pheromones**. Pheromones are released into the air. For example, the queen controls the worker bees in the colony by releasing a pheromone that prevents workers from developing into other queens.

Honeybees also communicate by using visual signals. The Austrian biologist Karl von Frisch (who along with Tinbergen and Lorenz shared a Nobel Prize in 1973) discovered that honeybees perform a very elaborate dance that tells other worker bees in the hive about food they have found. During the dance, the direction and speed the bee moves, the amount it shakes its body, and how it buzzes while dancing all provide information about a food source. The food's direction and distance from the hive are communicated in the dance. Other bees use this information to find the food source.

An insect society is very structured. An individual's position in the society is determined from birth. A worker bee never gets to be the queen.

Inside the Mind of an Infant

What could be more fascinating than trying to learn about the mind of a baby? During their first year, infants gain physical skills; they also take their first steps in being able to think and to make sense of the world around them. This field of study is called infant cognition. It includes the study of the increasing abilities of infants to perceive, recognize, categorize, and remember things in their environment.

In one research project, various patterns were shown to infants. (See Figure 12-11.) By measuring the amount of time the infants looked at each pattern, scientists were able to conclude that infants under 1 year of age look longer at concentric forms (top row) than at non-concentric forms (bottom row), and longer at curved lines than at straight lines.

Why? What do these results mean? What later developments occur in the mind of an infant? These and many other questions are being asked and examined by researchers studying infant cognition.

Figure 12-11

Drones remain useful until the queen has been fertilized; then they are often driven from the hive. On the other hand, societies of vertebrates are less structured and offer an individual a chance to change position. You might say that in these societies there are more chances for "job promotion."

Chickens show one type of behavior commonly found in vertebrate societies. This behavior determines who is the most important member of a particular society. With chickens, it is called the **pecking order**. In a flock of hens, one hen is the most dominant. This hen is able to peck at all the other hens. The second hen is able to peck at all the other hens except the first one. A third hen can peck at all the others except the first two, and so on down to the lowest hen, who is pecked at by all the other hens and has no hen below it to peck. Not a very good position to be in at all! The pecking order determines which hens get to eat first and which hens are able to use the best sites for making a nest. The feathers of hens at the top of the pecking order are always clean; these hens are always well groomed. They move around the yard with confidence. The hens at the bottom of the pecking order often have dirty feathers, are not well groomed, and often appear nervous or frightened. Other animals, such as wolves, also have highly structured societies similar to that of chickens.

Many vertebrates have another social behavior, one that involves defending the area in which the animals live. The defended area is called a **territory**. Territorial animals, those that defend their territories, include many bird species, wolves, deer, monkeys, crabs, and fish. To define their territory, many birds sing out a warning song and even attack other birds that cross the borders into their area. It is interesting to note that a bird will prevent other birds of the same species from entering its territory while ignoring other kinds of birds. This is because it competes with members of its own species for the same types of food and nesting sites.

Pets can also be observed displaying territorial behavior. Even cats or dogs, which spend most or all of their time inside their owners' homes, can become very upset when another cat or dog, a stranger, dares to appear outside the window of their territory. The question that needs to be asked is: Should humans be added to the list of territorial animals? The number of military conflicts that have occurred as people fight over land may suggest that the answer is yes.

Other forms of social behavior include **schooling**, in which animals travel together in groups, and **migration**, in which groups of animals travel long distances, often during specific times of the year. Dolphins and whales are very social animals. These marine mammals form groups, or pods, that range in size from about a dozen animals to several hundred individuals. Dolphins migrate over large areas in search of food. Some whales migrate thousands of kilometers to reach their feeding grounds. (See Figure 12-12 on page 268.) Fish, such as tuna, and birds, such as geese and ducks, are also migratory. Tuna swim for months around the Atlantic Ocean, traveling thousands of kilometers from one seasonal feeding place to another. In the northern hemisphere, geese and ducks fly from north to south every winter, and back again the next spring. These migrations often cover thousands of kilometers.

Social behaviors are generally not learned. Animals are born with these

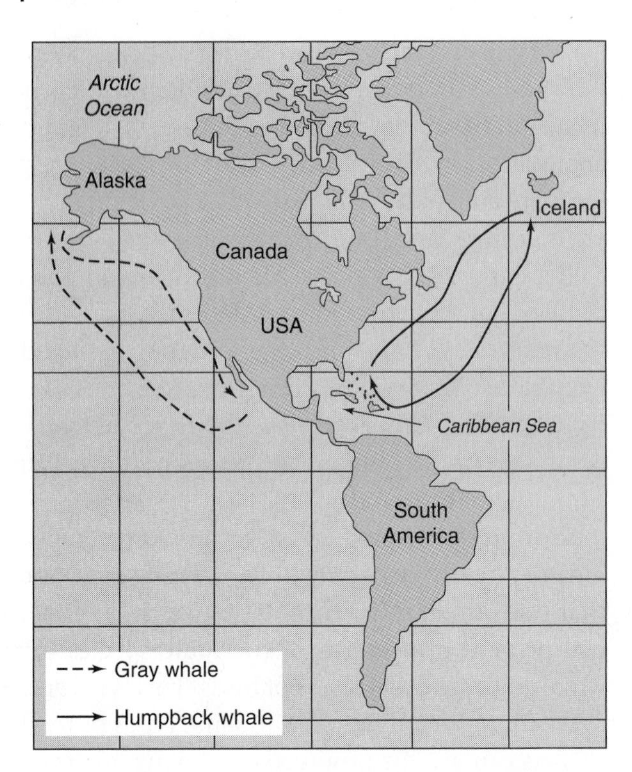

Figure 12-12 The migration routes of two whale species between their feeding and breeding grounds.

behaviors. Insect societies, pecking orders, territoriality, schooling, and migration are usually seen as characteristic behaviors of certain species. All members of a species have the characteristic behavior of the species. While the behavior has not been learned, it has evolved along with other characteristics. Therefore, the conclusion is that behavior, just like any other characteristic that has been selected naturally, provides the species with some survival advantage.

THE HUMAN MIND: WHO ARE YOU?

The human brain is the only structure on Earth that tries to understand itself. Are you who you are because of this mass of tissue, your brain? Is your brain, that vast collection of neurons and neurotransmitters, and other brain tissues, the same as your mind? These questions are certainly difficult to answer. However, biologists are confident that we will someday understand a great deal about perception, knowledge, learning, memory, and consciousness by understanding the structure and function of the brain.

A great deal is already understood about individual neurons, their structure and function, and how they are connected to each other. However, it seems that much of how the brain works depends on large networks of neurons that act together as systems. These systems are very complex and are not well understood. There are so many things the human brain can do. One type of thinking involves being able to group, or categorize, things. For example, we recognize that a saw, a hammer, and a screwdriver are all tools, and that a sweater does not belong in this group. It is believed that a network of neurons is responsible for allowing us to place the correct objects in a group. Another kind of brain activity occurs when you continue to look at an object in front of you while you turn your head to the side. Keeping your eyes focused on the object while your head turns requires very complex activity within your brain and within the neurons that connect your eyes to your brain. It has been suggested that understanding how very complicated computers function could help us understand how the human brain functions. However, is anyone intelligent enough to build a computer as complex as the brain? Not up to now.

Another puzzling question concerns the brain and sleep. Becoming tired and falling asleep is a basic part of living. But why? Neurons in the brain are very busy even while we sleep, almost as busy as when we are awake. While sleeping, our brains are really not at rest. And yet most people like to sleep. There must be a good reason why the brain needs sleep. This is another question yet to be answered.

Finally, what about memory? You hear a song, and suddenly you remember where you were that summer long ago when you first heard the song, and whom you were with. You also remember all those feelings of that time long ago, when that song was important to you. How is this memory stored in your brain? Does it exist in a certain place within the brain? Is the memory made up of organic molecules that are now stored in your brain? Does the memory consist of special connections of neurons in particular patterns?

Sometimes it seems as if there are more questions than answers in understanding the brain and the human mind. How exciting it is to be learning the answers as science explores this fascinating area of research.

LABORATORY INVESTIGATION 12
How Do We Learn?

INTRODUCTION

Learning is a complex process in all organisms. We can begin to understand the process of learning by identifying some of the methods by which humans learn. They are as follows:

- Repetition—observing and copying the actions of others
- Trial and error—repeating the efforts until a solution is found
- Memorization—a way to increase our body of knowledge
- Reasoning—drawing conclusions based on past experiences

In this investigation, you will complete various tasks in order to study these methods of learning.

MATERIALS

Letter "L" puzzle pieces in an envelope, word lists *A*, *B*, and *C* on separate pieces of paper (from the *Teacher's Manual*)

PROCEDURE

Part A: The Letter "L" Puzzle

1. Work in pairs. Take an envelope that contains the pieces of the letter L puzzle.
2. One member of the pair acts as a timekeeper. That person times how long it takes for the other person to solve the puzzle. The solver begins to arrange the puzzle pieces into the letter L when the timer says "go." The timer looks away from the person solving the puzzle.
3. Reverse roles and repeat step 2.
4. Repeat this process until each person has solved the puzzle three times.
5. Prepare a bar graph that shows the elapsed time of each trial on the vertical axis and the trial number on the horizontal axis.

Part B: Word Lists

1. Work by yourself. Carefully read through list *A* three times. Put this list away. On a blank piece of paper, write down all the items on the list that you can remember. Try to write the words in the same order as they appeared on the list.
2. Repeat step 1 with list *B*.
3. Repeat step 1 with list *C*.
4. Determine your score for each list. Count one point for each item in the list you remembered. For the purposes of calculating your score, the order of the items recalled does not count.

INTERPRETIVE QUESTIONS

1. What evidence was there that learning occurred in the activities in parts *A* and *B*? What were the primary methods by which learning occurred in each of these activities?
2. Discuss how the learning activities in this investigation are related to the skills and habits needed to become a successful learner.

CHAPTER 12 REVIEW

Answer these questions on a separate sheet of paper.

VOCABULARY

The following list contains all of the boldfaced terms in this chapter. Define each of these terms in your own words.

behavior, cerebellum, cerebral cortex, cerebrum, conditioned response, conditioned stimulus, fixed action patterns, habit, habituation, imitative learning, imprinting, medulla oblongata, migration, pecking order, pheromones, reflex, schooling, sensory perception, societies, territory

PART A—MULTIPLE CHOICE

Choose the response that best completes the sentence or answers the question.

1. Behavior *a.* is shaped by natural selection *b.* can be learned *c.* can be inborn *d.* all of these.
2. The largest part of the sensory cortex receives information from the *a.* face and neck *b.* torso *c.* arms and legs *d.* hands and face.
3. While staying overnight with a cousin in a busy city, you are kept awake by the noise of traffic. But your cousin can sleep due to *a.* sensory perception *b.* fixed action patterns *c.* habituation *d.* imprinting.
4. Which of these is *not* an example of a sensory receptor? *a.* the roof of a snake's mouth *b.* the hairs on your head *c.* a dog's nose *d.* a fish's lateral line
5. Although baby California condors are raised by humans, they are fed with a condor hand puppet so they will not *a.* imprint on humans *b.* engage in imitative learning *c.* form bad habits *d.* form a conditioned stimulus.
6. In humans, mechanical receptors are involved in the sensing of *a.* sound *b.* temperature *c.* light *d.* chemicals.
7. Which is an example of social behavior? *a.* People live in communities. *b.* Birds migrate toward the equator during the winter. *c.* Sardines swim in large schools. *d.* All of these.
8. The part of the brain that evolved most recently is the *a.* medulla oblongata *b.* pons *c.* cerebellum *d.* cerebrum.
9. If your tropical fish cluster around the top of the tank when you draw near, they are showing a *a.* conditioned response *b.* fixed action pattern *c.* result of imprinting *d.* pecking order.

10. A reflex *a.* is a learned behavior *b.* may not involve the brain *c.* is a conditioned response *d.* is imitative learning.
11. How much of the human body's oxygen supply is used by the brain? *a.* 1 percent *b.* 5 percent *c.* 25 percent *d.* 50 percent
12. Which symptom might you see in a person who has had a stroke on the right side of the brain? *a.* difficulty moving the right hand *b.* paralysis on the left side of the face *c.* difficulty recognizing the meaning of words *d.* all of these
13. On a playground, young children never tease the most popular kid, only the most popular kid can tease the second-most popular kid, and everyone teases the least popular kid. This behavior is an example of *a.* a conditioned stimulus *b.* imprinting *c.* habituation *d.* a pecking order.
14. In Pavlov's famous experiment, a ringing bell was *a.* a conditioned stimulus *b.* a conditioned response *c.* a fixed action pattern *d.* habituation.
15. The best explanation for a normally calm dog growling and running around frantically when a stranger or another dog enters the front yard is *a.* habituation *b.* territoriality *c.* imprinting *d.* fixed action patterns.

PART B—CONSTRUCTED RESPONSE

Use the information in the chapter to respond to these items.

16. What is the structure shown in the diagram? Identify the parts labeled A through E and briefly describe their functions.

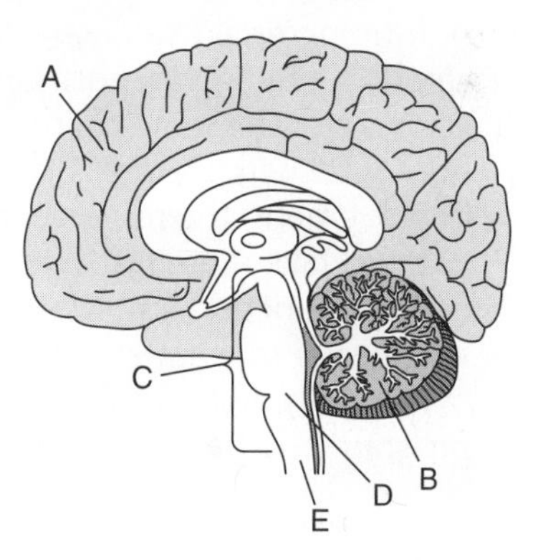

17. Predict what would happen if part D were damaged in an accident.
18. Describe what happens in the body in the knee-jerk reflex.
19. Why is it very difficult to understand how the human mind works?
20. Crayfish have an organ called a *statocyst* that consists of a granule inside a chamber lined with tiny hairs. If the granule is replaced by a tiny piece of iron, and a magnet is held over the crayfish, the crayfish will flip onto its back. Explain the crayfish's behavior.

PART C—READING COMPREHENSION

Base your answers to questions 21 through 23 on the information below and on your knowledge of biology. Source: *Science News* (March 15, 2003): vol. 163, p. 173.

Brain Training Aids Kids with Dyslexia

A reading-improvement course for children with dyslexia appears to go to their heads.

After completing the course, 20 grade-schoolers diagnosed with this reading disorder not only improved their speech and reading skills, but showed signs of increased activity in key brain areas as they read, according to a study in the Mar. 4 *Proceedings of the National Academy of Sciences.*

The children with dyslexia and another 12 kids with no reading problems, all ages 8 to 12, performed daily tasks on a computer for nearly 1 month.

The children were asked to match speech sounds to written consonants and vowels, and they practiced related skills. Before and after the course, the researchers used a functional-magnetic-resonance-imaging scanner to measure blood-flow changes throughout each child's brain that were uniquely linked to the identification of rhyming letters. This task taps into the ability to decode sounds associated with different letters, a crucial element of reading.

After training, only children with dyslexia exhibited substantial gains in reading and speech comprehension as well as blood-flow surges—a sign of increased cell activity—in several brain areas previously implicated in reading, report Elise Temple of Cornell University and her coworkers. The same children displayed elevated brain activity in attention and memory areas that probably contributed to their reading improvement, the researchers say.

It will take more research to determine how long the course's effects on brain function last. Moreover, Temple acknowledges, it's not known whether or to what extent other literacy programs influence the brains of kids with dyslexia. Two of Temple's colleagues designed the training program and have a financial interest in it.

21. State what developments, in addition to improved reading, occurred in the children after completion of the reading improvement course.
22. Explain the tasks given to the children, as well as the task linked to the area of the brain being studied, that were related to reading.
23. Describe two conclusions about the research that could not be made without further study.

Excretion and Water Balance

After you have finished reading this chapter, you should be able to:

Discuss the importance of the kidneys, skin, and liver to excretion and to homeostasis.

Compare and contrast the process of excretion in different organisms.

Describe the structures and functions of kidneys in humans.

Day after day, day after day,
We struck, nor breath nor motion;
As idle as a painted ship
Upon a painted ocean.

Water, water, everywhere,
And all the boards did shrink;
Water, water, everywhere
Nor any drop to drink.

Samuel Taylor Coleridge,
The Rime of the Ancient Mariner

Introduction

Imagine yourself in the middle of a calm sea. No winds blow. Your sailing ship is unable to move. You have been at this place in the ocean for weeks. Food and water have been used up. You and your fellow sailors are becoming very thirsty. In every direction, stretching as far as the horizon, all you can see is water. And yet, as the Ancient Mariner said in the poem by Coleridge, there is not a "drop to drink." How could this be? (See Figure 13-1 on page 276.)

Ocean water is salty. When dried, ocean salt is almost chemically identical to table salt. Salt, a solute, is only one of many substances that can be dissolved in water. This is true of the water in the human body. The

Figure 13-1 You can be sailing in the middle of an ocean and still not have a "drop to drink," since seawater is salty, and we can drink only freshwater.

human body is about 70 percent water. This water contains many types of solutes, including table salt. The amounts of these solutes in our bodies must remain constant day by day, and hour by hour. We have seen how regulating our internal body temperature is an important example of homeostasis. An even more important function of homeostasis is maintaining a constant internal chemical environment. Drinking salt water disrupts the careful balance that our bodies need internally. The results could be very dangerous, perhaps even fatal. We cannot drink salt water, only freshwater.

WATER COMPARTMENTS

In the body, water is found in three areas, or compartments:

- Plasma, the liquid portion of the blood without the blood cells, makes up about 7 percent of body fluid.
- The intercellular fluid and lymph make up about 28 percent of body fluid.
- The intracellular compartment, the fluid located inside cells, makes up about 65 percent of body fluid.

The real meaning of the cell theory is that we are only as healthy as our cells are. As you might expect, the chemical contents of our cells are therefore regulated carefully. We have seen that all body cells are surrounded by pools of intercellular fluid, which is mostly water. Because the level of chemicals in cells must remain within very narrow limits, it is essential that the composition of the intercellular fluid remains within narrow limits, too. How is the chemical composition of the intercellular fluid regulated? As you know, capillaries deliver blood to every cell in the body. Capillaries exchange materials with the intercellular fluids that surround each cell. It is the job of the blood, and more specifically the blood plasma, to control the levels of substances in the fluid that surrounds each cell. So, one of the most important jobs of homeostasis is regulating the chemical composition of blood plasma. (See Figure 13-2.)

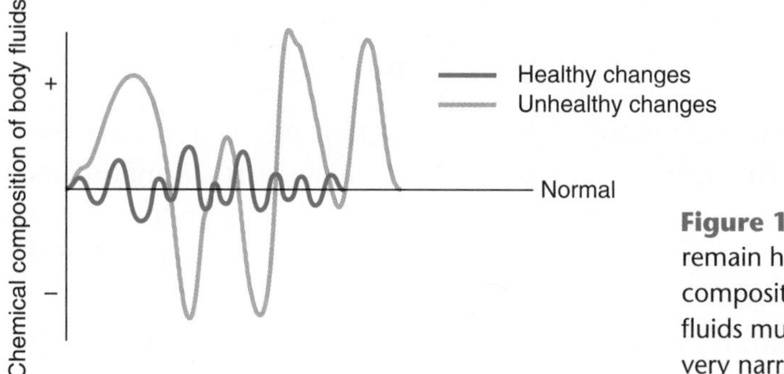

Figure 13-2 If we are to remain healthy, the composition of our body fluids must remain within very narrow limits.

In terrestrial vertebrates, including ourselves, the organ primarily responsible for regulating the chemical composition of plasma is the **kidney**. Approximately 25 percent of the total blood flow in the body passes through the kidneys at any moment, even though the kidneys make up only 1 percent of the body's total weight. When kidneys do not work properly, it results in blood that does not have the proper balance of chemicals. In time, this can cause death, unless the kidneys' functions are taken over by mechanical means. Let us now examine what the kidneys do and learn how they function.

THE KIDNEYS

A single bag of potato chips contains enough salt to kill you. Of course it doesn't, because your body does not allow this salt to be added directly to your body's fluids. One of the jobs of the kidneys is to maintain the proper levels of both salt and water in the body. Improper levels of salt cause a number of potential problems. The sodium and chloride ions of salt can disrupt normal chemical activity. For example, if for some reason your body did increase the level of salt in its fluids, your heart might no longer beat properly. Salt also affects the rate at which osmosis occurs. Water diffuses across a cell membrane toward an area of higher salt concentration. Therefore, the job of regulating both salt and water levels is often called maintaining a *water balance* in the body. This is one of the functions of the kidneys.

The kidneys are also part of the sanitation system of the body. All metabolic activities in the body produce wastes. These wastes are toxic and must not be allowed to accumulate in the body. So another important function of the kidneys is to carefully select the chemical wastes for removal while keeping useful nutrients. This process of ridding the body of metabolic wastes is called **excretion**.

WASTES THAT NEED TO BE REMOVED

You have already learned that cellular respiration produces carbon dioxide and water. The waste, carbon dioxide (CO_2), is released from the body primarily through gas exchange that occurs in the respiratory system. Carbon dioxide leaves the blood and enters the air in the alveoli of the lungs. Other wastes produced during metabolic activities include water from condensation synthesis and a variety of mineral salts from other metabolic processes.

However, by far the most important and potentially dangerous metabolic wastes produced in the body contain the element nitrogen. These **nitrogenous wastes** result when amino acids, the building blocks of proteins, are broken down. The main nitrogenous waste produced during this process is **ammonia** (NH_3). (See Figure 13-3.) Ammonia is a toxic compound that can kill living cells. Open a bottle of household cleanser that contains ammonia and you will smell its unpleasant characteristic odor.

All animals produce ammonia in their cells as they metabolize proteins. How do they get rid of this toxic waste? For an ameba, and some aquatic animals, ammonia diffuses into the water through the cell membrane. Other animals cannot get rid of ammonia so easily. These animals must change ammonia into a less harmful substance. Birds, terrestrial reptiles, and insects change ammonia into solid crystals of **uric acid**. This uric acid is mixed with solid, undigested wastes from the digestive system and passed out of the body. Any person trying to avoid being hit by bird droppings is trying to steer clear of these wastes. Nitrogen is a good fertilizer for plants, however. In some places, where large numbers of birds roost, these nitrogen-rich wastes accumulate in vast amounts. In the past, these droppings were shoveled into bags and sold for high prices as a potent plant fertilizer. Today, these natural sources of nitrogen have been replaced with nitrogen from ores that have been processed in factories.

Mammals have evolved a different solution for dealing with ammonia

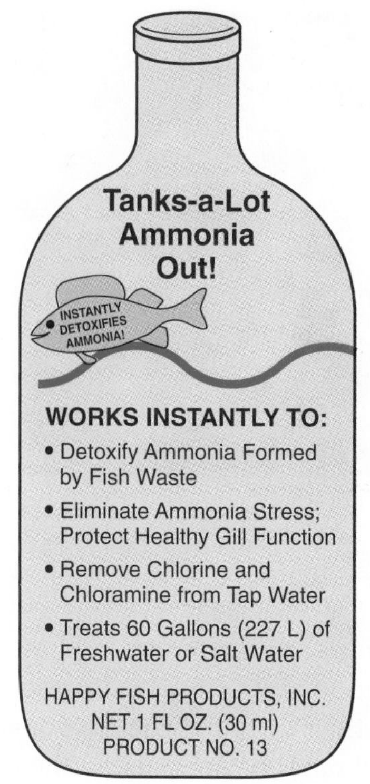

Figure 13-3 All animals produce ammonia—a toxic compound—when they metabolize amino acids. Ammonia that is formed by fish waste in an aquarium must be chemically treated to prevent damage.

wastes. In mammals, ammonia is mostly changed into **urea**, not uric acid. Unlike uric acid, which does not dissolve in water, urea is very soluble in water. To be removed from the body, urea must be dissolved in water. The ammonia, now in the form of urea, is diluted and removed from the body in **urine**.

EVOLUTION AND EXCRETION

A variety of solutions to the problem of maintaining a constant internal chemical environment can be seen in different organisms. We have seen that single-celled organisms, such as an ameba and a paramecium, rid themselves of wastes by simple diffusion across their cell membrane. However, these protists have evolved a structure that handles the problem of maintaining water balance. These protists live in freshwater, where the level of solutes inside them is higher than in the external freshwater. As a result, water tends to diffuse into these organisms. A unique organelle, the **contractile vacuole**, takes in this excess water in the cell and pumps it out of the organism and back into the environment. In this way, the proper level of water is maintained in the cell. Because the diffusion of

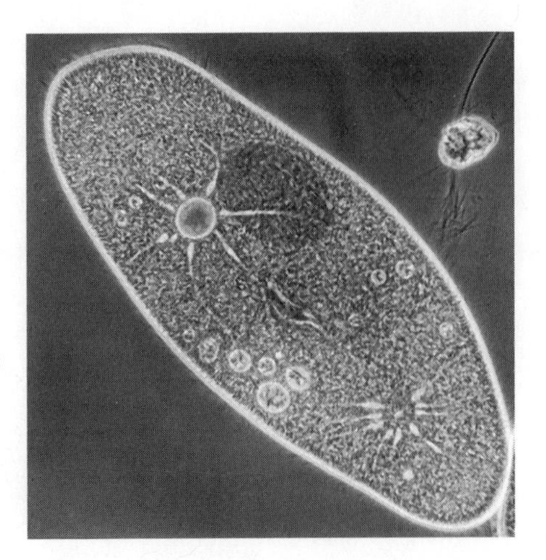

Figure 13-4 The paramecium rids itself of wastes by diffusion across its cell membrane. The contractile vacuole aids it in osmoregulation (see star-shaped organelle in upper half of paramecium).

water is called osmosis, the process of maintaining water balance is called **osmoregulation**. (See Figure 13-4.)

Some multicellular animals, such as sponges, sea anemones, and sea stars, have no special organs for osmoregulation and excretion. Instead, the exchange of water and wastes between their cells and the environment occurs by diffusion through the membranes of the cells on their body surfaces.

Most other animals, from earthworms to crabs to polar bears, have some type of internal organ that contains tubelike structures to carry out chemical regulation. By examining this structure in earthworms, we can learn about the kidneys in humans.

One thing you notice when you examine an earthworm is its segments —often, more than 100 of them. Each segment contains a pair of twisting tubes, each called a **nephridium**. The tube has an opening into the previous segment. The earthworm's body fluids pass into the opening. As these fluids move through the nephridia (tubes), useful materials are absorbed out of the tubes and into the capillaries that surround the tubes. The useful materials stay in the earthworm's blood. The wastes remain in the nephridia and pass out to the external environment through openings in the earthworm's body wall. (See Figure 13-5.)

The waste gas CO_2 diffuses directly from capillaries just below the earthworm's moist skin cells into the air between particles of soil. This reminds us of the process of excretion that occurs in the alveoli in the human lungs. During and just after a soaking rain, earthworms crawl out of their burrows in the ground onto the soil's surface. When the soil is saturated with water, an earthworm cannot get rid of waste gases. It cannot take in oxygen either. If it remains in wet soil, it might suffocate.

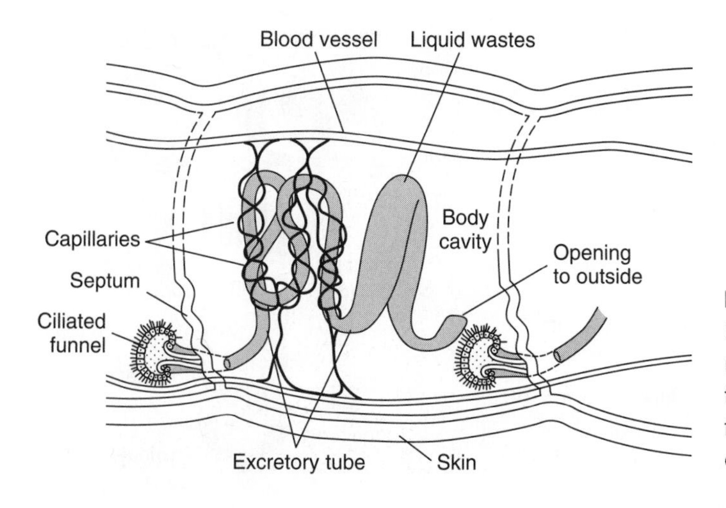

Figure 13-5 Nephridia tubes remove wastes from the body fluids of the earthworm.

Having seen the structure of the nephridia tubes in earthworms, we can now examine how the human kidneys work.

Check Your Understanding

What are some ways in which an excretory system helps organisms maintain homeostasis?

THE HUMAN KIDNEY: STRUCTURE AND FUNCTION

The kidneys are the most important part of the excretory system in humans. There are two kidneys, located in the lower, rear portion of the abdominal cavity. (See Figure 13-6 on page 282.) Every boxer knows which punches are illegal—a punch to the lower back area, on either side of the spinal column, can damage a kidney and is therefore banned.

Large blood vessels bring blood to the kidneys. Other blood vessels leave the kidneys with the newly cleansed blood. As blood passes through the kidneys, metabolic wastes are removed, and the correct balance of salt and water in the blood is maintained. As a result, the kidneys produce urine. Urine leaves each kidney through a tube, the **ureter**. The two ureters constantly drip urine into the **urinary bladder**. Here the urine is stored until it is convenient to pass it from the body. Urine leaves the body through the **urethra** when instructions from the brain are given.

A close look at the kidney's structure gives some idea of how this important organ works. If you examined a section of the kidney, you would see three large tubes at the center of the concave surface. One of these tubes is the *renal artery,* which brings blood to the kidney from the

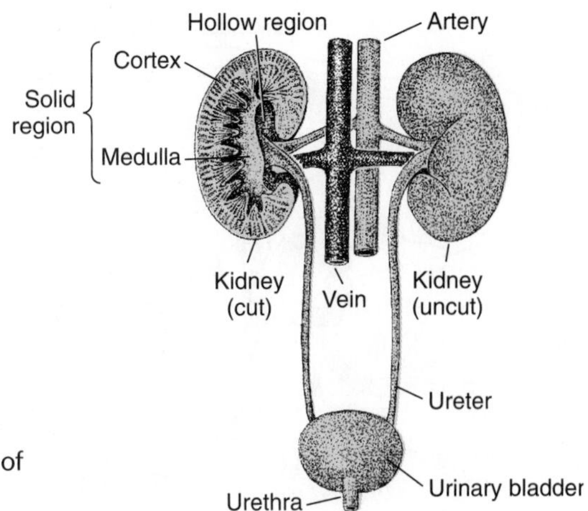

Figure 13-6 Human kidneys are located in the lower, rear portion of the abdominal cavity.

body; another is the *renal vein,* which returns blood to the body; and the third tube is the ureter, which connects each kidney to the urinary bladder. The inner portion of the kidney is hollow. This is where the urine is collected. The outer portion of the kidney consists of the medulla and the cortex. (See Figure 13-7.)

If you examined some tissue from the kidney under a microscope, you would see that the blood vessels go through the medulla and into the cortex. In the cortex are the beginnings of many tiny, individual tubular structures that extend into the medulla. These structures are called **nephrons**. It is in the nephrons that the real work of the kidney occurs. There are about a million nephrons in each kidney. So you have to jump in even closer for a microscopic view of an individual nephron.

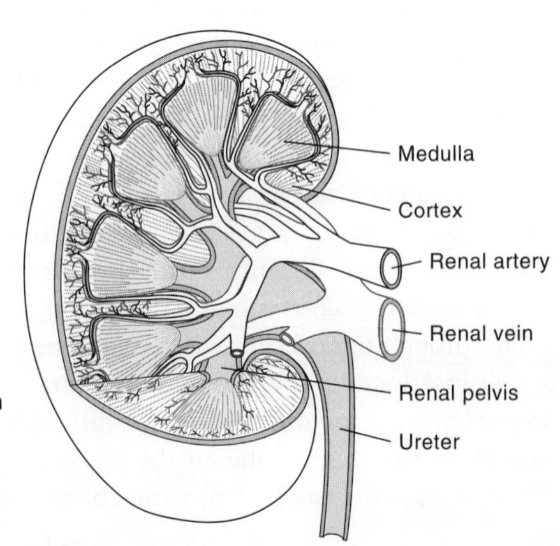

Figure 13-7 The outer portion of the kidney consists of the medulla and cortex. The inner portion, where urine collects, is hollow.

Under a microscope, you can observe the structure of a nephron. It is a tube whose walls are a single cell layer thick. You can see the complex network of capillaries wrapped around each nephron. Blood is cleansed, water balance is maintained, and urine is produced through the exchange of materials between the capillaries and the nephron.

The processes of excretion and chemical regulation begin in the nephron as blood from the renal artery enters a tight bundle of capillaries called the **glomerulus**. Inside these capillaries the blood is under high pressure. Filtration of the blood occurs here. Twenty percent of the liquid material of the blood, including the chemicals in the fluid, is forced out of the glomerular capillaries into a cuplike structure of the nephron that surrounds them. Almost all that remains behind in the capillaries are the relatively large blood cells and protein molecules. The cuplike structure is called **Bowman's capsule**. It is named after Sir William Bowman (1816–1892), an English physician who gained fame because of his work on how the kidneys function. (See Figure 13-8.)

Is this filtration sufficient to clean the blood and return it to the body? Definitely not, for two important reasons. First, so much fluid is forced out of the blood here that we would have to drink 7.5 liters of liquid every hour—180 liters per day—to replace the lost liquid. We would do little else than drink and urinate all day! Second, the process of filtration does not sort materials. Harmful wastes *and* valuable nutrients alike are forced out. Amino acids, sugar, salts, vitamins, and other nutrients are removed from the blood at this point. Therefore, the process of filtration alone is not enough to maintain homeostasis.

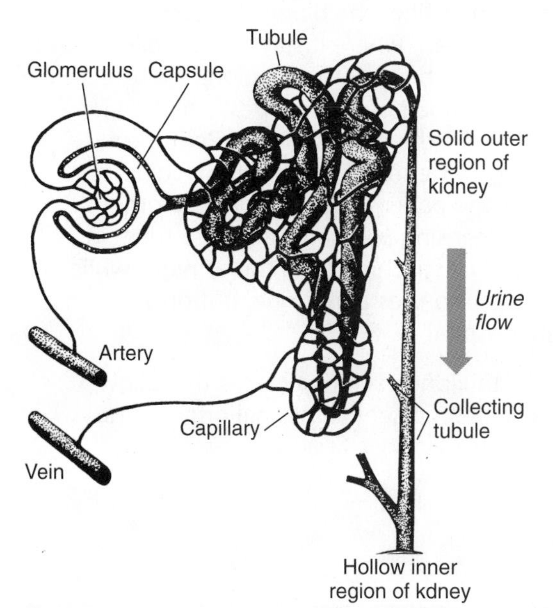

Figure 13-8 The structure of the nephron.

Living with Kidney Failure

Several years ago, a young woman named Jeanette fell and ruptured her patella, the knee bone. Surgery was planned to repair it, but this was abruptly cancelled when blood tests showed that Jeanette was in kidney failure. A few decades ago, this would have meant certain death. Now, however, Jeanette had options. The one she chose was at-home hemodialysis.

Kidney failure, now called *ESRD* (*e*nd-*s*tage *r*enal *d*isease), leads to a buildup of wastes in the blood, higher concentrations of salt and water, and unbalanced pH levels. Homeostasis is seriously disrupted. Without kidney function, life cannot continue. Fortunately, however, treatments have been developed to artificially cleanse the blood. The treatments involve the exchange of substances across membranes in a process called *dialysis*. In the 1960s, when the first artificial kidney was developed, a patient had to be connected to a dialysis machine for 8 to 10 hours at a time, several times a week. There were also serious side effects from the treatment, such as severe infections and inflammation around the heart.

The many options that Jeanette and other ESRD patients today can choose from are remarkable. The two basic choices are hemodialysis and peritoneal dialysis. With hemodialysis, the patient's blood passes through tubes into a machine next to him/her, where materials are exchanged between the blood and fluids in the machine. The patient has the option to undergo hemodialysis at home, like Jeanette did, or in the hospital, and at different times during the week. Some people have dialysis treatment every day, others two or three times a week.

The other kind of treatment, peritoneal dialysis, amazingly uses the lining of the patient's own abdomen to filter the blood. A cleansing solution is transported into the abdomen, where it stays for a while. As the person's blood passes in capillaries through the solution, the blood gets cleansed. Then the solution is drained out, and a fresh solution is put into the patient's abdomen to start the process again. Peritoneal dialysis can be done in several ways. One method uses no machine; it just goes on continuously, even as the patient walks around. Another way uses a machine to move the cleansing solution through a tube, in and out of the person's abdomen. The machine does this at night while the person sleeps. Another method also uses a machine, but only at certain times, and usually in the hospital.

So there are choices for people with kidney failure, choices that allow those who no longer have the use of an essential internal organ—the kidney—to go on living.

The next stage, reabsorption, occurs as the **filtrate** passes through the nephron tubule. By the process of active transport, glucose, salts, and other valuable nutrients are moved back from the filtrate into the blood in the capillaries. As these solutes move out of the filtrate, the concentration of water inside the nephron tubule becomes higher than outside it. As a result, by osmosis, water automatically moves out of the filtrate. In fact, almost 99 percent of the liquid in the filtrate is reabsorbed back into the blood. As a result, the quantity of urine we release is usually between one and two liters a day.

The final stage of kidney function occurs in the last portion of the nephron tubule. Secretion of additional wastes from the blood, including urea, occurs here. This process of secretion is one of the final steps in maintaining homeostasis of the fluids in the body. Regulating the pH of the blood is one more way the kidneys maintain homeostasis. In this case, if the blood becomes too acidic, hydrogen ions are removed from the blood. The reverse occurs if the blood is too basic.

The result of all this activity in the kidneys is that the blood returns through the renal vein to the body, newly cleansed of wastes and properly balanced with salts and water. In addition, urine that contains wastes, as well as any excess salts and water, is collected and removed from the body.

REGULATION OF THE EXCRETORY SYSTEM

The purpose of the excretory system is to maintain homeostasis by regulating the chemical composition of the blood and in turn of all the body's cells. At times, both the endocrine system and the nervous system help to carry out this extremely important job.

Perhaps, on occasion, you have eaten an entire bag of salty potato chips. As we have said, the salt in that one bag, if it all got into your blood, could be dangerous. However, all the salt in the bag of potato chips does not get into your blood. Why? As soon as the sodium level in your blood starts to rise, the adrenal glands—the endocrine gland located on top of each kidney—respond by secreting less of the hormone **aldosterone**. This hormone, when it reaches the kidney, causes nephrons to reabsorb salt from the urine into the blood. Smaller amounts of the hormone result in less reabsorption of salt. Therefore, after you have eaten the salty chips, more salt is excreted in your urine. (See Figure 13-9 on page 286.)

Regulation of the level of water in the body is very important. Consider when you have been exercising a great deal, sweating a lot, and drinking little to replace your water loss. Your body has been losing water. The result is that the volume of your blood decreases, since it consists

Figure 13-9 The adrenal glands help regulate the amount of salt in our blood: Sodium level rises; adrenal glands secrete less aldosterone; kidneys reabsorb less salt from urine into blood; more salt is excreted in urine; homeostasis is maintained.

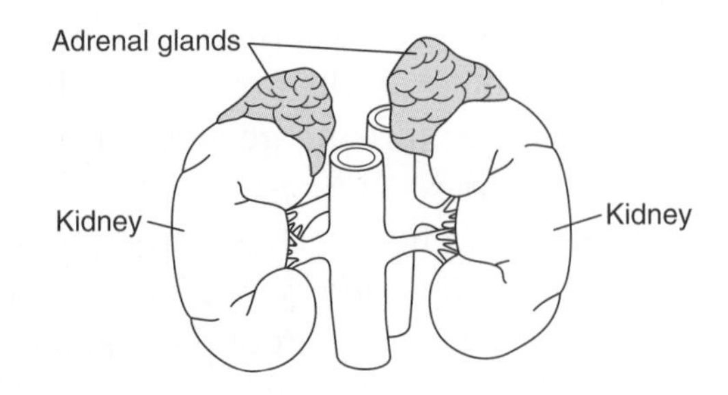

mostly of water. It is the blood and not your cells that loses the water. Our bodies "know" that it is absolutely necessary to keep the right amount of water inside our cells in the intracellular compartment.

However, once again the lower volume of the blood acts as a stimulus to the endocrine and nervous systems. This time the hypothalamus in the brain detects the lower blood volume. It directs the posterior pituitary gland to secrete **antidiuretic hormone** (ADH). When ADH reaches the kidneys, it increases the reabsorption of water from the urine back into the blood. The blood volume returns to normal. Meanwhile, solutes, including urea and urobilin—a yellow-colored waste—become more concentrated in the urine. The urine becomes darker. The darker color of the urine shows that the body is automatically readjusting itself by not releasing as much water.

Why do we feel thirsty? To be aware of this feeling, the cerebral cortex of the brain must be involved. It is the hypothalamus once again that does this job. In addition to releasing ADH when blood volume is low, the hypothalamus sends a message to the cerebral cortex. This message is the feeling you know as thirst. So you drink more water and, once again, homeostasis is maintained.

HOMEOSTASIS AND THE SKIN

The skin is the largest organ of the body. It is made up of a variety of different types of cells and tissues. (See Figure 13-10.) There are two separate layers in the skin: the outer layer, or **epidermis**, and the inner layer, or **dermis**. If we examine the structure of these layers, we can see how they function on behalf of homeostasis. It should be no surprise that the organ that surrounds the body has important roles in keeping the conditions that exist inside us fairly constant.

The epidermis is a very thin layer. New skin cells are formed at the bottom of this layer by cell division. These cells eventually push their way toward the surface, replacing the dead skin cells there. This process occurs

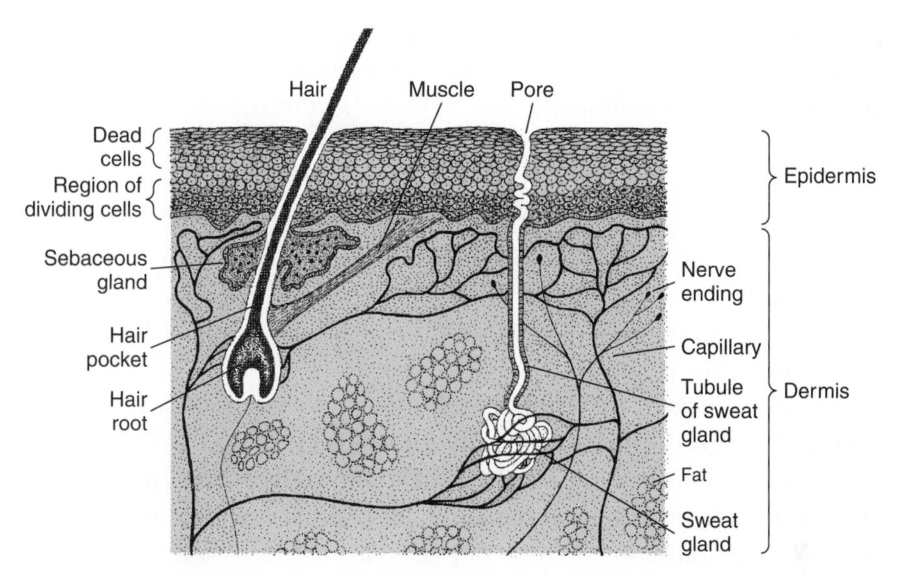

Figure 13-10 The skin, the largest organ of the body, helps us maintain homeostasis.

all the time. We constantly leave behind a trail of dead skin cells wherever we have been. These epidermal cells are very tough. They form a protective layer that prevents bacteria and harmful chemicals from entering, and water from leaving, our body. Beneath this layer are cells that produce the pigments that give skin its color.

The dermis is a thick, complex layer beneath the epidermis that contains a variety of structures. There are sensory nerve endings that can detect temperature and touch; hairs; and capillaries that help regulate body temperature (as discussed in Chapter 10). There also are muscles attached at the bottom ends of hairs. These muscles are able to cause the hairs to straighten, a response to fear in some animals. Some animals also conserve heat by causing their hairs to stand up. This increases the amount of air between the hairs, which insulates the body against heat loss. Early humans evolved in warm, tropical climates, where it was a disadvantage to be covered with hair. Thus, less body hair was a trait that was selected for, and humans lost the fur that most other mammals have. When these little muscles in our skin contract, the result is simply what we call goose bumps. In the dermis, there also are **sebaceous glands**, which keep skin and hair soft by releasing oils. Finally, there are **sweat glands**. Perspiration from sweat glands contains some urea, so they can be considered excretory structures. The main role of perspiration, however, is to assist in regulating temperature by cooling the body through evaporation.

Beneath the dermis is a layer of fatty tissue that many people wish were missing. However, fatty tissue contributes to homeostasis by acting as an insulator. It prevents the loss of body heat to the environment.

HOMEOSTASIS AND THE LIVER

No theme on homeostasis and excretion would be complete without talking about the liver. The liver is the largest organ in the abdominal cavity. It is involved in homeostasis by assisting most of the important systems of the body. (See Figure 13-11.)

The liver helps in excretion by removing nitrogen from waste amino acids and turning it into urea. The liver adds this urea to the blood. When the blood reaches the kidneys, this urea is removed in the nephrons. Red blood cells live for about 90 days. The liver breaks down old red blood cells. The components of these old cells are recycled to make new red blood cells. Chemical poisons are also made harmless by the liver. This is called **detoxification**. For example, alcohol—considered a poison by the body—is detoxified in the liver. Drinking excess alcohol makes the liver work very hard. Continued heavy drinking can, in time, cause liver disease.

A clear example of homeostasis is the need to keep blood sugar levels constant. The liver stores excess glucose in the form of the polysaccharide glycogen. The liver releases stored glycogen into the blood when glucose is needed.

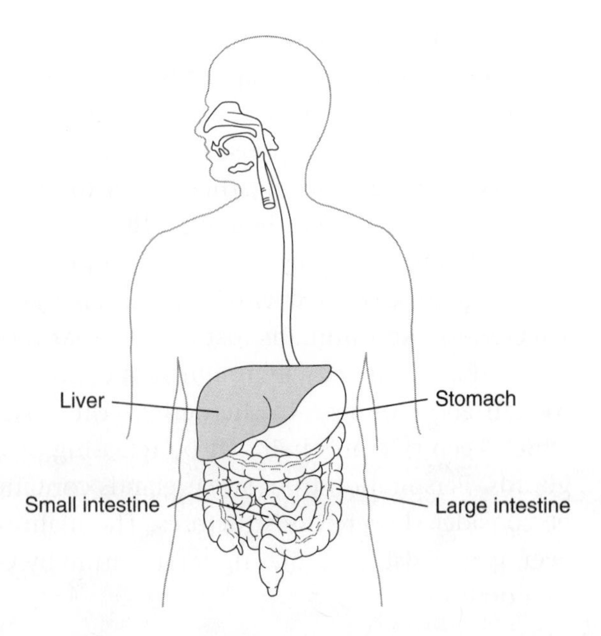

Figure 13-11 The liver turns amino acid wastes into urea, breaks down old red blood cells, and detoxifies chemical poisons such as alcohol.

The list of jobs of this 1.4-kilogram chemical factory, the liver, goes on. In Chapter 8, we saw how the liver helps the digestive process by making bile, the substance used by the body to digest fats. The liver packages fats for transport in the blood, controls the level of cholesterol in the blood, and stores vitamins A, D, and E. Finally, the liver helps maintain water balance between the blood and the intracellular fluid by making plasma proteins, important solutes in the blood.

WHEN THINGS GO WRONG: DISEASES OF THE EXCRETORY SYSTEM

One of the most common diseases of the kidney is nephritis, inflammation of the kidney. One way this disease can be detected is by a microscope examination of the urine. Cells and chemicals that normally should not be in the urine are indicators of the disease. Nephritis occurs most often in children and adolescents. Nephritis is caused by a bacterial infection and can usually be treated with antibiotics. Occasionally, however, nephritis lasts a long time and can lead to kidney failure.

Kidney stones occur when a chemical compound that contains the mineral calcium builds up in the kidney. In some cases, an improper diet leads to kidney stones. However, in most cases, the cause of kidney stones is unknown. A kidney stone can produce a great deal of pain when it travels through the ureter. Once the kidney stone is in the urinary bladder, it often leaves the body unnoticed. If the stone is too large to be passed, surgery to remove it, or the use of shock waves to break it into smaller pieces, may be necessary.

High levels of the nitrogenous waste uric acid in the blood causes gout, which occurs more often in men. The uric acid forms sharp crystals that produce severe pain in joints, usually in the big toe and sometimes in the anklebones. Treatment of gout includes drinking a lot of water, eating a special diet, and sometimes taking medication to lower the level of uric acid in the blood.

As was mentioned, liver disease can occur when too much alcohol is consumed. The liver needs to work extra hard to metabolize alcohol. Some by-products of this breakdown of alcohol, such as fats, accumulate in the liver. This causes scarring of the liver cells, called **cirrhosis**. Cirrhosis causes more of the liver cells to stop working. As the liver shuts down, homeostasis can no longer be maintained. Cirrhosis of the liver is the ninth leading cause of death in the United States.

LABORATORY INVESTIGATION 13
How Can We Test for the Presence of Certain Substances in "Urine"?

INTRODUCTION

The wastes produced by our cells are filtered from the blood in the kidneys. Then these wastes are eliminated from the body in urine. The kidneys also reabsorb many materials that must be returned to the blood. The kidneys are excretory organs that carry out the process of homeostasis in the body. During medical examinations, physicians often test urine for the presence of substances that may indicate a disease. Urine is also tested to determine if a person has used an illegal or controlled substance. The chemical testing of urine is called urinalysis. In this investigation, you will analyze mystery substances to determine what they contain.

MATERIALS

Artificial "urine" samples, test tubes, test-tube holder, test-tube rack, heat-resistant 500-mL beaker, tap water, hot plate, graduated cylinder, 10% acetic acid, Benedict solution, silver nitrate, biuret solution, safety goggles

PROCEDURE

CAUTION: *Wear safety goggles during this laboratory investigation.*

1. Your teacher will give you eight "urine" samples. Label eight clean test tubes the same way these samples are labeled. Place the labeled test tubes in your rack.
2. Prepare a hot-water bath by filling the beaker half full with tap water and placing it on the hot plate. Turn on the hot plate.
3. *Test for phosphates:*
   Place 10 mL of each of the two phosphate-urine samples into your two appropriately labeled test tubes. Look at the top of the liquid in the tube as you add five drops of acetic acid. Repeat for the other sample. Record your observations.
4. *Test for glucose:*
   Place 5 mL of each of the two glucose-urine samples into your appropriately labeled test tubes. Add five drops of Benedict solution to each test tube. Place both tubes in the hot-water bath for two minutes. Observe and record any color changes.

5. *Test for chlorides:*
   Place 5 mL of each of the two chloride-urine samples into your appropriately labeled test tubes. Carefully add three drops of silver nitrate to each test tube. Observe the surface of the solution in each tube and record any changes you observe.
6. *Test for albumin (protein):*
   Place 10 mL of each of the two albumin-urine samples into your appropriately labeled test tubes. Add five drops of biuret solution to each test tube. Observe and record the results.
7. Make a data table of your results.
8. Obtain an unknown "urine" sample from your teacher. The unknown samples will vary; some may test positive for one or more substances, some may test negative for one or more substances. Design and carry out an investigation to determine the substances present in your sample.

INTERPRETIVE QUESTIONS

1. Human blood contains molecules of glucose and albumin, along with phosphates and chlorides. Urine normally contains phosphate and chloride ions. What does this information indicate about the functions of the kidneys?
2. Why may the presence of glucose in the urine be a cause of concern?
3. What may be indicated by the presence of albumin in the urine?

CHAPTER 13 REVIEW

Answer these questions on a separate sheet of paper.

VOCABULARY

The following list contains all of the boldfaced terms in this chapter. Define each of these terms in your own words.

aldosterone, ammonia, antidiuretic hormone, Bowman's capsule, cirrhosis, contractile vacuole, dermis, detoxification, epidermis, excretion, filtrate, glomerulus, kidney, nephridium, nephrons, nitrogenous wastes, osmoregulation, sebaceous glands, sweat glands, urea, ureter, urethra, uric acid, urinary bladder, urine

PART A—MULTIPLE CHOICE

Choose the response that best completes the sentence or answers the question.

1. The organ in vertebrates that is primarily responsible for regulating the chemical composition of plasma is the *a.* lymphatic system *b.* liver *c.* kidney *d.* hypothalamus.
2. Which of these is *not* a solute in blood? *a.* water *b.* salt *c.* urea *d.* carbon dioxide
3. Which of these is *not* one of the water compartments in the body? *a.* lymph *b.* intracellular fluid *c.* plasma *d.* urinary bladder
4. The process of maintaining water balance is called *a.* thermoregulation *b.* osmoregulation *c.* excretion *d.* metabolism.
5. Excretion is best described as the process of *a.* perspiring *b.* ridding the body of metabolic wastes *c.* regulating the water balance in the body *d.* producing urine.
6. An ameba produces nitrogenous wastes in the form of *a.* uric acid *b.* urea *c.* ammonia *d.* carbon dioxide.
7. The excretory structures in earthworms are *a.* nephridia *b.* kidneys *c.* urinary bladders *d.* contractile vacuoles.
8. The nitrogenous waste that is least soluble in water is *a.* uric acid *b.* urea *c.* ammonia *d.* carbon dioxide.
9. Gout occurs when *a.* too much alcohol is consumed, causing liver damage *b.* uric acid crystals form in the joints *c.* calcium compounds build up in the kidneys *d.* bacteria infect the kidneys, causing inflammation.
10. When a paramecium (a one-celled organism) is put into distilled water, structures at its "front" and "tail" ends start to pulsate

rapidly. These structures are probably *a.* lysosomes *b.* hearts *c.* sodium pumps *d.* contractile vacuoles.

11. The functional units of the human kidneys are the *a.* nephridia *b.* nephrons *c.* glomeruli *d.* renal medulla.
12. The thin, outermost layer of the skin is the *a.* endoderm *b.* dermis *c.* epidermis *d.* cortex.
13. Overactive sebaceous glands cause *a.* acne *b.* gout *c.* kidney failure *d.* gallstones.
14. Nephritis is *a.* an inflammation of the kidneys *b.* a scarring of liver cells *c.* an excretory organ in a crayfish *d.* a type of nitrogenous waste.
15. The dermis is involved with *a.* detecting temperature and pressure *b.* causing goose bumps in response to cold or fear *c.* regulating body temperature *d.* all of these.

PART B—CONSTRUCTED RESPONSE

Use the information in the chapter to respond to these items.

16. Identify the structures labeled *A* through *L* in the diagram.

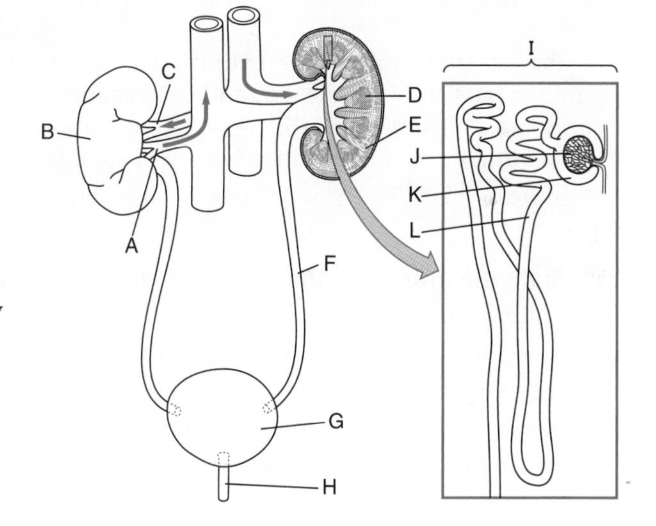

17. Where does the process of reabsorption take place? Why is this process important to maintaining homeostasis?
18. How do aldosterone and ADH contribute to homeostasis?
19. Compare the three types of nitrogenous wastes in animals. How might the excretion of uric acid be an adaptation to life on dry land?
20. Explain how the liver helps to maintain homeostasis.

PART C—READING COMPREHENSION

Base your answers to questions 21 through 23 on the information below and on your knowledge of biology. Source: *Science News* (March 1, 2003): vol. 163, p. 142.

Good Taste in Men Linked to Colon Risks

Men with exceptionally good taste may pay for it in health risks.

About 25 percent of people have extra taste buds on their tongues. They live in "a neon taste world" instead of a "pastel" one, as Yale University researcher Linda Bartoshuk puts it (*SN:* 7/5/97, p. 24).

There may be a nasty consequence to the benefit. Among men over 65, intense tasters have significantly more colon polyps than other tasters, according to Bartoshuk and Marc Basson of Wayne State University of Detroit. Extra polyps suggest an extra risk of colon cancer.

Bartoshuk speculates that sensitive tongues lead these men astray in food choices. Supertasters often cringe at intense vegetable flavors, and the supertasting seniors eat fewer vegetables than do their counterparts with normal taste sensitivity. The supertasters also tended to weigh more. Low-vegetable diets and extra weight both raise the risk of colon cancer.

Ear infections may exacerbate this cancer risk. The nerves from the tongue pass through the ear, and ear infections distort neural mechanisms so that the tongue increases its sensitivity to fat. Bartoshuk has found that among men with a history of ear infections, supertasters are especially likely to be very overweight.

21. State the difference that has been discovered among men in their ability to taste foods.
22. State the health risk associated with the men that have increased taste sensitivity.
23. Describe two hypotheses that might explain the health risk faced by the men with sensitive tongues.

14

Disease and Immunity, Wellness and Fitness

After you have finished reading this chapter, you should be able to:

Compare and contrast the various factors that cause disease.

Identify the body's defenses against disease-causing organisms.

Describe the structures and functions of the human immune system.

And it is well to superintend the sick to make them well, to care for the healthy to keep them well, also to care for one's own self, so as to observe what is seemly.

Hippocrates, *Precepts*

Introduction

In the past, many people believed that evil spirits were the cause of human illness. If someone became ill, the treatment prescribed was often horrific. The patient might be beaten, tortured, or starved. One technique used in some cultures involved drilling a hole in the ill person's skull to allow the evil spirit to leave. If a sick person didn't die from the disease, he or she might die from the "cure." (See Figure 14-1 on page 296.)

How different is our understanding of disease today! We have achieved this understanding from the careful thinking, experiments, and observations of many people over a long period of time. Twenty-four hundred years ago, the physician and teacher Hippocrates lived on the small Greek island of Kos. Through his work, Hippocrates did a great deal to move medicine away from superstitions. For this, Hippocrates has been called the "father of medicine." An article, "Airs, Waters, and Places," written by Hippocrates—or one of his students—discussed how disease, rather than coming from the gods, may have been related to the weather, drinking water, and winds in the town. The idea that diseases have understandable

Figure 14-1 In an ancient medical procedure, the Incas carved a hole in the patient's head, attempting to cure an illness.

causes was becoming part of medical knowledge. It followed logically that once the causes of diseases were known, sensible treatments could be offered. The need to understand the causes of disease and to discover successful treatments for diseases continues.

DISEASE: A LACK OF HOMEOSTASIS

Homeostasis, the theme we have been studying, emphasizes the need for organisms to maintain a carefully controlled internal set of conditions, a dynamic equilibrium. Maintaining these conditions—including pH, temperature, water and salt balance, and levels of CO_2 and O_2—allows an organism's cells to function normally. Living organisms allow changes within very definite limits to occur. Changes outside normal limits disrupt homeostasis, producing illness, disease, and possibly even death.

There are many reasons why the body can be pushed beyond its normal limits. These reasons, or factors, are often the causes of disease, causes that ancient peoples did not understand. An inherited defect in a genetic trait might be a cause of disease. The disruptions of homeostasis in such a disease would be caused, in a sense, by a factor inside the body. Many other diseases result from some influence outside the body, in the environment.

FACTORS THAT CAUSE DISEASES

Diseases may be caused by one of the following factors, or by a combination of several of these factors.

- **Inheritance.** Defective genetic traits can be passed from parents to offspring. Often the parents may not have the disease, but both may carry a single allele (the form of a gene) for the disease. It is the combination of these two defective alleles in the child that gives him or her the disease.

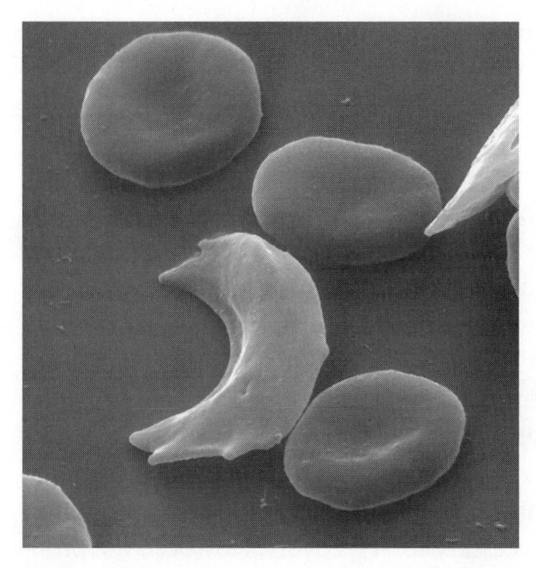

Figure 14-2 This scanning electron micrograph of red blood cells (magnified more than 7000 times) shows the distorted shape of a cell with flawed hemoglobin. The cause, sickle-cell anemia, is an inherited disease that disrupts homeostasis by obstructing blood flow in capillaries.

A well-known example of an inherited disease is sickle-cell anemia. In this condition, hemoglobin, the protein that carries oxygen in red blood cells, is flawed. As a result, the red blood cells may, at times, get twisted out of shape and resemble a sickle, a crescent-shaped tool used to cut grass. (See Figure 14-2.) In the human body, homeostasis is disrupted when the sickle cells obstruct capillaries. The normal movement of blood is interrupted. In addition, the flawed hemoglobin is unable to carry as much oxygen as is normal. At present, sickle-cell anemia cannot be cured. However, problems that arise from the disease are successfully treated.

- **Microorganisms**. Microorganisms that cause disease are called **pathogens**; they include certain fungi, bacteria, protozoa, and viruses. Some diseases caused by microorganisms may be passed, in a variety of ways, from one person to another. These are called **infectious diseases**. (See Figure 14-3.) Microorganisms, or microbes, most often enter the body through respiratory pathways, the digestive system, or the urethra. Infections may also occur through breaks in the skin. Some

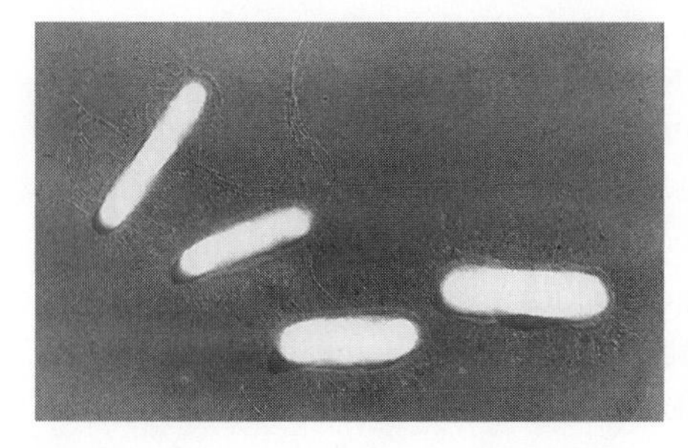

Figure 14-3 Certain microorganisms, such as these rod-shaped bacteria, can cause infectious diseases.

diseases are more easily transmitted from one person to another than other diseases.

Tuberculosis is a disease caused by an infectious microorganism, a rod-shaped bacterium. The bacteria that cause tuberculosis travel in the tiny water droplets released into the air when an infected person coughs or sneezes. The bacteria enter the body of another person through the nose or mouth. Once inside the body, the tuberculosis bacteria can infect any tissue, but most often they infect the tissues of the lungs. A person may carry the bacteria in his or her body and not show symptoms of the disease. However, tests can determine if a person has been exposed to the bacteria that cause tuberculosis, even if the person shows no symptoms of this disease. A person with a positive test for the tuberculosis bacteria, or a person who has the disease, is treated with antibiotics. Along with antibiotics, rest and time are necessary for a recovery. However, the problem of antibiotic resistance (discussed in Chapter 1) applies to tuberculosis, too. Some types of tuberculosis have developed that are resistant to the usual course of antibiotics. As a result, new drugs must be produced all the time.

- **Pollutants and poisons.** Chemical agents present in the environment may upset the body's normal functioning and produce disease. These pollutants include coal dust, asbestos, lead, phosphorus, mercury, polychlorinated biphenyls (PCBs), and many others. For example, when asbestos fibers enter the respiratory system, they cause asbestosis, a disease of the lungs. Years later, cancer in the lungs and chest may result from the inhalation of asbestos fibers. Poisons such as the chemical element arsenic and the toxin from the bacterium *Clostridium botulinum* are extremely toxic. They quickly cause death. Amazingly, however, both—in very tiny doses or in different forms—are now being used to treat and cure many diseases and ailments. Botox is helping with some 40 ailments, such as cerebral palsy and Parkinson's, as well as troubles like migraines, excessive sweating, and facial wrinkles.
- **Organ malfunction.** A disease may develop when one or more of the body's organs malfunction. When an organ such as the liver, lung, heart, stomach, or kidney does not function properly, serious effects on the body result. As you saw in Chapter 13, problems in the nephrons of the kidneys may lead to kidney failure. This causes the disease uremia. Waste products that should have been removed by the kidneys begin to build up in the blood.

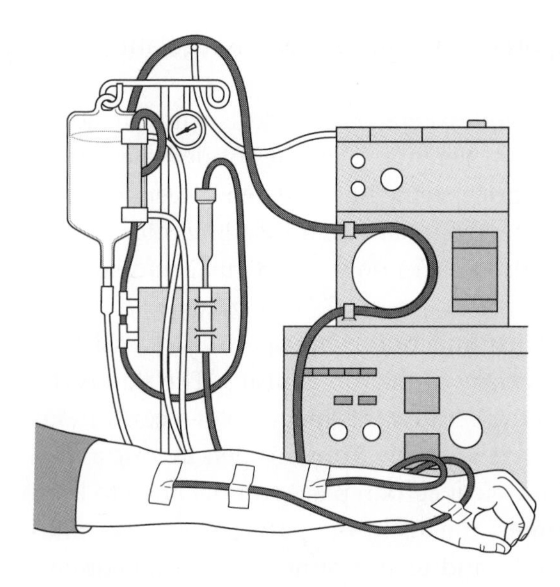

Figure 14-4 Dialysis machines can take the place of kidneys, but only temporarily.

Today, technology offers hope to people with kidney failure. **Dialysis**, featured in Chapter 13, is a process in which a person's blood is pumped through an artificial kidney to be cleansed, removing the wastes from the blood. (See Figure 14-4.) But dialysis provides only temporary help. Kidney transplant operations offer a more permanent treatment.

- **Harmful lifestyles**. The way one lives can also be an important factor in causing disease. Specifically, tobacco, alcohol, and drugs in the body can disrupt homeostasis, producing illness. In addition, overeating, not exercising, having unsafe sexual experiences, and living with stress can lead to certain diseases. **Hypertension**, or high blood pressure, is one such disease. Hypertension involves an increased pressure on the walls of arteries. Untreated, hypertension can lead to heart attacks, strokes, and damage to the kidneys, nervous system, and eyes. While medications are used to treat severe hypertension, a less stressful lifestyle combined with a diet that is low in sodium may help lower a person's blood pressure. Moderate exercise, on the advice of a physician, is also an important treatment for hypertension.

THE BODY'S DEFENSES AGAINST DISEASE

Our bodies are surrounded by microorganisms trying to get into us. Some of them succeed, through the nose, through cuts in our skin, or along with the food we eat. Many of these microorganisms cause serious problems if they survive and reproduce inside us without challenge. Controlling these microscopic invaders is as important to homeostasis as is regulating body temperature and chemistry.

The methods by which the body keeps out, or deals with, invading microorganisms are called lines of defense. The lines of defense can be either nonspecific or specific. A nonspecific line of defense keeps out any microorganism. It does not matter what particular invading microorganism it is. A specific line of defense attacks only a particular microorganism, one type at a time.

The first line of defense against infection is nonspecific. It consists of physical barriers that block the entry of microorganisms. The skin is the main physical barrier in our body. Because it is made up of a continuous layer of flat, tough cells, it protects the body from invaders as long as it is undamaged. Fluids are also released at certain places on the body to keep out microorganisms. For example, mucus in the passages of the respiratory system, saliva in the mouth, and tears around the eyes all contain substances that kill microorganisms. The strong acid in the stomach is very effective in killing microorganisms in the food we eat.

A second line of defense is present when microorganisms get through our physical barriers. We have all experienced a cut or scrape on the skin. In time, the injured area becomes warm, reddened, and perhaps swollen with pus. What is happening? The events described are called **inflammation**. (See Figure 14-5.) When the injury occurs, chemicals are released by the damaged tissues. In the body, these chemicals act like an alarm. They cause an increase in the blood flow to the site of the injury. Responding to the alarm, special white blood cells that can attack invaders arrive.

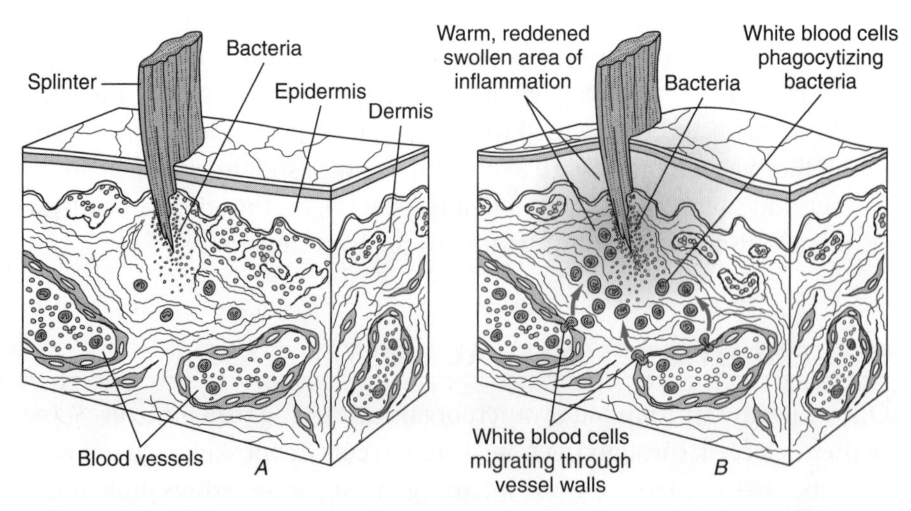

Figure 14-5 Inflammation is the body's second line of defense against disease.

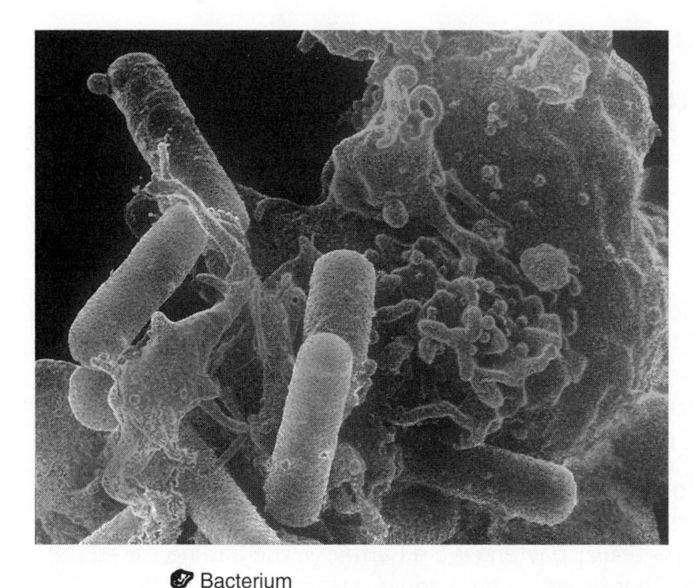

Figure 14-6 During inflammation, phagocytes engulf and destroy microorganisms, which prevents more serious infection.

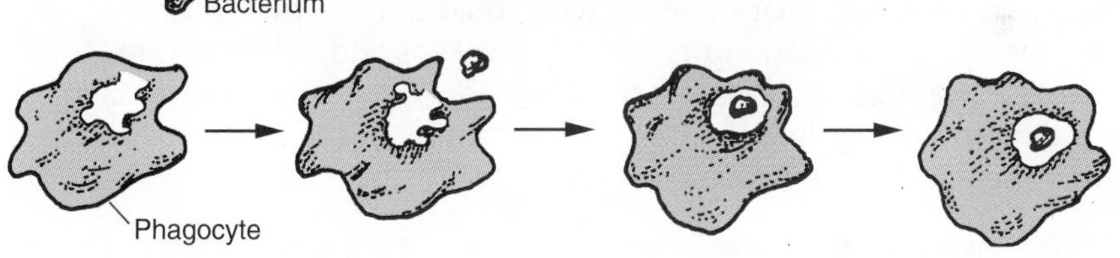

These cells are phagocytic. Like an ameba, phagocytes surround and engulf microorganisms, destroying them by ingesting them. All of this activity helps to prevent a more serious infection from developing. The redness, warmth, and pus are signs that the body is healing itself. (See Figure 14-6.)

Vertebrates have evolved a very important system that attacks specific invaders. This is the **immune system**. The immune system knows who the "bad guys" are. The immune system goes after very specific invaders to try to keep them from disrupting normal body functions.

THE IMMUNE SYSTEM

An English physician, Edward Jenner, attempted a very famous and very risky experiment in 1796. (If he were alive today, Jenner probably would have been prevented from carrying out this kind of research.) Jenner noticed that people who worked with cows did not usually contract the deadly disease smallpox. He therefore intentionally infected an eight-year-old boy with cowpox, a mild disease similar to smallpox. Jenner suspected that cowpox in some way protected people from catching smallpox. His

Antibiotics, Infections, and You

Antibiotics are used to treat infections in people and animals. Due to the enormous success of antibiotics, their use is very common worldwide. When we are ill, we have come to expect quick, effective treatment with antibiotics. Physicians often prescribe antibiotics at the earliest sign of an infection.

One result of the widespread use of these important medicines is a growing number of antibiotic-resistant strains of bacteria. Some scientists have suggested the alarming possibility of infections that will not be treatable by the antibiotics we have. Already, one disease, tuberculosis—which was largely under control—has reappeared in a strain that is much more difficult to treat with antibiotics.

Recently, scientists became alarmed when they found—in food being given to chickens—bacteria that are resistant to the most powerful antibiotics. Even though those particular bacteria were harmless, the finding raised the disturbing possibility that these bacteria could pass on their antibiotic resistance to disease-causing bacteria in chickens and, ultimately, in humans. One possible reason that such drug-resistant bacteria are being found more frequently is the heavy use of antibiotics to promote health and growth in farm animals.

This is an issue for everyone to be aware of and concerned about. Science has provided us with a group of wonder drugs to treat diseases that once killed many people. However, we must be thoughtful and wise in the use of antibiotics. The laws of nature—in this case, the process of natural selection that produces resistance to antibiotics—can never be ignored.

observation that people who worked with cows often came down with cowpox, but rarely if ever contracted smallpox, formed the basis of his hypothesis. After the boy recovered from cowpox, Jenner deliberately injected the boy with smallpox. The boy did not get sick! Did cowpox protect the boy from smallpox? How could a previous illness protect a person from getting sick again?

We now know that the immune system defends our bodies against very specific invaders. Each invader—usually a bacterium or virus—has specific protein molecules attached to its surface. Each such molecule is called an **antigen**. It is these molecules that are detected by the body's immune system.

When the immune system detects an antigen, it produces **antibodies** —the molecules that an individual produces as a defense against disease. Antibodies provide this defense by binding to the antigens. Once this occurs, the invader can be destroyed by the body. As it turns out, the cowpox and smallpox antigens are almost identical. After he was injected with cowpox, the boy's immune system made antibodies against the cowpox antigen. Later, when he was injected with smallpox, the boy's body was ready with a defense. The smallpox virus was destroyed with the help of the antibodies that the boy had made against cowpox antigens. That is why he did not get sick. (See Figure 14-7.)

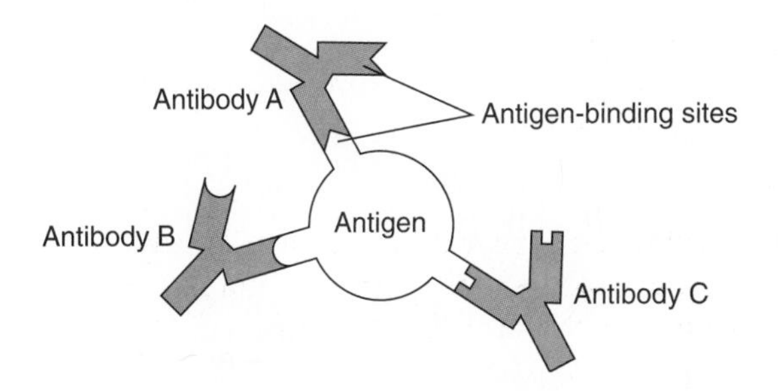

Figure 14-7 Antibodies, part of the immune system, bind to antigens—specific protein molecules on an invader's surface.

This technique came to be known as **vaccination**. *Vacca* is the Latin word for "cow." Today, vaccines offer protection against a number of diseases. People are now given harmless antigens in a vaccine, which cause the body to produce antibodies.

Check Your Understanding

How is the immune system more "specific" in its defense against infection than the defenses presented by the skin and by inflammation reactions?

B CELLS AND T CELLS

The immune system also includes B cells and T cells, actually two types of **macrophages**. These macrophages are kinds of white blood cells that are produced in bone marrow, the thymus gland, the spleen, the lymph nodes, and the tonsils. (See Figure 14-8 on page 304.) Macrophages are the cells (phagocytes) we have already mentioned, which move to infected areas to engulf and digest invading microorganisms. B cells are the ones that respond to specific antigens by beginning to produce antibody proteins that will bind only with that antigen.

As time goes on, the body contains many different types of B cells, each producing antibodies for one specific antigen. After having been invaded once by an antigen, some special B cells that recognize that antigen remain

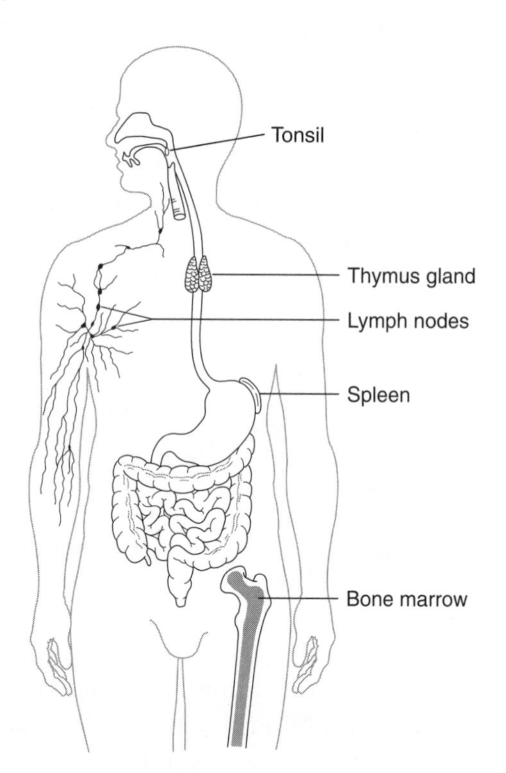

Figure 14-8 Macrophages are produced in bone marrow, the thymus gland, the spleen, the lymph nodes, and the tonsils.

in the body for the rest of your life. These are called memory B cells. Because they are already present in the body, you instantly start making antibodies the moment you encounter the same invading microorganisms again. That is why individuals usually do not get measles or chicken pox a second time. The immune system remembers the first exposure to the disease and is ready! This type of protection is called **active immunity**. (See Figure 14-9.)

Passive immunity is related to active immunity. In passive immunity, a person is injected with a large quantity of the correct already formed antibodies to a particular antigen. These antibodies protect the body from a disease only for as long as the antibodies remain in the body. However, the body is passive, uninvolved in this protection. It did not make the antibodies and, more important, does not "remember" how to make them. You therefore have protection for a limited time from passive immunity.

Antibodies are good at recognizing antigens on invaders only when the invaders are in the fluids in your body but not inside your cells. The problem is that many bacteria and all viruses quickly get inside body cells. Once inside your cells, bacteria and viruses begin to multiply, to really make you sick. To make matters worse, the antibodies cannot find the

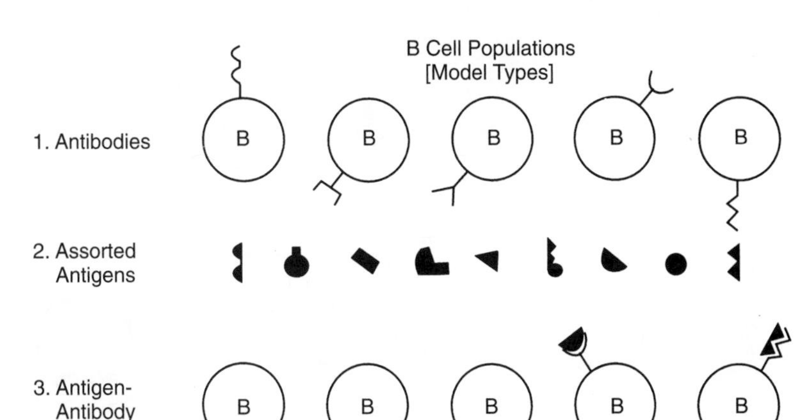

Figure 14-9 In active immunity, memory B cells in the body from a previous exposure can instantly make antibodies when they encounter the same antigen.

invaders once they are inside cells. It is for this reason that the immune system has T cells. One type of T cell is called **cytotoxic**, or **killer, T cells**. Through protein receptors on their surface, they can recognize cells in the body that have been infected with invading microorganisms. This recognition occurs when an antigen present on the surface of an infected cell binds specifically with a receptor protein projecting from the cell membrane of a T cell. Then the killer T cells punch holes in the membranes of the infected cells, sometimes injecting poison into them. The infected cells are killed and the invaders in these cells are destroyed. (See Figure 14-10.)

In addition, another important type of T cell acts as a helper. Helper T cells assist both B cells and killer T cells. Without helper T cells, the other members of the immune system family cannot do their job. Just

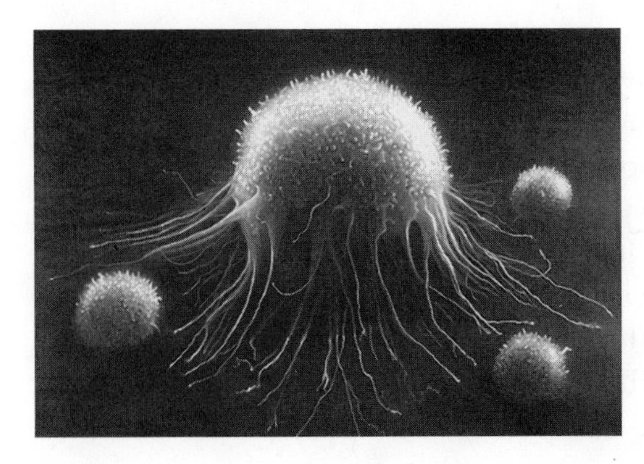

Figure 14-10 Killer T cells can recognize cells in the body that have been infected by invading microorganisms.

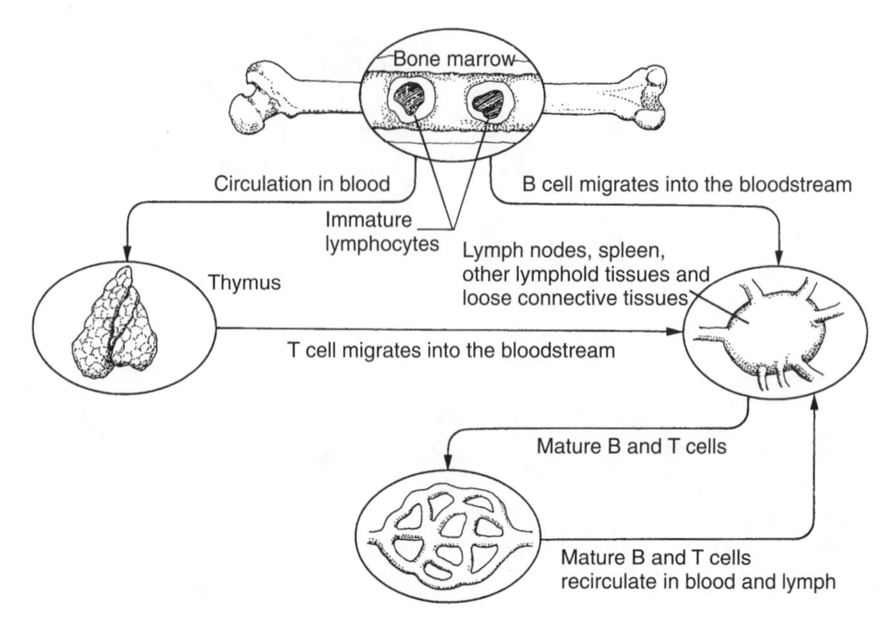

Figure 14-11 Mature B and T cells are white blood cells that circulate through the body to fight infection.

how important helper T cells are is shown by the fact that they are the cells destroyed by the human immunodeficiency virus (HIV), which results in the disease called AIDS. (See Figure 14-11.)

WHEN THINGS GO WRONG: DISEASES OF THE IMMUNE SYSTEM

The immune system helps maintain the internal dynamic equilibrium necessary for life. However, the immune system can become out of balance. It can be overactive or underactive, and in either case the body's equilibrium is upset.

Allergic reactions result from overactivity of the immune system. The body responds inappropriately to common substances such as dust, mold, pollen, or certain foods. The immune system begins making a special type of antibody to these substances, which under normal conditions would not stimulate the immune system. These antibodies cause cells in the body to release substances, including **histamines**, which cause many allergic symptoms, such as extra fluid in the nasal pathways, difficulty breathing, or inflammation (hives). Allergies are often treated with **antihistamines**, drugs that stop the release of histamine. Some severe allergic reactions may be life threatening. These reactions may need other types of medications. Allergies should be treated by a physician.

Sometimes the immune system begins to attack normal body tissues. These are called **autoimmune diseases** and are very serious. Autoimmune diseases include myasthenia gravis, rheumatic fever, lupus erythematosus, and rheumatoid arthritis. (See Figure 14-12.) In all autoimmune diseases, the body is literally rejecting its own tissues. A similar kind of rejection also often occurs when an organ is transplanted from one person to another. Medications that keep the organ recipient's immune system from attacking and rejecting the newly transplanted organ must be taken.

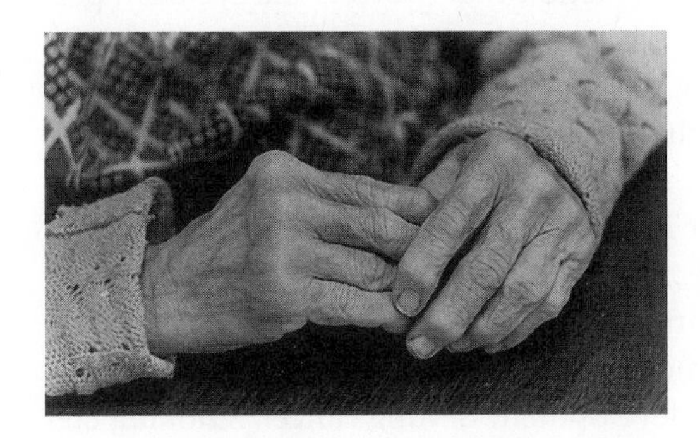

Figure 14-12 These hands have joints that are affected by rheumatoid arthritis, an autoimmune disease.

Recent studies seem to indicate that the process of inflammation, which protects us when we are young, may actually contribute to crippling diseases when we get older. For example, only three out of ten heart attacks occur in people whose arteries have narrowed. Researchers now suspect that many heart attacks are caused when a rupture develops in the wall of an artery—brought on by overactive immune system cells causing an inflammation. Sudden clotting occurs, the artery gets blocked, and a heart attack or a stroke may occur.

Disease also occurs when the immune system is underactive instead of overactive. These are called **immunodeficiency diseases**. AIDS, acquired immunodeficiency syndrome, is an immunodeficiency disease. AIDS is caused by a virus that can be transmitted from one person to another. (This disease is not inherited; that is why it is called "acquired.") AIDS develops when the human immunodeficiency virus, HIV, destroys a person's helper T cells. The body is no longer able to protect itself from diseases that may attack it.

On rare occasions, an individual is born without a functioning immune system. Such children have SCIDS, *s*evere *c*ombined *i*mmuno*d*eficiency *s*yndrome, and they have almost no defense against diseases.

Sadly, life expectancy is short for such persons, but progress is being made. Early detection of this condition, followed by bone marrow transplants, has resulted in significant improvements in some children.

Drugs that are used to kill cancer cells in the body can also interfere with the immune system. This is called depressed immunity and can cause complications in the treatment of cancer.

DISEASES AND CHOICES: REDUCING RISKS

Certain automatic behaviors, such as simple reflexes, reduce the risk of infection. An eye's blink is a reflex action that protects the eye from particles or objects that could harm it. Conscious thoughts and actions can reduce the risks of contracting certain diseases. Some of the choices we make can have important influences on our health. Factors that cause diseases may be controllable or uncontrollable. When you think about reducing the risk of disease, it is necessary to think about what risks are under your conscious control.

For example, if there is an inherited tendency in one's family to develop lung cancer, this cannot be controlled. On the other hand, risk factors can also influence the development of lung cancer. Smoking cigarettes has been identified as a risk factor that can cause lung cancer. Smoking or not smoking is a choice people make; thus, it can be controlled. The damaging effects of cigarette smoking include a much higher risk of developing not only lung cancer but also heart disease and the painful, fatal disease emphysema. What a person needs to ask is: With what is known about the risks of cigarette smoking, do the risks outweigh the benefits of the activity? Should I begin to smoke? Should I continue to smoke?

Reducing risks can also include avoiding exposure to sources of infectious microorganisms. These sources may include polluted water, contaminated food, animals, or people with contagious diseases. However, it is necessary to be informed about the risks. For example, breathing air in the same room as a person with tuberculosis over a long period of time may result in your being infected. On the other hand, sharing a room with a person infected with HIV, even shaking hands, will not make you become infected. However, contact of your body fluids with his or her body fluids, by your blood mixing with his or her blood, sharing a needle, or having sexual intercourse, may certainly cause you to become infected. Reducing the risk of contracting a particular disease is often a matter of behavior and choices.

WELLNESS AND FITNESS

What is wellness? **Wellness** is defined as a lifestyle, that is, how one lives. How one lives is a result of many factors. Some factors are outside of one's control. The inherited characteristics that a person has from birth are one such factor. Other factors are within one's control. It is mostly these controllable factors that determine how you live.

Wellness involves all the various components that make up one's life. These components are called dimensions. The dimensions of wellness include:

- emotional wellness—maintaining good mental health, a positive attitude, high self-esteem, and a strong self-image
- physical wellness—maintaining good nutrition, getting regular exercise, and getting adequate sleep
- social wellness—having positive interactions with and enjoying being with others
- intellectual wellness—having a sense of curiosity and a strong desire to learn
- environmental wellness—maintaining a way of life that protects the environment and minimizes harm to one's surroundings
- cultural wellness—being aware of one's own cultural background while respecting the diversity and richness of the cultures of others
- spiritual wellness—paying attention to personal values and beliefs

What is fitness? **Fitness** is more specifically about one's own body. Fitness has been defined as the ability to carry out normal activities while having enough energy and strength left over to meet an unusual challenge. A fit person can walk to work and then climb eight flights of stairs to the office without suffering physical discomfort. A person who is not fit runs out of breath very early in the climb up the stairs. (See Figure 14-13 on page 310.)

How does the body of a fit person differ from the body of a person who is not fit? Remember that for the body to function, it always needs matter and energy. To be fit, there must be an optimum relationship among the different tissues in the body for matter and energy to be used most efficiently. One way of comparing tissues is to look at the ratio of muscle (matter) to fat (stored energy). The amount of fat in one's body is not as important as the percentage of fat and the amount of muscle. Too little fat limits one's energy; too much fat can cause many life-threatening

Figure 14-13 Fitness is the ability to perform normal activities yet have enough energy and strength to meet an unusual challenge.

conditions, such as heart disease. The ideal body fat percentage for men ranges from 15 percent to 18 percent and for women from 20 percent to 22 percent.

How does a person develop physical wellness and physical fitness? Eating properly is most important. A person's daily diet should include the proper amount of carbohydrates, fats, proteins, vitamins, minerals, and water (see Chapter 8).

An individual requires a certain number of calories in the diet. For example, the average teenage boy needs about 3000 calories and the average teenage girl needs about 2000 calories per day. The number of calories a person needs is also affected by his or her body size, age, general level of activity, and physical health.

For physical wellness and fitness, it is also very important to exercise regularly. Exercise includes aerobic activities, which concentrate on the respiratory and circulatory systems, and strength-training activities, which improve one's muscles and skeletal system. A combination of both kinds of activities should be a normal and regular part of one's lifestyle.

Finally, adequate rest is essential to good health. Being well rested helps a person feel less stress during the day. The reduction of stress is one of the most important ways of promoting wellness.

LABORATORY INVESTIGATION 14
How Does an Earthworm Respond to a Change in Temperature?

INTRODUCTION

In order to stay healthy, and thus alive, all organisms must maintain relatively constant internal conditions, even when the environment around them changes. In this investigation, you will study the circulatory system of an earthworm as it experiences changes in the temperature of its environment. It is easy to observe the rate of blood flow in an earthworm by studying the pulsing of the dorsal blood vessel, which can be seen through the earthworm's skin.

MATERIALS

Paper towels, rectangular pan, live earthworm, clock or watch with second hand, tap water, laboratory thermometer, ice, warm-water bath

PROCEDURE

1. Place an earthworm on a paper towel lining the inside of the pan.
2. Find the dorsal blood vessel, located in the middle of the worm's dorsal surface. Observe the pulsing of the blood vessel. Count the number of beats in one minute.
3. Propose a hypothesis that explains the relationship of the pulse rate of the dorsal blood vessel to the temperature of the worm's environment.
4. Expose the worm to at least five different temperatures, from nearly freezing to warm. Keep the temperature below 45°C, however. For each temperature, mix hot and cold water, then pour a few centimeters of water into the pan. It is not important to mix the water to an exact temperature.
5. Let the worm adjust to the new temperature for at least three minutes. Then take the temperature of the water and count the number of times the earthworm's dorsal blood vessel pulses in one minute. Pour off the water and let the earthworm breathe for at least three minutes.
6. Repeat with water of a different temperature. After collecting your data, let the worm rest again.

7. Plot the data on a graph, with the independent variable (the temperature) on the *x*-axis and the dependent variable (pulses) on the *y*-axis. You may also enter data from other students and you should share your data with them. Study the graph for a pattern and, if possible, draw a best-fit curve for the data points.

INTERPRETIVE QUESTIONS

1. Why was it necessary to wait for several minutes each time before you took another pulse rate?
2. Why was it necessary to pour off the water to let the worm breathe?
3. Describe any relationship you see in the data on the maintenance of body temperature in the earthworm.
4. What are some possible sources of error in this investigation?
5. Did your data support your hypothesis?

CHAPTER 14 REVIEW

Answer these questions on a separate sheet of paper.

VOCABULARY

The following list contains all of the boldfaced terms in this chapter. Define each of these terms in your own words.

active immunity, allergic reactions, antibodies, antigen, antihistamines, autoimmune diseases, cytotoxic (killer) T cells, dialysis, fitness, histamines, hypertension, immune system, immunodeficiency diseases, infectious diseases, inflammation, macrophages, passive immunity, pathogens, vaccination, wellness

PART A—MULTIPLE CHOICE

Choose the response that best completes the sentence.

1. Diseases that are contagious, or "catching," are said to be *a.* infectious *b.* hereditary *c.* viral *d.* nonspecific.
2. Fitness includes *a.* being aware of your cultural heritage *b.* interacting positively with other people *c.* the ability to carry out normal activities and to meet unusual challenges *d.* having a healthy lifestyle.
3. Passive immunity *a.* occurs after a person has had a disease *b.* wears off after a period of time *c.* is induced by injecting antigens into the body *d.* involves the response of memory B cells.
4. An example of an inherited disease is *a.* tuberculosis *b.* AIDS *c.* chicken pox *d.* sickle-cell anemia.
5. Phagocytes help to protect the body by *a.* forming a tough physical barrier *b.* containing acids that kill microorganisms *c.* eating invading microorganisms *d.* producing antigens that bind to disease-causing microorganisms.
6. Diseases can be prevented by *a.* performing the correct religious ceremonies *b.* keeping windows tightly shut at night *c.* reducing exposure to harmful chemicals *d.* scaring the diseases out of sick people, so the diseases do not spread.
7. Dialysis is used to treat *a.* sickle-cell anemia *b.* uremia *c.* antibiotic-resistant tuberculosis *d.* severe combined immune deficiency syndrome.
8. Hypertension is often associated with *a.* a high-sodium diet *b.* asbestos *c.* inflammation *d.* histamines.

9. Thrush, a yeast infection of the mouth, is rarely seen except in infants, people with AIDS, and people who have received organ transplants. Thrush is thus a sign of *a.* autoimmunity *b.* allergies *c.* inherited disorders *d.* immune deficiency.
10. Histamines *a.* provide passive immunity *b.* cause allergic symptoms *c.* bind to antigens on microorganisms *d.* cause autoimmune responses.
11. B cells *a.* manufacture antibodies *b.* recognize cells that have been infected by invading microorganisms *c.* stimulate other macrophages to fight infection *d.* provide passive immunity.
12. Smoking increases your risk of developing *a.* cancer *b.* emphysema *c.* heart disease *d.* all of these.
13. Edward Jenner contributed to medical science by *a.* showing that diseases were caused by natural rather than supernatural forces *b.* developing a vaccine against smallpox *c.* discovering killer T cells *d.* demonstrating that juvenile diabetes is an autoimmune disease.
14. AIDS can be transmitted by *a.* sneezing *b.* shaking hands *c.* unprotected sexual intercourse *d.* all of these.
15. Because the antigens on strep bacteria are similar to those on human tissues, an untreated strep infection may result in the production of antibodies that attack the heart valves. This is an example of *a.* an immunodeficiency disease *b.* a severe allergic reaction *c.* a disease-promoting lifestyle *d.* an autoimmune disease.

PART B—CONSTRUCTED RESPONSE

Use the information in the chapter to respond to these items.

16. On a separate sheet of paper, complete the following table. What is described in the table?

DIMENSION	DESCRIPTION
	Being aware of your own cultural background and respecting the culture of others
Environmental	
	Interacting positively with others
Emotional	
Spiritual	
	Obtaining good nutrition, regular exercise, and adequate sleep
Intellectual	

17. Select three dimensions from the table and describe how you can improve them in your own life.
18. List three nonspecific defenses of the body and explain how each protects against disease.
19. Relate disease to homeostasis.
20. How does a vaccine prevent disease?

PART C—READING COMPREHENSION

Base your answers to questions 21 through 23 on the information below and on your knowledge of biology. Source: *Science News* (February 1, 2003): vol. 163, p. 78.

As Population Ages, Flu Takes Deadly Turn

The annual toll of influenza has risen dramatically since the late 1970s, according to an analysis of U.S. death statistics. One major factor is the advancing average age of the population. Another is the increasing prevalence of virulent strains of the flu virus.

Influenza is typically not a direct cause of death, but researchers at the Centers for Disease Control and Prevention in Atlanta, estimated the disease's contribution to mortality by noting seasonal fluctuations in deaths that might have resulted from underlying flu infections. Bacterial pneumonia, for example, can be a fatal consequence of severe flu.

Such calculations suggest that influenza claimed more than 68,000 lives on average during each of the last three flu seasons of the 1990s, William W. Thompson and his colleagues report in the Jan. 8 *Journal of the American Medical Association* (*JAMA*). That's well up from about 16,000 annual deaths attributable to flu during a similar period 2 decades earlier.

People over age 65 are nearly 100 times as likely to die from flu than people 5 to 50 years old are, and the efficacy of flu vaccinations wanes in older adults.

Responding to the new findings in the same issue of *JAMA*, David M. Morens of the National Institutes of Health in Bethesda, Md., urges physicians to get annual flu shots, in order to avoid transmitting the virus to patients. They should also encourage their patients, especially older ones, to get the shots, he says. "Even an imperfect vaccine, used optimally, can prevent many thousands of deaths," says Morens.

21. State two explanations for the increasing number of deaths from influenza in the United States.
22. Explain the connection between bacterial pneumonia and the flu.
23. Describe the actions that doctors are being advised to take to help limit the number of deaths from the flu.

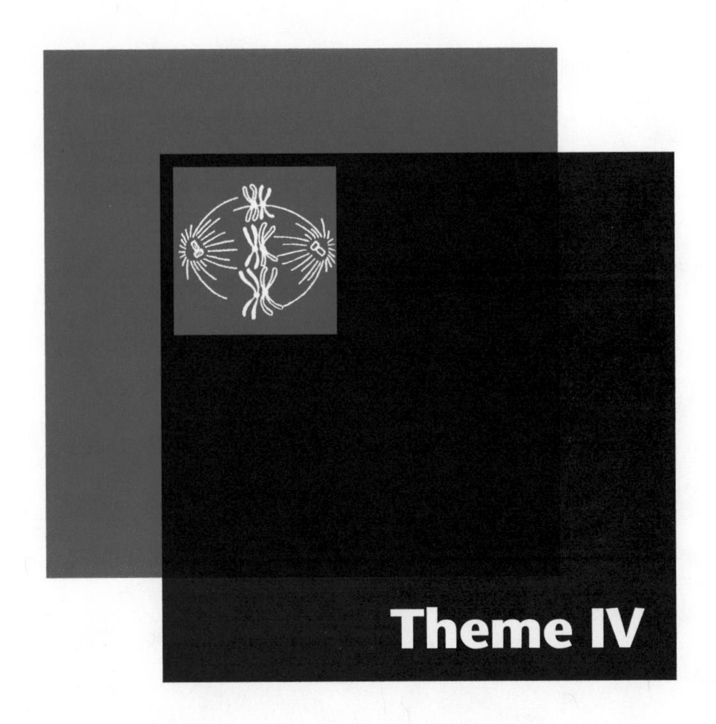

Theme IV

Reproduction, Growth, and Development

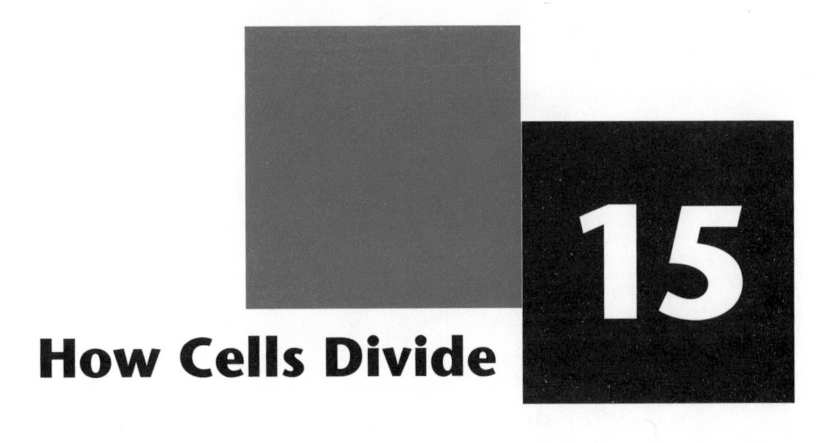

How Cells Divide

After you have finished reading this chapter, you should be able to:

Discuss the importance of cell division to reproduction, growth, and development, and the negative effect of uncontrolled cell division.

Describe the stages of mitosis and the process of cytokinesis in plant and animal cells.

Compare and contrast different methods of asexual reproduction.

Life is a whim of several trillion cells to be you for a while.

Anonymous

Introduction

Life goes on. Each of us, as an individual, plays a part in the great ongoing process of life on Earth. Each individual organism—as varied in form as a mosquito, an oak tree, or a rhinoceros—is part of this process that began 3.5 billion years ago and continues today.

We have examined what it takes for an individual to remain alive in a constantly changing environment. Now, our focus changes. Staying alive is important for every individual organism. But this is not sufficient to maintain the process of life on Earth. No individual organism lives forever. Every organism has a normal life span—the length of time between when its life begins and when it ends. (See Figure 15-1 on page 320.) So, for life to continue on Earth, individuals must reproduce new individuals.

A group of individuals of the same species living in a particular place makes up a population. Individuals may come and go, but a population continues. For example, a population of squirrels has existed in New York City parks for hundreds of years. Individual squirrels are born, live, and die. However, no squirrel that was alive 100, or even 50, years ago is alive

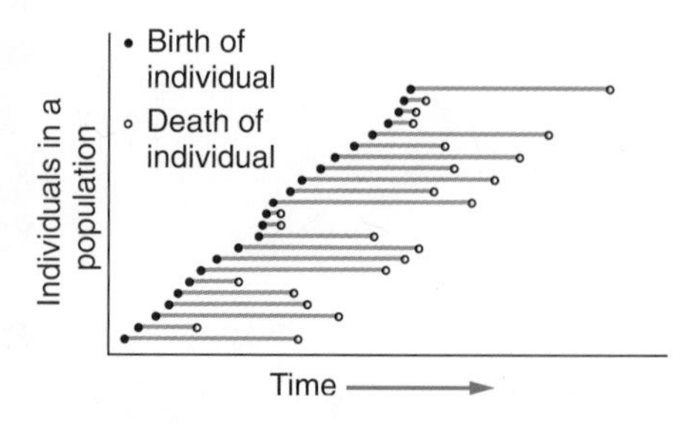

Figure 15-1 Life spans in a population.

today. The squirrel population of New York City will probably continue to exist far into the future. The continuity of life requires **reproduction**, the ability of a species to produce more of its kind. It is reproduction within populations that allows species to survive. It is reproduction that allows life on Earth to continue.

THE LIFE OF A CELL

Every cell has a life of its own. This is as true for single-celled organisms as it is for each of the billions and trillions of cells that make up the bodies of plants and animals—including ourselves. Each cell has a beginning, a period of growth, and then an ending. The series of events that occurs during the life of a cell is called the **cell cycle**. In the first stage of the cell cycle, the cell begins to grow in size. Organic materials such as amino acids and sugars and inorganic materials such as water are moved into the cell. The cell increases in size by adding these materials to itself. The cell also increases the number of its parts. For example, its mitochondria divide in two to make more mitochondria. If it is a plant cell, the same thing happens to its chloroplasts. (See Figure 15-2.)

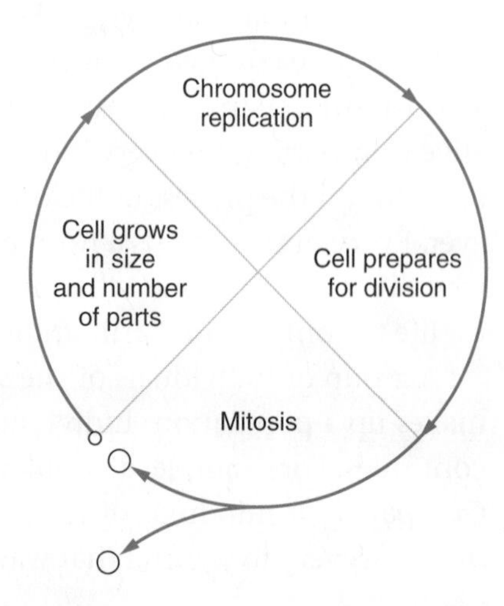

Figure 15-2 The cell cycle—the series of events in the life of a cell.

During the next period in the cycle, the cell stops getting larger. Now, the genetic material in the cell—the set of instructions received from the previous cell—duplicates. The genetic material is the building plan, similar in some ways to the set of blueprints used to build a house. The genetic material contains all the information about how the cell is to be built and how it functions. The genetic material is made up of the chemical called **DNA**, deoxyribonucleic acid. (You will study DNA in detail in Chapter 19.)

In cells that are reproducing, DNA is found in "packages" known as **chromosomes**. Bacterial cells may have a single chromosome. A goldfish has 94 chromosomes in each cell of its body. The number of chromosomes in other organisms varies. A fruit fly has eight chromosomes, a cabbage plant has 18, and a human has 46. The chromosome number is specific for each type of organism. The exact chromosome number must be maintained for the species to continue. This means that as cells reproduce, the new cells must have the same number of chromosomes as did the original cells.

During this middle stage in the life cycle of a cell, the genetic material in the cell is duplicated. This process is called **replication**. It is the most important stage in preparation for reproduction of the cell. Following this stage, some additional cell growth occurs. What is growing here is material needed for the final big event: cell division. This is the way a single cell reproduces: It divides into two cells. (See Figure 15-3.)

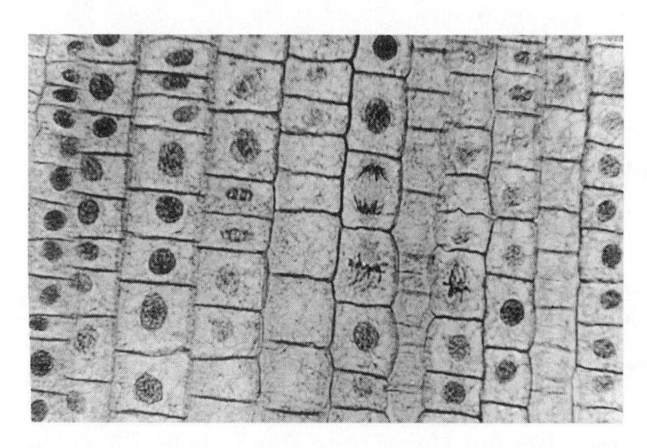

Figure 15-3 Several cells can be seen here preparing for cell division. When cells reproduce, the new cells must have the same number of chromosomes as did the original cells.

CELL DIVISION

During cell division, the genetic material must be equally divided. When a eukaryotic cell divides, it must send one copy of each of its chromosomes to each of the new cells. In addition, the cytoplasm and other cell parts must be divided between the two cells.

The division of the chromosomes occurs first. This division happens during a sequence of events called **mitosis**. Following mitosis, the cytoplasm of the cell divides. This process is called **cytokinesis**. Following cytokinesis, each of the new cells has a complete set of chromosomes, just like the original cell. The new cells are called *daughter cells*. The cell they came from, which no longer exists, is the *parent cell*.

WHY MUST CELLS DIVIDE?

Ants are small and elephants are large. Does that mean that ant cells are small and elephant cells are large? Absolutely not! Larger organisms are large because they have more cells, not bigger ones. As a baby, you had blood cells that were exactly the same size as your blood cells are now. The only difference is that now you have a lot more of them.

Why don't cells grow larger and larger? Why must cells divide rather than grow beyond a certain size? Recall how important the cell membrane is in the life of the cell. All materials that enter or leave the cell must pass through the cell membrane. Cells that are too large do not have enough cell membrane in relation to the size of their cytoplasm. The cell membrane is not large enough to permit enough materials to enter and leave the cell in order for the cell to survive. Therefore, rather than increasing beyond a certain size, all cells have built-in instructions to divide instead.

THE STAGES OF MITOSIS

Mitosis, we have said, is the sequence of events that produces and separates a cell's chromosomes into two identical sets. The preparation for mitosis is made through the replication of the chromosomes. Let's look more closely at a chromosome. After replication, a chromosome is double-stranded. Each strand of the chromosome is actually an extremely long, twisting molecule of DNA, usually combined with some proteins. The two identical DNA strands are called **chromatids**. They are joined by a structure known as the **centromere**. All the genetic information in one chromatid has been replicated in the other chromatid. (See Figure 15-4.)

Mitosis consists of several different stages. The stage of the life cycle before the actual beginning of mitosis is called **interphase**. Even though this stage is not considered part of mitosis, the cell is far from inactive. Growth of the cell and replication of the genetic material occur during interphase. Interphase is also the longest phase of a cell's life. Mitosis actually begins with **prophase**. During prophase, the long, twisted ribbons of genetic material become shorter and thicker. These are the chromosomes

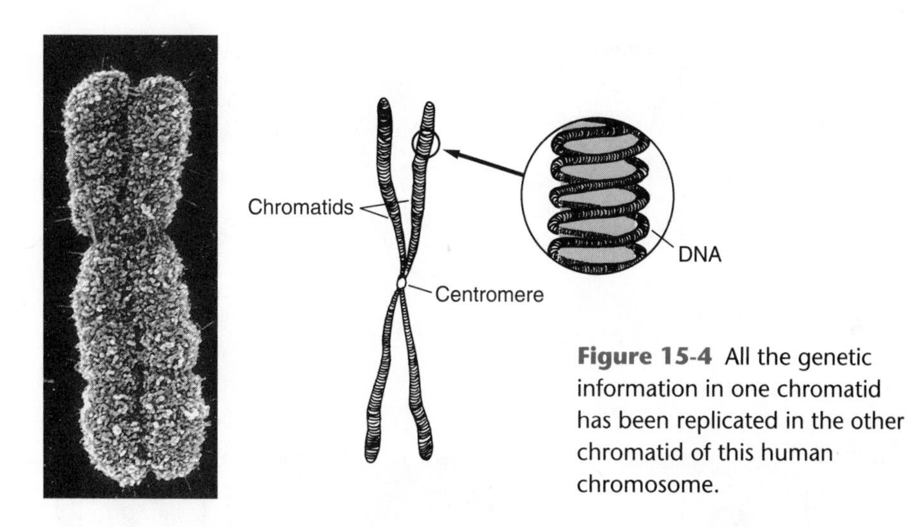

Figure 15-4 All the genetic information in one chromatid has been replicated in the other chromatid of this human chromosome.

that are visible with a microscope. In animal cells, **centrioles**, structures important in cell division, move to opposite ends of the cell. Until now, the genetic material in all eukaryotic cells has been stored inside the nuclear membrane. During prophase, the nuclear membrane breaks down. The double-stranded chromosomes are able to move freely around in the cytoplasm. (See the table below.)

The next stage of mitosis involves the movement of the chromosomes. Cells that are dividing take on a very organized appearance. Mitosis has been called the "dance of the chromosomes." This dance becomes clear if you use a microscope to observe cells as they begin mitosis. First of all, a series of fine, highly organized fibers begins to appear. The fibers stretch from each end of the cell toward the middle. These fibers, called **spindle** fibers, are made of protein. When fully arranged, the spindle is shaped like a football.

Metaphase is the next stage in mitosis. During this stage, the fibers of the spindle connect to the centromere in the middle of each double-stranded chromosome. In this manner, the chromosomes are lined up in a single file across the middle of the cell.

THE STAGES OF MITOSIS

Interphase	Prophase	Metaphase	Anaphase	Telophase
Growth of the cell and replication of its genetic material	Nuclear membrane breaks down; chromosomes become visible	Spindle fibers appear; chromosomes line up across middle of cell	Double-stranded chromosomes separate; identical chromatids move apart	Chromosomes move to opposite ends; nuclear membranes reappear; cell pinches in

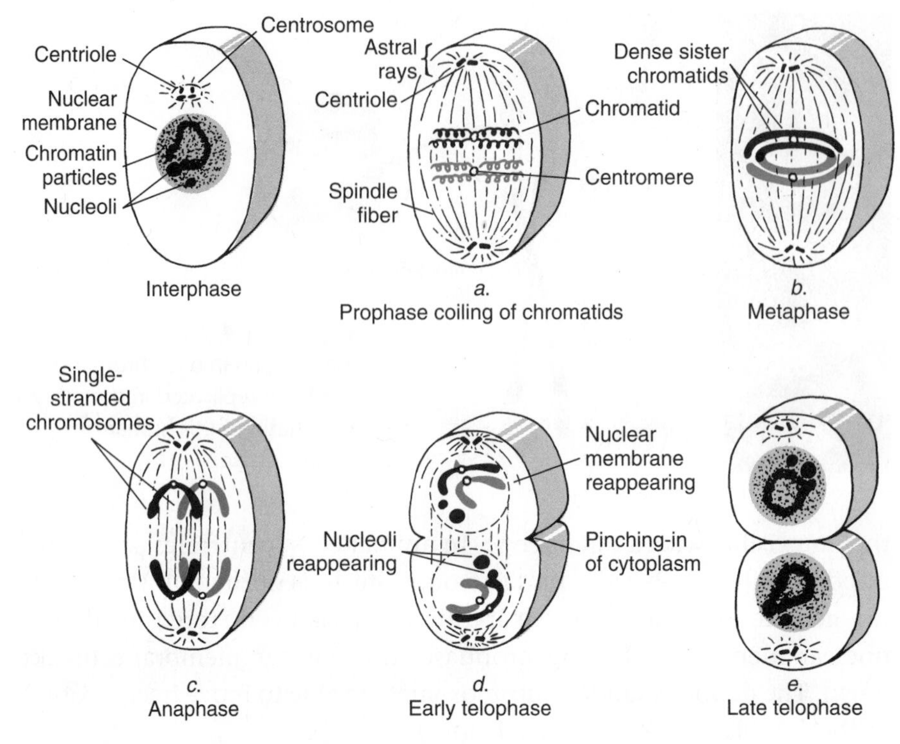

Figure 15-5 The stages of mitosis.

And then it happens! The centromeres that connect the two identical chromatids in each chromosome divide. As a result, the chromatids begin to move apart. The spindle fibers direct the movement of the identical chromatids away from each other. Each chromatid now becomes a separate single-stranded chromosome. This stage of mitosis is called **anaphase**.

The final stage of mitosis is **telophase**. The new chromosomes are pulled toward opposite ends of the dividing cell. There are now two identical, complete sets of chromosomes, one set at each end of the cell. A nuclear membrane forms around each set of chromosomes. Now, the final event in cell division begins. (See Figure 15-5.)

CYTOKINESIS: ONE CELL BECOMES TWO

For a cell to divide, it must produce two identical sets of chromosomes during mitosis. But that is not enough. To complete the division, the cytoplasm of the parent cell must be split in two. Through this process, called *cytokinesis*, two new daughter cells are produced.

The process of mitosis is essentially the same in plant and animal cells. This is one of the great pieces of evidence for evolution. The steps

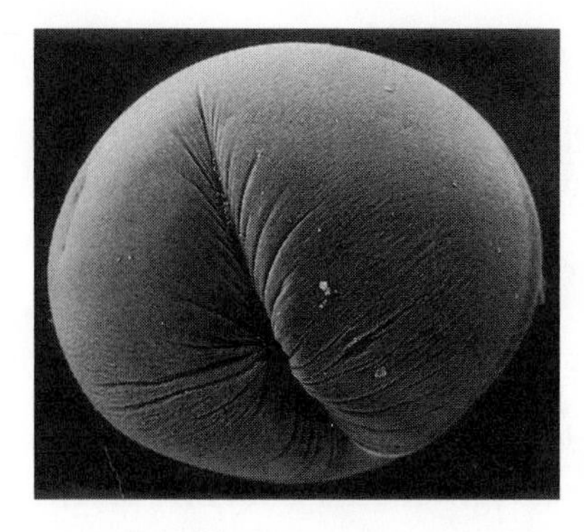

Figure 15-6 The final stage of mitosis—the animal cell "pinches in" to split into two. Each new daughter cell has some of the cytoplasm from the parent cell and an identical set of chromosomes.

involved in mitosis evolved long before organisms became specialized as plants or animals. However, differences can be seen between plants and animals during cytokinesis.

While watching animal cells under the microscope, you will see the cell membrane begin to change as mitosis concludes. The cell membrane begins to tighten in a band around the middle of the cell. This is called "pinching in." The process seems similar to pulling the drawstring tight on a plastic bag or sweat pants. When the pinching in is complete, the result is two new animal cells. Each new daughter cell has some of the cytoplasm from the parent cell and an identical set of chromosomes. (See Figure 15-6.)

A plant cell divides in a very different manner. The process of separation does not begin at the outside. Instead, a dividing wall begins to grow in the middle of the cell, from the inside out. This dividing wall is called a **cell plate**. It is made of tough polysaccharides. When the cell plate has completed growing, it extends outward to both sides of the cell. It will become a cell wall. The result is two new plant cells. Every growing plant goes through this process again and again as its cells divide. (See Figure 15-7.)

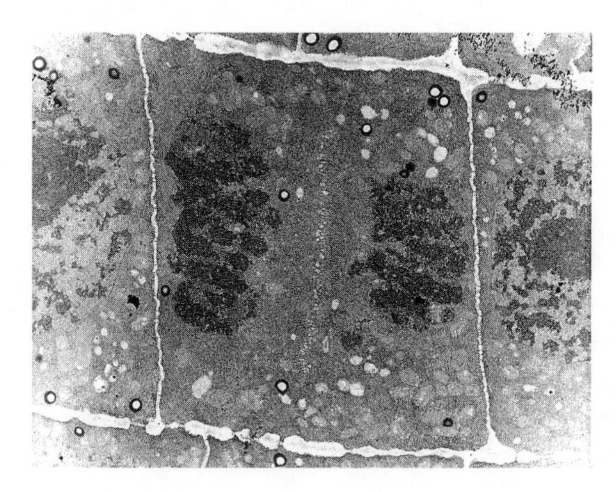

Figure 15-7 When a plant cell divides, a cell plate forms in the middle of the cell, growing from the inside out.

Check Your Understanding

Why is it important that the chromosomes of a cell replicate before cell division occurs? When does replication occur?

MAKING NEW INDIVIDUALS: ASEXUAL REPRODUCTION

Mitosis and cytokinesis produce two new daughter cells from one parent cell. The daughter cells are identical. They are also genetically identical to the parent cell. But have new individuals been produced? Well, sometimes.

Certainly this is the case if the original parent cell was a single-celled organism. An ameba, through mitosis and cytokinesis, becomes two new identical organisms. Reproduction in an ameba involves only one parent. This is **asexual reproduction**. (See Figure 15-8.) The asexual reproduction that occurs in single-celled organisms, including bacteria and protists, is known as **binary fission**, which means "splitting in two." (For reproduction to be called *sexual,* it must involve two parents.)

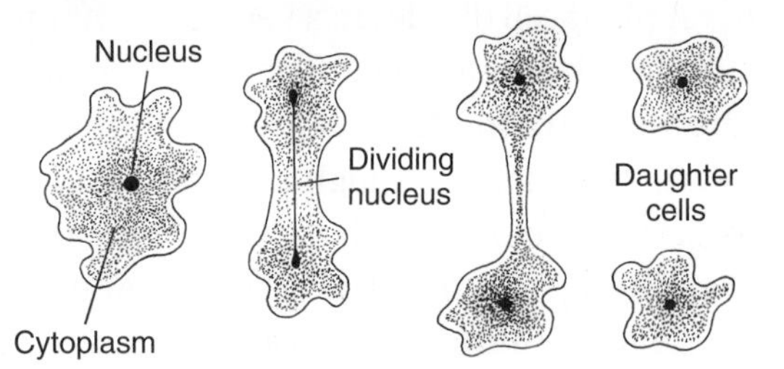

Figure 15-8 Asexual reproduction, as in the ameba, requires only one parent.

Plants have a variety of types of asexual reproduction. In each type, a plant or a part of the plant reproduces itself through mitosis. As a result, the offspring are identical to the parent plant. For example, strawberry plants send out horizontal stems across the soil. These horizontal stems, called runners, touch the surface of the soil at a new place. At that spot, an entirely new, identical plant with roots and leaves begins to grow. (See Figure 15-9.)

Parts of many plants can be cut off and used to start new individuals. For example, a leaf from an African violet can be placed in soil to begin a new plant. The underground bulbs of daffodils produce new bulbs. All the plants that result are identical to the "parent" plant. Even potatoes can reproduce asexually. Each "eye" on a potato, the point where a shoot

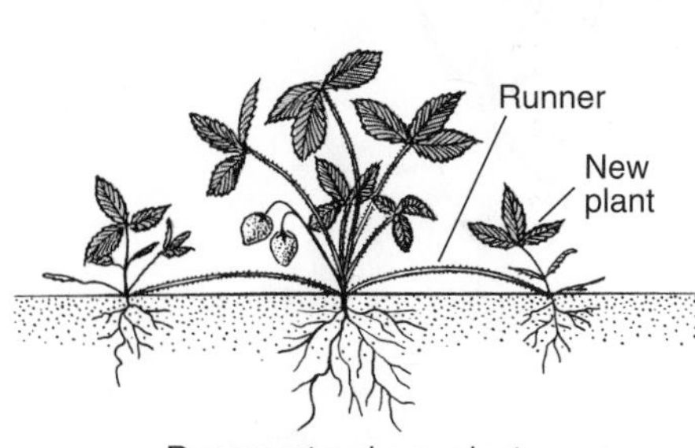

Figure 15-9 Strawberry plants can reproduce asexually by means of runners.

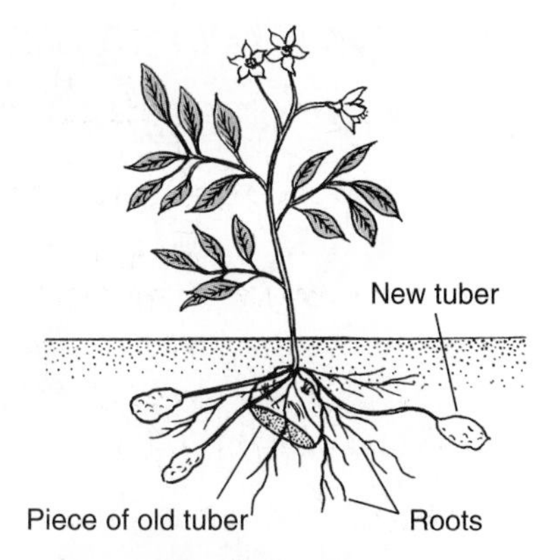

Figure 15-10 Potatoes can reproduce asexually. Each "eye" on a potato can develop into a new potato plant.

begins to grow, can develop into a new potato plant. All these activities are the result of mitotic cell division. (See Figure 15-10.)

Can animals also reproduce asexually? In some cases, yes. The flatworm, planaria, is able to reproduce from parts of itself. If a planaria is cut into two or three sections, each section can grow into a new worm, replacing the missing sections. If an arm of a sea star is broken off, the arm can sometimes grow into a new sea star. The sea star that lost the arm will regrow one. This process of growth, or regrowth, of an animal from a part is known as **regeneration**. (See Figure 15-11.) Some invertebrates have this ability. However, most vertebrates cannot do this. It is not possible, as you know, for you to regenerate a part of yourself. A severed finger can, in some cases, be reattached; but your body cannot grow a new one.

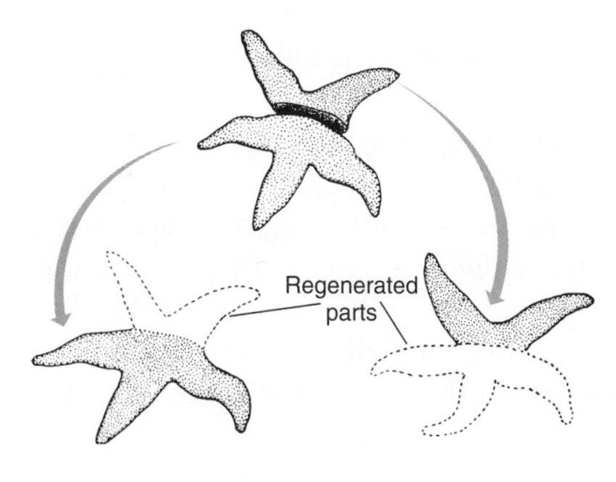

Figure 15-11 Some animals, such as the sea star, can reproduce asexually through the process of regeneration.

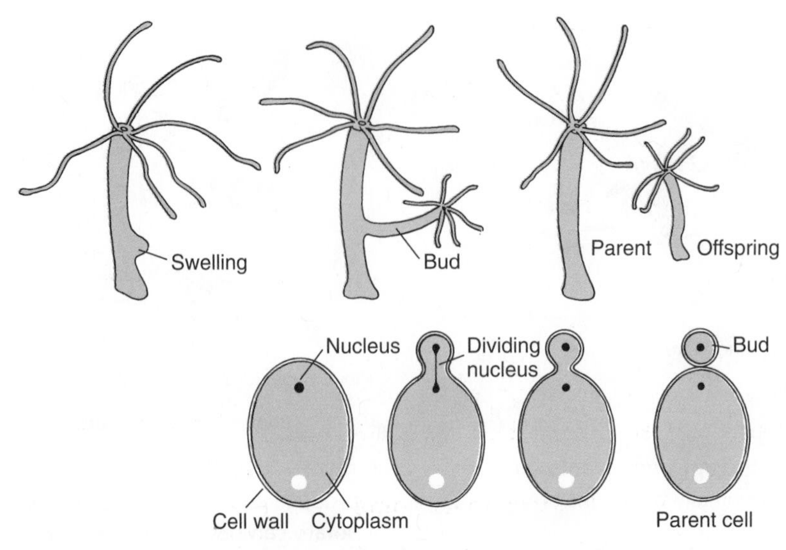

Figure 15-12 Both the hydra and the yeast can reproduce asexually by budding.

Other organisms, such as single-celled yeasts, sponges, and hydra, can produce offspring by **budding**. During the process of budding, a new small individual begins to grow out of the side of the parent. The cells that form this new individual result from mitotic cell division. In time, the bud, large enough to live on its own, breaks free of the parent. (See Figure 15-12.)

THE RATE OF CELL DIVISION

When does a cell divide? How long does it take for one segment of cell division to begin and end? Do all types of cells divide at the same rate? What controls the speed at which cells go through the cell cycle? The answers are very important to the process of growth and development in organisms. The answers may also provide the keys that help unlock the secrets of cancer—actually a disease caused by cells whose cycle of growth is out of control.

Every multicellular organism is made up of various types of tissues and cells. For example, the human body contains blood tissue, skin tissue, muscle tissue, bone tissue, and nerve tissue. Controlling the rate at which cells of each particular kind of tissue divide is a necessary part of homeostasis. Red blood cells have a relatively short life span, and we need an enormous number of them. As a result, the cells that develop into red blood cells divide quickly. To maintain the correct number of red blood cells in our body, about 2.5 million new red blood cells are made each second in the bone marrow. (See Figure 15-13.) Bone cells, on the other hand, divide much more slowly. An even more extreme example is nerve cells, which have almost completely lost the ability to divide. It was long thought that the number of brain cells you have does not increase as you mature. New research is questioning this idea. However, we do know for certain that it is very difficult to repair or replace damaged nerve cells when the spinal cord is damaged.

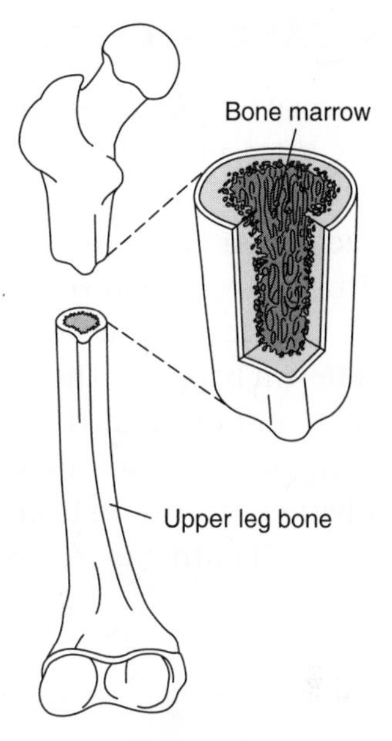

Figure 15-13 Millions of new red blood cells are made each day in the bone marrow.

If conditions in the body change, homeostasis requires that cells change the rate at which they divide. For example, a skin cell normally takes about 20 hours to complete its cell cycle. However, your skin cells speed up their rate of division if you cut yourself. The cell cycle shortens and your cut skin heals faster. Liver cells normally do not divide. However, if part of the liver is removed by surgery, the cells in the remaining part start dividing until the liver returns to its normal size.

How do cells know when to divide and when not to divide? In biology, this is one of the most important current areas of research. Scientists have found internal and external controls for cell division. Substances inside the cell control when, and how rapidly, the cell divides. In addition, external environmental factors affect cell division. Environmental factors include changes in temperature, pH, and the amounts of available nutrients. The presence of other cells is another factor that affects cell division. For example, normal cells keep dividing until they touch other cells. Then cell division stops. This is called *contact inhibition*; it prevents cells from getting overcrowded.

Hormones are another type of external control of cell division. In Chapter 11, you learned that hormones are chemicals produced in one part of the body that affect cells in another part of the body. For example, human growth hormone, from the anterior pituitary gland, controls growth throughout the body. In particular, it stimulates cell division in the long bones of the arms and legs. As the rate of cell division increases, you grow and your clothes no longer fit!

CANCER: CELL DIVISION OUT OF CONTROL

Normal cells from vertebrates can also be grown outside the body. By giving them nutrients and the correct temperature and pH, the cells will begin to divide in a smooth glass dish. They will divide until they touch each other and cover the bottom of the dish. Then cell division stops. The cells remain alive and very "well behaved."

On occasion, however, scientists have observed that something very different happens. The cells divide, cover the dish, touch each other, and then continue to divide. The cells begin to pile up in the dish, crowding each other, and still they divide. It seems that nothing will stop the cell division. It is like an uncontrolled "blob" in a science fiction movie. Only when all nutrients are used up does the growth cease. (See Figure 15-14.)

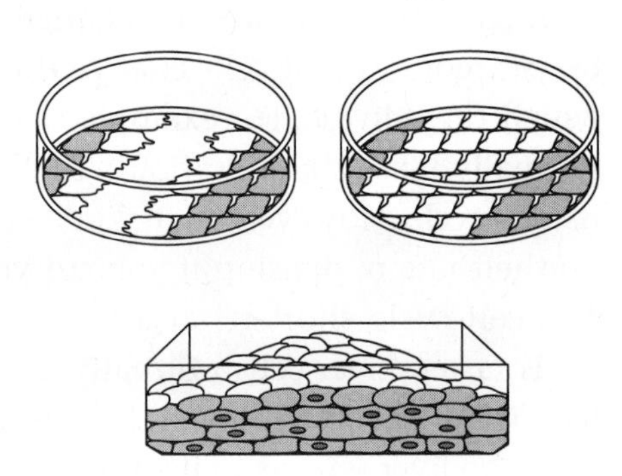

Figure 15-14 When normal cells are grown in a glass dish, they continue to divide until they touch each other. Occasionally the cells continue to divide even after they touch, resulting in the type of uncontrolled cell growth seen in cancer.

This phenomenon is the type of cell behavior that characterizes the disease cancer. Cancer results from uncontrolled cell division. Cancer cells seem to no longer follow the rules. They do not recognize the signals that control normal cell division. When this happens in the body, the growing mass of cells may become a **tumor**. The tumor steals energy and nutrients from other normal tissues. The results can be deadly.

Tumors steal energy from surrounding tissues by growing many blood vessels to bring nutrients to them. Some of the most exciting cancer research is now aimed at finding substances that will stop this growth of blood vessels (a process called *angiogenesis*). Two of these substances, angiostatin and endostatin, have in fact recently been found to destroy tumors in mice. Will a cure for cancer in humans actually be found? There is hope that the answer may indeed be yes. However, only time—and further research—will tell.

Humans are not the only organisms that develop cancer. Frogs, chickens, mice, and even plants can develop cancer. In fact, almost all multicellular organisms can develop cancer. You may have seen a tree with a

Cancer Treatment in the Twenty-First Century

Non-Hodgkin's lymphoma (NHL) is a cancer of the lymph system. This system collects intercellular fluid from throughout the body, returning it in tubes to the bloodstream. The tiny lymph vessels join together to eventually form large ones that empty into veins in the neck. Enlargements along these lymph vessels are known as lymph nodes. These nodes, or glands, are involved in the body's defenses against diseases. However the lymph system is also the site for NHL cancer—one of the few cancers that is occurring with greater frequency. No one knows why the incidence of NHL is increasing, but it now accounts for more than 4 percent of cancer deaths in this country.

Chemotherapy, the traditional use of drugs to treat cancers such as NHL, was developed during the twentieth century. These anti-cancer drugs use a variety of methods to attack cancer cells: by attacking DNA; by shutting down protein synthesis; or by stimulating the immune system. In the mid-1990s, trials began for the use of a very different type of drug—monoclonal antibodies. These drugs are actually designer-made antibodies that have been produced to find and attack cell-surface targets that exist only on cancer cells. The monoclonal antibody drugs are therefore referred to as *targeted drugs*; they search out the cancer cells. The cutting-edge capability of these twenty-first-century drugs is to attach radioactivity or some other cancer-fighting drug to the monoclonal antibody. The targeted drug will go find the cancer cells, deliver its deadly payload, and then kill the cancer cells. This treatment is now being used against NHL with some success.

Doctors currently stress that the best approach is to use both methods—twentieth-century and twenty-first-century cancer treatments. Well-respected experts are optimistic about the chances for real progress in the years ahead. For example, Dr. Andrew Zelenetz, chief of the lymphoma services at Memorial Sloan Kettering Cancer Center in New York City has said, "This is a very exciting time. We didn't have new important agents for the treatment of lymphoma for many years. Now we're seeing the emergence of these targeted therapies that are very exciting, and in fact, we're starting to see the emergence of other chemotherapeutic agents that actually have activity in lymphoma. We're entering a new era where we have both the traditional tools as well as these new targeted tools, and we're going to be seeing more of them coming down the pike. There are a number of new agents that are in development that are being tested that I think have real promise."

Hopefully these new cancer-fighting agents will be developed in time to fight the increase in incidence of NHL and other potentially deadly cancers.

Figure 15-15 Even trees can develop cancers. The tumors on this tree were caused by a virus.

strange, large swelling partway up its trunk. Most likely, that swelling was a cancerous tumor. (See Figure 15-15.)

Sometimes, after the dividing cells form a tumor, nothing else happens. Cells do not break off from the tumor and travel to other places in the body. The tumor causes no further damage and is said to be **benign**. What is dangerous is when cancer cells have the ability to break away from the tumor. These cells may get into the blood system. The blood can carry the cells to other parts of the body where they can start new tumors. The spread of a cancer in this way is known as **metastasis**. A tumor that metastasizes is said to be **malignant**. Malignant cancers can lead to death. (See Figure 15-16.)

Uncontrolled cell growth can occur in many different types of cells. As a result, there are different types of cancer. There is skin cancer, breast cancer, prostate cancer, lung cancer, leukemia (a cancer of the blood), and

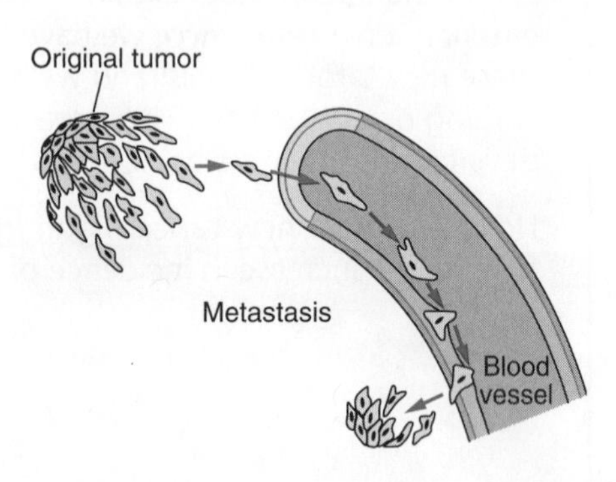

Figure 15-16 Metastasis is the spread of cancer cells through the body.

many others. It does not seem that there is a single cause for all types of cancer. Currently, a great deal of research is being done to try to learn the causes of each type of cancer.

Even though there are differences, it is quite certain that all types of uncontrolled cell division involve the cells' genetic instructions. In Chapter 20, you will learn much more about how these genetic instructions work in cells. We have already seen that these instructions, the genes, are made of DNA. Internal and external factors that cause cancer do so by damaging or changing the DNA. For example, some skin cancers result from DNA damage caused by exposure to ultraviolet (UV) rays that reach Earth from the sun. Physicians recommend limiting exposure to the sun and encourage the use of creams or lotions that contain chemicals to block UV rays from reaching the skin. Tobacco smoke contains several types of chemicals that are known to act as **carcinogens**. These carcinogens cause mutations in the DNA of lung cells, which result in the formation of cancer. It seems that breast cancer, rather than resulting from external factors, sometimes results from internal causes. Some physicians suggest that there are hereditary factors (genes) that make a person more likely to form cancers in breast tissue. This would explain why women in certain families are more likely to develop this disease. If the cancer has also occurred in relatives, it is likely that an inherited factor is the cause.

No matter what the cause of cancer is, it has been said that one of the most important ways to reduce the risk of cancer is to maintain good health habits. The immune system constantly attacks not only invading cells but also abnormal cancerous cells from our own body. Much of the time, the immune system is successful. It destroys cancer cells before they can develop and cause problems. In fact, many biologists describe cancer as being the result of the rare times when abnormal cells get past the immune system. It is no surprise that many people who die from AIDS, a disease of the immune system, actually die from some type of cancer. The patient's damaged immune system is not able to protect the person from cancerous cells. A healthy body, meaning a healthy immune system, is therefore one of the best protections against cancer.

LABORATORY INVESTIGATION 15
How Do Plant Cells Change During Cell Division?

INTRODUCTION

As plants grow, new cells are produced by cell division. Cell division includes the process of mitosis, the rearrangement and regrouping of chromosomes in cells. The stages in plant cell division can be observed in cells from the tips of roots. In this activity, you will examine prepared slides of root-tip cells. To make this kind of slide, a thin slice of tissue is cut from the tip of a root of an onion plant. This tissue is stained, and a permanent slide is made. Through a microscope, you can observe the cells trapped in various stages of mitosis. Your task is to make observations and conclusions about how mitosis occurs in plants.

MATERIALS

Compound microscope, prepared slide of onion root tips, lens paper

PROCEDURE

1. Look at the slide of the onion root tip. How many root tips are on your slide? Look closely at one of the root tips. Decide which end of the root is the bottom and which end was cut from the end of the root. Make a labeled drawing of your observations.
2. Now observe the slide under the low-power objective on your microscope. Observe the cells at the cut end and at the rounded end of the root. List the differences you observe in the two areas.
3. Observe the end of the root where the cells are showing the most evidence that mitosis is occurring. Which end did you choose? Why?
4. Change to the high-power objective. Choose five cells that are in different stages of mitosis. Draw these cells. Label as many structures as possible. Number the drawings from the earliest stage of mitosis to the latest.
5. If time permits, survey the cells throughout the zone where mitosis is occurring. Record the number of cells in each of the stages you numbered in step 4. Prepare a data table of your results.

INTERPRETIVE QUESTIONS

1. Which part of the root, the round end or the cut end, contains the youngest cells? What evidence do you have to support this conclusion?
2. In step 4, how did you decide the order of the stages you observed?
3. Once mitosis is completed, how are the daughter cells similar to the parent cell? How are they different?
4. From your survey of the cells in step 5, draw a conclusion about the length of time it takes to complete each of the five stages observed in step 4. Explain your reasoning.
5. Use your knowledge of the structure and functions of chromosomes to explain why it is necessary for mitosis to produce a complete, new, and identical set of chromosomes for the cells produced by cell division.

CHAPTER 15 REVIEW

Answer these questions on a separate sheet of paper.

VOCABULARY

The following list contains all of the boldfaced terms in this chapter. Define each of these terms in your own words.

anaphase, asexual reproduction, benign, binary fission, budding, carcinogens, cell cycle, cell plate, centrioles, centromere, chromatids, chromosomes, cytokinesis, DNA, interphase, replication, malignant, metaphase, metastasis, mitosis, prophase, regeneration, reproduction, spindle, telophase, tumor

PART A—MULTIPLE CHOICE

Choose the response that best completes the sentence or answers the question.

1. How many chromosomes are found in a human body cell?
   a. 8 *b.* 18 *c.* 46 *d.* 94
2. The duplication of the genetic material in a cell is called
   a. reproduction *b.* replication *c.* binary fission *d.* metastasis.
3. Bacteria reproduce by *a.* binary fission *b.* budding
   c. metastasis *d.* meiosis.
4. The process by which the cytoplasm of a cell divides is called
   a. mitosis *b.* meiosis *c.* cytokinesis *d.* regeneration.
5. Under normal circumstances, which body cells divide most quickly? *a.* bone marrow cells *b.* liver cells *c.* bone cells
   d. brain cells
6. Chromosomes line up in a single file across the middle of the cell during *a.* anaphase *b.* metaphase *c.* prophase *d.* telophase.
7. A tumor that spreads is said to be *a.* benign *b.* malignant
   c. carcinogenic *d.* leukemic.
8. Which of the following statements is correct? *a.* A chromosome consists of two chromatids joined by a centriole. *b.* A chromatid consists of two chromosomes joined by a centriole.
   c. A chromatid consists of two chromosomes joined by a centromere. *d.* A chromosome consists of two chromatids joined by a centromere.
9. A group of individuals of the same kind living in a particular place is a *a.* population *b.* genus *c.* crowd *d.* species.

10. Geckos, a type of lizard, can detach their tails to fool predators. The gecko later grows a new tail. This is an example of *a.* asexual reproduction *b.* binary fission *c.* budding *d.* regeneration.
11. The series of events that occurs during the life of a cell is called *a.* mitosis *b.* metastasis *c.* binary fission *d.* the cell cycle.
12. During cell division in a plant cell, you would *not* expect to see a *a.* centriole *b.* spindle *c.* cell plate *d.* chromosome.
13. A twig from a willow tree placed in a bucket of water will soon sprout roots and can be planted to produce a new tree. This is an example of *a.* asexual reproduction *b.* sexual reproduction *c.* budding *d.* binary fission.
14. The genetic material is made up of *a.* centromeres *b.* DNA *c.* NADP *d.* proteins.
15. Cell division is affected by *a.* contact inhibition *b.* hormones *c.* nutrient availability *d.* all of these.

PART B—CONSTRUCTED RESPONSE

Use the information in the chapter to respond to these items.

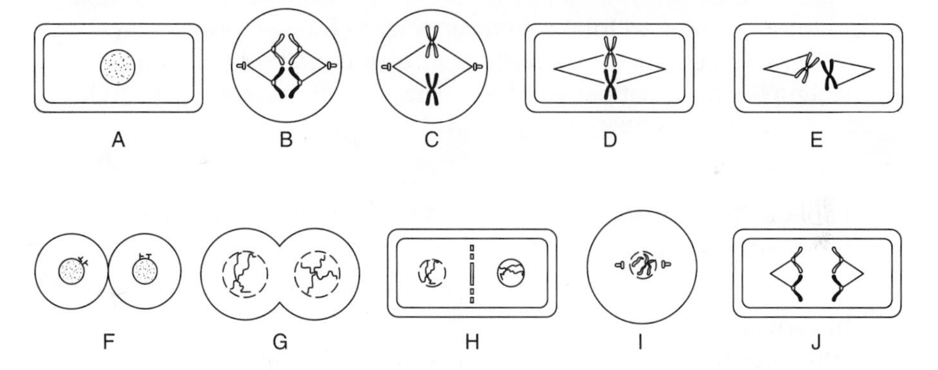

16. The diagram shows the stages of two versions of a process in the incorrect order. Identify the process and put the stages in the proper sequence.
17. Name the stages of the process and briefly describe the major events associated with each stage.
18. Although movies like *The Blob* can be entertaining, they rarely portray cell biology accurately. Explain why a giant human-eating ameba is not biologically possible.
19. What is cancer? What are some possible causes of cancer? What are some ways of reducing the risk of cancer?
20. Why is asexual reproduction important to people who grow plants?

PART C—READING COMPREHENSION

Base your answers to questions 21 through 23 on the information below and on your knowledge of biology. Source: *Science News* (December 14, 2002): vol. 162, p. 382.

Zapping Bone Brings Relief from Tumor Pain

By unleashing radio waves inside bone, researchers have stopped intractable pain in people with cancer that has spread to their skeletons.

Tumors that form inside bone when cancers spread can be especially painful. The new technique, called radio-frequency ablation, unleashes energy via a needle inserted into bone to reach the edge of the tumor. The radio waves create intense heat that kills nearby tumor cells within about 10 minutes, says study coauthor Matthew R. Callstrom of the Mayo Clinic in Rochester, Minn.

Targeting the surface where the tumor meets the bone seems critical, he says. "Our thought is that nerve fibers in that area—where tumor cells are eroding bone—are the pain generators," he says. Bone itself appears unaffected by the procedure.

The researchers treated 62 patients in whom conventional cancer therapy had failed. Of these, 59 reported significant pain relief, and 28 said they experienced total pain relief at some times, Callstrom says.

"We're not curing cancer with this treatment," he says. "But we're affecting the pain that patients have. The most important [concern] for all these patients is their quality of life."

21. Explain why researchers are using a new technique on cancer tumors inside bone.
22. State the process by which radio waves are being used to treat pain in cancer patients.
23. Why are the surfaces where tumors meet the bone considered important in this study?

Meiosis and Sexual Reproduction

After you have finished reading this chapter, you should be able to:

Explain the importance of reduction division to sexual reproduction.

Describe what happens to chromosomes during Meiosis I and Meiosis II.

Discuss methods of sexual reproduction in plants and animals.

And here's the happy bounding flea,
You cannot tell the he from she.
The sexes look alike, you see;
But she can tell, and so can he.

Roland Young, *The Flea*

Introduction

For almost all animals, it takes two to reproduce—a male and a female. This is **sexual reproduction**. Even most plants use this method of reproduction to make more of their own kind. Sexual reproduction is very important in understanding living things. It also plays a significant role in the process of evolution. To understand why this is so, we must look, as always, at individual cells. Within cells, we must closely examine the chromosomes. (See Figure 16-1 on page 340.)

IT'S ALL ABOUT CHROMOSOMES

Each of our cells contains chromosomes. The word *chromosome* means "colored body." Chromosomes are microscopically small. They were discovered only in the late 1800s as microscope lenses were improved. With a microscope, chromosomes can be observed when the cell is stained with a dark dye. The chromosomes absorb the dye and appear to be colored, thus giving them their name.

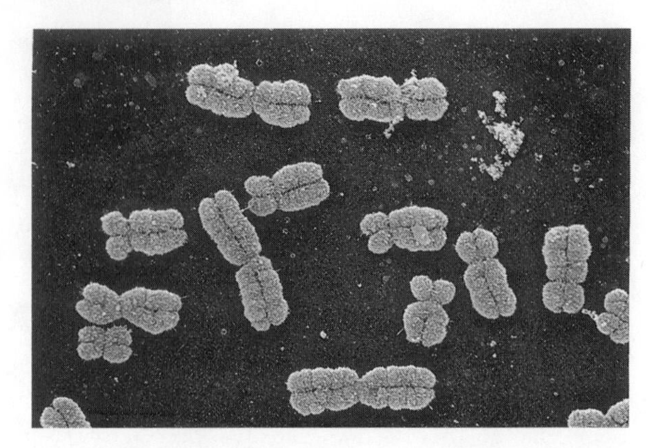

Figure 16-1 Human chromosomes.

Chromosomes have the two most important jobs in the world. They contain the inherited information that has been passed along from the beginning of life on Earth. It is this information that determines an individual's characteristics. In humans, this information makes us who we are. The chromosomes also contain the "know-how" that keeps our cells running correctly. This is what keeps us and all organisms alive.

Why is sexual reproduction all about chromosomes? About 105 years ago, it was observed—again with the newly improved microscopes—that when a sperm cell and an egg cell unite during sexual reproduction, it is the nucleus from each cell that joins. In particular, it is only after the nucleus from the sperm cell enters the egg cell and fuses with its nucleus that the development of a new organism begins. (See Figure 16-2.) What

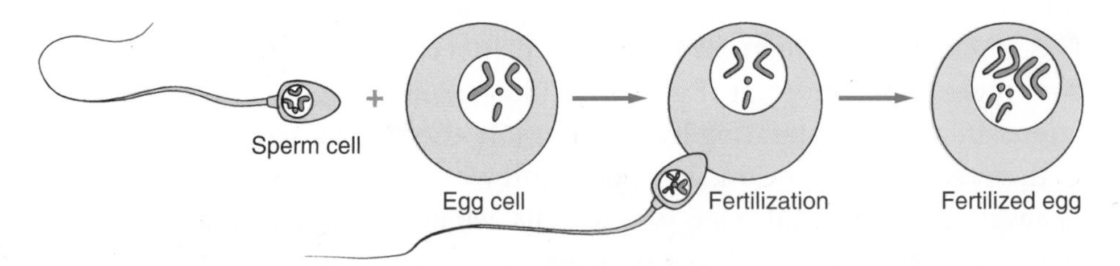

Figure 16-2 Sexual reproduction involves the joining of chromosomes from a sperm cell and an egg cell.

does the nucleus contain? The chromosomes, of course. So sexual reproduction is all about joining chromosomes from two individuals, usually those from a male individual (the father) with those from a female individual (the mother). Put this way, it does not sound very romantic, but that's biology.

How many chromosomes are found in a human body cell? In the past

100 years, guesses based on microscopic observation have varied from 8 to 50 chromosomes per cell. Finally, T. C. Hou, a young Ph.D. student at the University of Texas, accidentally found a way to separate the chromosomes and count them accurately. Hou found that human body cells each contain 46 chromosomes.

Right away, a problem becomes obvious. Your body cells require a chromosome number of 46. That means that the first cell from which each of us came, the cell resulting from the combination of a sperm and an egg, must have had 46 chromosomes. The fertilized egg divided by mitosis many, many times to become you. After each mitotic division, the number of chromosomes in each body cell is equal to that of the parent cell. Therefore, every body cell now in you still has 46 chromosomes. The problem is, how can a cell from each of your parents, a sperm and an egg, combine to make a new cell with 46 chromosomes?

Arithmetic tells us there is only one way. Both the sperm and the egg must have only 23 chromosomes—half the number of chromosomes from the normal number of 46. And indeed this is the case. In this chapter, you will study the type of cell division that produces sperm and egg cells, which have that reduced number of chromosomes. (See Figure 16-3.)

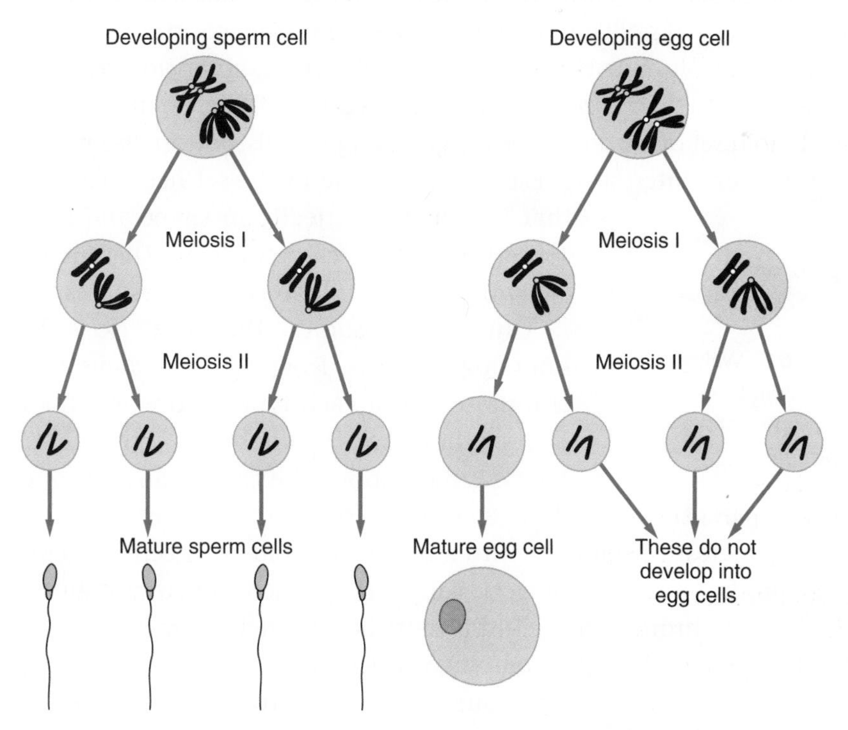

Figure 16-3 Sperm cells and egg cells have half the normal number of chromosomes for their species.

GAMETES

The sperm and egg cells, or sex cells, are called **gametes**. In the process of sexual reproduction, the nuclei of the gametes join together. This fusion of the nuclei is called **fertilization**. The resulting cell, a fertilized egg cell, is called a **zygote**.

Each gamete, as we have said, has exactly one-half the normal number of chromosomes. Cells that have half the normal number of chromosomes are said to be **haploid**. In humans, the haploid number is 23. The zygote and all body cells that come from the mitotic division of the zygote have two sets of chromosomes in them, one from each parent. These cells are **diploid**. In humans, the diploid number is 46.

Haploid gamete cells must be made by a special type of cell division that reduces the chromosome number by one-half; so, when fertilization occurs, the normal diploid number of chromosomes for the species is maintained. This type of cell division is called **meiosis**.

A CLOSER LOOK AT CHROMOSOMES

Grasshoppers played an important role in helping us understand the next step in the story of sexual reproduction. In 1903, Walter Sutton, a graduate student at Columbia University in New York City, observed grasshopper cells to study their chromosomes. (See Figure 16-4.) By looking closely at the diploid set of chromosomes in grasshopper cells, Sutton discovered something very interesting. Each chromosome in the set had a partner that matched it perfectly in shape and size. The chromosomes came in pairs. These pairs are known as **homologous chromosomes**.

Figure 16-4 Walter Sutton studied the chromosomes of grasshoppers.

Our chromosomes therefore exist in homologous pairs. Essentially, we have two chromosomes of each type. And where does each of these two chromosomes come from? That question should be easy to answer: one from each parent. Now we have a clear idea of the job that must be performed by meiosis. Beginning with a normal diploid body cell, gametes must be produced through meiotic cell division. Each gamete contains a haploid set of chromosomes. And it must be an exact set, meaning one and only one of each of the homologous chromosomes.

Today it is a standard procedure to examine the chromosomes in human cells. White blood cells are used. Certain diseases can be detected in unusual patterns in chromosomes. While undergoing mitosis, white

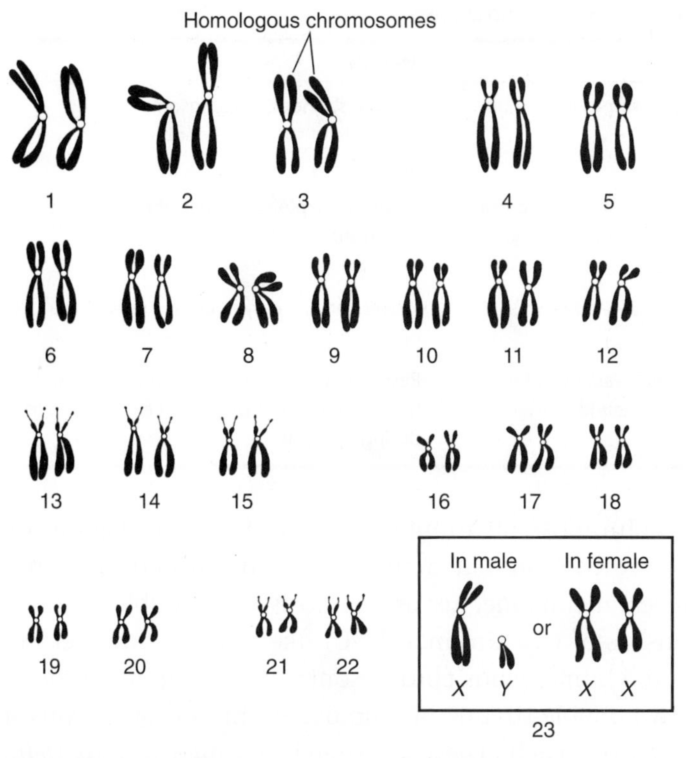

Figure 16-5 A karyotype is prepared by arranging pictures of matching chromosomes by size and shape.

blood cells are collected and their chromosomes are photographed under magnification. Pictures of the chromosomes are enlarged, cut out, and arranged by size and shape. This process of arranging pictures of chromosomes by their size and shape is called preparing a **karyotype**. In Chapter 22, you will see that it is often very important to study the karyotype of chromosomes in a developing fetus's cells. (See Figure 16-5.)

MEIOSIS: REDUCING THE CHROMOSOME NUMBER

Mitosis and meiosis take place during cell division, and in some ways these two processes are similar. Chromosomes replicate before either process begins. However, the results of mitosis and meiosis are very different. When mitosis is completed, the chromosome number remains the same as the original parent cells. When meiosis is completed, the chromosome number is half the original number. (See table on page 344.)

How does this happen? Meiosis actually involves two separate cell divisions that take place one after the other. Meiosis I is the first cell division in this process. It is called **reduction division** because it is during this

SOME DIFFERENCES BETWEEN MITOSIS AND MEIOSIS

Mitosis	Meiosis
Double-stranded chromosomes line up in middle of the cell in single file.	Double-stranded chromosomes line up in middle of the cell in double file (i.e., in tetrads).
Results in diploid number of chromosomes in daughter cells (i.e., the normal species number).	Results in haploid number of chromosomes in daughter cells (i.e., half the species number).
Process occurs in all cells of the body.	Process occurs only in certain cells of the sex organs.
Results in very few genetic variations because the chromosomes retain their identity.	Results in many genetic variations because the chromosomes break and exchange parts during synapsis.

process that the chromosome number is halved. At the beginning (Prophase I), we see the familiar double-stranded chromosomes not enclosed by a nuclear membrane. Just as in mitosis, each double-stranded chromosome consists of two chromatids. Unlike mitosis, however, we also begin to see the homologous chromosomes pairing up in a process called *synapsis*. Two double-stranded homologous chromosomes consist of four chromatids, two each. These are therefore called **tetrads** (*tetra* means "four"). (See Figure 16-6a.)

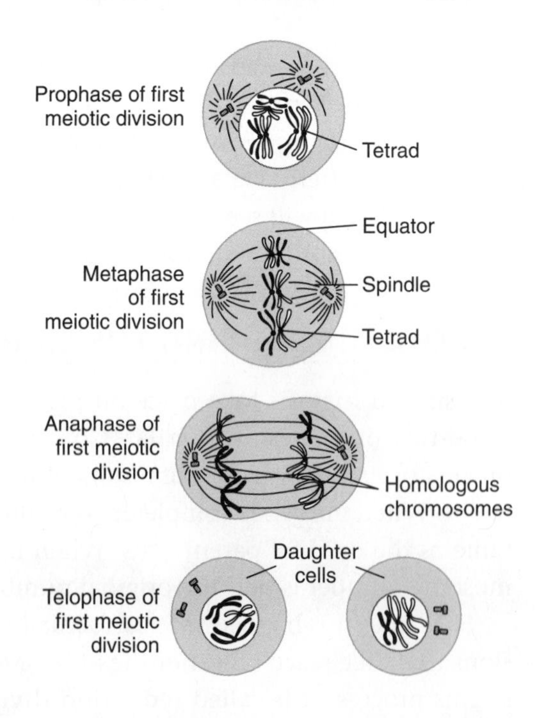

Figure 16-6a Meiosis I.

Now the chromosomes line up (Metaphase I). However, this time they are not in single file. They are in homologous pairs. In Anaphase I, the paired chromosomes separate, with the homologous chromosome (consisting of two chromatids) from each pair moving to opposite ends of the cell. Finally, in Telophase I, we have two new cells, each with only one of each homologous chromosome. For humans, if we began with the 46 double-stranded chromosomes, we now have two cells, each having 23 double-stranded chromosomes.

The second division, Meiosis II, also passes through four stages. In Meiosis II, the chromosome number is not being reduced. Rather, the chromatids from each double-stranded chromosome are being separated. Meiosis II in humans begins with two cells that have 23 double-stranded chromosomes each. It ends with four cells, each with 23 single-stranded chromosomes. (See Figure 16-6b.)

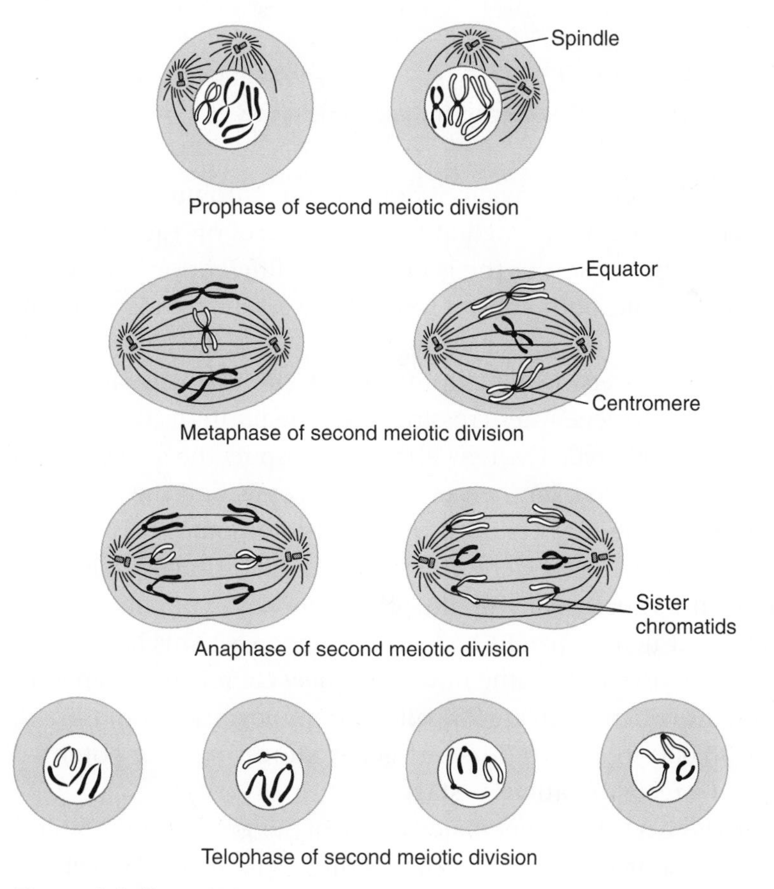

Figure 16-6b Meiosis II.

MEIOSIS I: THE SOURCE OF OUR DIFFERENCES

With the exception of identical twins, children in the same family are never exactly alike. Differences occur in eye color, hair color, hair texture, height, nose shape, ear size, and many other characteristics. Why is this so, if the children were born of the same parents? The explanation for these differences lies in the details of Meiosis I.

During Prophase I, tetrads are formed. The chromatids of homologous chromosomes are very close to each other. In fact, they often overlap and actually exchange pieces in a process called **crossing-over**. (See Figure 16-7.) Why is crossing-over important?

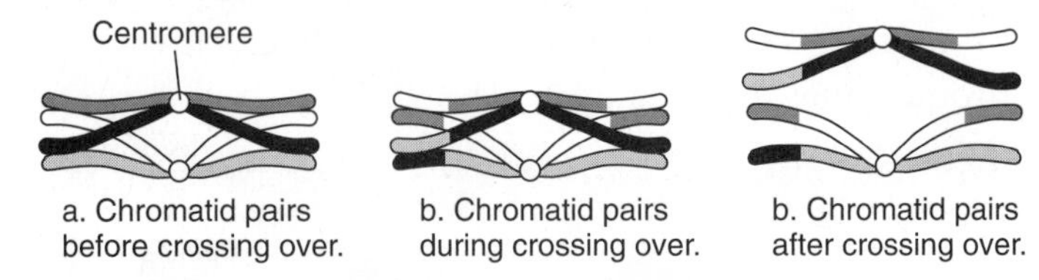

Figure 16-7 In crossing-over, the chromatids of homologous chromosomes overlap and exchange pieces.

Suppose we are looking at a pair of homologous chromosomes from a plant. The same genes are located at the same places on each chromosome. However, the version of the gene may be different on each chromosome. For example, both chromosomes have a gene for plant height and a gene for flower color. But chromosome A has the genes for tall plants and white flowers, while chromosome B has the genes for short plants and yellow flowers. After crossing-over, the positions of genes on a chromosome are altered. The new arrangement puts the genes for tall plants and yellow flowers on the same chromosome, and the genes for short plants and white flowers on the same chromosome. This is obviously a new combination of genes. As a result of crossing-over, genetic **recombination** has occurred. The offspring may have combinations of genetic traits that neither parent had. (See Figure 16-8.) This happens in all meiotic cell divisions. Imagine how many new combinations are possible in humans, when there are 23 pairs of homologous chromosomes, with about 30,000 to 50,000 different genes altogether. The potential number of gene recombinations is staggering!

However, there is yet another major source of our genetic and physical differences. This source of change occurs during Metaphase I. We get one of each homologous chromosome from each parent: the maternal chromosome from Mom and the paternal chromosome from Dad. Imagine the

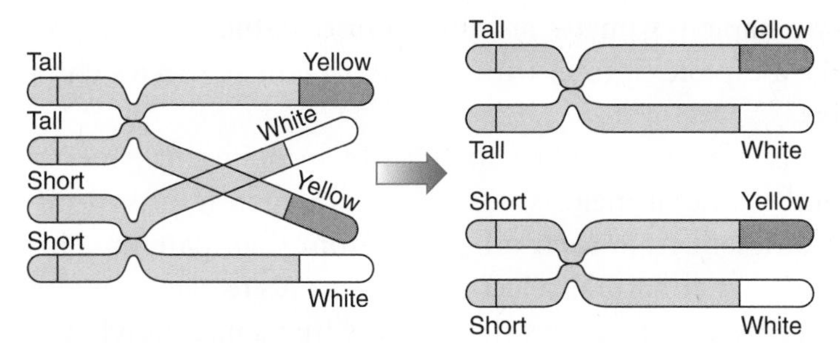

Figure 16-8 Genetic recombination is the result of crossing-over.

maternal chromosomes were green and the paternal ones yellow. (Of course, they have no color unless they are artificially stained.) Nevertheless, the green maternal chromosomes could all line up on the left side, the yellow paternal ones on the right side. This would put only green maternal chromosomes in one daughter cell and only yellow paternal chromosomes in the other daughter cell. However, there is no rule as to how they line up. Any number of maternals may be on one side; any number of paternals on the other. Each time meiosis occurs, the lining up can be different. This is known as **independent assortment**. The number of possible different assortments is huge. In fact, with 23 chromosome pairs in humans, there are more than eight million (2^{23}) different possible arrangements. Since this happens during meiosis in both parents, the chances of two identical children being born at different times within one family are one in 70 trillion ($2^{23} \times 2^{23}$). You have a much better chance of winning a lottery, and you know how small that chance is! (See Figure 16-9.)

In conclusion, meiosis has two all-important functions. First, meiosis maintains the normal species chromosome number by preparing haploid

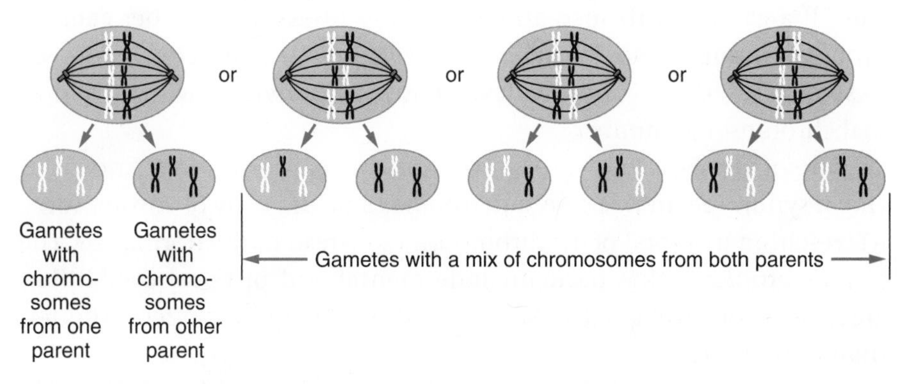

Figure 16-9 Each time meiosis occurs, the chromosomes line up in a different arrangement.

gametes. The two haploid gametes are able to fuse during sexual reproduction to make a diploid zygote, with the characteristic species chromosome number.

Second, meiosis increases genetic variability by recombining genes in eggs and sperm. Because of meiosis, sexual reproduction results in offspring that are different from each other and from their parents. This genetic variation is what natural selection acts on. A greater variety of characteristics in offspring increases the chances that some individuals will be better suited than others to survive in a particular place and time. As natural selection acts on the varied offspring, generation after generation, the species evolves.

Check Your Understanding

Why does crossing-over occur only between homologous chromosomes? How does this lead to genetic variation?

MEIOTIC MISTAKES

The dance of the chromosomes that occurs in cell division—especially during meiosis—is a wonderfully complex sequence of events. However, it does not always proceed correctly. For example, in Meiosis I, a pair of homologous chromosomes occasionally fails to separate; and in Meiosis II, sometimes a double-stranded chromosome does not separate. Either of these situations creates gametes with an incorrect number of chromosomes. A gamete may have an extra chromosome because it has both members of a homologous pair of chromosomes, instead of only one. Or a gamete may be one chromosome short, having neither member of a homologous pair. These differences are called **chromosomal abnormalities**. If a gamete with such an abnormality fuses with another gamete, problems occur. In most instances, the zygote fails to develop. However, in some cases, the zygote does develop into an individual with an abnormal chromosome number.

One of the best-known examples of this is the disorder known as Down syndrome. In this case, a person has an extra copy of chromosome 21, resulting in a total of 47 chromosomes instead of the normal 46. The serious problems that occur include mental and physical disabilities, greater risks of developing leukemia and heart disease, and often a shorter-than-normal life span.

At one time, Down syndrome children rarely lived beyond age 10. Today, there is significant potential for development in Down syndrome children. In fact, the popular 1990s television program *Life Goes On*

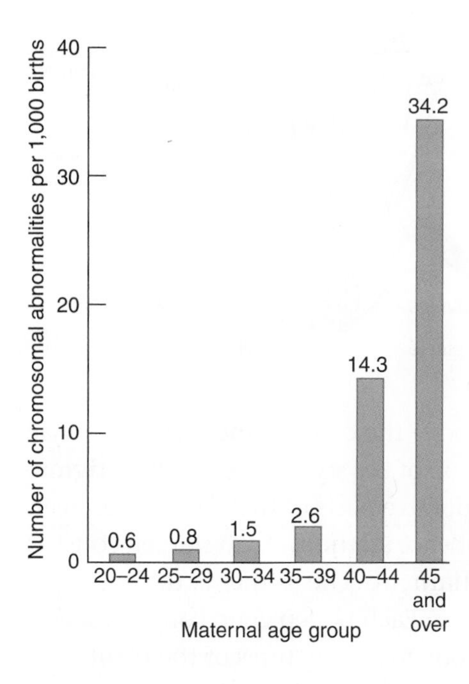

Figure 16-10 As maternal age increases, especially after 40, the possibility of giving birth to a child with a chromosomal abnormality increases.

starred Chris Burke, an actor with Down syndrome. By the year 2002, at age 37, Burke was recording music and making appearances in such shows as *The Commish* and *Touched by an Angel.*

The age of the mother affects the chances of having a child with Down syndrome. For a mother between the ages of 20 and 24, the chances are 6 in 10,000 births. For a mother who is 45 years old or over, the chances increase greatly, to 342 in 10,000 births. Early detection of a chromosomal abnormality such as this is one of the main reasons for preparing a karyotype of fetal cells early in a woman's pregnancy. (See Figure 16-10.)

THE SEX LIFE OF FLOWERING PLANTS

Of the many types of plants on Earth, flowering plants are the group that has evolved most recently. Most types of plants reproduce sexually. As in animals, the male gamete (sperm) joins together with the female gamete (egg) to produce a zygote. The zygote then begins to grow into the new plant. What is special about flowering plants is that the place where all this happens is very visible. The location is often brightly colored, beautifully shaped, and sweet smelling. Flowers are the parts of plants where sexual reproduction occurs. In fact, the parts that make up flowers include the sex organs of the plants. Let's take a closer look. (See Figure 16-11 on page 350.)

A typical flower contains a central structure called the **pistil**. The pistil consists of three parts. The base of the pistil is the **ovary**. The ovary

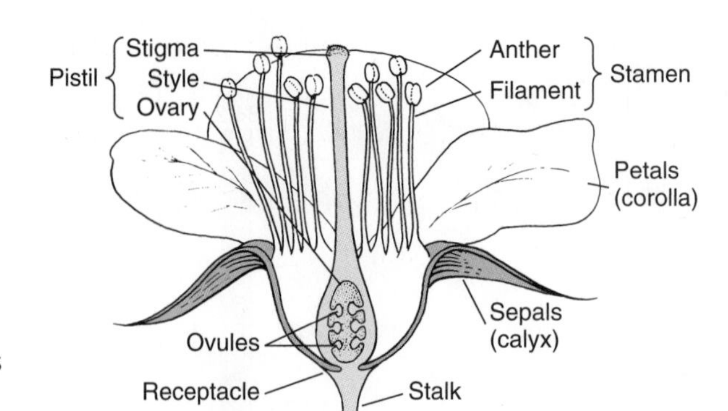

Figure 16-11 In plants, sexual reproduction takes place in the flowers.

contains eggs cells, as in animals. Above the ovary is the **style**. The style is usually long and thin. Located on top of the style is the "sticky" **stigma**. The pistil contains the female reproductive parts of the plant. Often seen arranged around the pistil are a number of **stamens**. Each stamen consists of an **anther** supported by a thin **filament**. The anther is the source of countless numbers of tiny pollen grains. Each pollen grain contains sperm cells. The stamens are the male reproductive structures of the plant.

Why is the stigma sticky? It must catch and hold the pollen grains that land on it. Each pollen grain begins to grow a tube down through the style. Eventually, a sperm nucleus from the pollen grain reaches and fertilizes an egg cell in the ovary. The sperm and egg cells fuse, fertilization occurs, and a zygote is the result. (See Figure 16-12.)

But why is the rest of the flower so showy? Anything that increases the chances of fertilization occurring—which involves getting pollen grains to the sticky stigma—is a survival advantage to the plant species. One of the best methods of getting pollen grains to the stigma is through the action of insects. Many flowers produce nectar, a nutrient-rich, sweet-smelling fluid. Insects such as bees are attracted by the nectar, since it is a good

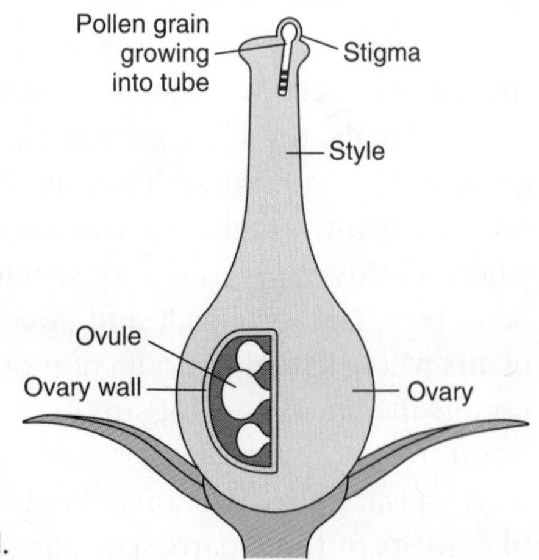

Figure 16-12 The pollen grain grows a tube down through the style until it reaches the ovary and fertilizes an egg cell.

source of energy. While gathering it, the bees carry pollen from the anther to the stigma. Other types of flowers have structures that allow wind to blow pollen from one flower to another. There are many different types of flowers and methods of **pollination**. In all cases, however, the same process, sexual reproduction, is about to occur after the transfer of pollen from an anther to a stigma. The next time you look at a flower, it may be of interest to remember that these are the reproductive organs of a plant. The ultimate purpose of a flower is to make more of its own kind.

SEXUAL REPRODUCTION: ON THE INSIDE OR THE OUTSIDE?

Various methods of sexual reproduction occur in plant and animal species. However, whatever the method, sexual reproduction always involves fertilization—the fusing of nuclei from gametes; and development—the growth of the zygote into a new individual. One of the main differences in the types of reproduction involves the location of the events. As strange as it may seem, both fertilization and development may occur either inside or outside the bodies of the reproducing organisms. Let's see how.

Single-celled algae and many plants live in water. Gametes, both sperm and eggs, meet in the open water. Fertilization occurs, and the zygote begins to develop. These events that occur in the environment, and not inside the organism, are known as external fertilization and external development. (See Figure 16-13.)

Although plants first evolved in water, many species eventually adapted

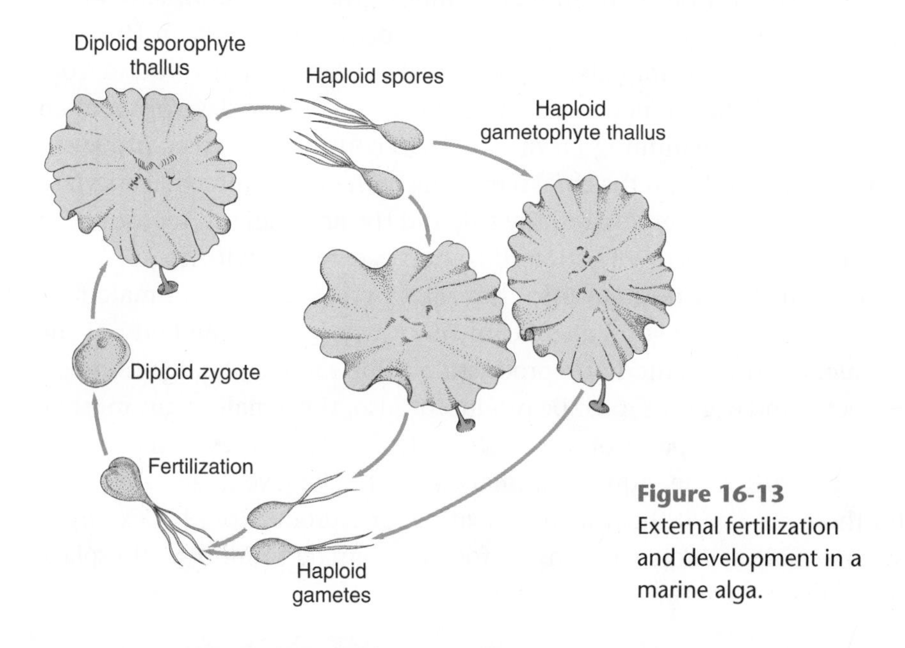

Figure 16-13 External fertilization and development in a marine alga.

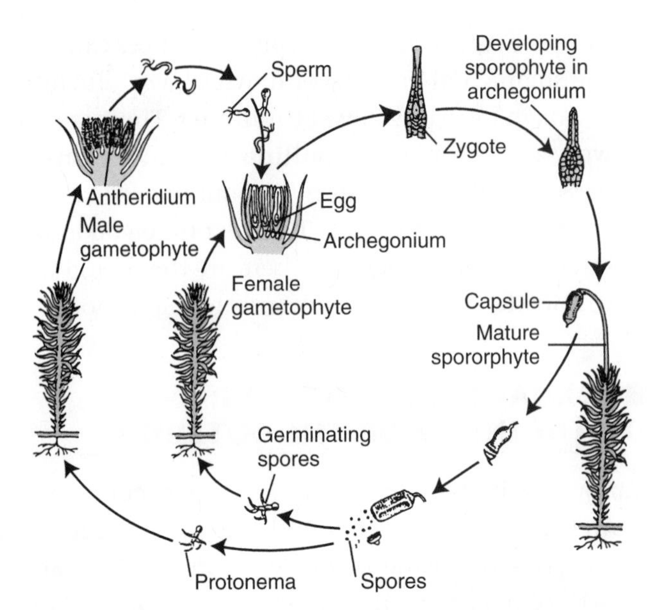

Figure 16-14 Internal fertilization and development in a moss plant.

to life on land. One of the important changes that occurred was the evolution of internal fertilization and development. For example, a moss plant consists of a small bundle of leaves that live on the moist forest floor. Egg cells develop inside a round structure on the plant. Sperm cells are attracted to this structure and swim to it. They enter it and fertilize the egg. The zygote develops into a new moss plant. (See Figure 16-14.)

External fertilization in animals requires a watery environment. Many invertebrates simply release their eggs and sperm into the water. This happens with the coral animals that make up the beautiful reefs around tropical islands. When a sperm and an egg unite, the zygote develops into an immature, free-swimming **larva**. The larva continues to grow and eventually settles down on the coral reef along with other adult coral animals.

Two groups of vertebrates, the fish and the amphibians, reproduce in water. In most species of fish, fertilization is external, with egg cells and sperm cells being released into the water by the female and male fish, respectively. The zygotes also begin their lives outside the body of the female. For this method of reproduction to be successful, large numbers of sperm and egg cells must be produced. Also, the female's eggs must be fertilized by the sperm of a male from the same species—preferably a strong, healthy one. Aquatic animals will often behave in amazing ways for these reasons at the time of spawning, or reproduction. For example, several male salmon will closely follow a female salmon to the place

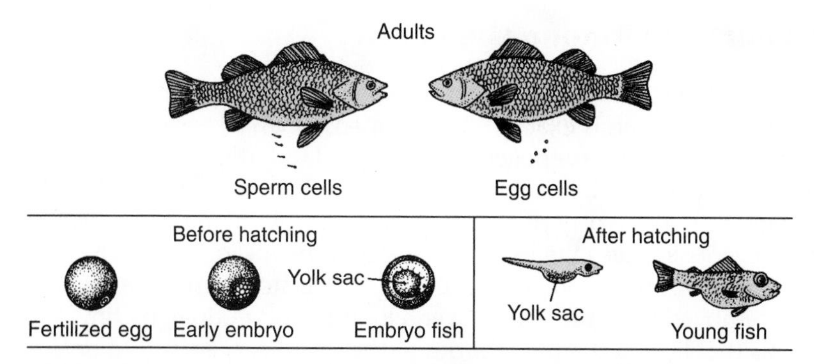

Figure 16-15 In most species of fish, fertilization and development are external.

where she prepares a nest in the river bottom. The males fight each other until the strongest one remains. The male is ready to release his sperm into the water the moment the female lays her eggs. On the river bottom, the fertilized eggs begin to develop. (See Figure 16-15.)

Amphibians also release their eggs into the water, where they are fertilized and develop into adults. Amphibians usually provide no parental care. However, there is one species of tropical frog in which the males carry the developing embryos as tadpoles in their stomachs. During this time, the male's stomach does not release digestive juices. Once the tadpoles have developed and are ready to live on their own, the male opens his mouth and out hop the young frogs!

Similarities and differences can be seen in the patterns of sexual reproduction in the vertebrates that reproduce on land. Gametes need moisture to meet and fuse. Reptiles and birds make use of the fluids inside their bodies for fertilization. Fertilization is internal, within the body of the female. The male and the female must mate for the sperm to be deposited in the female. Once fertilization occurs, the fertilized egg is prepared for development on land. A watertight membrane, the **amnion**, forms around the fertilized egg. (See Figure 16-16.) A protective shell also forms, the egg is laid, and the development of the new organism now occurs externally.

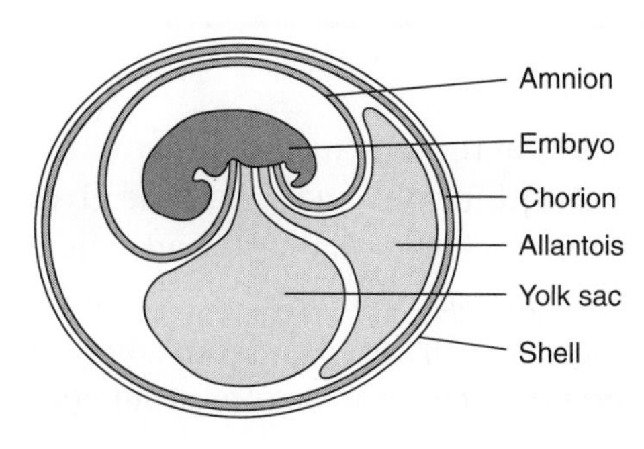

Figure 16-16 In birds, fertilization is internal. The amnion, a watertight membrane, forms around the fertilized egg, which is then covered with a shell. Development takes place externally.

Group Courtship Among the Corals

It seems impossible to imagine all organisms of a single species reproducing at the same time. But this is exactly what happens when all trees of a species simultaneously release pollen into the wind to fertilize the female flowers of that species. Still, in animal species, simultaneous reproduction of many individuals is a rare occurrence. However, one of the most spectacular underwater events involves the mass spawning of the millions of small organisms that make up a coral reef. A coral reef is a stony structure made of minerals removed from the water, over a long period of time, by tiny coral animals that live on the outer edges of the reef. Reefs are found only in the clear, warm, shallow waters of the tropics. During their lifetime, coral organisms remain in one place. They catch and remove food particles from the water that surrounds them. One good place to observe the mass spawning of coral animals is the Flower Gardens National Marine Sanctuary in the Gulf of Mexico.

One of the most amazing things about a mass spawning is its timing. It involves all coral animals in an area at the same time. The coral animals release packets of eggs and sperm simultaneously. This is called the *synchronous release of gametes.* Even more amazing is that this release of gametes happens only once a year, and at exactly the same time each year! At Flower Gardens, you must be diving on the eighth night after the full moon in August, between the hours of eight and eleven, to see this. What you see is the water teeming with brightly colored eggs intermixing with countless packets of sperm. To see this again, you need to come back one year later, again eight nights after the August full moon, and at the same time in the evening.

Scientists believe that mass spawning increases the chances of successful fertilization in three ways. First, with so many eggs and sperm in the water at the same time, fertilization is more likely to occur. Second, with gametes from different colonies of one species being released at the same time, cross-fertilization between different colonies is more likely. This increases the genetic variation among the offspring. Finally, with so many fertilized gametes in the water at once, the amount lost to predation is limited.

This is "love on the rocks," and it happens once every year.

(In Chapter 18, you will look more closely at the structure of the amniotic egg.) The eggs of some snakes develop within the female, and she gives "birth" to live young. Turtles simply bury their eggs in the ground, where they hatch several weeks later. Alligators prepare elaborate nests for their eggs. During incubation and after the eggs hatch, the alligator parents care for their offspring. Most birds prepare nests and provide careful supervision of the eggs by sitting on them, keeping the eggs warm until they hatch. The parents then care for the young, bringing food to and providing protection for the hatchlings.

One final pattern of sexual reproduction takes place only in mammals. Fertilization occurs internally. Males often fight with one another for the chance to mate with an available female or females. With many mammals, mating occurs only at a specific time of the year. What is the big difference between mammals and most other animal groups? It is that the development of the zygote occurs within the body of the female. The food for the developing embryo comes entirely from the body of the mother. A structure, the **placenta**, has evolved to bring nutrients to the developing baby and to remove wastes. After birth, continuing nourishment of the baby mammal occurs through its nursing on milk provided by the mother's mammary glands. Of course, there are exceptions even among the mammals. *Marsupials*, such as kangaroos, are mammals that do not have a placenta. Instead, they give birth to very undeveloped embryos, which continue to develop and receive nutrition from mammary glands within the mother's pouch. The platypus and the echidna are unusual mammals that lay eggs, which develop outside of the female's body. Yet they also nurse their offspring after they hatch.

Two ways to compare patterns of sexual reproduction are in terms of the number of eggs and the amount of parental care provided. External fertilization and development are risky. Eggs and sperm may not meet and, even if they do, the fertilized eggs may easily be destroyed before they can develop. Large numbers of eggs increase the chances of reproduction and the chance that some offspring will survive. Internal fertilization, while more complex, increases the chances of reproductive success and survival; so, fewer eggs are produced. Providing parental care also protects the developing embryo. The most complete form of protection, of course, is allowing the embryo to develop within the body of the female. (See table below.)

SEXUAL REPRODUCTION IN VERTEBRATES

Vertebrate	Fertilization	Development	Number of Eggs	Parental Care
Fish	Most external	Most external	Many	None, usually
Amphibians	External	External	Many	None, usually
Reptiles	Internal	Most external	Not very many	Little
Birds	Internal	External	Few	Much
Mammals	Internal	Most internal	Few	Much

LABORATORY INVESTIGATION 16
What Are the Relationships Among the Structures of Flowers, Fruits, and Seeds?

INTRODUCTION

Because a flower is the reproductive organ of higher plants, someone once suggested that a "flower is a plant's way of making more flowers." In this investigation, you will observe and compare the structures of flowers, fruits, and seeds.

MATERIALS

Gladiolus flowers, scalpels, hand lens, pea pods, presoaked lima beans

PROCEDURE

1. Observe the gladiolus flower. Describe the appearance of the sepals and the petals. Carefully remove the sepals and petals. Leave the rest of the flower intact and draw it in your notebook.
2. Use a scalpel to cut along the length of the ovary. Use a hand lens to observe the internal structures. Remove a stamen and rub the anther on your finger. Describe what you observe.
3. Carefully open the pea pod. Observe, draw, and label the contents.
4. Obtain a lima bean that has been soaked in water. Locate the tiny opening along the inner edge of the bean and the scar from the point of attachment to the pod. Draw what you observe.
5. Carefully remove the seed coat from the lima bean. Gently separate the halves of the bean to expose the embryo inside. Use a hand lens to observe the embryo. Draw and label what you see.

INTERPRETIVE QUESTIONS

1. Explain the relationship of a plant's fruit and seeds to the structure of its flower.
2. What is the purpose of a seed? How is a seed's structure related to this function? What are the functions of a fruit?
3. Why are tomatoes, peppers, and cucumbers actually fruits and not vegetables, as many people believe?

CHAPTER 16 REVIEW

Answer these questions on a separate sheet of paper.

VOCABULARY

The following list contains all of the boldfaced terms in this chapter. Define each of these terms in your own words.

amnion, anther, chromosomal abnormalities, crossing-over, diploid, fertilization, filament, gametes, haploid, homologous chromosomes, independent assortment, karyotype, larva, meiosis, ovary, pistil, placenta, pollination, recombination, reduction division, sexual reproduction, stamens, stigma, style, tetrads, zygote

PART A—MULTIPLE CHOICE

Choose the response that best completes the sentence or answers the question.

1. Reduction division is another term for *a.* mitosis *b.* Meiosis I *c.* Meiosis II *d.* Prophase II.
2. How many chromosomes are found in a normal human egg cell? *a.* 8 *b.* 16 *c.* 23 *d.* 46
3. Down syndrome results when an individual *a.* has an extra copy of chromosome 21 *b.* is exposed to harmful chemicals before birth *c.* is missing a chromosome *d.* has an extra piece of chromosome 17.
4. New combinations of genes result from *a.* budding *b.* mitosis *c.* asexual reproduction *d.* sexual reproduction.
5. An example of a gamete is a *a.* sperm cell *b.* zygote *c.* stigma *d.* maternal chromosome.
6. Which is a male reproductive structure in a flower? *a.* ovary *b.* stigma *c.* style *d.* stamen
7. Crossing-over may take place during *a.* Anaphase I *b.* Metaphase II *c.* Prophase I *d.* Telophase II.
8. Because of independent assortment, *a.* gametes can receive different mixes of maternal and paternal chromosomes *b.* maternal chromosomes are separated from paternal chromosomes during reduction division *c.* mothers over the age of 35 are at greater risk of having a child with Down syndrome *d.* organisms with fewer offspring are more likely to give more parental care to each offspring.

9. A multicellular, immature stage in the life cycle of an organism that is unlike the adult form is *a.* a zygote *b.* a gamete *c.* an anther *d.* a larva.
10. Which of these animals would you expect to have internal fertilization? *a.* goldfish *b.* butterfly *c.* coral *d.* frog
11. In humans, the diploid number is *a.* 8 *b.* 16 *c.* 23 *d.* 46.
12. Homologous chromosomes *a.* match in shape and size *b.* stay together during Anaphase I *c.* come from the mother only *d.* come from the father only.
13. In flowers, what structure produces pollen? *a.* ovary *b.* anther *c.* filament *d.* stigma
14. Which of these mammals nourishes the developing embryo through a placenta? *a.* kangaroo *b.* duckbill platypus *c.* echidna *d.* dog
15. Tetrads line up along the center of a cell during *a.* Meiosis II *b.* fertilization *c.* Metaphase I *d.* Prophase II.

PART B—CONSTRUCTED RESPONSE

Use the information in the chapter to respond to these items. (Note: Two X chromosomes indicate a female; an X and a Y indicate a male.)

16. What is shown in the diagram? How is it used to understand chromosomal abnormalities in humans? 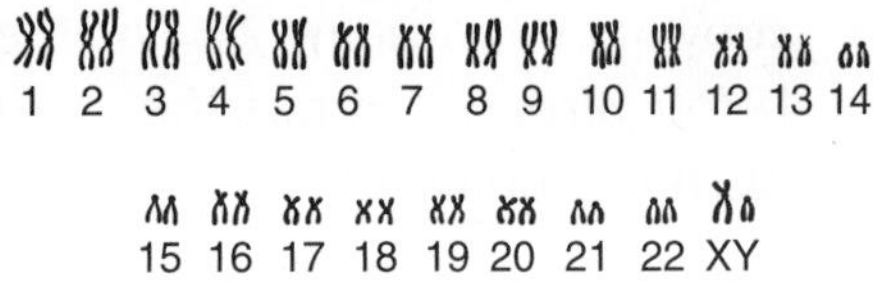
17. Examination of the diagram shows a *a.* male with Down syndrome *b.* male without Down syndrome *c.* female with Down syndrome *d.* female without Down syndrome.
18. Compare and contrast mitosis and meiosis. Be sure to consider where in the body these processes occur as well as the steps and results of each process.
19. What is the significance of crossing-over?
20. Discuss the general evolutionary trends in vertebrates regarding fertilization, number of eggs, location of development, and amount of parental care.

PART C—READING COMPREHENSION

Base your answers to questions 21 through 23 on the information below and on your knowledge of biology. Source: *Science News* (December 21, 2002): vol. 162, pp. 189–190.

Novel Enzyme Provides Sperm's Spark of Life

Biologists may have finally found what they call the "spark of life," a molecule in sperm that triggers a fertilized egg to begin developing.

Immediately after a sperm penetrates an egg, several waves of calcium flow out of the egg's stores of the ion. These calcium surges set off development of the fertilized egg. For more than a century, biologists have speculated that sperm must contain something that liberates this calcium. Several egg-activating factors have been proposed, but none has withstood scrutiny.

Because of its calcium-releasing role in some other cells, an enzyme called phospholipase C (PLC) was among the suspects. None of the known versions of PLC fits the bill as an egg activator, however.

Now, in the Aug. 15 *Development*, F. Anthony Lai of the University of Wales College of Medicine in Cardiff and his colleagues report the discovery of a new form of PLC that's present only in sperm. Moreover, when injected into an unfertilized egg, the enzyme stimulates calcium surges identical to those caused by sperm. This enzyme may provide a seemingly natural means of activating eggs in cloning or other forms of artificial reproduction, the scientists suggest.

Given the history of this issue, the role of the new PLC must be verified "10 times over," cautions Sergio Oehninger of the Jones Institute for Reproductive Medicine in Norfolk, Va.

21. Explain where the molecule called "spark of life" is found and what it does.
22. Describe the events that occur as soon as a sperm enters into an egg cell.
23. State the discovery that may explain why a fertilized egg begins developing.

17

Human Reproduction

After you have finished reading this chapter, you should be able to:

Describe the structures and functions of the male and female reproductive systems.

Explain the importance of hormones to the functioning of reproductive systems.

Discuss issues that affect reproductive health: contraception, infertility, and STDs.

Where did you come from, Baby dear?
Out of the everywhere into the here.

George Macdonald, *"Baby,"*
At the Back of the North Wind

Introduction

Everyone loves the news stories about the taxi driver who delivers a baby for the woman who could not get to the hospital on time. At some point each of us asks ourselves with awe for the first time, "Where did *I* come from?"

It has been said of the body that it is "fearfully and wonderfully made." How true this is of the reproductive systems on which the survival of our species depends. In this chapter, you will study the systems of the human body responsible for our being here: the male and the female reproductive systems.

THE MALE REPRODUCTIVE SYSTEM

The male reproductive system has two main functions: It produces haploid male gametes, the **sperm** cells; and it must be able to deposit the sperm cells it produces inside the female. (As described in Chapter 16,

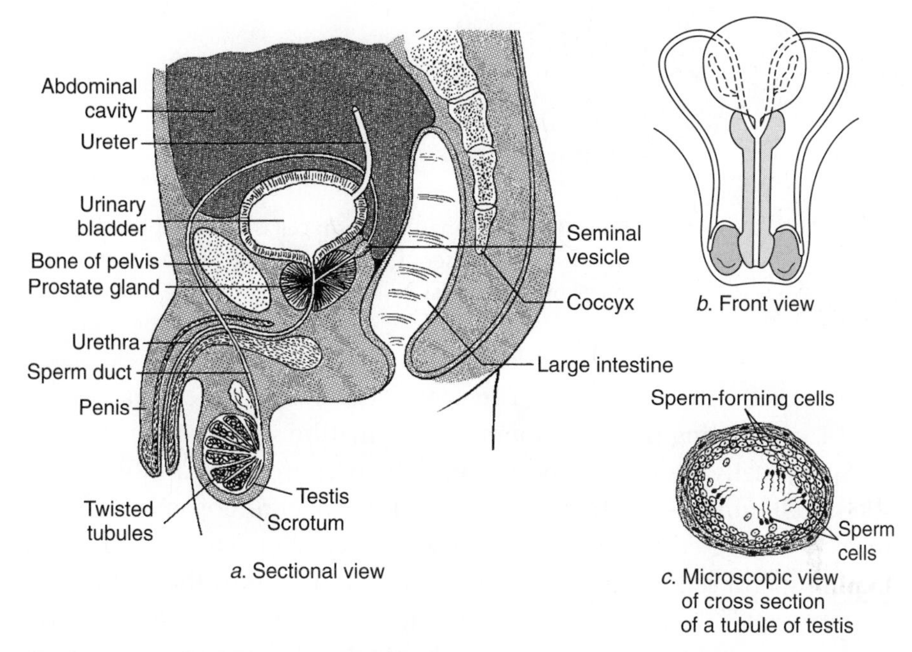

Figure 17-1 The male reproductive system.

fertilization in all mammals, including humans, occurs internally.) In addition, the male reproductive system provides a pathway for the removal of urine. (See Figure 17-1.)

The first function, sperm cell formation, occurs in the two **testes**. The formation of sperm requires a temperature that is a few degrees cooler than the temperature of the rest of the body. How is this temperature reduction produced? The testes are suspended within a sac called the **scrotum**. Because it is not located within the body cavity, the temperature of the scrotum is a few degrees lower than normal body temperature. The scrotum is an adaptation that has evolved to increase the chances of producing healthy sperm.

Inside the testes are a great many tiny tubes, or tubules. In fact, if all the tubules in a single testis were laid end-to-end, they would reach from one end of a football field to the other, seven times over! As normal diploid cells move through these tubules, they undergo meiotic cell division. The production of haploid gametes occurs in a process called **gametogenesis**. The formation of sperm cells is called *spermatogenesis*. Nowhere else in the male's body does meiotic cell division occur. In the tubules, the cells go through two divisions that produce four haploid **spermatids** from each diploid cell. The spermatids are actually immature sperm and are pushed from the tubules into the **epididymis**, a tubule about 6 meters

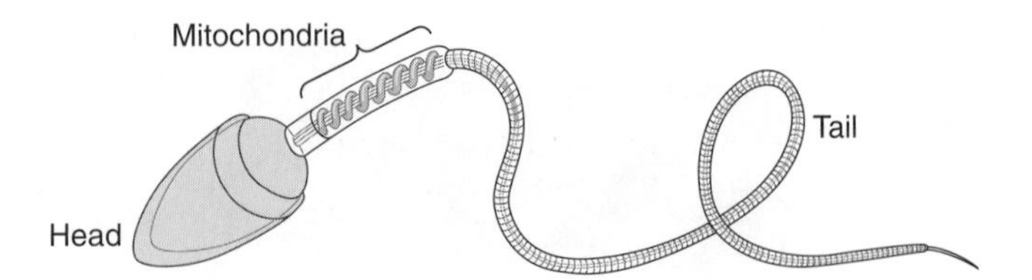

Figure 17-2 The mature sperm cell is well adapted to deliver the haploid set of chromosomes from the male to the egg cell of the female.

long packed into each testis. The spermatids remain there for several weeks, completing their development into mature sperm cells.

Sperm cells are highly specialized cells that are able to move. Each sperm cell must be able to deliver a haploid set of chromosomes from the male to an egg cell in the female reproductive tract. The structure of a mature sperm cell is well adapted to its function. Almost the entire head of the sperm is the haploid nucleus, the all-important genetic information that is delivered to the egg. Attached to the head of the sperm is a long tail that propels the sperm along. Also present are large numbers of mitochondria that produce ATP, which yields the energy the sperm use to propel themselves to the egg. (See Figure 17-2.)

From the epididymis, sperm move into the **vas deferens**, the beginning of their pathway out of the body. As sperm move through the vas deferens, a number of glands add fluids. Sperm and these added fluids make up the **semen**. In fact, most of the semen is not sperm. Instead, it consists mostly of fructose, a sugar that provides an additional source of energy for the sperm. Prostaglandins are also found in the semen. Prostaglandins are chemicals that cause muscle contractions to occur in the female reproductive system. These contractions help move the sperm toward the egg. In addition, semen has an alkaline pH. This high pH neutralizes the acid in the male's urethra and in the female reproductive system that might damage the sperm.

The male reproductive system is adapted for internal fertilization. The **penis** is a structure that has evolved to deposit sperm safely within the female's body. The tube that passes through the penis, the urethra, has two functions. It is the pathway for sperm out of the body as well as the pathway for urine to leave the body after having been stored in the urinary bladder. At the time of sexual excitement, a valve closes. This valve prevents the bladder from releasing urine. At the same time, nerve impulses from the brain cause blood vessels in the penis to relax. Blood rushes in and fills empty spaces in the spongy tissue in the penis. The penis enlarges and becomes rigid. The result is an erection. Strong muscular contractions occur in a sequence of events called an orgasm. Semen is forced from the body during **ejaculation**. The semen in a single ejaculation contains up to 300 million sperm. Any one of these sperm cells could fertilize an egg and lead to the birth of a new human.

THE FEMALE REPRODUCTIVE SYSTEM

Three important functions are performed by the female reproductive system. First, gametes (eggs) are produced in the ovaries. Second, a pathway is provided for sperm cells to reach an egg. Third, the female reproductive system provides a temporary home for the developing embryo. (See Figure 17-3.)

In females, gametogenesis produces haploid egg cells in the **ovaries**, a pair of reproductive organs. The production of egg cells, a process called *oogenesis,* actually begins in a woman's body before she is born. Approximately 2 million cells in the female fetus have already begun the first phases of meiosis in the immature ovaries by the time of birth. Nothing further happens to these cells until the female reaches puberty. In an adult male, sperm production occurs all the time; an average of 30 million sperm are produced each day. In a female, all potential eggs are present when she is born. Throughout her reproductive life, a female releases only a few hundred of these eggs. Usually only a single egg matures and is released each month. During egg formation, meiosis produces four cells after two divisions. However, only one mature egg is made. Along with this single mature egg, three other smaller structures are formed. Because

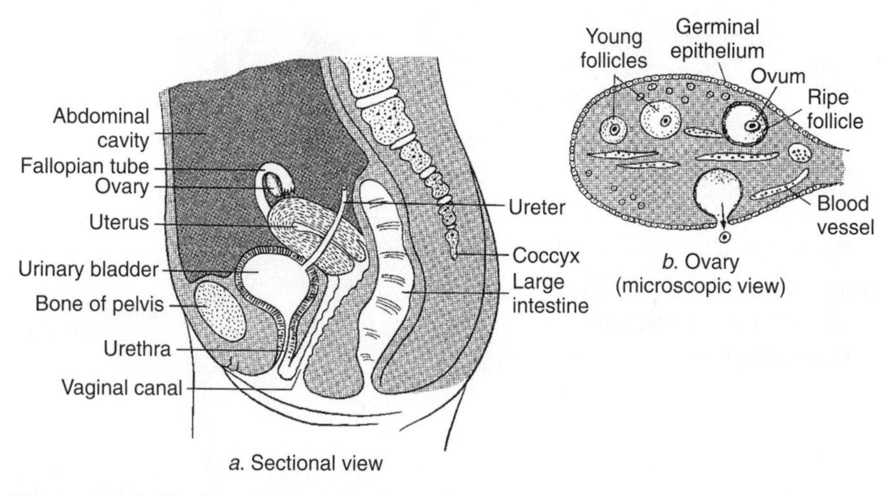

Figure 17-3 The female reproductive system.

an egg cell needs a relatively large amount of raw materials, nutrients stored in the cytoplasm are not divided equally between the egg and the three other structures produced during meiosis. Instead, just the large, single egg is packed with the nutrients needed to nourish the embryo right after fertilization.

This development of egg cells occurs within the ovary once a month. One mature egg cell is released from one of the ovaries. The ovaries contain **follicles.** A follicle is a hollow bundle of cells with an egg inside. The follicle helps the egg mature. As the egg matures, the follicle enlarges and fills with fluid. The greatly enlarged follicle moves to the edge of the ovary and bursts, releasing the egg and the fluid. This event is **ovulation**, the release of an egg from the ovary. (See Figure 17-4.)

The egg cell gets swept into the **oviduct** (or fallopian tube), a long tubular structure found next to each ovary. If fertilization occurs, the sperm usually joins the egg in the oviduct. The egg continues to move along the oviduct to the **uterus**, a pear-shaped organ with thick muscular walls. If the egg cell was fertilized, the embryo becomes attached to the inside wall of the uterus and continues to develop. If fertilization did not

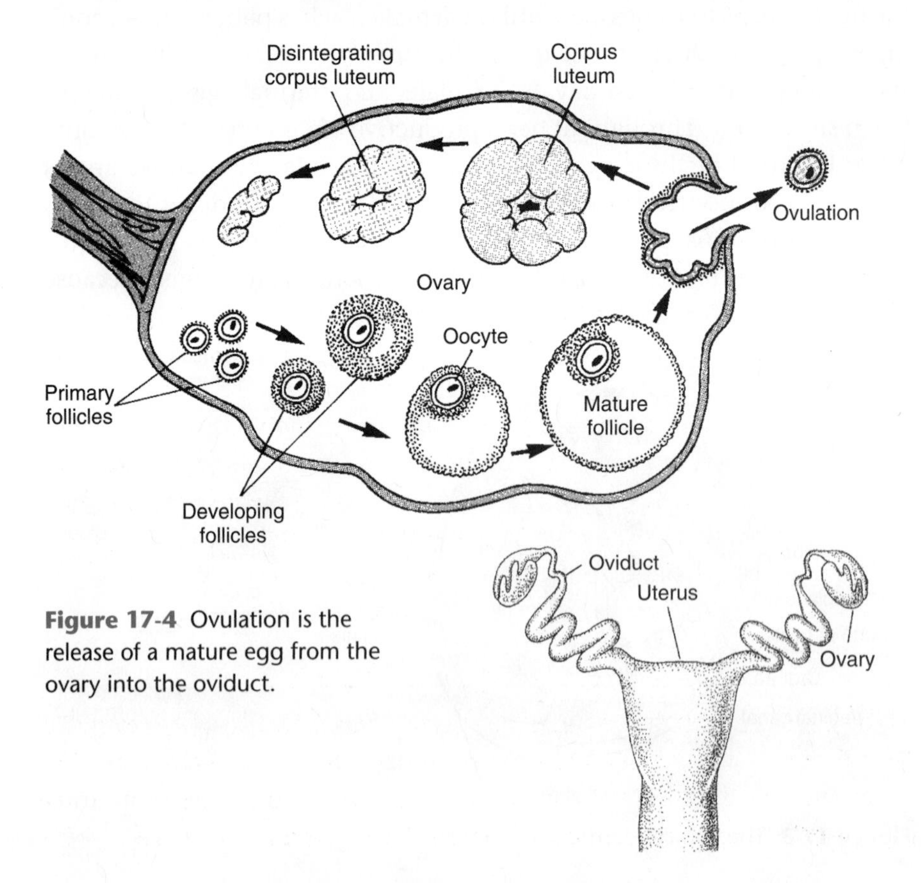

Figure 17-4 Ovulation is the release of a mature egg from the ovary into the oviduct.

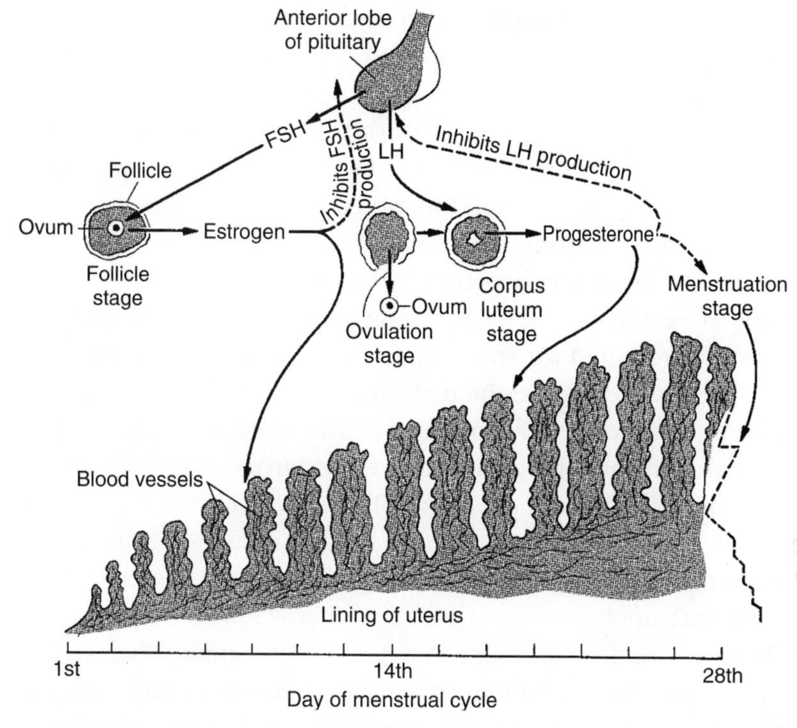

Figure 17-5 Ovulation occurs in the middle of the menstrual cycle.

occur, the egg cell breaks down within 24 hours of ovulation and is passed from the body, along with some blood and tissue from the uterine lining. (See Figure 17-5.)

At the lower end of the uterus is the **cervix**, a narrow opening. It is through the cervix that the sperm traveled on their way to the egg cell. Connecting the cervix to the outside of the body is the **vagina**. The vagina is made up of muscular tissue. It is into the vagina that sperm are ejaculated from the penis. Also, the vagina is the birth canal, through which the infant passes as it leaves the mother's body during childbirth.

Unlike in males, the reproductive pathway in females is not combined with the pathway for excretion. The urethra, through which urine passes from the urinary bladder, ends at an opening near the vagina.

HORMONES AND SEXUAL REPRODUCTION

During one's life, many changes and events occur in the body to make sexual reproduction possible. Hormones coordinate these changes. The main endocrine gland in charge of producing these hormones is the anterior pituitary gland in the brain. The pituitary gland is controlled by the hypothalamus, a part of the brain. This connection suggests that, in some way, thoughts stimulate the release of hormones in the brain that affect the reproductive system.

Two main hormones, **follicle-stimulating hormone** (FSH) and **luteinizing hormone** (LH), are secreted from the anterior pituitary. These hormones were first named for their functions in females. Today, we know that these hormones have important functions in males, too.

The effects of a hormone depend not on the hormone itself but on the hormone's target tissue. In males, the target tissues for FSH and LH are the testes. FSH is required for sperm formation to occur. LH causes the testes to produce **testosterone**, the main male sex hormone. The effects of testosterone include the development of the male sex organs before and after birth. Without the presence of testosterone, the fetus would develop into a female. Around the age of 11, the level of testosterone suddenly increases in a boy's body. As a result, sperm production begins. This event is the beginning of **puberty**. During puberty, the penis and the testes begin to mature.

Testosterone also affects various other tissues in the male. Testosterone causes pubic and underarm hair growth, causes changes in body proportions, increases muscle tissue, and deepens the voice. Testosterone may influence the development of acne and the beginning of body odors from bacteria attracted to sweat glands. Because none of these characteristics are directly related to sexual reproduction, they are called secondary sex characteristics. In males, the level of testosterone in the body remains much the same for about 40 years after puberty. After that time, the level of testosterone gradually begins to decrease.

In females, FSH and LH from the anterior pituitary stimulate the release of the major sex hormones, **estrogen** and **progesterone**, from the ovaries. Estrogen and progesterone are produced in the ovaries. The onset of puberty in females occurs somewhat earlier than in males. At about age 10, the levels of estrogen and progesterone increase dramatically, causing the uterus, vagina, and ovaries to mature. Secondary sex characteristics, including the growth of pubic and underarm hair, breast development, acne, and body odors, also are influenced by estrogen and progesterone. In addition, a monthly cycle of events, known as the **menstrual cycle**, begins. Remember, in males, sperm production occurs all the time after puberty. In females, the menstrual cycle occurs every month after puberty. Part of this cycle includes the release of an egg cell (or egg cells) from the ovaries. (See Figure 17-6.)

If you take a close-up look at the ovary, you can see that an egg develops in its own follicle. The development of the egg is stimulated by

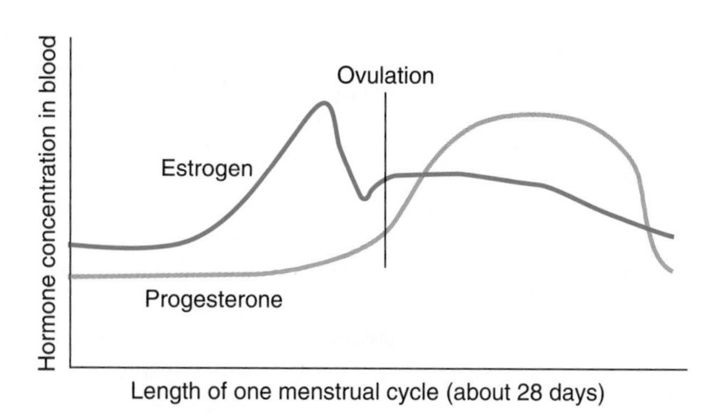

Figure 17-6 The menstrual cycle and ovulation are controlled by the release of hormones.

FSH. The follicle releases estrogen which, in turn, stimulates a sudden release of LH from the anterior pituitary. The release of LH happens about 14 days into the menstrual cycle and causes ovulation—the release of the egg—to occur. Eggs can be fertilized for up to 10 to 15 hours after ovulation.

In addition to ovulation, another critical function occurs during the menstrual cycle. The woman's body must be prepared in case fertilization occurs. Everything must be ready to nurture the developing embryo. During the first two weeks of the cycle, estrogen causes the lining of the uterus to thicken. There is also an increase in the amount of blood that flows to this area.

During the second half of the cycle, after ovulation has occurred, the follicle (now without the egg cell) develops into a yellow-colored body, the **corpus luteum**. The corpus luteum begins to secrete progesterone, the pregnancy hormone, which prepares the uterus for an embryo. Sperm can survive in the female for up to 72 hours after sexual intercourse. Therefore, sexual intercourse a few days before or a few days after ovulation may result in pregnancy. If pregnancy occurs, the embryo becomes attached to the inner lining of the uterus. The growing tissue then begins to release more hormones to keep everything in the right condition. However, if fertilization does not occur, the continued preparations in the uterus are unnecessary. The body realizes this near the end of the four-week period of the menstrual cycle. At this time, the level of LH decreases, the corpus luteum breaks down, and the level of progesterone decreases. Without this hormone, the uterine lining no longer remains intact. Instead, the lining breaks down. The built-up tissue along with some blood and the unfertilized egg are released from the body. This flow of blood, called **menstruation**, lasts for about four days. Then the levels of FSH again begin to increase. A new follicle releases an egg. The cycle continues. (See table on page 368, which describes the menstrual cycle.)

THE FOUR PHASES OF THE MENSTRUAL CYCLE

Follicle Phase	Ovulation	Corpus Luteum Phase	Menstruation
Pituitary gland secretes FSH and LH into blood → follicle grows and matures within ovary → estrogen levels rise and uterine lining thickens. (10–14 days)	Pituitary gland secretes burst of FSH and LH → follicle ruptures → mature ovum is released from ovary into the oviduct (where it can be fertilized). (3–4 days)	Ruptured follicle (corpus luteum) continues to release estrogen and starts to release progesterone → uterine lining fully ready to receive fertilized ovum. (2–3 days)	If unfertilized, ovum passes through uterus → corpus luteum disintegrates, so it releases less estrogen and progesterone into blood → uterine lining is shed and discharged. (3–7 days)

In most women, the menstrual cycle continues for at least 40 years from puberty. Between the ages of 45 and 55, the levels of hormones change; the menstrual cycle becomes less regular and eventually stops. This stage, called **menopause**, marks the point at which a female is no longer capable of reproducing. Menopause is a normal occurrence in all women; however, the effects of menopause vary widely from one woman to another. In men, by contrast, sperm production continues throughout life, although the number of healthy sperm likely declines with age.

Check Your Understanding

How is it true that "the effects of a hormone depend not on the hormone itself but on the hormone's target tissue"—specifically in the case of FSH and LH in both males and females?

SEXUAL REPRODUCTION IN OTHER MAMMALS

Only human females and some other higher primates have a regular monthly menstrual cycle. Humans, however, are able to mate at any time. Females of almost all other mammal species will mate only at specific times during a year. These times are known as **estrus** or "heat." It is only during estrus that egg cells are released for fertilization. Estrus occurs in wolves and deer once a year, in dogs usually twice a year, in cows and horses once a month, and in rats and mice every few days.

Ovulation, mating, and fertilization occur on a specific timetable. In many animals, such as deer or sheep, estrus occurs during the autumn. This mating pattern evolved through natural selection and allows the development of the embryo to occur during the winter. The young are born in early spring, when food is more plentiful. The newborn then have plenty of time to grow before the next harsh winter sets in.

Human females are among the few animals that allow mating at times when no eggs have been released. In other words, sexual intercourse in humans can occur even when fertilization is not possible. It has been suggested that this sexual pattern evolved in the human species in order to develop strong bonds between males and females. Such ties between the sexes are the basis of the human family unit. Is there a connection among biology, behavior, and the strength of family ties?

BIRTH CONTROL

Over the years, a wide variety of methods have been developed to reduce the chances of becoming pregnant. These methods are called birth control or **contraception**.

Birth control helps a couple plan if and when to have a child. However, birth control gained even more importance in the late twentieth century for two additional reasons. The world's human population is growing faster than at any other time in history. For example, in 1800, Earth supported a population of almost 1 billion people. By the year 2000, the human population reached 6 billion. If the current rate of growth continues, that number will double within the next 40 years. Today, the highest rates of population growth are in the poorest, least-developed countries. Population growth is a matter of great concern throughout the world and is of critical concern in certain countries. As a result, family planning, including the use of various birth control methods, is being arranged not only by individual couples but also by government agencies.

In countries such as India, Indonesia, Mexico, and China, the population growth rate is very high. The governments in these countries are concerned about their ability to feed and care for an ever-increasing population. China has given the most attention to family planning. In 1970, China began an official "one child–one family" policy. In each community in China, government agencies work with couples to reach this goal. The rate of population growth has declined in China, but other social problems are developing. For example, how will a country take care of its older population if there are fewer younger people working to support them? Other countries are watching China closely to see if it succeeds in meeting its population growth goals. Indeed, because China has almost one-quarter of Earth's total human population, the success of family planning in China is an important matter for the entire world. (See table on page 370, which lists the most populous countries.)

WORLD'S MOST POPULOUS COUNTRIES IN 1996

Country	Population in Millions	Percent of World's Total Population
China	1,232	21.4
India	945	16.4
United States of America	269	4.7
Indonesia	200	3.5
Brazil	161	2.8
Russian Federation	148	2.6
Pakistan	140	2.4
Japan	125	2.2
Bangladesh	120	2.1
Nigeria	115	2.0

The second reason why contraception has gained importance is due to disease. As we will see at the end of this chapter, many serious diseases can be passed between people during sex. Some birth control methods are also important ways to reduce the transmission of certain diseases.

Over time, various methods of birth control have been developed. Some methods are more effective than others. The most effective means of birth control is **abstinence**, that is, not having sexual intercourse. Abstaining from sexual intercourse prevents sperm from reaching an egg and also limits the spread of sexually transmitted diseases.

Other birth control methods work by interfering with the process of reproduction. The two most effective methods are both forms of sterilization. In males, the operation called a vasectomy cuts and closes the vas deferens. This operation prevents sperm from being included in the fluids that are ejaculated. The male's experience of orgasm is not affected. In females, tubal ligation surgery cuts and closes off the oviducts. Fertilization cannot occur because the sperm and the egg cannot meet. These two operations are nearly 100 percent effective at preventing a pregnancy from occurring. In some cases, these operations can be reversed and the ability to produce a child is restored. These two procedures have no effect on the experiences felt during sexual intercourse.

Barrier methods such as condoms and diaphragms also block the movement of sperm. A condom is a sheath that covers the penis. Usually made of latex, condoms when used properly are also highly effective in preventing the transmission of disease during sex. The possibility of contracting serious, life-threatening diseases has increased the need for individuals to practice "safe sex." A diaphragm is a latex barrier placed over the cervix, the entrance to the uterus. A diaphragm prevents sperm from reaching the area of the female's reproductive tract where fertilization of an egg is possible. The use of condoms and diaphragms can be combined with the use of a spermicide, a chemical substance that kills sperm. Using a spermicide with a condom or a diaphragm greatly reduces the chances that an unwanted pregnancy will occur.

Birth control pills contain hormones that prevent ovulation. The estrogen and progesterone in these pills keep the levels of these two hormones in the female body high. As a result, FSH and LH are not released from the anterior pituitary. Therefore, oogenesis and ovulation—which depend on these two hormones—do not occur. In 1990, another method of delivering hormones that prevent ovulation was approved. In this method, thin strips of material that release hormones are placed under the skin. The strips remain effective for about five years, when they can be replaced. The strips can be removed at any time. After their removal, the levels of hormones in the blood return to normal. The ability to become pregnant is restored. There is also an injectable hormone that prevents pregnancy from occurring. (See Figure 17-7.)

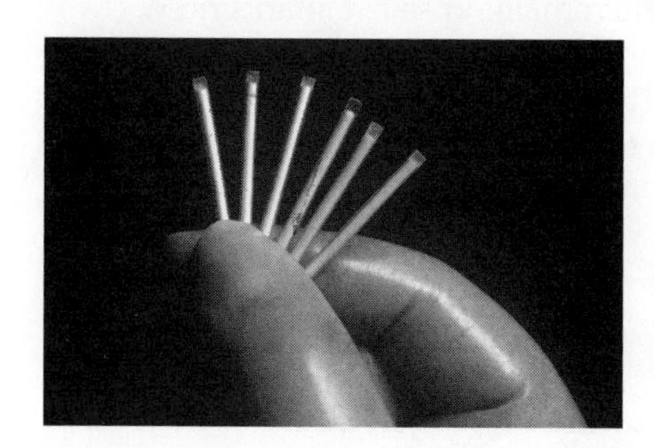

Figure 17-7 Contraceptive implants for women, which release a synthetic progesteronelike hormone, can be effective for five years.

Pregnancy will not occur, of course, if sexual intercourse does not happen. Although some methods of birth control are almost as effective as abstinence, other methods are not. Whichever methods are used must be carefully considered by both individuals in a sexual relationship. If a woman becomes pregnant, both partners have to be able to deal with the emotional, financial, and medical responsibilities.

WHEN THINGS GO WRONG: INFERTILITY

Approximately 10 percent of couples in the United States who are trying to have a child are unable to do so. This condition is called **infertility**. In males, infertility is most often due to a lower-than-normal number of healthy sperm. Occasionally, the cause of this problem is simple to

Texas Woman Makes History by Delivering Octuplets

On December 8, 1998, Nkem Chukwu gave birth to a 1-pound, 6-ounce girl. At the time of the baby's birth, doctors discovered that the mother was carrying seven additional fetuses. At this point, the doctors were able to stop her labor for two more weeks. They wanted to give the tiny fetuses more time to develop. On December 20, the remaining seven infants were born by cesarean section at St. Luke's Episcopal Hospital in Houston, Texas. For the first time in recorded history, a woman had given birth to eight surviving babies.

According to the *Guinness Book of World Records,* the largest recorded multiple birth was nine babies born in Sydney, Australia, in 1971. However, only six of the nine survived. Three other octuplet births have been recorded—one each in Britain, Spain, and Turkey. In one of these cases, six children survived; in the other two, no children survived.

Although they were born in Nigeria, Nkem Chukwu and her husband are American citizens who now live in Houston. Mrs. Chukwu took fertility drugs to help her become pregnant. She was offered the option to reduce the number of fetuses she carried to increase the remaining infants' chances of surviving, but she refused. One week after they were born, the smallest baby, a girl, died. However, the remaining seven infants grew stronger, left the hospital after five months, and by December 2000 were described by their pediatrician, Dr. Patti Savrick, as "normal, healthy two-year-olds." Surprisingly, a healthy baby girl was born to Mrs. Chukwu in October 2002, joining her seven famous four-year-old brothers and sisters.

remedy. For example, a lack of zinc in the diet is one cause of infertility. Additional zinc can increase healthy sperm production.

In females, infertility is usually due to the lack of regular ovulation or to a blocked oviduct. Hormones known as fertility drugs can be given to stimulate ovulation. These drugs often cause several eggs to be released during ovulation. Today, it is more common for triplets, quadruplets, and even quintuplets to be born than it was in the past. The use of fertility drugs has increased the number of multiple births.

If the oviduct is blocked, a couple may attempt **in vitro fertilization**. The term *in vitro* means "in glass." Eggs and sperm are placed in a laboratory dish. If fertilization occurs, several embryos are placed in the woman's uterus. Even with this method, pregnancies occur no more than 20 percent of the time. The name "test-tube babies" has been used to describe children born through this method.

It was reported that Mrs. Chukwu had taken two hormone-based fertility drugs—follicle-stimulating hormone and human chorionic gonadotropic hormone—to produce multiple eggs. The occurrence of this woman giving birth to eight babies raises some serious questions about the use of fertility drugs. In this case, the eight babies were born very prematurely. They spent only six-and-one-half months developing instead of the normal nine months. The beginning of their lives involved intravenous feeding, living in closely monitored incubators, and breathing through a tube attached to a respirator. Once the eight premature babies got past the grave health risks that could occur in the first few days, they had to face the risks of life-threatening infections due to their almost nonexistent immunity at birth.

Fertility experts frequently express concerns about multiple births such as this. Dr. Alan DeCherney, chairman of the Department of Obstetrics and Gynecology at UCLA, says, "The risk here is tremendous with these kids, of death as well as of severe neurological impairment. So there's just no reason to do this." He pointed out that doctors are able to limit the number of eggs that become fertilized in the womb. Another specialist in reproductive medicine, Dr. Jirain Konialian, said it was "reckless" for a doctor not to monitor egg production more closely to reduce the chances for such a high-number multiple birth. Clearly, there are some people who see the birth of octuplets as a wondrous event, and others who view this situation as a consequence of reckless medical practice that can endanger the lives of the babies and even the mother, too.

Embryos from in vitro fertilization can also be frozen and used at a later time. They can be returned to any woman's body, not only to the woman who produced the eggs. The recipient of the embryo—if it is not the woman who initially produced the eggs—is known as a surrogate mother.

However, a number of very difficult ethical and legal questions must now be dealt with by society. For example, a court decision in England in 1996 to destroy all unclaimed frozen embryos that were more than five years old created much controversy. In the late 1980s, a woman in New Jersey who acted as a surrogate mother decided she wanted to keep the baby. The couple who provided the egg and sperm asked the courts to give them custody of the baby. Who was right?

Many infertile couples are choosing not to use any of these new techniques to have a baby. They are, instead, finding the satisfaction they had hoped for by adopting a child.

WHEN THINGS GO WRONG: SEXUALLY TRANSMITTED DISEASES

Sexual intercourse provides an opportunity for body fluids from one person to be passed to another person. As a result of this contact, it is relatively easy for certain disease-causing microorganisms to be passed from one person to another. An infection that occurs in this way is called a **sexually transmitted disease** (STD).

By far the most dangerous STD is AIDS (*A*cquired *I*mmune *D*eficiency *S*yndrome.) AIDS now kills more people than any other disease contracted through sexual activity. AIDS is spreading rapidly throughout the world. It is becoming truly devastating in Africa, where 2.4 million AIDS deaths, 85 percent of the world's total, occurred in 1999. Medicines that slow down the progression of the disease in infected people have been developed. Today, people with AIDS must follow a complicated schedule taking expensive medications. However, no cure or vaccine for AIDS is currently available, nor, unfortunately, is one expected to be available for a long time. The available medicines are costly and generally not available in poor countries.

Another STD, gonorrhea, is usually treatable with antibiotics. However, certain new strains of the bacteria that cause gonorrhea have developed resistance to many antibiotics. Physicians are concerned that more new strains, resistant to all antibiotics, may develop in the future. (This is the same problem that was seen with tuberculosis in Chapter 14.) More than 1 million cases of gonorrhea are reported each year. One very important problem with this disease is that people infected with gonorrhea, particularly females, may show no symptoms. Unaware that he or she has the disease, a person may unknowingly infect a sex partner. In most cases, however, gonorrhea produces a painful release of pus from the genitals. The pain is usually severe enough to cause a person to see a physician for treatment. If untreated, gonorrhea can cause infertility.

Like gonorrhea, chlamydia is caused by a bacterium. The symptoms of a chlamydia infection are similar to the symptoms of gonorrhea, but treatment for this disease requires different antibiotics from the ones commonly used to treat gonorrhea. Diagnosis by a physician is important to insure that the proper antibiotics are used. It is estimated that more than five million Americans become infected with chlamydia each year.

A further danger posed by gonorrhea and chlamydia is the risk, if the mother is infected, of infecting a newborn as it passes through the birth canal. Both diseases can cause blindness in infants. In the United States, every baby born in a hospital is given antibiotic eyedrops immediately after birth to prevent possible infections.

Before modern antibiotics were developed, syphilis was one of the most terrible diseases in humans. The infectious bacterium that causes syphilis destroys nerve tissue, including the brain. Over time, if left untreated, syphilis often caused blindness, madness, and, eventually, death in the people it infected. Today, accurate diagnosis and antibiotic treatment have made this disease far less common in most parts of the world. The early stage of this disease involves a chancre—a hard, painless sore—on or near the genitals or on the lips or hands. The later stages of syphilis affect nervous tissue deep inside the body, so the effects of syphilis are not easily seen. If a mother is infected, her babies can become infected during birth, with tragic results. Since the later stages of this disease are not outwardly visible, it is important that syphilis be detected early. It was for this reason that all persons who wanted to marry in the United States once needed to take a blood test before they could get a marriage license.

Genital herpes is a common STD caused by a virus. Because it is a viral infection, herpes cannot be treated with antibiotics. Genital herpes causes open sores in the areas of the male and female sex organs. Since these open sores are filled with viruses, they are very infectious. In time, the sores heal, but the person remains permanently infected with the virus. The viral sores can return again and again throughout an infected person's life. This disease can also be passed to an infant as it is being born. The dangers to a newborn are much greater than to an adult. Infected infants can become blind; they may also suffer brain damage. In infants, this disease may be fatal. A new drug, acyclovir, can prevent the virus from infecting a person, if it is used the first time someone is infected.

The effects of STDs are real and very dangerous. Early detection and treatment are vitally important. However, some STDs are still untreatable today. Abstaining from sexual relations ensures that a person will not become infected with one of these potentially life-threatening diseases. "Safe sex"—a better term might be "safer sex"—includes the use of condoms and maintaining a monogamous relationship as important preventative steps. If symptoms of an STD occur, it is important to be treated promptly by a physician. It is also important to inform those with whom one has had sex that they also need to be seen by a physician. Keep in mind that several important STDs cannot be cured at this time. Learn how to protect yourself, and the people you care for, from contracting an STD. Responsible behaviors are important and may save lives.

LABORATORY INVESTIGATION 17
What Are Some Stages in the Development of Human Embryos?

INTRODUCTION

In humans, the embryonic development of organs occurs in the early stages of pregnancy. During its development, an embryo may encounter risks from faults present in its genes. It may also be harmed by problems from its mother, such as an inadequate diet, the use of alcohol, drugs, and tobacco, and other toxins or infections. Therefore, an understanding of the stages of development that occur during pregnancy is important. Pregnancy and human embryonic development are the topics of this investigation.

MATERIALS

NOVA video *The Miracle of Life*, handouts "Cutouts of Stages of Human Development" and "Stages of Human Development" chart (from the *Teacher's Manual*), scissors, blank paper, tape

PROCEDURE

1. Your class will view the video *The Miracle of Life* before you begin this laboratory activity.
2. In groups of two or three, study the handout "Cutouts of Stages of Human Development." Cut out each diagram and tape it to a sheet of paper. As a group, prepare a description of each diagram. Write your description beneath the diagram. Now study the diagrams and place them in order from the earliest to the latest. Number them from 1 to 6.
3. Join another group and share your findings.
4. Your teacher will now hand out the "Stages of Human Development" chart. Compare the sequence your group developed with the actual sequence as shown in this handout. How does the sequence you developed compare?
5. Read and study the following questions. Discuss them with your group before you present your answers to the class.

INTERPRETIVE QUESTIONS

1. How is the knowledge of the prenatal stages of human development used by physicians to monitor a pregnancy?
2. How does a sonogram help monitor a woman's pregnancy?
3. What are some problems that may occur during pregnancy that can cause health problems in babies?
4. How does amniocentesis monitor developmental problems?
5. What role does prenatal care—including counseling and proper diet—play in ensuring a healthy pregnancy?
6. Each of the following may affect prenatal development. Find out the effects of two of them and report your findings to the class: smoking crack; hereditary disorders; the use of nicotine and caffeine by the mother while pregnant; fetal alcohol syndrome; diseases such as measles, rubella, or AIDS.

CHAPTER 17 REVIEW

Answer these questions on a separate sheet of paper.

VOCABULARY

The following list contains all of the boldfaced terms in this chapter. Define each of these terms in your own words.

abstinence, cervix, contraception, corpus luteum, ejaculation, epididymis, estrogen, estrus, follicle-stimulating hormone (FSH), follicles, gametogenesis, infertility, in vitro fertilization, luteinizing hormone (LH), menopause, menstrual cycle, menstruation, ovaries, oviduct, ovulation, penis, progesterone, puberty, scrotum, semen, sexually transmitted disease (STD), sperm, spermatids, testes, testosterone, uterus, vas deferens, vagina

PART A—MULTIPLE CHOICE

Choose the response that best completes the sentence or answers the question.

1. Egg formation in human females starts *a.* before the individual is born *b.* shortly after birth *c.* at the onset of puberty *d.* at the beginning of the menstrual cycle.
2. Sperm formation occurs in the *a.* ovaries *b.* testes *c.* penis *d.* semen.
3. FSH is required for *a.* testosterone production in males *b.* egg formation in females *c.* follicle maturation in females *d.* progesterone production in females.
4. Sperm cells finish maturing in the *a.* sperm-bearing tubules in the testes *b.* epididymis *c.* vas deferens *d.* urethra.
5. Which is *not* an adaptation for helping the sperm reach the egg? *a.* mitochondria in the sperm cells *b.* the acidic environment of the female reproductive tract *c.* prostaglandins in semen *d.* the penis
6. On which day of the menstrual cycle does ovulation typically occur? *a.* day 1 *b.* day 10 *c.* day 14 *d.* day 28
7. What is the only 100 percent effective form of contraception? *a.* condoms *b.* birth control pills *c.* vasectomy *d.* abstinence
8. Fertilization usually occurs in the *a.* oviduct *b.* uterus *c.* vagina *d.* follicle.
9. Meiosis in females produces *a.* four spermatids *b.* four eggs *c.* one egg and three small structures *d.* two eggs and two small structures.

10. The corpus luteum produces the hormone *a.* progesterone *b.* estrogen *c.* testosterone *d.* luteinizing hormone.
11. In most mammals, egg cells are released only during *a.* menopause *b.* estrus *c.* orgasm *d.* in vitro fertilization.
12. Sexually transmitted diseases include *a.* AIDS *b.* syphilis *c.* chlamydia *d.* all of these.
13. LH causes *a.* the testes to produce testosterone *b.* the ovaries to release estrogen *c.* ovulation *d.* all of these.
14. The release of the egg into the oviduct is called *a.* ovulation *b.* menstruation *c.* ejaculation *d.* contraception.
15. Which of the following is the most useful in preventing STDs? *a.* birth control pills *b.* spermicide *c.* diaphragm *d.* condom

PART B—CONSTRUCTED RESPONSE

Use the information in the chapter to respond to these items.

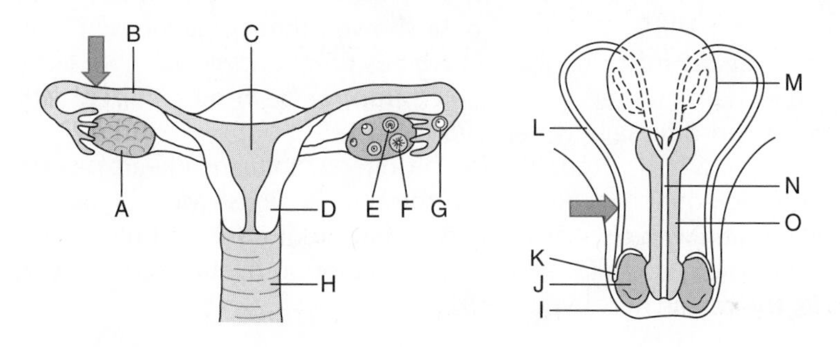

16. Identify the structures labeled A through O on the diagrams.
17. What would be the effect of cutting and tying off structures B and L at the arrows? What are these surgical procedures called?
18. Briefly describe the events in gamete production in humans.
19. Briefly describe five STDs. What are some ways to avoid contracting STDs?
20. Why might a fertility specialist recommend that a would-be father switch from wearing briefs to wearing boxer shorts?

PART C—READING COMPREHENSION

Base your answers to questions 21 through 23 on the information below and on your knowledge of biology. Source: *Science News* (March 8, 2003): vol. 163, p. 157.

Miscarriages Foretell Heart Trouble

A woman's experiences in childbearing may presage her risk of heart disease, according to new research. Women who spontaneously lose one or more fetuses in pregnancy are about 50 percent more likely than other women to later suffer ischemic heart disease, in which constricted or obstructed blood vessels choke the flow of blood to the heart.

Elective abortions don't appear to influence women's risk for ischemic heart disease, Gordon C.S. Smith of Cambridge University in England and his colleagues say.

The researchers reached these conclusions after analyzing data on all 129,290 women in Scotland who delivered their first live baby from 1981 through 1985. Additional data showed that those women who had had an early miscarriage in a previous pregnancy were more likely than other new mothers to have died from or been hospitalized with ischemic heart disease between 1981 and 1999.

The loss of a fetus probably doesn't directly influence heart disease risk, Smith and his colleagues say in the Feb. 22 *British Medical Journal*. Rather, women with circulatory defects that predispose their blood vessels to become blocked face an elevated risk for both fetal loss and heart disease, the researchers hypothesize.

21. Explain what health risk in women may be connected to the loss of a fetus during pregnancy.
22. How was the research conducted that investigated the relation between miscarriages and heart disease?
23. Explain why there may be a connection between miscarriages and heart disease.

Growth and Development

After you have finished reading this chapter, you should be able to:

Outline the main events that characterize embryonic development.

Describe the events that occur during human prenatal development.

Discuss the growth and development stages of the human life cycle.

A hen is only an egg's way of making another egg.

Samuel Butler, *Life and Habit*

Introduction

The world was stunned in February 1997 when newspapers such as *The New York Times* printed in headlines: "Scientist Reports First Cloning Ever of Adult Mammal: Researchers Astounded." British scientists had succeeded in creating a lamb, named Dolly, which lived for six years, using DNA from a body cell of an adult sheep. But students were still reading in textbooks, "**Cloning** uses the information present in one of your cells to build a whole new you . . . It has not been achieved in mammals . . . Such a procedure is not likely to be possible for human beings for a long time if, in fact, it ever becomes possible." (See Figure 18-1 on page 382.)

Yet by 2001, a group of scientists announced their plans to proceed, despite intense controversy, within the next two years to try to clone human babies to help infertile couples with no other means to have their own children. Others predicted that cloning technology would begin to be used by fertility clinics by about 2005.

When you read about scientific advances such as these, you should keep in mind that science is not *only* a subject you take in school. Science is an activity that occurs when we learn about the world around us. This world includes ourselves and how our bodies stay alive and grow. One of

Figure 18-1 Dolly, the first mammal cloned from the DNA of an adult cell.

the great beauties of science is that it is a never-ending search for knowledge. This knowledge has the ability to change what we are able to do almost every day. Some of these developments cause intense political, ethical, and religious controversies. In this chapter, you will study how organisms grow and develop. It is necessary for each of us to have such knowledge. We need to be able to understand the rapid, profound changes that science is making in our world.

EMBRYONIC DEVELOPMENT: FROM THE BEGINNING

Embryonic development is the sequence of events that gradually changes a zygote into a functional organism. Most of the instructions that control this series of events are in the genetic material—that is, the chromosomes—of the zygote. In addition, the environment that surrounds the zygote can have profound effects on the results. Much of the process of embryonic development is the same, whether the organism is a sea urchin, a frog, a chicken, a mouse, or a human. A related series of events also occurs in the embryonic development of plants.

The beginning of embryonic development occurs as the egg is fertilized. A zygote forms at the moment the cell membrane of the egg and the sperm join. This event makes it possible for the haploid nuclei of the two cells to fuse. The moment of fusion has another significant consequence. A complex series of changes occurs on the outside of the newly fertilized egg that prevents another sperm from entering. Regardless of how many sperm reach the egg, only one—the first one—gets in. (See Figure 18-2.)

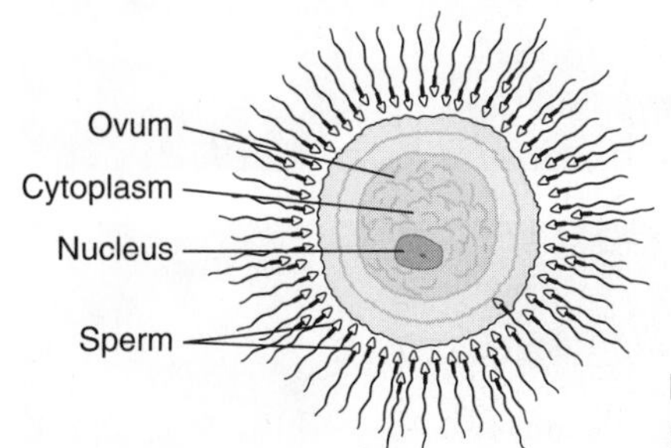

Figure 18-2 Embryonic development begins after the egg is fertilized.

The changes that begin to occur after fertilization, and that will continue until death, are known as **development**. The most dramatic developmental changes—growth and differentiation—occur early in the life of an organism. Through growth, the organism becomes larger as the number of its cells increases. Differentiation occurs as these cells begin to develop their own specific structures and functions. We increase in size because our bodies are made up of many cells. However, we stay alive because our cells differentiate into more than 200 types, including blood, skin, muscle, and bone cells.

Compared to a typical cell in an animal, a fertilized egg has an enormous amount of cytoplasm. Once it joins with a sperm, the fertilized egg, or zygote, immediately begins to divide. All divisions of a zygote after fertilization are mitotic cell divisions. The diploid number of chromosomes is maintained at each division. Therefore, the cells in the adult body still have the same complete diploid set of chromosomes found in the fertilized egg cell. The first series of mitotic cell divisions is called **cleavage**. The overall original size of the zygote does not change, but the number of cells in the zygote increases. (See Figure 18-3 on page 384.)

At the end of cleavage, the zygote has become a ball of cells called a **blastula**. The blastula consists of a single layer of cells that is hollow (no cells) in the middle and filled with liquid. It is important to remember that all the cells and tissues of the adult organism will eventually develop from the cells of this embryo. You might wonder when the cells in an embryo "know" what part of the animal they will become. This is one of the basic kinds of questions asked by embryologists, biologists who study

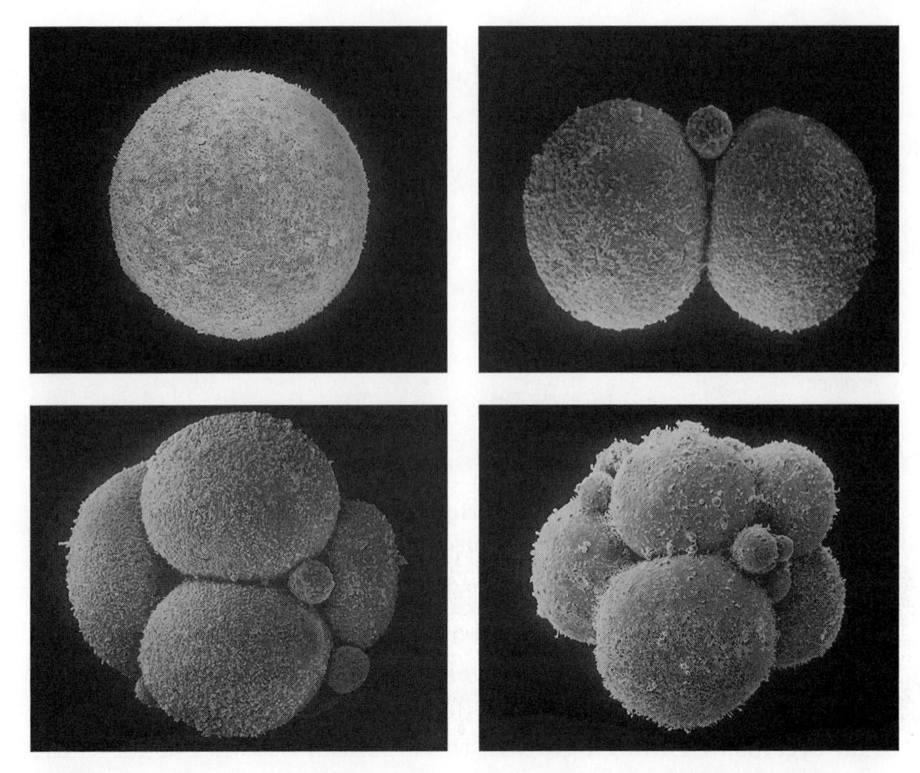

Figure 18-3 Scanning electron micrographs of the mitotic cell division of a fertilized human egg up to the eight-cell stage—the early cleavage of the zygote.

the development of embryos. Simple experiments with frog embryos have provided a great deal of information. After one mitotic division, a frog embryo consists of two cells. A dye placed in one of the two cells will be found in one half of the frog's body. Therefore, even at this early stage of development, these two cells "know" which side of the frog they will become. In addition, if these two cells are left to develop together, they will produce one frog. If they are separated, each cell will develop into a frog. The result: two frogs. More interesting questions arise from these findings. How do these two early cells "know" if they are alone or together? How do they "know" to become one-half a frog when they remain together, or a whole frog if they are separated? The science of embryology tries to answer these and other similar questions.

Although the blastula appears to have little organization, careful experiments have shown that it already has been determined which parts of the future adult each of the blastula cells will become. One part will become the eye, another the ear, another the leg, another the digestive system. By pinpointing these parts on the blastula, scientists can create a **fate map** that shows what each part of the blastula is destined to become.

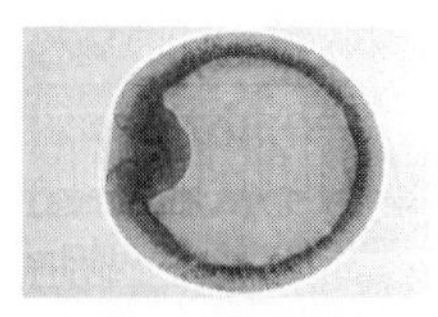
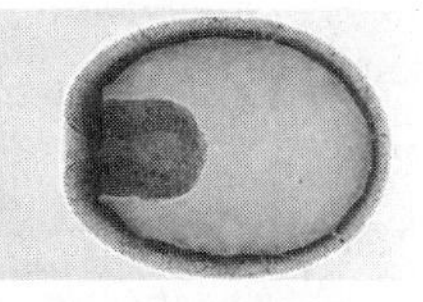
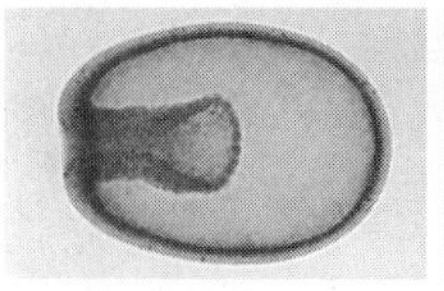
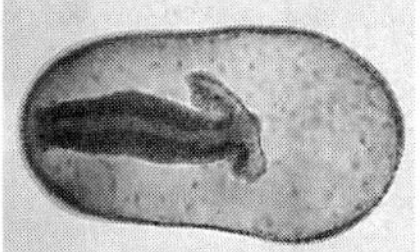

Figure 18-4 The process of gastrulation.

How can a structure as "simple" as a blastula develop the organization that will, in time, produce arms, legs, a stomach, a heart, and a nervous system? The process following cleavage begins to develop this organization.

During this next stage, called **gastrulation**, cells in the blastula begin to move. They change position in a highly regulated fashion. Imagine poking a tennis ball with your finger. If you poke hard enough, an indentation occurs. By pushing harder with your finger, the outer wall of the tennis ball is made to fold in. This folding-in resembles gastrulation. (See Figure 18-4.) As gastrulation continues, an inner layer of cells forms. The initial single-layer structure becomes a two-layered structure, called a *gastrula*. The outer layer of cells is called the **ectoderm**, and the inner layer is called the **endoderm**. By the end of gastrulation, a third layer of cells has formed between these two layers. This middle layer is the **mesoderm**. (See Figure 18-5.)

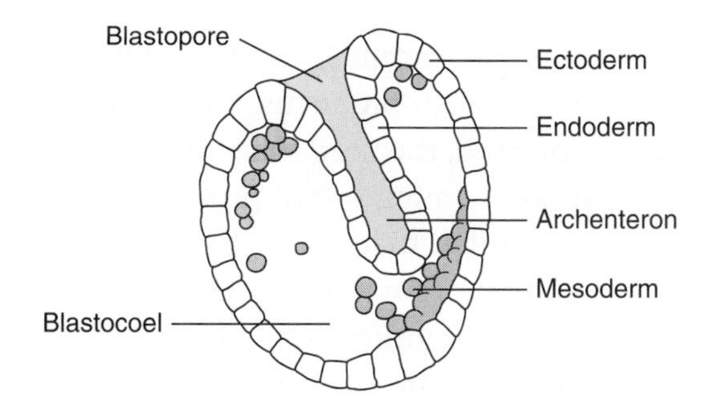

Figure 18-5 The mature gastrula has three germ layers: the ectoderm, mesoderm, and endoderm.

The endoderm, mesoderm, and ectoderm are the three embryonic germ layers. In vertebrates, the endoderm develops into the liver, pancreas, digestive system, and respiratory system. The ectoderm becomes the epidermis of the skin and the nervous system. The mesoderm gives rise to all other body parts, including the muscles, bones, circulatory system, excretory system, and sex organs. (See Figure 18-6 on page 386.)

The next significant step that occurs is the beginning of the nervous system. Perhaps because of its central role in coordinating activities of the organism, the nervous system begins development early. For whatever reason, after the three germ layers have formed, the ectoderm cells

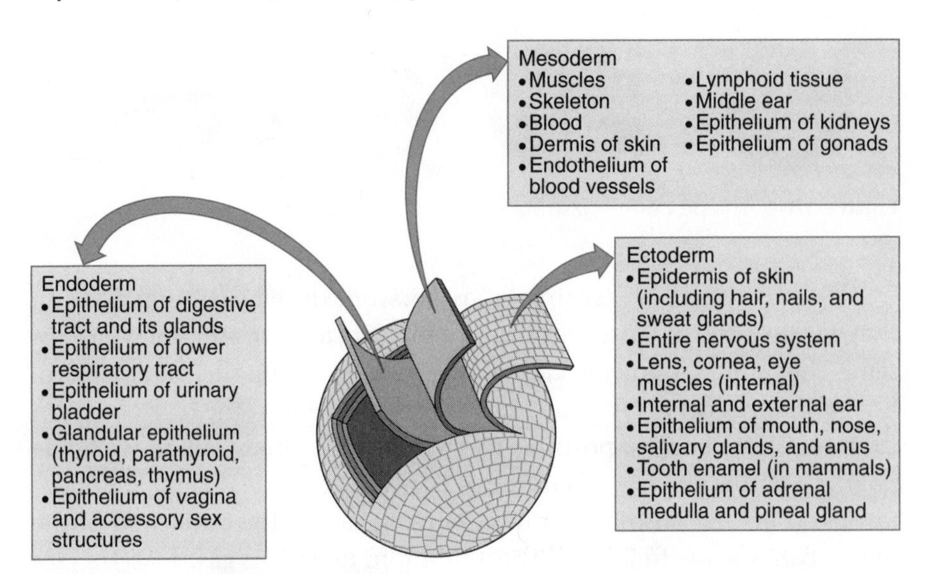

Figure 18-6 In vertebrates, the three embryonic germ layers are each destined to develop into specific body parts.

along the dorsal surface of the embryo begin to grow into a long hollow tube. This **neural tube** becomes wider at one end. Eventually this wider part develops into the brain. The long and narrow end of the neural tube develops into the spinal cord.

All eggs develop in a fluid environment. For animals that live in water, such as a sea urchin or a frog, this presents no problems. However, for reptiles, birds, and mammals that live on land, the need for a watery environment for embryonic development poses an important problem. If we examine a bird embryo, we can see the solution that first evolved perhaps 300 million years ago. A series of membranes form around but outside of the embryo. Included in these extraembryonic membranes are the **chorion**, **yolk sac**, **allantois**, and **amnion**. The chorion and allantois combine to help the embryo exchange gases and remove wastes. The yolk sac surrounds the yolk, the food source for the embryo. Most important, the amnion surrounds the embryo, protects it, and holds in a solution of salt water not too different in composition from ocean water. Because of the amnion, the embryo of land animals develops in water just as all aquatic animals do. (See Figure 18-7 and the table below it.)

At this point in development, the rest of the embryo's cells are still quite similar to each other. However, now the process of cellular changes begins. As cells start to develop into muscle cells, skin cells, blood cells, or other tissue types, they begin to organize into organs. Cell differentiation is occurring. All of the cells contain the same genetic information, but each type of cell uses this information differently.

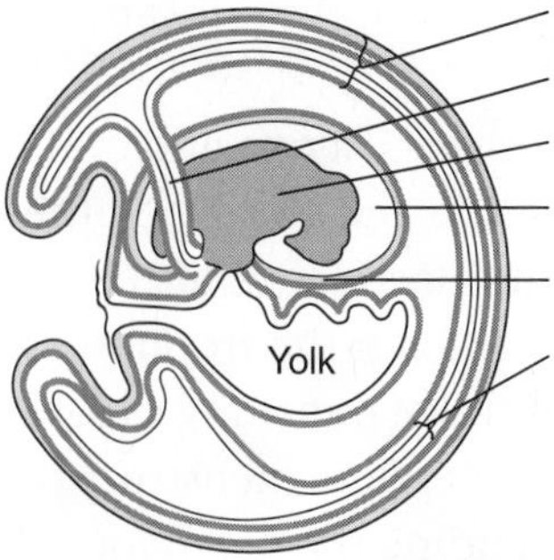

Figure 18-7 Extraembryonic membranes surround the embryos of birds.

THE EXTRAEMBRYONIC MEMBRANES

Chorion	Allantois	Yolk Sac	Amnion
Combines with allantois to help embryo exchange gases and remove wastes.	Combines with chorion to help embryo exchange gases and remove wastes.	Surrounds the yolk, which is the food source for the developing embryo.	Holds in a saltwater solution that surrounds and protects the developing embryo.

In most vertebrates, by the time birth occurs, all the major structures of the animal have been formed. The development that occurs after birth is mostly confined to an increase in size as the animal develops into its adult form.

Check Your Understanding

What are the three embryonic germ layers of the gastrula? Why are they important?

HUMAN DEVELOPMENT

Early stages of development are amazingly similar in all animals and are especially similar in all vertebrates. We have learned about how the human body forms by observing the development of sea urchins, frogs, chicks, and mammals such as mice. We will now examine some of the specific events that occur in human development.

As you read in Chapter 17, human egg cells are released about once a month from an ovary. The egg is fertilized in the oviduct and continues to move down this tube. The first cell division, which is the beginning of cleavage, occurs in the oviduct. This first division occurs about 36 hours after fertilization. Twenty-four hours later, the second division occurs. Twelve hours after this division, the third cell division occurs. Now, after three days and three cell divisions, the embryo consists of eight cells. The divisions continue, and by the end of six days the embryo—located in the uterus—is called a **blastocyst**.

There are two main groups of cells in the blastocyst. At one end of the blastocyst, a group of cells will actually become the embryo. The other group, the outer layer of cells, is called the **trophoblast**. Once inside the uterus, the trophoblast touches the uterine lining. Then the trophoblast secretes a hormone that tells the corpus luteum to keep making progesterone. This is necessary to protect the embryo. Otherwise, menstruation would occur; the uterine lining would fall away, ending the pregnancy. A pregnancy test actually detects the hormone secreted by the trophoblast.

Now an important and pivotal series of events occur. First, the trophoblast secretes enzymes that allow it to "eat" its way into the uterine lining so that it is completely embedded there. Then the cells of the trophoblast begin to divide rapidly. These cells invade the lining of the uterus. The trophoblast is the layer of cells around the embryo that connects the embryo to its mother. This connecting process is known as **implantation**. (See Figure 18-8.)

After implantation, a series of extraembryonic membranes form. These are similar to the membranes you saw in the chicken egg. The same types of membranes are also present in reptiles. The similarities in these extraembryonic membranes are clear evidence that reptiles, birds, and mammals evolved from a common ancestor. In humans and other mammals, the extraembryonic membrane on the outside is the chorion. It connects to the uterine wall. The membrane on the inside, the amnion, contains the **amniotic fluid** that surrounds and protects the embryo.

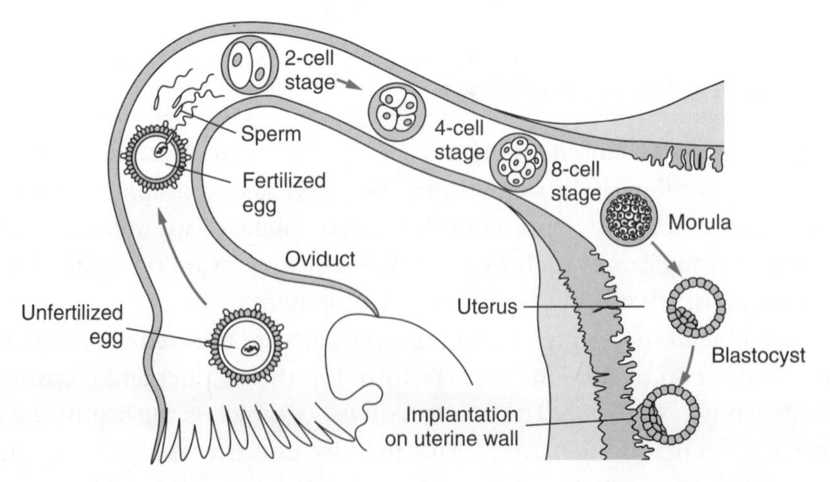

Figure 18-8 In the development of a human embryo, the fertilized egg travels through the oviduct and implants on the wall of the uterus.

In the next step, the chorion begins to grow into the mother's tissues. Blood vessels develop in this fetal tissue. At the same time, tissue with its own blood vessels grows from the uterine lining of the mother. This newly connected fetal and maternal tissue becomes a structure called the **placenta**. In the placenta, the fetus's blood vessels and the mother's blood vessels are close enough to each other to allow the diffusion of substances between them. The mother's blood and the fetus's blood do not mix. However, nutrients and oxygen can move from the mother's blood into the fetus, and wastes from the fetal blood can move into the mother's blood. The **umbilical cord** connects the placenta to the fetus. We can all point to the place on our bodies where, when each of us was once a fetus, the umbilical cord was attached—the navel, or belly button. (See Figure 18-9.)

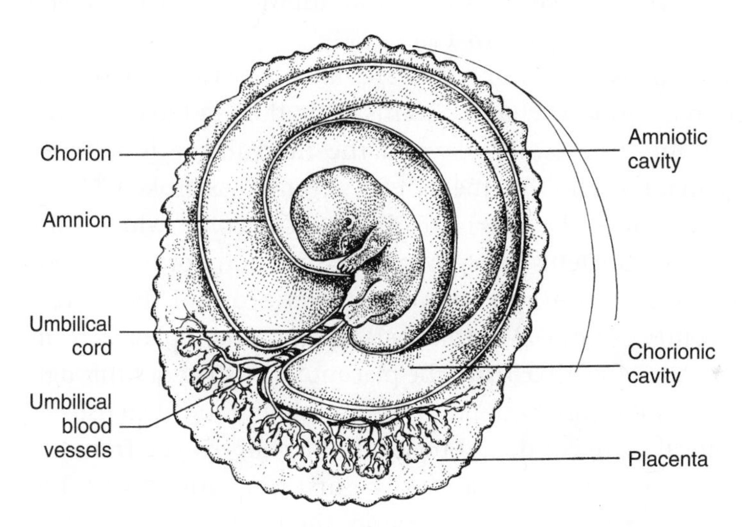

Figure 18-9 The umbilical cord connects the placenta to the fetus.

Gastrulation begins a few days after implantation. After one month, the embryo is barely 0.5 centimeter long. Yet, by this time, the nervous system, circulatory system, lungs, liver, and several other organs have begun to develop. During the second month, the beginnings of fingers and toes and a face with eyes can be observed. After the third month, the embryo is called a **fetus**. During the third month, the fetus is mostly growing in size, not developing new structures. One exception is that now the genitals are forming and the gender of the fetus can be observed. This is the end of the first three months, or the first **trimester**.

The second trimester begins at the fourth month. Now we observe the fetus doing things—sucking its thumb, swallowing, and kicking. By the fifth month, the mother can feel these kicking movements. Also, the fetal heartbeat can be heard for the first time. Most organs are developed by the end of the second trimester and can function at least partially. However, a fetus has very little chance of surviving outside the mother on its own before the seventh month.

During the last three months of pregnancy, the fetus increases in weight a great deal. By birth, the fetus is five to six times heavier than at the beginning of the third trimester. The most important change that occurs during the third trimester is the development of the brain. For the brain to develop normally, the fetus must receive a high level of protein from the mother. Thus, it is essential that the mother have a protein-rich diet herself at this time. By the end of the ninth month, the fetus is able to control its own breathing. Being able to use its fully developed lungs to breathe on its own is the key reason why an infant is ready to survive outside of its mother. (See Figure 18-10.)

And then it happens. For reasons that are not yet clearly understood, hormone levels in the mother change. Events begin that lead to the baby's birth. First, the amniotic sac breaks and the fluid inside is released through the vagina. It is said that the mother's "water has broken." Muscular contractions cause the cervix to dilate, or become wider. This increase in size allows the fetus to begin to be pushed out, normally head-first. More contractions continue to move the fetus out of the uterus. After the baby comes out of the birth canal, the placenta separates from the uterus. Now called the **afterbirth**, the placenta is pushed out through the vagina by some final contractions. Those assisting in the birth usually clamp the umbilical cord, cut it, and tie it off. Without oxygen from the mother, the baby's lungs fill with air for the first time. Along with the first gulps of air comes that wonderful sound, the loud strong cries of a newborn baby.

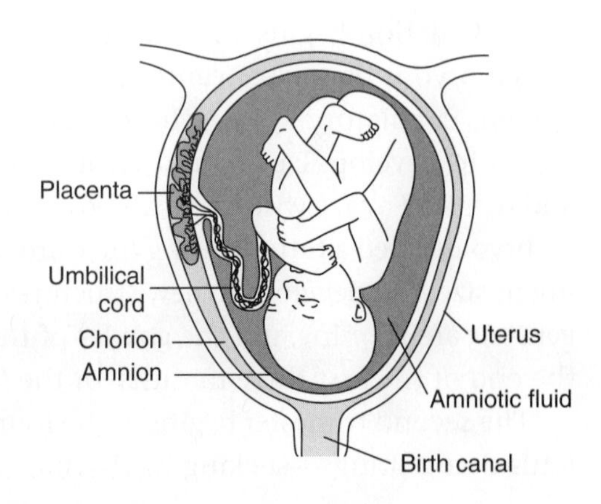

Figure 18-10 A developed fetus, ready for birth.

FACTORS THAT CONTROL GROWTH

Embryonic development has been compared to fixing an airplane while it is flying. In this sense, the process is very different from just building a machine. A machine is assembled, checked all over, and finally switched on. A developing embryo is "on" from the beginning. It is being built while it is running, that is, when it is alive. Thus it is somewhat like fixing a plane while it is already flying—but much more complicated.

What controls and coordinates human growth and development? Obviously, this kind of control and coordination is not easy, and the question is not a simple one to answer. In fact, this is one of the main questions being investigated by biologists. We already know part of the answer. Earlier, you studied the sequence of events called the cell cycle. Each time it divides, a cell proceeds through the steps in the cell cycle. A developing embryo is a vast collection of dividing cells. These cells are rapidly becoming specialized into specific types of tissues. For the embryo to take shape, for organs to develop and begin functioning, the growth of each cell type must be carefully timed and controlled. Biologists have learned that cells respond to substances present in the environment around them. These substances, usually proteins, are known as **growth factors**. Each tissue type has a specific growth factor produced by nearby cells. Because growth factors work in one specific area of the embryo, they are called local signaling molecules. Our future understanding of how an embryo develops will no doubt show that a complex interaction of many different growth factors exists in an embryo.

THE DANGERS THAT FACE A FETUS

It has often been said that there is no place as safe as a mother's womb, or uterus. Surrounded by a watery cushion, kept warm and nourished, a fetus does have its safety provided by the mother. It is true that most infections that may make a mother ill cannot cross over the placenta and into the fetus. It is not true, however, that the fetus is entirely safe. Dangers can intrude into the small world of the fetus.

Some forms of radiation can pass through the tissues of the mother and into the fetus. X rays, for example, can affect a fetus. The fetus is a place of tremendous cellular activity. Cells are dividing, growing, and changing. Dividing cells are easily damaged by X rays. Damage to even a few cells early in fetal development can produce disastrous effects. If the genetic material in one cell is altered, then every one of the millions of cells that develop from that damaged single cell will contain the same

Longer Telomeres Mean a Longer Cell Life

You know that chromosomes are made up of long chains of DNA, which determine all the traits of an organism. Scientists have learned that the end section of a chromosome—called the *telomere*—plays an important role in the life of a cell.

Normal human cells divide only a limited number of times. Each time a cell divides, it gets a little older. Research has suggested that when a cell divides, the telomeres in its chromosomes become a little shorter. The telomere is now thought of as a kind of molecular clock that keeps track of, or controls, the age of cells. It did not take long for researchers to begin to think that if the telomeres could somehow be kept from getting shorter, a cell could continue to divide forever. And possibly, because their cells would never get older, the person such cells belonged to would also never age. In effect, an ageless telomere would become a molecular fountain of youth.

Normally, the telomeres become shorter and shorter with each cycle of cell division. A sufficiently short telomere is believed to tell a cell to stop dividing. A key enzyme that can also stop the shortening process is *telomerase*. The telomerase enzyme reverses this process by adding DNA to the telomeres at the chromosome ends.

In support of this hypothesis, researchers found that telomerase remains active in most immortal cell lines. These are types of cancerous cells that keep on dividing in an uncontrolled manner. The importance of telomerase therefore has become even greater. Not only may the absence of telomerase lead to cell aging, but its presence may lead to cancer.

One other source of evidence about the role of telomerase is found in human sex cells. Telomerase is normally active in the human cells that give rise to sperm and eggs. This makes sense because these cells are expected to keep on replicating without limit.

Will further research provide a means to keep cells from aging through the action of telomerase, and yet without also becoming cancerous and dividing uncontrollably? These are significant questions to answer and an important area for future research.

defective genetic message. As a result of exposure to radiation, fetuses may have organs that do not work. Sometimes, the fetus cannot even survive. (See Figure 18-11.)

Some infectious microorganisms from the mother can enter the fetus. The virus that causes rubella, or German measles—a mild illness in the mother—can severely damage the fetus during the first trimester. A type of bacteria, streptococci, may not even make the mother ill; but in the fetus, permanent brain damage, blindness, and even death may result from a streptococcal infection.

Chemicals taken in by the mother also can affect a developing fetus. Some chemicals may cause enormously harmful effects. During the 1950s, a new drug called thalidomide was introduced as a tranquilizer in Europe. Unfortunately, this medicine had profound side effects. As a result, thousands of babies were born with severely deformed limbs because their mothers took thalidomide while they were pregnant. No one knew that thalidomide would produce this type of change in these infants—until it was too late.

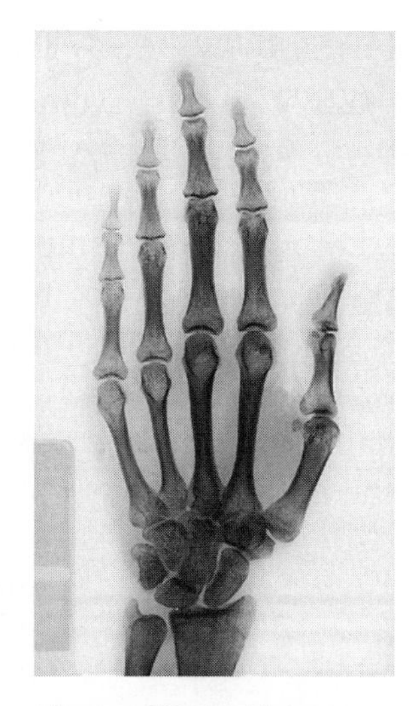

Figure 18-11 X rays, which penetrate through soft tissues to show bones, can be harmful to a developing embryo.

Other drugs can also affect fetal development. Heroin, LSD, cocaine, and alcohol can endanger a fetus during pregnancy. Many serious effects result. Babies can be born addicted to these drugs. Mental retardation can occur as a result of the alcohol that may pass from the mother's blood into the fetus.

And, finally, imagine what might happen to a fetus if, while it is developing, it is regularly denied the oxygen that its growing tissues need. This is what occurs every time a pregnant woman smokes. Cigarette smoke contains carbon monoxide, which replaces oxygen in the mother's blood. Less oxygen in the mother's blood means less oxygen for the fetus. A child whose mother smoked cigarettes during her pregnancy usually has a smaller birth weight, is more likely to die shortly after birth, or may have health problems for the rest of his or her life.

THE HUMAN LIFE CYCLE

Throughout this chapter, we have concentrated on the development that occurs before birth. But life does not remain the same after birth. Our life is one of continuous change. In fact, humans pass through a predictable series of stages, just as every individual cell does. For us, this series of changes is called the human life cycle.

The first stage, which consists of all development before birth, is the **prenatal** period. All development after birth is **postnatal**. For the first two years of life—the period of infancy—the individual is a newborn and then a toddler. During this time, body weight more than triples. Fat stored in the body begins to decrease. The muscular and nervous systems develop rapidly, as does coordination. A newborn cannot lift up its head on its own. Yet a two-year-old has learned to walk, run, and even talk.

Changes in the body's physical proportions continue from infancy to adulthood. For example, the head of a newborn is much larger in proportion to the rest of its body than the head of an adult. The human brain is extremely important to the success of our species. Therefore, it is necessary to begin life as a human with a relatively large head. On the other hand, the arms and legs of a newborn are much shorter in proportion to the rest of the body than in an adult. (See Figure 18-12.)

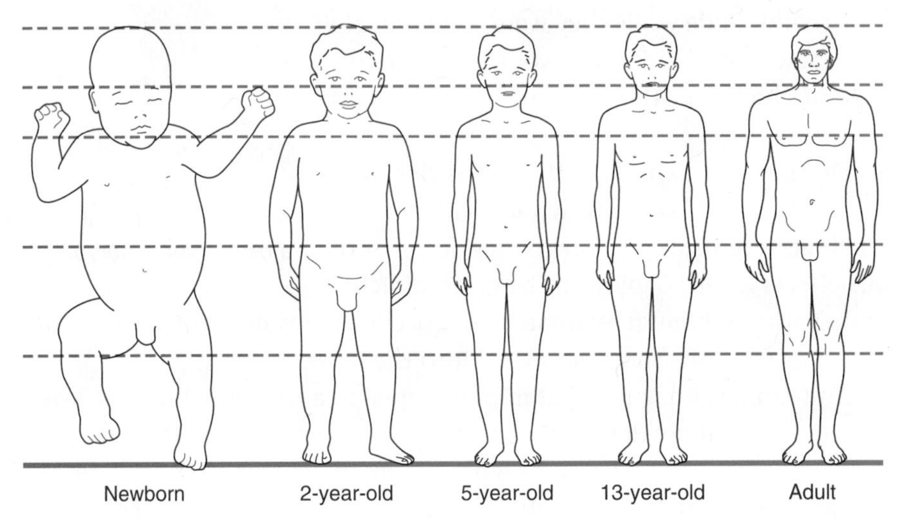

Figure 18-12 The changing body proportions of a human—the head of a newborn is much larger in proportion to the rest of its body than the head of an adult.

The stage known as childhood continues from infancy to puberty. Growth is fairly rapid during this period, but it is not as dramatic as growth during infancy. The child begins to look less like a baby and more like a preadolescent. In most children, the chubby face and potbelly present as an infant disappear. By 10 years of age, the child has developed fairly good muscular coordination and is able to complete a number of complicated tasks.

A spurt in height occurs first in girls, between the ages of 10 and 12. This growth spurt occurs a little later in boys, usually between the ages of 12 and 13. The teenage years—the period of **adolescence**—last from approximately 13 to 19 years of age. During this period, rapid intense physical growth occurs, including sexual maturation.

Some changes that begin in childhood are not completed until early or middle adulthood. For example, bones continue to grow and harden into adulthood. Throughout adulthood, the body needs to spend more and more energy on repairing tissues that are already there, rather than on their development.

Eventually, the process of maintenance and repair cannot keep the body running properly. Throughout life, cells are dying and being replaced. However, by older adulthood, many dead cells are not replaced. Tissues begin to wear down. This stage of life can be called old age, and it ends for all people—as for other organisms—in death.

EXTENDED LABORATORY INVESTIGATION 18

How Do Plants Know Which Way Is Up and Which Way Is Down?

INTRODUCTION

Unlike animals, plants cannot move from place to place. In their environment, however, plants must orient themselves to have the best chance of survival. Light is the most obvious factor that plants react to in their environment. Plants usually grow toward the sunlight. This response and other responses that plants show to environmental stimuli are called tropisms. It was from the study of phototropism—the movement of a plant toward light—that our knowledge of plant hormones and their effects on plant growth began. Plants are also able to respond to the effects of gravity. The plant shoot must be able to grow upward and the roots must grow downward. The response to gravity is called gravitropism, and it is the subject of this extended investigation.

MATERIALS

Clear plastic cups, thick absorbent paper (blotting paper), scissors, corn seeds, bean seeds, water

PROCEDURE

1. Prepare four cups in the following manner. Cut a piece of absorbent paper to the same height as the cup. Roll up the strip and place it in the cup. The paper should unroll and be held in position by the walls of the cup. Place several of the same kinds of seeds between the paper and the walls of the cup. Place the seeds midway between the top and the bottom of the cup. Leave room between the seeds. You should have a circle of seeds all at about the same height. Pour water into the bottom of the cup. The water should move up the paper. In time, the paper will be uniformly moist. Keep a small reserve of water in the bottom of the cup as this investigation proceeds. It will keep the seeds moist as they grow.
2. Place the four cups in a dark place to eliminate light as an environmental growth factor.
3. Record your observations in a log book as the plants begin to grow. Make careful observations of the growth of the roots and shoots from the seedlings.

4. Repeat the investigation. Again position the seeds in a circle around the cup. But this time, make sure the seeds have the same orientation. Do this by noting any structural features in the seeds. In one of your cups, place the seeds horizontally (lying on their sides) instead of vertically.
5. Place the cups away from the light and carefully note the direction of growth of the seed's roots and shoots.
6. Continue your investigations to explore how seedlings determine which direction is up and which is down. Listed below are several variables you might like to test.
   - Carefully reposition the seedlings (turn them upside down or on their side) after they have grown for a period of time.
   - At various times after they have begun to grow, cut the tips of the growing shoots or roots and observe the seedlings to determine the effects of these actions.
   - Try various combinations of the above methods of repositioning and cutting.
   - Compare the growth of monocot (corn) seeds with dicot (bean) seeds.
7. Prepare a research report of your experiment. Include a hypothesis, procedure, observations, and conclusions.

CHAPTER 18 REVIEW

Answer these questions on a separate sheet of paper.

VOCABULARY

The following list contains all of the boldfaced terms in this chapter. Define each of these terms in your own words.

adolescence, afterbirth, allantois, amnion, amniotic fluid, blastocyst, blastula, chorion, cleavage, cloning, development, ectoderm, embryonic development, endoderm, fate map, fetus, gastrulation, growth factors, implantation, mesoderm, neural tube, placenta, postnatal, prenatal, trimester, trophoblast, umbilical cord, yolk sac

PART A—MULTIPLE CHOICE

Choose the response that best completes the sentence or answers the question.

1. For most organisms, growth primarily involves *a.* an increase in cell size *b.* an increase in the number of cells *c.* specialization of cells *d.* cell migration.
2. The process in which DNA is taken from an adult animal and used to create a genetically identical animal is *a.* cloning *b.* replication *c.* gastrulation *d.* implantation.
3. The sequence of cell divisions and movements that change a zygote into a multicellular organism is called *a.* differentiation *b.* blastulation *c.* gastrulation *d.* embryonic development.
4. In a gastrula, the inner layer of cells is the *a.* ectoderm *b.* endoderm *c.* epiderm *d.* mesoderm.
5. The neural tube eventually develops into the *a.* backbone *b.* notochord *c.* brain and spinal cord *d.* esophagus.
6. Which of these best illustrates the concept of differentiation? *a.* A tadpole loses its tail as it matures. *b.* Cleavage results in the formation of a blastula. *c.* A child is taller in third grade than in second grade. *d.* A mature red blood cell lacks a nucleus, mitochondria, and other organelles.
7. The outermost extraembryonic membrane is the *a.* chorion *b.* yolk sac *c.* allantois *d.* amnion.
8. The part of the blastocyst that helps it to implant itself in the uterine lining is the *a.* embryonic disc *b.* trophoblast *c.* amnion *d.* blastula.

9. A blastula is most similar in shape and structure to *a.* a bowling ball *b.* a soccer ball *c.* a partially deflated, dented basketball *d.* two Frisbees glued together at the rims.
10. The mesoderm gives rise to the *a.* digestive and respiratory systems *b.* yolk sac *c.* nervous system *d.* circulatory and skeletomuscular systems.
11. The membrane that holds the watery environment in which the embryo develops is the *a.* allantois *b.* amnion *c.* trophoblast *d.* mesoderm.
12. Materials are transported between the fetus's blood and its mother's via the *a.* umbilical cord *b.* yolk sac *c.* allantois *d.* placenta.
13. The human embryo is called a fetus after *a.* implantation occurs *b.* the mother can feel it move *c.* the first trimester is completed *d.* the second trimester is completed.
14. In what way is it certain that maternal cigarette smoking damages the fetus? *a.* Exposure to carcinogens increases the risk of cancer later in life. *b.* Chemicals in cigarettes cause limb defects. *c.* Oxygen deprivation stunts growth and causes nervous system damage. *d.* The baby is born addicted to nicotine.
15. The afterbirth consists of the *a.* umbilical cord *b.* events of postnatal development *c.* amniotic sac *d.* placenta.

PART B—CONSTRUCTED RESPONSE

Use the information in the chapter to respond to these items.

16. What does this diagram show? What is structure A called?
17. What is structure B called? What is the process that changes structure A into structure B?

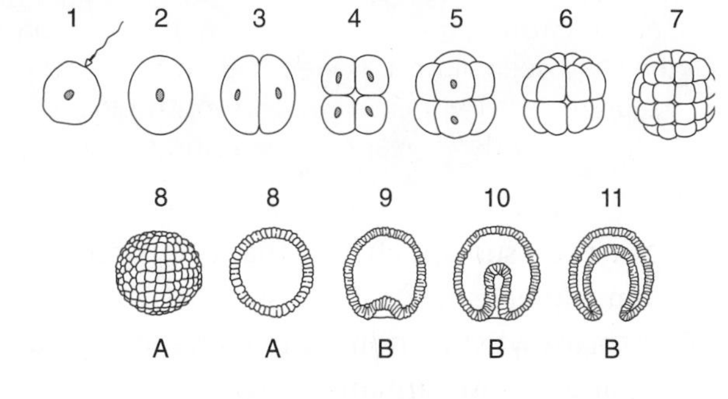

18. What controls embryonic growth and development?
19. Describe major stages in postnatal human development.
20. Why is it important for a woman to know early on she is pregnant?

PART C—READING COMPREHENSION

Base your answers to questions 21 through 23 on the information below and on your knowledge of biology. Source: *Science News* (November 9, 2002): vol. 162, p. 301.

Memory Grows Up in 1-Year-Olds

The second year of life may be particularly memorable. Around the time of their first birthday, children make dramatic advances in remembering simple events for 4 months after witnessing them, a new study finds. This memory breakthrough depends on a proliferation of neural connections in memory-related brain structures known to develop as infants approach age 1, propose Harvard University psychologists Conor Liston and Jerome Kagan.

The researchers recruited 12 babies and toddlers at each of three ages: 9 months, 17 months, and 24 months. Children watched an experimenter both perform and describe three action sequences. In one sequence, for example, the experimenter said "Clean-up time!" while wiping a table with a paper towel and then throwing the towel into a trash basket.

Kids in the two older groups watched four demonstrations of each action sequence, and 9-month-olds saw six repetitions. After each presentation, the experimenter encouraged children to imitate what they had just seen.

Four months later, the youngsters—then ages 13 months, 21 months, and 28 months—were asked to reenact each set of actions with the same materials after hearing the same verbal descriptions.

The children now 28 months old correctly performed a majority of previously observed actions, usually in their original order, Liston and Kagan report in the Oct. 31 *Nature*. The 21-month-olds reenacted what they had seen almost as well as their older peers did. Far fewer signs of accurate recall appeared in 13-month-olds, the only participants who had been under 1 year of age during initial memory trials.

21. State the sudden change that occurs in a child's ability to remember as he or she passes the first birthday.
22. Explain what is thought to occur in the brain that results in the change in how infants remember.
23. Describe the design of the experiment that was used to conduct memory research on 1-year-olds and 2-year-olds.

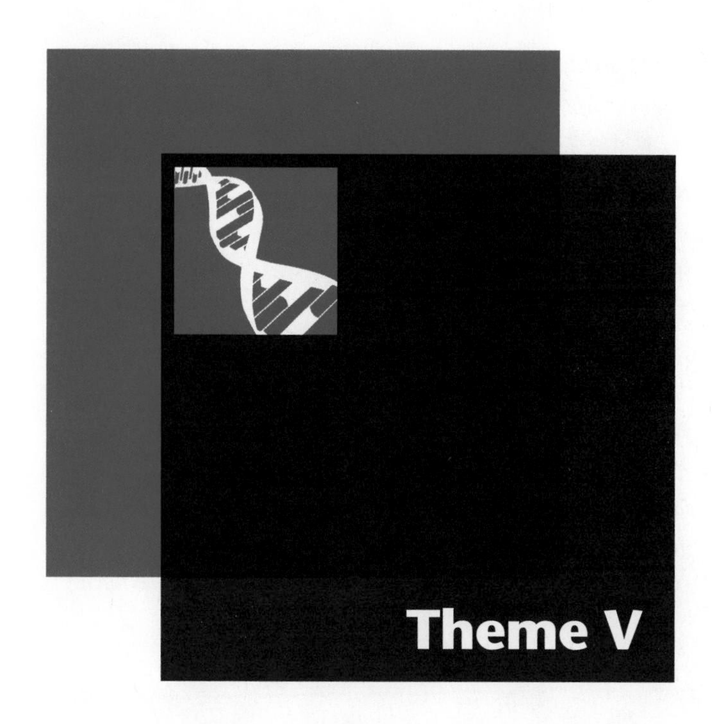

Genetics and Molecular Biology

DNA Structure and Function

After you have finished reading this chapter, you should be able to:

Describe the basic structure of a chromosome, the DNA molecule, and the nucleotide subunits.

List the four nitrogenous bases and explain how they are paired.

Discuss the process of DNA replication and explain how errors in replication may cause mutations.

There is no substance as important as DNA.

James Watson, *Recombinant DNA: A Short Course*

Introduction

Deoxyribonucleic acid, or **DNA**, has been called the molecule of life. Scientists have known of DNA's existence for about 100 years, but they have come to understand it only since the early 1950s. In the relatively short time since then, scientists have learned that DNA contains the information on which all life depends.

During the lifetimes of our parents and our grandparents, scientists discovered the secret of what was quietly, invisibly making life possible on this planet for the past 3.5 billion years. DNA is the master molecule that controls all of life. Indeed, life as we know it would not exist without DNA. You will learn about it in this chapter.

DNA: A BRIEF HISTORY

We have become very familiar with the chromosomes contained in the nucleus of every eukaryotic cell. Chromosomes carry the genetic information that is passed on from generation to generation. Chromosomes determine the genetic traits of every living organism—color of skin or

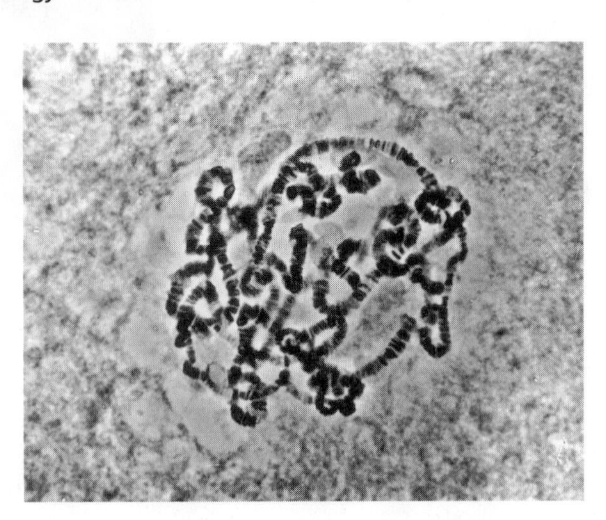

Figure 19-1 Fruit fly chromosomes.

fur, size of ears, number of legs or wings, height, and countless other characteristics. (See Figure 19-1.) By the early 1920s, scientists knew that chromosomes were made up of two substances, DNA and proteins. The question asked from 1923 on was: Is the genetic material—the substance that determined the traits in all living things—the DNA or the protein?

DNA, actually nucleic acid, had first been identified in 1868 by a Swiss physician, Johann Friedrich Miescher. Nucleic acid was thought to be a simple molecule that contained only four different parts. On the other hand, many different kinds of proteins were known to exist. Proteins are made up of an almost infinite number of combinations of the 20 different amino acids. Because of the great number of different proteins, many scientists thought proteins were the genetic material—the source of the great diversity of traits—found in living things.

The search for the substance that was responsible for determining traits in living things began. Some interesting experiments, mostly using bacteria and viruses, were performed. One set of experiments by a British researcher, Frederick Griffith, found that a substance from dead pneumonia bacteria could change or transform harmless pneumonia bacteria into pneumonia-causing ones. He called the substance a "transforming factor." (See Figure 19-2a.) Oswald Avery, at Rockefeller University in New York City, conducted very careful chemical analyses on the transforming factor. In 1944, Avery concluded that DNA was the transforming factor. At that time, the importance of Avery's discovery remained unrecognized. (See Figure 19-2b.)

It took one more set of crucial experiments to demonstrate that DNA was the substance that determined which traits were inherited. In 1952,

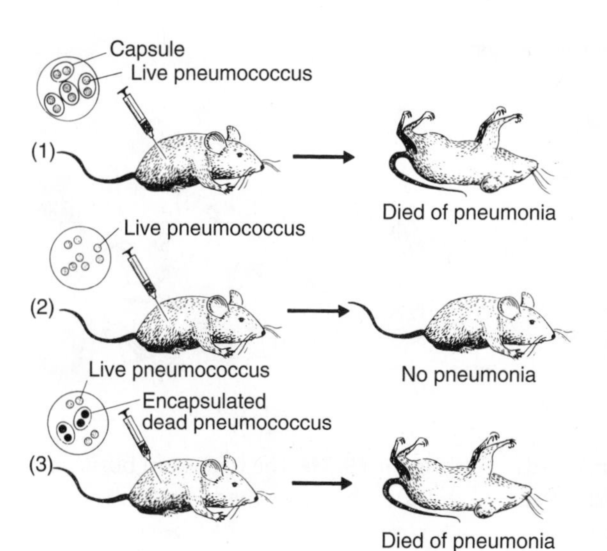

Figure 19-2a Griffith's experiment. A substance from dead, coated, pneumonia-causing bacteria could transform live, uncoated, harmless pneumonia bacteria into coated, harmful, pneumonia-causing bacteria.

Alfred Hershey and Martha Chase, working at the Cold Spring Harbor Laboratories on Long Island in New York, used bacteria and the viruses that infect bacteria in a now-famous experiment. Hershey and Chase knew that the viruses they used reproduced by injecting viral genetic material into bacterial cells; the rest of the virus—consisting of its protein coat—remained attached to the outside of the bacteria. The viruses then

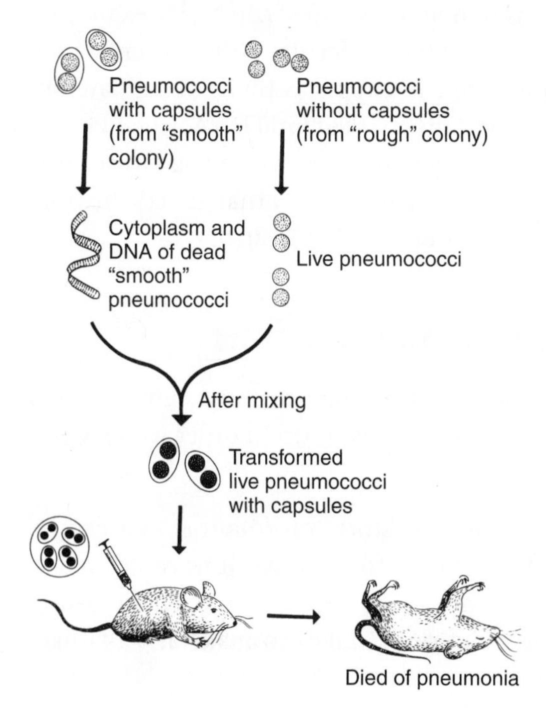

Figure 19-2b Avery's experiment. He concluded that Griffith's "transforming factor" was actually the bacteria's DNA.

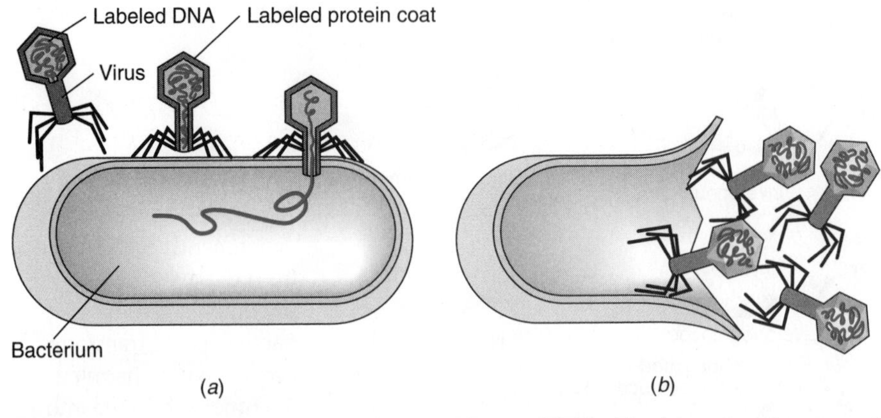

Figure 19-3a Viral DNA is injected into a bacterium, where it will direct the production of new viruses. The virus's protein coat remains outside the bacterium.

Figure 19-3b The bacterium bursts and releases the newly made viruses. The new viruses contain the labeled DNA, not the labeled protein.

used the material in the infected bacterial cells to make more viruses. (See Figures 19-3a and 19-3b.)

Hershey and Chase prepared two kinds of viruses. One kind of virus had a special radioactive marker on its DNA. The other kind of virus had a different radioactive marker on its proteins. Hershey and Chase infected bacterial cells with the two kinds of marked viruses. Later they examined the inside of the bacterial cells to see what molecules—the protein or the DNA—had entered the bacteria from the viruses as the genetic material. The answer was clear. It was the specially marked viral DNA. Hershey and Chase concluded, correctly, that the genetic material in the viruses was DNA. Except for a few unusual viruses, we now know that genetic instructions are carried in DNA in all living organisms on Earth.

GENETIC MATERIAL: A JOB DESCRIPTION

A substance that serves as the genetic material has the most significant job in the world: to carry on life itself. What must it do in order to carry out this all-important job?

- The genetic material must be able to store information that can be passed on from one generation of cells to the next. It must be able to store a lot of information. Look at a rosebush or at yourself and think about how many genetic instructions it takes to make an organism like the rose or like you.

- The genetic material must be able to make a copy of itself in order to pass its information on again and again.
- The genetic material must be strong and stable so it does not easily fall apart, causing perhaps harmful changes to its store of information.
- The genetic material must be able to mutate, or change, slightly from time to time. These changes allow a species to produce variations on which natural selection acts. This is the way evolution occurs.

It was concluded, in 1952, that DNA is the genetic material. At that time, scientists wondered if it would be possible to learn how DNA is built and how it functions in order to do these jobs. No one imagined that the very next year the answer would be found, and that this answer would revolutionize the science of biology.

THE WORLD LEARNS OF THE DOUBLE HELIX

Scientists are always looking for interesting projects to work on. In 1953, James Watson, a young researcher from the United States, went to Cambridge University in England. He joined Francis Crick, a physicist, because both of them were interested in learning more about DNA. Few other people were interested in studying DNA at that time. Many researchers suspected that trying to decipher the structure of DNA would turn out to be very dull work. And they thought that it would not be at all clear how DNA functions as the genetic material.

To learn more about the structure of DNA, Watson and Crick did not work the way most scientists do. They did not perform a long series of experiments in a laboratory. Instead, they used data already known about DNA molecules and made models of molecules with structures to fit the data. (See Figure 19-4 on page 408.)

It was known that DNA was made up of smaller nucleotide subunits. Each **nucleotide** consists of a five-carbon sugar called deoxyribose, a phosphate, and a nitrogenous base. A chemist at Columbia University, Erwin Chargaff, had discovered that the four types of nitrogenous bases in DNA actually formed two pairs. The amounts of adenine (A) and thymine (T) were always the same (A = T). The amounts of guanine (G) and cytosine (C) were always the same (G = C). Another critical piece of information that Watson and Crick had was from pictures taken when X rays were shot through crystals of DNA. Called X-ray diffraction patterns, these pictures had been taken by Rosalind Franklin and Maurice Wilkins. It was determined from their pictures that the DNA molecule had a spiral, or

Figure 19-4 James Watson (*left*) and Francis Crick shown in 1953 with their model of part of a DNA molecule.

helical, pattern—a pattern that looks like a spiral staircase or the threads of a wood screw. (See Figure 19-5.)

With this information, based largely on the work of other scientists, Watson and Crick used tin and wire to build a large model. Their model of DNA, in the shape of a double helix, rather than being dull, was about the most exciting discovery in the history of biology.

The easiest way to understand the double helix structure of DNA is to picture a ladder that has been twisted. The two sides of the ladder are parallel to each other. The steps of the ladder link the two sides to each other.

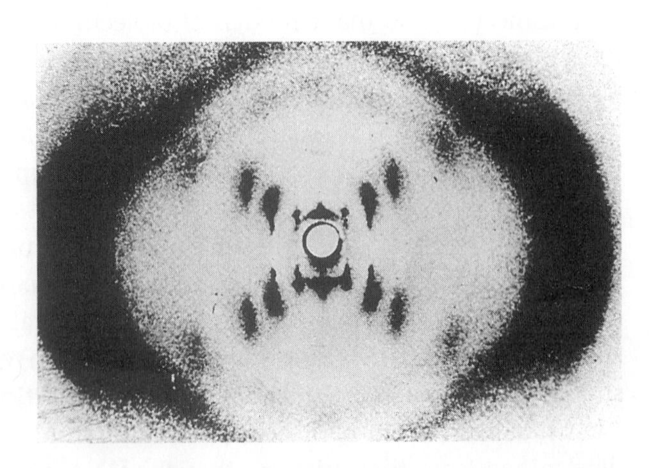

Figure 19-5 X-ray diffraction pattern of the DNA molecule.

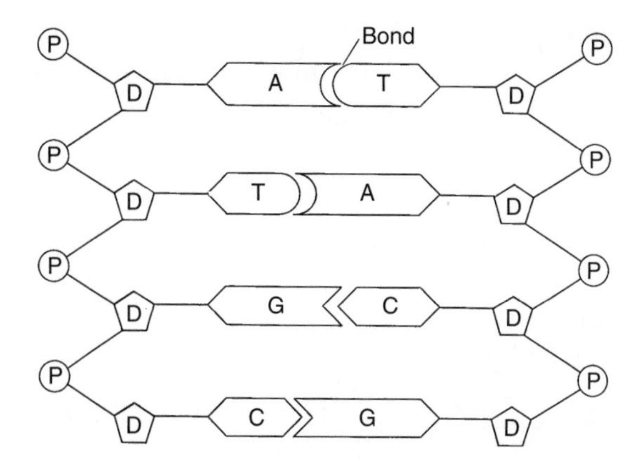

Figure 19-6 The structure of DNA.

In this model, the sides of the ladder are the sugar-phosphate backbone of the DNA molecule. Alternating deoxyribose sugar and phosphate molecules make up this backbone. Stretching between the two sides, from a sugar on one side to a sugar on the other, are pairs of molecules that contain nitrogen. These are the nitrogenous bases. One sugar, one phosphate, and one nitrogenous base make up a nucleotide. The nucleotides on one side of the molecule are bonded together to make one side, or strand, of the molecule. Other nucleotides are bonded together to make up the other side of the DNA molecule. (See Figure 19-6.)

Finally, how are the two sides of the molecule joined? The answer to this involves the nitrogenous base pairs. The Watson-Crick model showed that the only possible way all the parts could fit was for each large adenine base to be matched opposite a smaller thymine base. Similarly, a larger guanine had to be opposite a smaller cytosine. These nitrogenous bases were joined by fairly weak hydrogen bonds between them. It is now possible to realize the importance of Chargaff's discovery, A = T and G = C.

So a molecule of DNA consists of two strands, opposite each other, connected by matching nitrogenous base pairs. If we look at one strand, we can describe it in terms of the order or sequence of nucleotides. A DNA molecule may be very long. Because the nucleotides are in a long line, the order of the nucleotides is called a *linear sequence.* Imagine walking along a single strand of DNA. The bases in the nucleotides may occur in any order. The linear sequence on a short molecule of DNA might be A-T-T-G-A-C-C-G. Now imagine walking along the opposite strand starting at the same place. Opposite the A in the first strand is a T. Because we know the sequence of nucleotides in the first strand, we automatically know the sequence of nucleotides in the other strand. In this example, beginning with the T, it must be T-A-A-C-T-G-G-C. This is the key to how

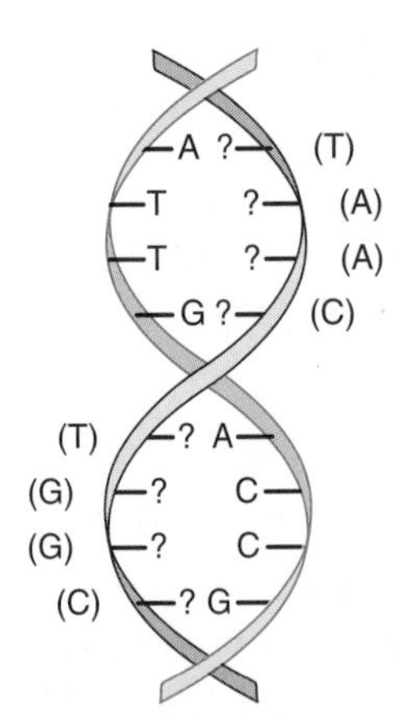

Figure 19-7 From the sequence of nucleotides on one strand of DNA, we can determine the sequence on the opposite strand.

the DNA copies itself. The copying of DNA is replication and it depends on the matching nitrogenous base pairs in the nucleotides of each strand. (See Figure 19-7.)

What is so important about the order of the nucleotides in a strand of DNA? Everything! The sequence of nucleotides *is* the information that the strand of DNA contains.

DNA: A LIBRARY OF INFORMATION

The words of any language are a means to store and transmit information. For example, all English words are made from combinations of the 26 letters in the alphabet. Many, many words can be made from these letters. An unabridged dictionary contains 450,000 words, all made from only 26 different letters. In some ways, the nucleotides in DNA are like the letters of the alphabet, only the DNA letters are "chemical" letters. Because there are only four letters—A, T, G, and C—in the DNA alphabet, scientists thought that DNA was too simple to contain the complex genetic information of life. But what is also significant in DNA is the sequence of the letters, not only the letters themselves. Using these four letters in long sequences, nature can create an almost unlimited variety of genetic messages. In fact, by creating messages only ten nucleotides long, it is possible to make more than one million different sequences or messages with the four nucleotides.

When you realize that human DNA consists of not ten but three billion pairs of nitrogenous bases, you can begin to imagine how much information can be stored in the DNA of our cells! All of the information for constructing our bodies, determining all of our characteristics or traits, and keeping our bodies running is stored in the linear sequences of nucleotides in our DNA. The same is true for every bacterial cell, insect, fish, bird, tree, and all other organisms on Earth.

Once scientists realized the importance of the linear sequence of DNA nucleotides, they were anxious to determine the sequence of nucleotides in a particular DNA molecule. In Chapter 2, you learned about one important reason for doing this. The evolutionary relationship of two organisms can be learned by comparing their DNA. The more similar their nucleotide sequences, the more recently the two organisms evolved from a common ancestor. (See Figure 19-8.)

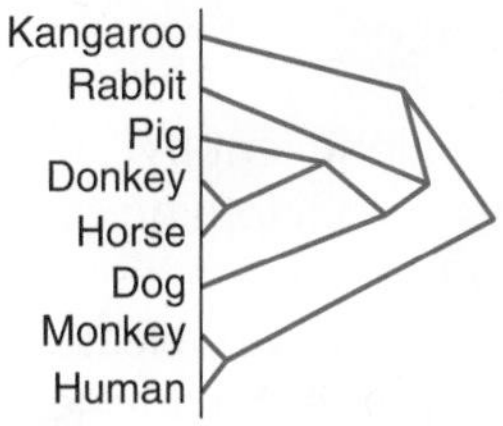

Kangaroo
Rabbit
Pig
Donkey
Horse
Dog
Monkey
Human

Figure 19-8 Evolutionary relationships for DNA closeness. The more similar their nucleotide sequences, the more recently the two organisms evolved from a common ancestor.

To make use of the genetic information stored in DNA, organisms must change the information into proteins. Proteins are made up of amino acids that are linked to each other. So, a protein is another linear sequence of subunits. In the next chapter, you will learn how the information stored in DNA gets expressed in the form of proteins.

Check Your Understanding

Why is the sequence of nucleotides in each strand of a DNA molecule so important? (Give more than one reason.)

MAPPING THE HUMAN GENOME

At one time, determining the sequence of nucleotides in a particular type of DNA was difficult and time-consuming. In the 1960s, it took seven years to determine the sequence of a DNA molecule with only 77 nucleotides. Now, like many other tasks, the analysis of DNA is automated. Laboratory equipment analyzes DNA quickly, and computers tabulate the results. Because of these technological advances, in the late 1980s molecular biologists began to plan for what they considered the most important biology investigation of all time: determining the entire nucleotide sequence of human DNA. In 1992, a worldwide effort—the **Human Genome Project**—began to analyze the three billion base pairs of human DNA. If printed on paper, the linear sequence of DNA contained in each of our cells would require 2000 books the size of this one. Molecular biologists all over the world are working together on this project, and they expect to finish before the year 2005.

In 1998, a well-known geneticist, along with a highly respected research company that makes machines to analyze DNA, announced that they planned to map the human genome faster and for less money than the government-sponsored Human Genome Project. They claimed that for only $200 million they would be done in just three years. The public may have been concerned that one private company could control so

Death-Row Inmate Cleared by DNA Evidence

Every time a prisoner awaiting a death sentence is proven innocent by DNA evidence and released, it makes the news. And it should. Nothing demonstrates the power of DNA technology better. Ray Krone owes his freedom, and probably his life, to this technology. In 2002, he was released from an Arizona prison after serving 10 years. During that time, Mr. Krone, who had served in the U.S. Air Force and worked as a letter carrier with no criminal record, was tried twice for the sexual assault and stabbing murder of a bartender in 1991. Mr. Krone was in the bar where the victim worked the night of the murder. The only evidence used to convict him was the similarity between the pattern of tooth marks on the victim, where she had been bitten, and Mr. Krone's teeth.

The first trial sentenced Mr. Krone to death, the second trial to a life sentence. Finally, after 10 years, DNA testing was done on saliva from bite marks found on the victim's clothing. Not only did the DNA *not* match that of Mr. Krone, but it *did* match that of a person serving time in another Arizona prison for an unrelated sex crime. The odds were 1.3 quadrillion (1,300,000,000,000,000) to 1 that it was this other man's DNA on the victim and not that of Mr. Krone or anyone else. A judge ordered the immediate release of Ray Krone when the DNA test results were announced.

much important information. However, most scientists agreed that the sooner the complete human genome was decoded, the sooner more research could be conducted to understand what it all means. By 2001, the first working draft sequence of the entire human genome was published.

Scientists agree that the Human Genome Project, described by some people as the effort to read the "book of humankind," is just the beginning of human genetic research. Only through this effort will we be able to understand ourselves on the molecular level, the most basic level of all. Indeed, scientists aim to someday understand all life-forms on this level. In 2002, six more model organisms were chosen to have their entire genetic codes spelled out. These six are the chimpanzee, the chicken, the honeybee, the sea urchin, a yeastlike protozoan, and a family of fungi.

CHROMOSOMES: PACKAGING THE DNA

Chromosomes were observed with microscopes long before anyone knew what they were made of. Now we know that chromosomes in eukaryotes

The DNA match was made possible because Arizona now has a database that contains a DNA profile of every prison inmate. In fact, every state in America now has such a database; and a national system, the National DNA Index System (NDIS), was started in 1998. By 2002, the one-millionth DNA profile had been entered into the computerized system. DNA evidence collected from any crime scene can now be quickly compared to that of any one of the million convicted offenders in the NDIS database. The system is quickly growing and the technology of DNA testing is rapidly improving. For example, a portable DNA testing kit is under development in Britain. It will be smaller than a suitcase and will be linked to the national DNA database of that country. It is expected that the crime scene evidence will be put in a solution and then placed inside the mobile unit. Silicon chip technology in the testing kit will then extract a DNA profile that will be sent to the national database via a laptop computer. The results may be returned in under an hour to the detective's palm-held computer. Saliva on discarded cigarette butts at crime scenes has already been used successfully to provide DNA profiles of suspects.

It is hoped that someday, thanks to this kind of technology, there will be no more wrongful convictions such as that of Mr. Krone, and more positive identifications of those who do deserve the jail time.

are made of proteins and DNA. Chromosomes are packages of DNA that seem to be held together by proteins. Why do organisms need packages of DNA?

Consider that DNA molecules are very, very long. A typical cell in the human body is much smaller than the period at the end of this sentence, and yet that single cell contains more than 2 meters of DNA. In addition, for DNA to do its job correctly, it cannot get tangled like a long piece of string thrown carelessly into a drawer. Chromosomes help maintain the DNA in the proper shape, untangled and ready for use. In a chromosome, the long double-helix DNA molecules get wound around protein molecules to form bundles. These bundles get looped together, and the loops, in turn, get coiled and folded together. This all works well to squeeze DNA into a very tight space and yet keep it well organized in order to do its job.

DNA REPLICATION: PASSING IT ON

To qualify as genetic material, DNA has to be able to replicate, or make a copy of, itself. This process of **DNA replication** occurs during the middle

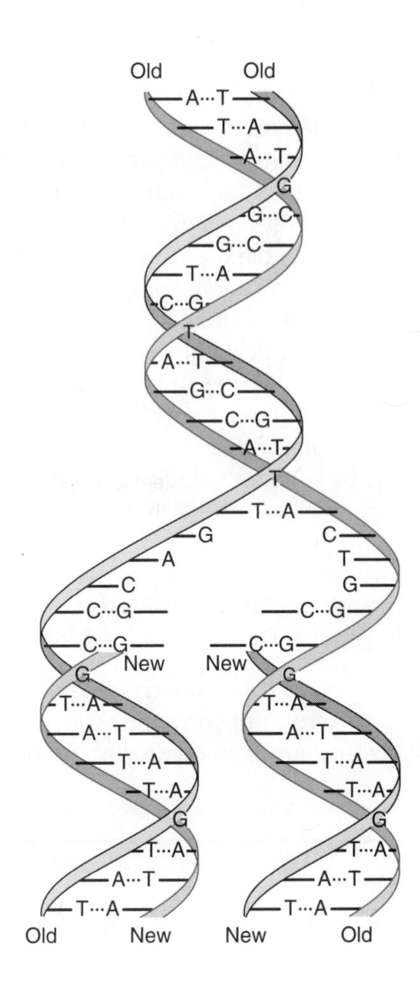

Figure 19-9 During DNA replication, the double helix unwinds, the DNA strands separate, and new complementary DNA strands form opposite each of the original strands.

of the cell cycle. What we already know about its structure is enough to explain how DNA replicates. (See Figure 19-9.)

To make a copy, you need an original, sometimes called a template. Because DNA is a double helix, it has templates built into it. To begin the process, the double helix unwinds. As with all metabolic activities, enzymes are needed for this process. Once the double-stranded molecule is untwisted, it begins to unzip, just like a zipper. Recall that the nitrogen base pairs A-T and G-C are connected by weak hydrogen bonds. However, at the correct moment, through the activity of an enzyme, these bonds begin to break apart. As the hydrogen bonds break, each strand of the DNA molecule becomes separate. Many free nucleotides float around in the cell. Specific enzymes match up these free nucleotides with the existing nucleotides in each DNA strand. Wherever a T is located on a strand, an A pairs to it; wherever a C is located, a G joins up, and so on. One by one, new nucleotides are joined together to make a new strand opposite

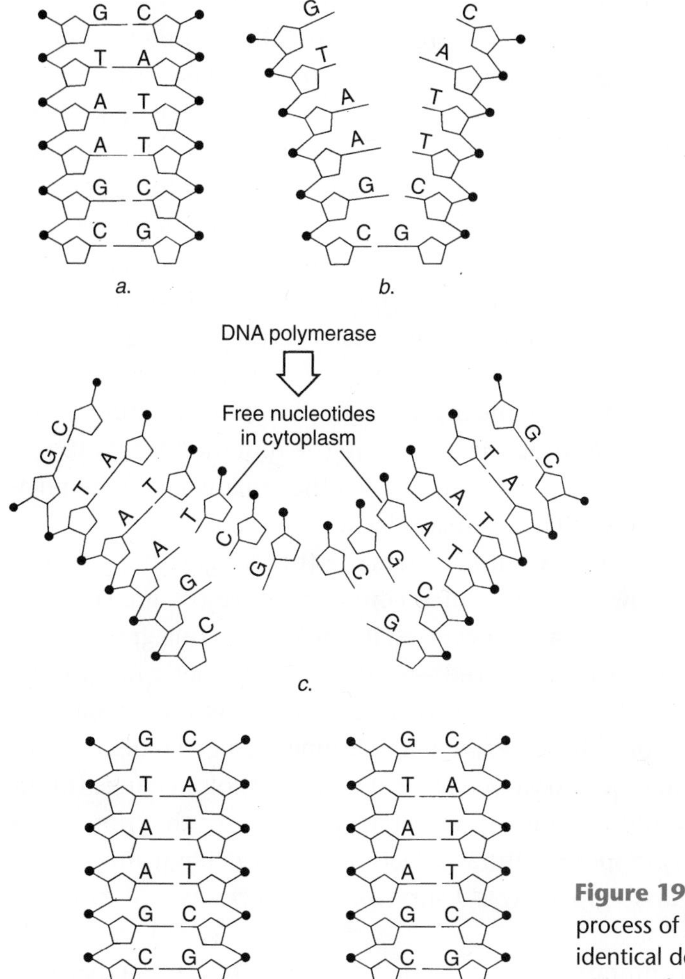

Figure 19-10 Through the process of DNA replication, identical double-stranded DNA molecules are formed.

each old strand. What determines the linear sequence of nucleotides in the new strands? The sequence of bases in the old strands. When replication is complete, two double-stranded DNA molecules are formed. Each molecule is made up of one old strand joined to a newly synthesized strand. How do the two new DNA molecules compare to the original one? They are identical. DNA replication has occurred. (See Figure 19-10.)

ERRORS IN DNA REPLICATION

In life, nothing is always perfect. That is true about DNA replication, too. The enzymes responsible for directing the correct pairing of nucleotides during DNA replication occasionally make mistakes. A nitrogenous base may be left out. Or the wrong base may be matched up. Sometimes an extra one is added. These mistakes produce errors in the linear sequence in one strand of the DNA molecule. Such an error is called a **mutation**.

From what we know about the replication process, once an error occurs in a DNA strand, it may be copied again and again. The mutation in the genetic material in one cell can easily be passed on to future cells.

Are these mutations good or bad? It seems a strange question to ask, because we assume that mistakes are always bad. However, what is obvious is not always true. A mutation is simply a change. Many changes in the genetic material are harmful. In fact, many of these changes make it impossible for the future cells, or even the entire organism, to continue living. Other mutations cause the organism to change in an unnoticeable fashion. Rather than harming the organism, the mutation seems to produce no effect. Sometimes the mutation gives the organism a sudden new advantage that other similar organisms lack.

Let's consider a simple example. Imagine that all grasshoppers in a green field were brown in color. Birds could easily see the grasshoppers and eat them. Then a mutation occurred in the DNA of one grasshopper. When that grasshopper reproduced—passing on its genetic mutation—the offspring with the mutation were green, not brown. Is this a good mutation or a bad one? Being a green grasshopper in a green field is good if it makes it harder for birds to see and eat you. This color change (mutation) to green would provide a survival advantage over the more easily seen brown grasshoppers. Natural selection would make it more likely that the green grasshoppers would survive. The species would evolve in terms of body color.

Not only can mutations in DNA be good, but they are actually an important source for the genetic variations necessary for natural selection to occur. Much of the evolution of life on Earth has depended on the chance occurrence of these mutations.

CORRECTING THE ERRORS

"Do not lie on the beach without using a sunscreen lotion with at least SPF 30." This is common advice given by doctors. Sun Protection Factor (SPF) 30 gives a person 30 times more protection from the sun than using no lotion at all. (See Figure 19-11.) You may wonder why we even need to be protected from the sun. Doesn't the light of the sun provide energy for life on Earth? Although the sun's energy is vitally important for life on Earth, sunlight also contains ultraviolet (UV) radiation. When the high-energy waves of ultraviolet light strike cells in a person's skin, the DNA in those cells is damaged. Mutations at specific places in the DNA can occur.

This damage occurs all the time. After all, the sun is shining on us much of the time, not just when we are at the beach, but also when we walk down the street or attend a ball game. In addition, other factors, including a variety of chemicals, tobacco smoke, and X rays, can cause mutations in DNA. All of these substances, including sunlight, are referred to as **mutagens**.

Figure 19-11 Using a sunblock helps to protect the DNA in skin cells from damage by ultraviolet radiation in sunlight.

Finally, as we have mentioned, mutations occur naturally or spontaneously in the cells of our bodies all the time. What keeps our bodies operating normally most of the time is our own built-in repair system. This system consists of a series of repair enzymes that detect damaged pieces of DNA. The damaged pieces are removed, and the DNA is repaired. The problem is that sometimes, if the exposure to the mutagen is too great, too much damage occurs. The repair enzymes are unable to fix the damage. Mutations go uncorrected. Mutations frequently produce cancers, although this disease may occur many years after exposure to the mutagen. That is the reason for using the SPF 30 lotion. Ultraviolet radiation from the sun today greatly increases the risk of skin cancer later in life. One of the three types of skin cancer, malignant melanoma, has the ability to metastasize or spread throughout the body. It can kill. A blistering sunburn early in life increases the risk of cancer years later. It is necessary to help the body protect itself by minimizing one's exposure to mutagens.

LABORATORY INVESTIGATION 19
How Do Molecules of DNA Replicate?

INTRODUCTION

In 1953, Francis Crick and James Watson discovered the structure of the DNA molecule. This discovery enabled other researchers to begin to determine how molecules of DNA function. Like many other polymers, a strand of DNA is made of many individual connected subunits. A molecule of DNA consists of two parallel strands. The subunits on one strand are connected to the subunits on the other strand. It is the connections between the two strands, which can be broken and reformed, that enable DNA to make a copy of, or replicate, itself. In this laboratory investigation, you will use paper "models" of DNA subunits and portions of DNA strands to learn how this remarkable chemical is constructed and can be copied.

MATERIALS

Diagrams for steps 1–4 (from *Teacher's Manual*), scissors, blank paper, tape

PROCEDURE

Step 1. Cut out the six nucleotides in the handout. Assemble them into a double strand of DNA that consists of three pairs of nucleotides that are joined together. (*Hint:* The phosphates attached to the sugars point up on one strand and down on the other strand. Biochemists say the two strands are "antiparallel.") Tape the pieces into place on a sheet of paper.

Step 2. Cut out the four nucleotides from the second handout. Match them up correctly according to rules for nitrogenous-base pairing. Remember that the hydrogen bonds (dotted lines) coming from one base must match up with the hydrogen bonds in its partner. Tape the pieces into place on a sheet of paper.

Step 3. Refer to the diagram for step 3. Use the rules that govern the pairing of bases. Write the complementary base on the unlettered strand that will bond with the base shown on the lettered strand.

Step 4. Refer to the diagram for step 4. Show the replication of the 14-base-pair DNA molecule. Write in the nitrogenous base pairs after the strands have separated. Then show all the base pairs for molecules *A* and *B*. Compare these two molecules to the original.

INTERPRETIVE QUESTIONS

1. From the results in step 2, determine the two reasons why each base can pair with only one of the three types of nitrogenous bases.
2. From the results of all four steps, write a paragraph that describes how DNA is a well-organized molecule. Explain why this organization makes it possible for DNA to be the genetic material for all organisms.

CHAPTER 19 REVIEW

Answer these questions on a separate sheet of paper.

VOCABULARY

The following list contains all of the boldfaced terms in this chapter. Define each of these terms in your own words.

deoxyribonucleic acid (DNA), DNA replication, Human Genome Project, mutagens, mutation, nucleotide

PART A—MULTIPLE CHOICE

Choose the response that best completes the sentence or answers the question.

1. DNA stands for *a.* double nucleus acid *b.* double nucleic acid *c.* diribonucleic acid *d.* deoxyribonucleic acid.
2. Chromosomes are made up of *a.* proteins and carbohydrates *b.* proteins and DNA *c.* DNA and phospholipids *d.* amino acids and phosphate.
3. Proteins are made up of *a.* nucleic acids *b.* amino acids *c.* fatty acids *d.* polysaccharides.
4. The process by which DNA makes a copy of itself is *a.* replication *b.* transcription *c.* translation *d.* duplication.
5. Why did it take a long time to determine that DNA was the "master molecule"? *a.* DNA seemed too simple. *b.* DNA was not isolated chemically until the 1950s. *c.* Only a few types of living things contain DNA. *d.* Proteins combine with DNA in confusing ways.
6. A substance that causes changes in DNA is called a *a.* carcinogen *b.* transformation factor *c.* mutagen *d.* replication error.
7. A nucleotide consists of *a.* a sugar, a phosphate, and an amino acid *b.* a sugar, a phosphate, and a nitrogenous base *c.* an amino acid, a phosphate, and a nitrogenous base *d.* an amino acid, a sugar, and a nitrogenous base.
8. How many amino acids are there? *a.* 20 *b.* 46 *c.* 81 *d.* 129
9. If a segment of DNA contains 10 adenines, it will contain *a.* 10 guanines *b.* 10 cytosines *c.* 10 thymines *d.* none of these.
10. Which scientist is associated with the discovery of bacterial transformation? *a.* Watson *b.* Franklin *c.* Chargaff *d.* Griffith

11. Cytosine pairs with *a.* adenine *b.* guanine *c.* thymine *d.* threonine.
12. Watson and Crick described DNA as a *a.* cross within a circle *b.* single spiral *c.* double helix *d.* branching chain.
13. Which of these is *not* a nitrogenous base? *a.* adenine *b.* cytosine *c.* guanine *d.* threonine
14. The backbone of the DNA molecule is made of *a.* sugar and nitrogenous bases *b.* nitrogenous bases only *c.* sugar only *d.* sugar and phosphate.
15. The sequence of nucleotides in DNA *a.* has little significance *b.* tells how a protein molecule should be made *c.* codes for specific types of simple sugars *d.* tells how a chromosome should be assembled.

PART B—CONSTRUCTED RESPONSE

Use the information in the chapter to respond to these items. Refer to the following diagram to answer questions 16 and 17.

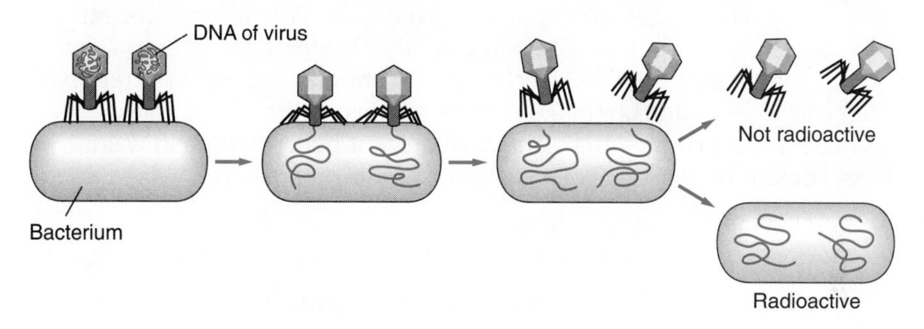

16. Identify the scientists who performed the experiment shown in the diagram.
17. What were the results of this experiment? What did the scientists conclude from these results?
18. What four characteristics must the substance that serves as the genetic material have?
19. Explain why mutations can have positive or negative effects.
20. Create a time line that shows important events in our understanding of DNA.

PART C—READING COMPREHENSION

Base your answers to questions 21 through 23 on the information below and on your knowledge of biology. Source: *Science News* (March 1, 2003): vol. 163, p. 141.

Worms May Spin Silk Fit for Skin

Silk cocoons could become puffs of valuable human proteins if a new bioengineering method developed by Japanese scientists pans out.

In the past few decades, various biotechnology research teams have devised way to mass-produce medically or industrially useful proteins by modifying the DNA of organisms. The animals create the proteins in their cells, milk, urine, or eggs (*SN:* 4/6/02, p. 213).

Now, Katsutoshi Yoshizato of Hiroshima University and his colleagues have genetically altered silkworms to produce a partial form of human collagen in their silk. Collagen is the structural protein in skin, cartilage, tendons, ligaments, and bones.

Given that silkworms worldwide annually spin about 60,000 tons of silk, the technique could lead to inexpensive, high-volume manufacture of collagen for artificial skin grafts. The method might also produce the blood-serum component albumin and other proteins, the scientists say.

In the January *Nature Biotechnology*, Yoshizato and his team report attaining concentrations of 0.8 percent collagen in the altered silkworms' cocoons. "If we raised the yield to 10 percent per total protein weight, we could produce it cheaply enough," Yoshizato predicts.

21. State the basic method by which biotechnology researchers have learned to make large quantities of useful proteins.
22. Explain why it would be valuable to produce large amounts of the protein collagen.
23. How might silkworms become involved in the inexpensive production of collagen?

Genes and Gene Action

After you have finished reading this chapter, you should be able to:

Explain how DNA is copied into RNA by the process of transcription.

Describe how mRNA builds a protein by the process of translation.

Discuss the importance of codons, anticodons, and peptide bonds to the process of translation and protein synthesis.

All evolution is nothing but a struggle among genes.

Stephen Jay Gould,
Hen's Teeth and Horses's Toes

Introduction

Genes are one thing we all get from our parents. We say that we hope we get "good genes." Now that DNA has been shown to be what makes up the genetic material, it is time to look more closely at genes. What is a **gene**?

In 1908, an English doctor, Sir Archibald Garrod, was explaining his new ideas about diseases to other doctors. As an example, Garrod told about a man who had a disease that made his urine black instead of yellow. All his life he had produced black urine. Most people have an enzyme that destroys the black substance in urine. Garrod believed that people who produce black urine lack the gene that makes the needed enzyme. His idea suggested that each gene is responsible for producing a single enzyme. Garrod was the first person to suggest what a gene is and how it works.

GENES AND PROTEINS

Every cell can be thought of as a "chemical factory." In every cell, many different chemical reactions are taking place at the same time to keep the cell alive. All the chemical reactions that occur in a cell make up its

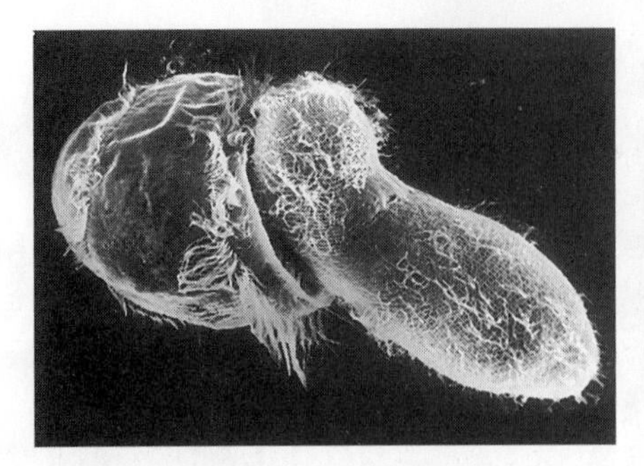

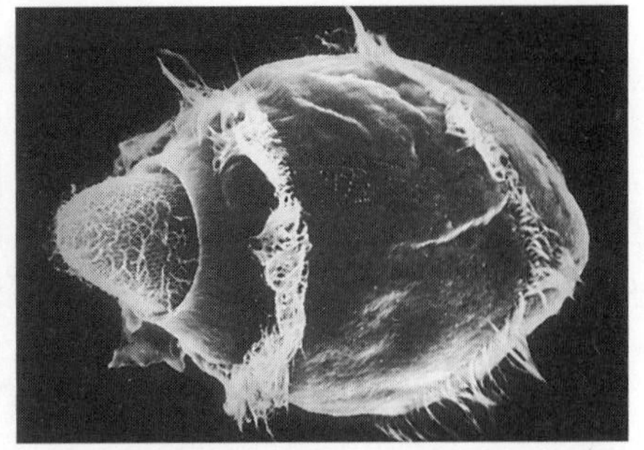

Figure 20-1 The genes of the didinium, a one-celled organism, contain the instructions for making the enzymes that will digest the paramecium it has caught.

metabolism. Scientists have learned that a specific enzyme controls each metabolic step, and that cells make their own enzymes. Where do the cells get the directions for making these enzymes? Garrod and other researchers have shown that the directions for making each enzyme are located in each cell's genes. (See Figure 20-1.)

Enzymes are proteins. Genes, which instruct cells to make enzymes, are therefore really packages of information that tell a cell how to make proteins. Proteins are polymers, or long chains, of amino acids. As you learned already, there are 20 different types of amino acids. Valine, phenylalanine, glycine, tryptophan, and serine are some of these amino acids. The order in which amino acids are joined determines which protein is made. Every different protein molecule has a unique sequence of amino acids. (See Figure 20-2.)

Genes are specific sections of DNA molecules that are made up of linear sequences of nucleotides. Proteins are linear sequences of amino acids. The big question is: How do cells use a linear sequence of nucleotides in DNA to build a linear sequence of amino acids for a protein? This question was answered by investigations carried out during the 1950s and

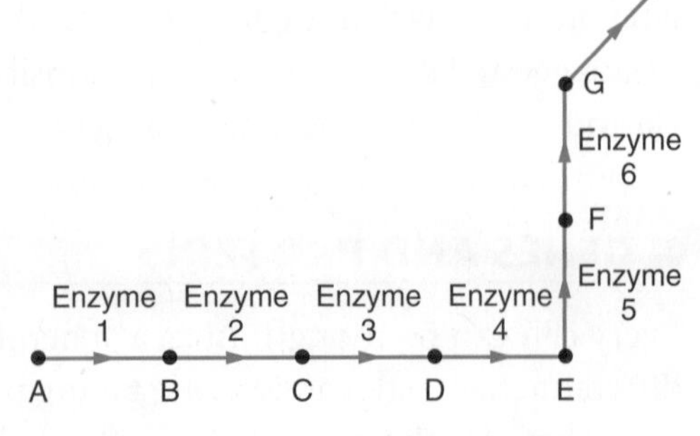

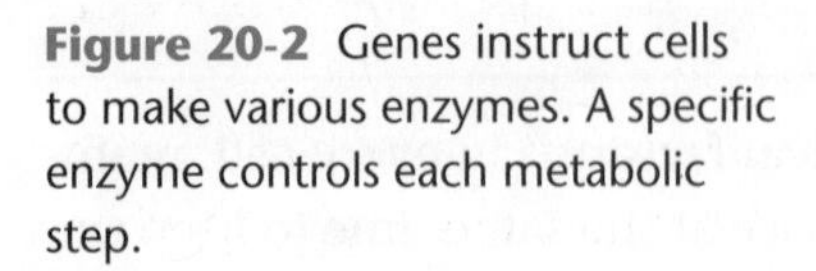

Figure 20-2 Genes instruct cells to make various enzymes. A specific enzyme controls each metabolic step.

1960s. This effort has been called "breaking the genetic code." In a real sense, the secret of life was no longer a mystery.

In eukaryotic cells, DNA is stored in the nucleus. The construction of cell proteins occurs outside the nuclear membrane. Protein synthesis occurs at the **ribosomes**. These small organelles are distributed throughout the cytoplasm in all eukaryotic cells. How does the genetic information in the DNA within the nucleus get to the ribosomes? A third type of molecule, **ribonucleic acid**, or **RNA**, works as a helper to transfer the information. A flow of information can be described. Information flows from DNA to RNA to protein. We can now examine in more detail how this process works. (See Figure 20-3.)

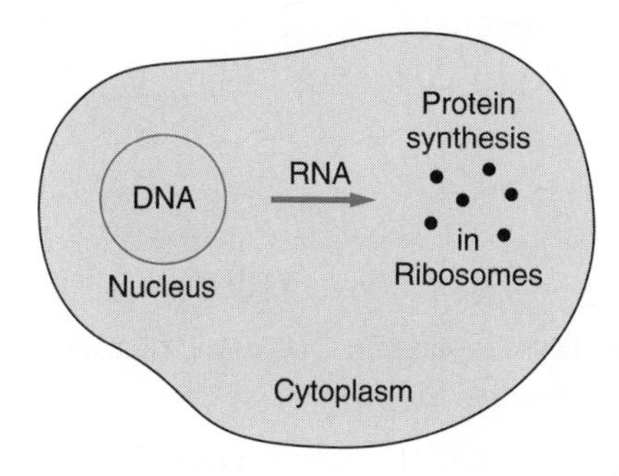

Figure 20-3 The flow of genetic information in a eukaryotic cell.

FROM DNA TO RNA

Each gene is a portion of a chromosome, in effect a portion of the nucleotide chain. To be able to move the information in the nucleotide sequence out to the ribosomes, a messenger is needed. An RNA molecule called *messenger RNA* does this job.

The structure of RNA differs from that of DNA. RNA contains the sugar ribose instead of deoxyribose. Because of this difference, the enzymes that work with RNA are able to distinguish it from DNA. RNA has four nucleotides, just like DNA, but not the same four. Like DNA, RNA contains adenine, cytosine, and guanine; but RNA substitutes uracil for thymine. Thus, the RNA nitrogenous base uracil pairs with adenine, as thymine does in DNA. Finally, RNA molecules usually consist of only a single strand. The single-stranded RNA can more easily interact with other molecules than can the double-stranded DNA molecule. (See Figure 20-4, and the table below it, on page 426.)

DNA is copied into RNA by a process called **transcription**. Transcription is similar to DNA replication. The DNA double helix opens up where

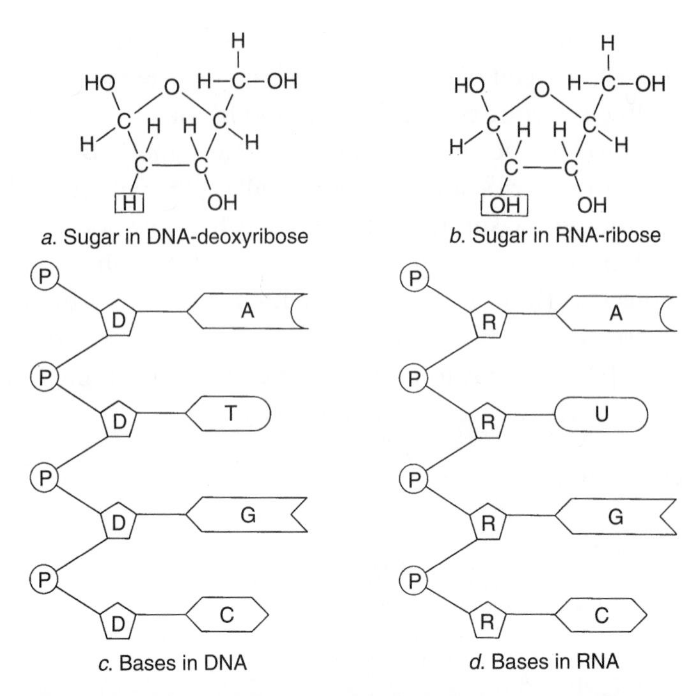

Figure 20-4 The bases in DNA are adenine, cytosine, guanine, and thymine. RNA substitutes uracil for thymine.

DIFFERENCES BETWEEN DNA AND RNA

Deoxyribonucleic Acid (DNA)	Ribonucleic Acid (RNA)
Is stored in the nucleus (of eukaryotic cells)	Moves out of nucleus to ribosomes in the cytoplasm
Contains the sugar deoxyribose	Contains the sugar ribose
Has four nucleotides: adenine, cytosine, guanine, and thymine	Has four nucleotides: adenine, cytosine, guanine, and uracil (substitutes for thymine)
Consists of a double-stranded molecule	Consists of a single-stranded molecule

a particular gene is located. Special enzymes begin to match up ribose-containing nucleotides (RNA) with the correct nucleotides in the DNA. (See Figure 20-5a.) The base-pairing rules are followed: A-U, T-A, G-C, and C-G. A messenger RNA, or mRNA, molecule is built. It has a nucleotide sequence that is complementary to the DNA strand it copied. This means that the mRNA molecule has the same nucleotide sequence as one strand of the original DNA, except where uracil is substituted for thymine. (See Figure 20-5b.) The newly formed mRNA separates from the DNA, and the double helix closes. The mRNA then moves out of the nucleus through pores in the nuclear membrane to the ribosomes in the cytoplasm.

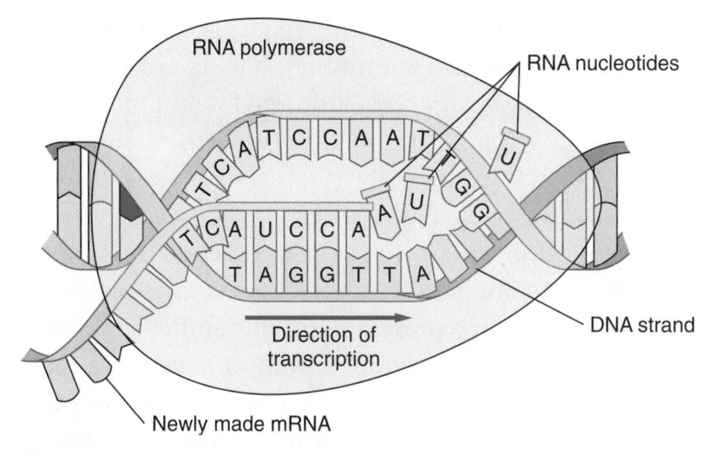

Figure 20-5a By a process called transcription, DNA is copied into RNA.

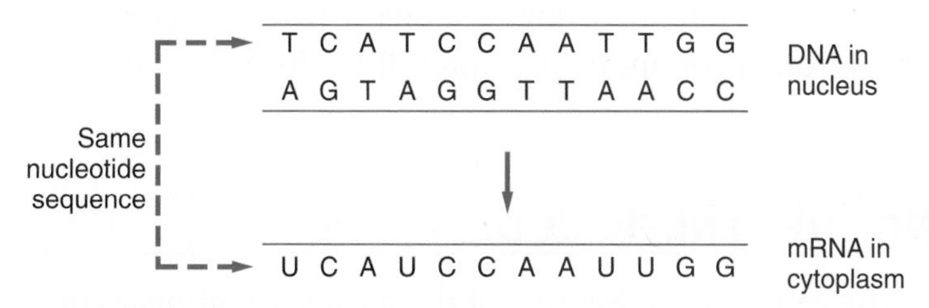

Figure 20-5b DNA and RNA nucleotide sequences.

FROM RNA TO PROTEIN

So far we have solved one problem. We have moved the genetic information, stored as a nucleotide sequence, from the nucleus to the cytoplasm by using mRNA. However, another problem still remains: how to use the nucleotide sequence in the mRNA molecule to build a protein with the correct sequence of amino acids. This problem involves a change of "language." Going from one spoken language to another is called translation. The genetic process that now occurs—using the mRNA molecule to build a protein—is also called **translation**. The nucleotide language of RNA must be translated into the amino acid language of proteins. This translation occurs at the ribosome. It is necessary to figure out how the four different nucleotides present in mRNA—A, C, G, and U—can be used to assemble 20 different types of amino acids in the correct order to form a particular protein. Clearly, one nucleotide cannot represent one amino acid. If the nucleotides were taken in pairs, there would be only 16 different pairs possible (4^2). Try it yourself: AA, AT, AC, AG, TA, TT, TC, TG, and so on. This still does not provide enough codes for the 20 amino acids. The information of the mRNA nucleotide sequence must make at least 20 different "words," one for each amino acid. If we take a sequence of three nucleotides at a time, we get 4^3 or 64 different combinations—plenty of combinations to code for 20 amino acids! Indeed, through several years of ingenious research in the early 1960s, this is exactly what scientists discovered.

Built into every living cell in the world is a **genetic code**. It is called the *triplet code*. Each different combination of three nucleotides makes up a word, called a **codon**. Each codon represents a specific amino acid. Each of the 20 amino acids has at least one codon, and most have more than one. (See Figure 20-6.) This genetic code is universal. For example, the codon CCU stands for proline in bacteria, worms, sharks, polar bears, trees, and humans. GCA stands for alanine. AGA means arginine in all organisms, and so on. All organisms on Earth use the same genetic code. This important similarity in living things is convincing evidence that all organisms evolved from a common ancestral cell life-form in Earth's far distant past.

CRACKING THE GENETIC CODE

The cracking of the genetic code was one of the most fascinating accomplishments in recent scientific research. It occurred about 10 years after Watson and Crick first revealed the structure of DNA. At that time, scientists were faced with the same kind of puzzle as government specialists who crack secret enemy military codes. However, the scientists needed more than pencils and paper to solve the genetic code. The tools they used to crack the genetic code turned out to be test tubes of chemicals and the inner contents of bacteria.

Marshall Nirenberg, at the National Institutes of Health, was the researcher who cracked the code. Nirenberg began by collecting the contents of bacterial cells in test tubes. By adding RNA to the test tubes, Nirenberg found that proteins were synthesized. It did not matter what organism the RNA came from. The test tubes contained ribosomes from bacterial cells plus everything else needed for protein synthesis. The ribosomes accepted and used the instructions for making proteins from the RNA, even if the RNA was from another organism.

Nirenberg decided to use these artificial protein factories in his test tubes. However, now he added a very special RNA. Another scientist, Severo Ochoa at New York University, had learned how to assemble nucleotides into a long strand of RNA. Ochoa had been able to make a

First Position	Second Position: U	Second Position: C	Second Position: A	Second Position: G	Third Position
U	UUU, UUC: Phe UUA, UUG: Leu	UCU, UCC, UCA, UCG: Ser	UAU, UAC: Tyr UAA: *Stop* UAG: *Stop*	UGU, UGC: Cys UGA: *Stop* UGG: Trp	U C A G
C	CUU, CUC, CUA, CUG: Leu	CCU, CCC, CCA, CCG: Pro	CAU, CAC: His CAA, CAG: Gln	CGU, CGC, CGA, CGG: Arg	U C A G
A	AUU, AUC, AUA: Ile AUG: Met	ACU, ACC, ACA, ACG: Thr	AAU, AAC: Asn AAA, AAG: Lys	AGU, AGC: Ser AGA, AGG: Arg	U C A G
G	GUU, GUC, GUA, GUG: Val	GCU, GCC, GCA, GCG: Ala	GAU, GAC: Asp GAA, GAG: Glu	GGU, GGC, GGA, GGG: Gly	U C A G

Figure 20-6 The mRNA amino acid triplet codes.

long RNA molecule that contained only uracil. The molecule was called "poly-U."

Nirenberg placed the poly-U into his test tube protein factories. After allowing time for protein synthesis to occur, he analyzed the results. What had been made? A polypeptide that contained a single amino acid, phenylalanine, repeated over and over again. Nirenberg and his colleague, Heinrich Matthaei, had sent the test tubes the following message: "uracil... uracil... uracil... uracil... uracil... uracil..."

The ribosomes in the test tube translated this message into: "phenylalanine... phenylalanine..." and so on. In effect, UUU-UUU = Phe-Phe. The first triplet code word had been discovered: UUU is the codon for phenylalanine. In addition, a method had been established to decipher the rest of the genetic code. By adding different RNA molecules to the test tubes, each of the codons was discovered. By 1965, all 64 possible nucleotide triplets had been decoded. A 3.5-billion-year-old secret, the genetic code, had been deciphered.

Check Your Understanding

Where does protein synthesis take place in a cell? How did these organelles help scientists decipher the genetic code?

TRANSLATION

All that remained to be discovered was what are the translators. What identifies the nucleotide triplets, that is, the codons that are lined up along the mRNA molecule? What matches these codons with the corresponding amino acids? That is, what translates the RNA code into a protein? A large group of hardworking molecules known as transfer RNA, or tRNA, does this job. A tRNA molecule has a "split personality." Located at one end of the molecule is a triplet of nucleotides called the **anticodon**. At the other end of the molecule, an amino acid is attached. (See Figure 20-7.) What would the "correct" amino acid be? The genetic code is needed to explain this. The table of triplet codes shows that the codon CCU stands for proline. If we use the base-pairing rules, we can determine that the anticodon—a triplet of nucleotides that will pair up with the codon—for CCU is GGA. Therefore, the tRNA molecule with proline at one end will have an anticodon of GGA at the other end. Similarly, for the codon GCA, there will be a tRNA molecule with an anticodon CGU on one end, and alanine as the amino acid on the other end.

The ribosomes are the organelles where the codes are read and the proteins are assembled. The mRNA molecule passes through a large groove in the middle of the ribosome. The ribosome's structure allows only one codon in the mRNA to be exposed at a time. The cell's cytoplasm is full of countless free-floating tRNA molecules. As the first codon in the mRNA molecule is exposed at the ribosome, the correct tRNA—that is, one with

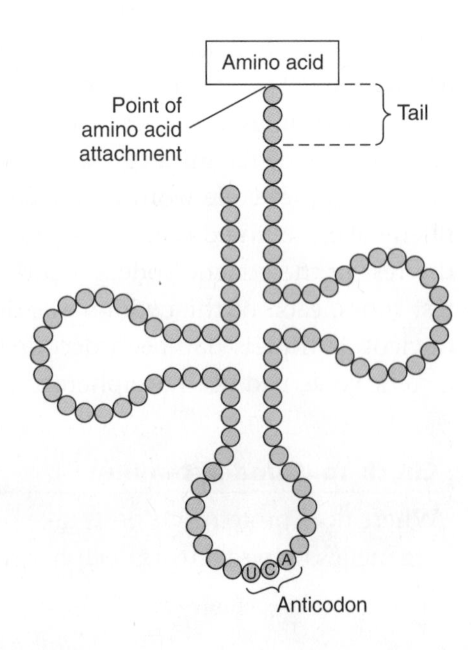

Figure 20-7 A tRNA anticodon.

a matching anticodon—lines up. It brings the amino acid it carries into place. This amino acid remains at the ribosome. The empty tRNA leaves the ribosome. The ribosome moves to the next codon on the mRNA chain. A tRNA with the matching anticodon delivers its amino acid to the ribosome. This second amino acid joins the first one. They are attached by a strong covalent bond known as a **peptide bond**. The second tRNA leaves without its amino acid. The cycle continues as the ribosome moves down the chain of mRNA. (See Figure 20-8.)

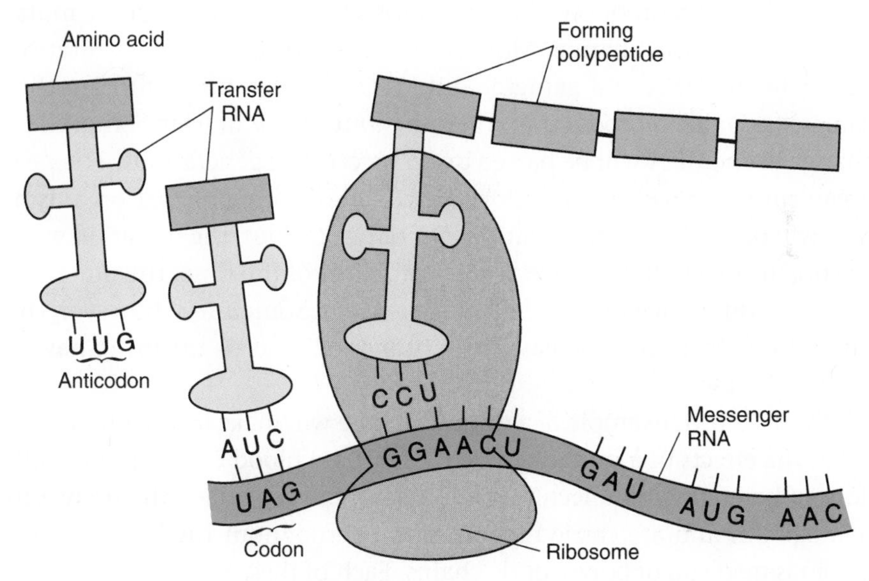

Figure 20-8 Model of translation at a ribosome.

Try to picture all of this in motion. A succession of tRNA molecules arrives at the mRNA in the ribosome. The correct tRNAs pair their anticodons with the exposed codons, bringing the correct amino acids into position. All this happens as the ribosome moves down the mRNA molecule, "reading" its message. The tRNA molecules leave, now empty. Also, the amino acids are being joined to each other, one at a time, in the correct order. A chain of amino acids is being formed, each amino acid linked by a peptide bond to the others. This chain of amino acids forms a **polypeptide**. When all the amino acids are finally linked, the polypeptide —by itself or in combination with other polypeptide chains and molecules—forms a protein. It is a specific protein, such as insulin or hemoglobin, with its own unique amino acid sequence based on the nucleotide sequence of mRNA. And that sequence was originally determined by the nucleotide sequence of the DNA in the cell's nucleus.

MUTATIONS: A CLOSER LOOK

In Chapter 19, we defined a mutation as a change in the nucleotide sequence in a DNA molecule. The possible effects of a mutation can now be explained in terms of what we know about protein synthesis.

The order of nucleotides in DNA determines the order of amino acids in proteins. In certain cases, a change, or mutation, in one nucleotide will change the triplet code, which in turn may make a change in an amino acid. If this change occurs in a body cell, then all other cells in an organism's body that reproduced through mitosis from that cell with the mutation will have the change. More important, however, if the mutation occurs in the DNA of a gamete, and if that gamete fuses with another gamete in sexual reproduction, then the mutation will be inherited. The change in the DNA will be passed on to succeeding generations. The new organism will have the mutation, as will all offspring of that organism. This will be an inherited condition. On rare occasions, the mutation gives the organism a better chance of survival. Most of the time, the mutation makes no difference in the life of an organism. Sometimes, however, the mutation is bad. In these cases, the individual and its offspring have a genetic disease.

A well-known example of a genetic disease will make it easier to understand the effects of a mutation. Each of our red blood cells contains millions of hemoglobin molecules. Molecules of oxygen attach themselves to hemoglobin and are carried in the blood throughout the body. Hemoglobin is made up of polypeptide chains. Each of these polypeptide chains contains 150 amino acids in a row. In order for the hemoglobin molecule to work properly, it must have a certain shape. The shape of the hemoglobin molecules depends on having the correct amino acids in the right order. Therefore, 450 nucleotides in the DNA—one triplet for each amino acid—must make up a gene for each polypeptide chain in hemoglobin. Sometimes, one of those 450 nucleotides is not correct. For example, instead of CTT at one location in his DNA, a person may have CAT. Then, instead of GAA on his mRNA, he has GUA. The genetic code has been changed. Instead of putting glutamic acid in place on an amino acid chain, this mutation inserts valine. (See Figure 20-9.)

One incorrect amino acid out of 150 seems like a very small change. But it is an important change, and it can be deadly. The disease caused by the single nucleotide change described here is sickle-cell anemia. With an incorrect amino acid in place in the polypeptide chain, the shape of a hemoglobin molecule changes. The red blood cells become long, curved,

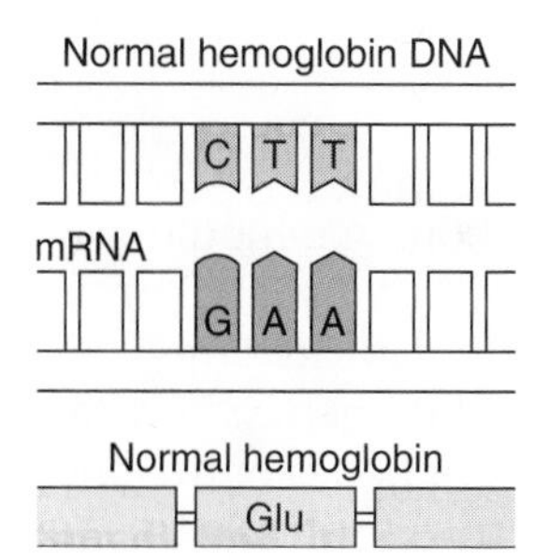

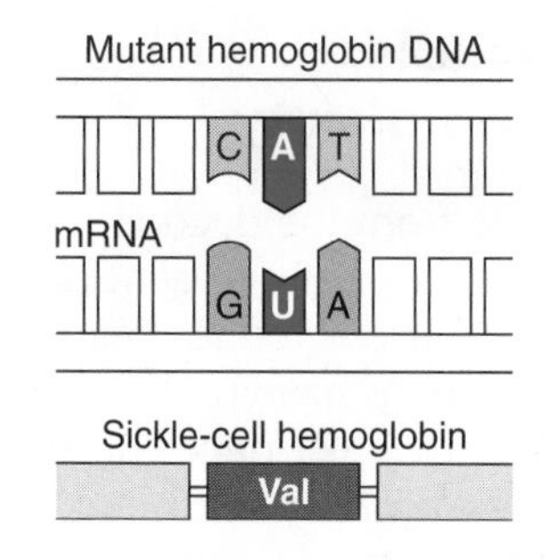

Figure 20-9 Just one incorrect nucleotide out of the 450 that make up the gene for hemoglobin is enough to cause sickle-cell anemia.

and twisted—like the grass-cutting tool, the sickle—instead of plump and round. These deformed cells get stuck in capillaries and cannot transport sufficient oxygen throughout the body. As a result, people with sickle-cell anemia tire easily. Other serious consequences may also occur. This disease can be passed onto succeeding generations. So a single change in the order of nucleotides in a DNA strand may have profound effects on a person and on that person's descendants.

The exact order of amino acids in proteins is essential for life. The fact that life goes on, with the synthesis of proteins occurring all the time, shows how precise the molecular machinery in cells usually is.

GENE EXPRESSION AND CELL DIFFERENTIATION

Chromosomes contain extremely long DNA molecules. Many genes are stretched out along these molecules. For example, it is estimated that there are 30,000 to 50,000 different genes on the 46 chromosomes in human cells. We know that following fertilization, every cell of the growing organism arises from the mitotic cell division of other cells. Through mitosis, every cell in our body has 46 chromosomes with the same DNA as the original fertilized egg cell.

But there are different types of cells in our bodies. We have skin cells, muscle cells, bone cells, nerve cells, blood cells, and so on. All of these cells have the same DNA in them. So why are these cells so different from each other? The answer is that only certain genes are used in certain cells. The use of the information from a gene is called **gene expression**. Proteins are synthesized only from genes that are expressed; these genes are turned on. All other genes in a cell are kept silent; they are turned off. This gives the cell its own structure, enzymes, functions, and physical characteristics. A muscle cell is able to contract, a nerve cell to transmit an impulse, and a skin cell to help form a flat protective layer. Creating

Human Genome's Riddle: Few Genes, Much Complexity

The first big surprise to arise from the decoding of the human genome was a matter of numbers. Where were all the genes? Rather than 100,000 or more genes, as scientists had predicted for years, the Human Genome Project has revealed that humans have perhaps only 30,000 genes. This can be compared to the fruit fly's 12,371 genes or the 19,098 genes of the tiny roundworm *Caenorhabditis elegans*. But it is not just a matter of human pride for people to consider themselves more complex. We *are* more complex. The little worm *C. elegans*, with over 19,000 genes, has a body of only 959 cells, of which 302 are neurons—the worm's "brain." The human body, built by perhaps only 50 percent more genes than the worm, has 100 trillion cells, with the brain alone containing 100 billion cells. So where does the complexity come from?

There are two main ways that scientists think human complexity has arisen. The first way concerns proteins. Proteins are the working parts of every cell, and it turns out that proteins themselves have different sections or domains in them. Ninety-three percent of the protein domains in humans are also in the worm and the fly. However, it seems that a lot of mixing and matching of these domains has occurred. Dr. Francis S. Collins, director of the genome institute at the National Institutes of Health said, "Maybe evolution designed most of the basic folds that proteins could use a long time ago, and the major advances in the last 400 million years have been to figure out how to shuffle those in interesting ways." The second ingenious way that evolution seems to have increased complexity is by dividing the genes themselves into several different segments, and using these segments in different arrangements to make different proteins.

There are many different ways of thinking about human complexity. One scientist has compared people to the machines created by them. Dr. Jean-Michel Claverie of the French National Research Center writes, "In fact, with 30,000 genes, each directly interacting with four or five others on average, the human genome is not significantly more complex than a modern jet airplane, which contains more than 200,000 unique parts, each of them interacting with three or four others on average."

the special types of cells through controlled gene expression is called **cell differentiation**. This is an essential process of life. Without cell differentiation, our bodies would be made up of only one type of cell. Your body might be simply 100 kilograms of red blood cells. That would mean life in a jar, which would not be much of a life at all! (See Figure 20-10.)

You may wonder what controls gene expression. So far, the best that

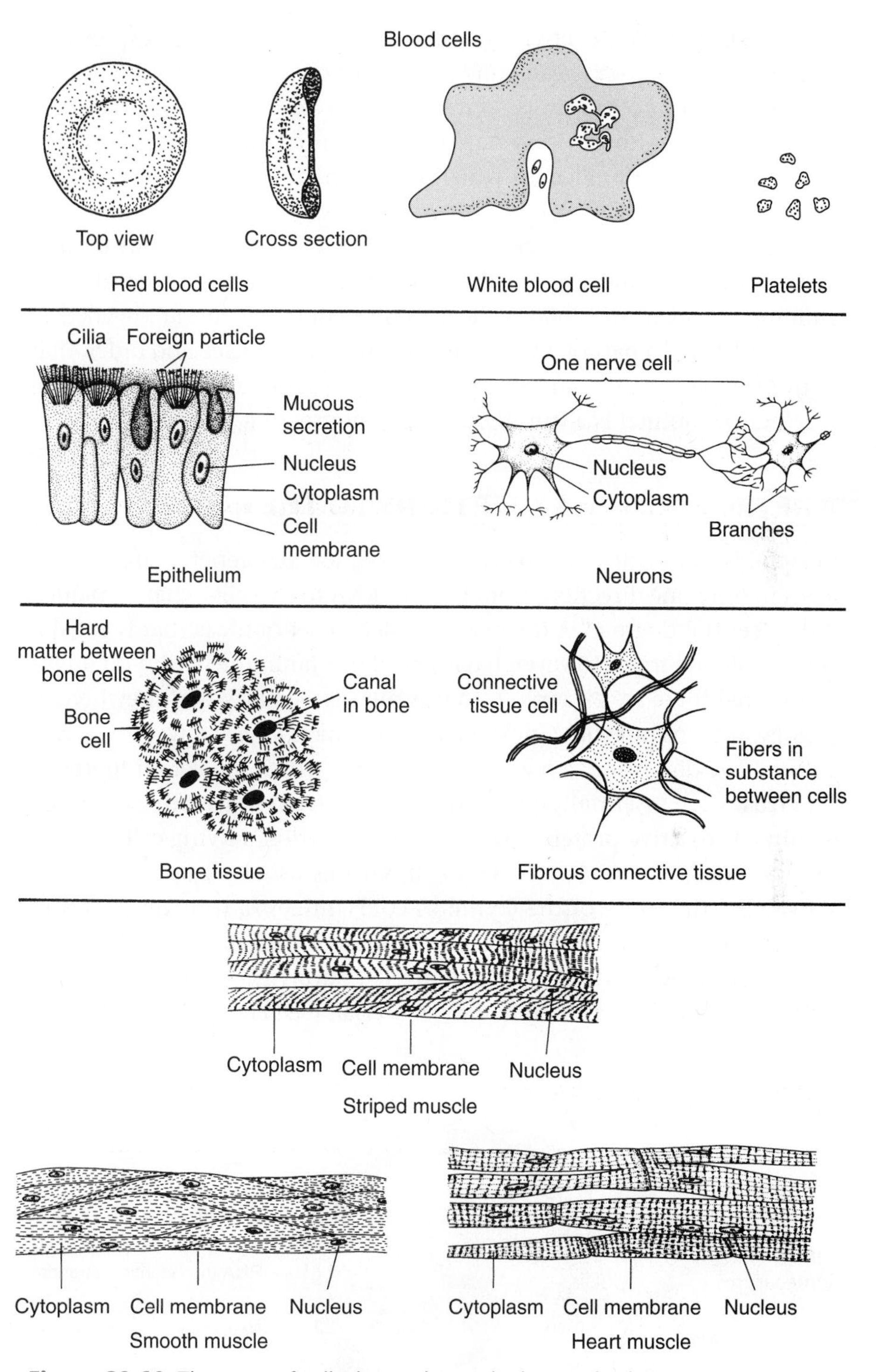

Figure 20-10 The types of cells that make up the human body.

scientists can say is that environmental factors influence gene expression. By environment, they mean conditions both outside and inside the cell. One of the most exciting areas of research today is the study of what these conditions are. Almost every day, new substances are being discovered that have some influence on which of the many genes in each cell are expressed.

In addition to this research, and just as fascinating, is the study of master control genes that regulate the activity of other genes responsible for building an organism's body. These master control genes are needed to put together each part of an organism in the correct place and order, that is, from "head to tail." Amazingly, the control genes are almost identical whether the animal is a fruit fly, a frog, a mouse, or a human.

RETROVIRUSES: EXCEPTIONS TO THE RULE

In 1956, Francis Crick was so certain of the idea that genetic information flows in only one direction, from DNA to RNA to proteins, that he named it the "central dogma." A dogma is an idea or set of ideas that is considered absolutely true. However, having read this far in the book, you might expect that there are exceptions to this rule, as we have seen elsewhere in biology. It turns out that Crick's central dogma is *almost* always true. Scientists have discovered a few rare exceptions, which are very important.

Viruses are essentially pieces of genetic material, usually contained within a protective protein coat. Viruses must enter a living cell in order to reproduce. When they enter a cell, viruses usually cause problems. Viruses are the cause of the common cold, influenza (flu), and diseases such as AIDS and herpes. Most viruses contain a DNA molecule. As would be expected, this is their genetic material. Surprisingly, however, some viruses contain RNA as their genetic material. (See Figure 20-11.)

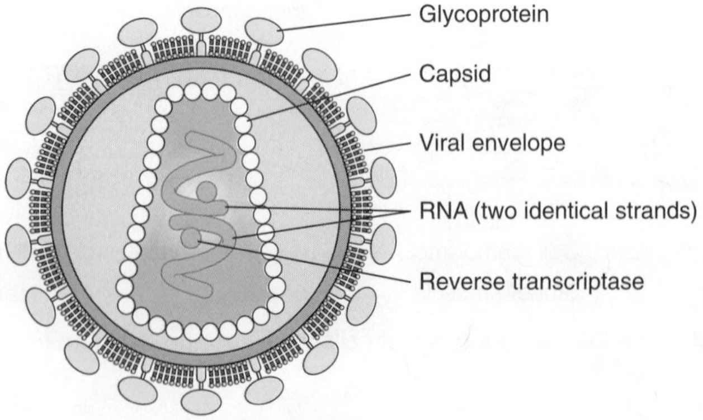

Glycoprotein
Capsid
Viral envelope
RNA (two identical strands)
Reverse transcriptase

Figure 20-11 Some viruses contain RNA as their genetic material. However, most have DNA as their genetic material.

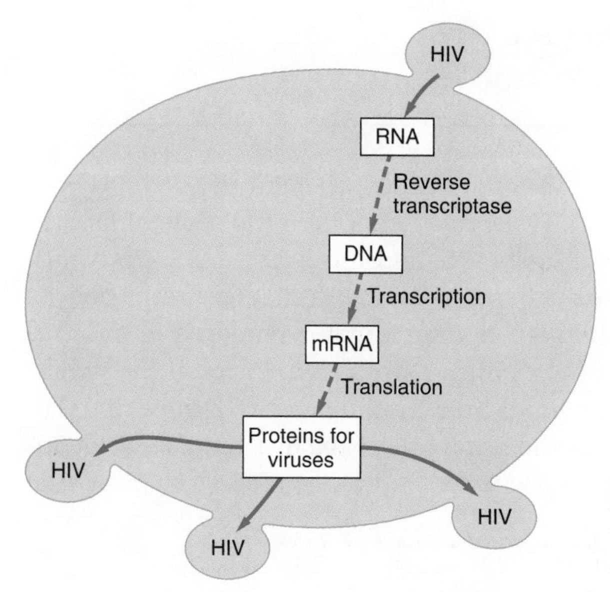

Figure 20-12 HIV, a retrovirus, changes its RNA code into a DNA code.

One of today's best-known viruses, HIV (human immunodeficiency virus), is an RNA virus. Therefore, HIV is called a **retrovirus.** (The prefix *retro-* means "backward.") In a retrovirus, the flow of genetic information at first goes in the reverse direction. That is, a retrovirus first changes its RNA code into a DNA code. It does this after infecting a cell, because the cell knows how to work only with DNA as the starting genetic material. Retroviruses carry with them the enzyme to make the process go in this opposite direction. It is called *reverse transcriptase.* As soon as HIV infects a cell, reverse transcriptase transcribes DNA from the virus's RNA. Then the viral DNA in the cell is used to build more copies of the virus. (See Figure 20-12.) One of the first medicines used to limit the effects of HIV in infected people was AZT. AZT interferes with the action of the enzyme reverse transcriptase. It is, however, not a cure for this disease. A person remains infected with HIV even while taking AZT.

You will study more about reverse transcriptase in Chapter 23. It is one of the most useful tools for molecular biologists who work with DNA.

LABORATORY INVESTIGATION 20
How Does DNA Direct the Construction of Proteins?

INTRODUCTION

Scientists consider DNA to be the most important molecule in biology because it contains the instructions that organisms use to build proteins. One protein molecule that performs important functions in the body is insulin. The protein insulin is a hormone that is produced in the pancreas. It lowers the level of blood sugar. Insulin also increases the storage of glycogen in muscles and in the liver. Insulin consists of three chains of amino acids. The gene for insulin was one of the first genes to be identified. In this investigation, you will study how the protein insulin is made from DNA. The codes for the first six amino acids in one of insulin's two chains are: Thr Lys Pro Thr Tyr Phe.

MATERIALS

Paper clips: 10 black, 10 white, 10 red, 10 green, and 10 yellow

PROCEDURE

1. In pairs or small groups, write down the DNA code for the six amino acids listed above. Find the codes for each amino acid in Figure 20-6.
2. Use the paper clips as DNA letters for each nucleotide. Make a model of a strand of DNA for the part of the insulin molecule listed above. Use the following key:

   black = adenine (A)
   white = thymine (T)
   red = cytosine (C)
   green = guanine (G)
   yellow = uracil (U)

3. Write your group's DNA code for the six amino acids of insulin on the chalkboard. Why might some of the codes be different from the others?

4. The gene for insulin may be found within a much longer DNA molecule. How does a cell know where the molecule says to start and stop reading the code for insulin? Study Figure 20-6 for help.
5. Add paper-clip letters to your insulin "gene" at the beginning and end to show the cell where to start and stop reading it.
6. Remember that in order to make a molecule of insulin, DNA must first be transcribed to RNA. Then the RNA must be moved out of the nucleus and translated into a protein at the ribosome. Use your model of insulin DNA to transcribe another paper-clip chain of messenger RNA. Use the key in step 2 and the base-pairing rule—with one exception. What is the exception?

INTERPRETIVE QUESTIONS

1. Why do you think that DNA is kept in the nucleus of the cell even though the information it contains must be transmitted to other parts of the cell?
2. Assume that the DNA codon for GGG mutates to GGT. Would this mutation have an effect on the synthesis of an insulin molecule?
3. Assume that the last DNA codon, AAG, mutates to AAT. Would this mutation have any effect on the synthesis of an insulin molecule?

CHAPTER 20 REVIEW

Answer these questions on a separate sheet of paper.

VOCABULARY

The following list contains all of the boldfaced terms in this chapter. Define each of these terms in your own words.

anticodon, cell differentiation, codon, gene, gene expression, genetic code, peptide bond, polypeptide, retrovirus, ribonucleic acid (RNA), ribosomes, transcription, translation, viruses

PART A—MULTIPLE CHOICE

Choose the response that best completes the sentence or answers the question.

1. The genetic material in retroviruses is *a.* polypeptides *b.* DNA *c.* RNA *d.* reverse transcriptase.
2. Proteins are made up of units called *a.* nucleic acids *b.* amino acids *c.* codons *d.* monosaccharides.
3. DNA is copied into RNA by a process called *a.* translation *b.* replication *c.* decoding *d.* transcription.
4. How many nucleotides make up a codon? *a.* 1 *b.* 2 *c.* 3 *d.* 4
5. Protein synthesis occurs in the *a.* ribosomes *b.* nucleus *c.* chromosomes *d.* anticodons.
6. A mutation *a.* is always harmful *b.* involves a change in the nucleotide sequence *c.* is never passed on to an organism's offspring *d.* results solely from the action of viruses.
7. One way RNA differs from DNA is that *a.* RNA is double stranded, while DNA is single stranded *b.* DNA contains uracil, while RNA contains thymine *c.* RNA is found only in the nucleus *d.* RNA contains ribose, while DNA contains deoxyribose.
8. Cell differentiation is the result of *a.* varying gene expression *b.* mutations *c.* external environmental factors *d.* different DNA in different cells.
9. Adenine in DNA pairs with *a.* adenine in RNA *b.* uracil in RNA *c.* thymine in RNA *d.* cytosine in DNA.
10. Viruses consist of *a.* eukaryotic cells *b.* prokaryotic cells *c.* an RNA coat surrounding a polypeptide core *d.* a protein coat surrounding a nucleic acid core.
11. Each amino acid is represented by *a.* one codon only *b.* one to six codons *c.* two codons *d.* two or more codons.
12. The scientist who figured out the first "word" in the genetic code was *a.* Marshall Nirenberg *b.* Severo Ochoa *c.* Francis Crick *d.* Archibald Garrod.

13. The sequence of nucleotides in mRNA is the same as that of the DNA strand that was *a.* transcribed *b.* not transcribed *c.* not transcribed, with the substitution of uracil for thymine *d.* not transcribed, with the substitution of cytosine for guanine.
14. The molecule that carries amino acids to the forming polypeptide is *a.* ribosomal RNA *b.* messenger RNA *c.* transfer RNA *d.* DNA.
15. If the codon is AUG, its anticodon is *a.* AUG *b.* ATG *c.* CGA *d.* UAC.

PART B—CONSTRUCTED RESPONSE

Use the information in the chapter to respond to these items.

16. Based on the information given in Figure 20-6 on page 429, what is the sequence of nucleotides in the gene (DNA) shown in the figure below? What amino acids do these nucleotides code for?

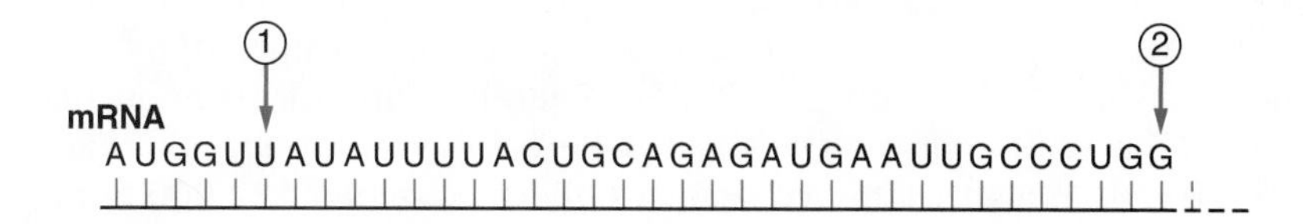

17. What happens to this mRNA after it leaves the nucleus? Name the anticodons on the tRNA molecules that interact with it.
18. What would happen if the nucleotide at arrow 1 mutated into adenine? What if the same thing happened to the nucleotide at arrow 2?
19. How is protein synthesis similar to constructing a building?
20. How did scientists "crack" the genetic code?

PART C—READING COMPREHENSION

Base your answers to questions 21 through 23 on the information below and on your knowledge of biology. Source: *Science News* (January 25, 2003): vol. 163, p. 54.

Transplanted Male Bone Marrow Makes Nerve Cells in Women and Girls

An unusual study of the brains of women and girls who had received transplants of bone marrow from men indicates that marrow cells can transform into nerve cells. Researchers found that each female brain had nerve cells containing a Y chromosome, presumably derived from the transplanted bone marrow.

Over the past several years, numerous research groups have reported that bone marrow, the source of a person's blood cells, can transform into cells of the skin, muscle, heart, liver, and even brain. These lab and animal studies have raised hopes that bone marrow or cells derived from it could repair hearts, cure neurological disorders, and treat many other medical conditions.

Some investigators, however, have challenged the bone-marrow results. The stakes are high because of the politicized debate over whether adult stem cells, such as those in bone marrow, are as promising a therapeutic tool as stem cells derived from embryos are.

In an upcoming *Proceedings of the National Academy of Sciences*, Eva Mezey of the National Institute of Neurological Disorders and Stroke in Bethesda, Md., and her colleagues report their analysis of the brain tissue of two girls and two women. Each had received a bone-marrow transplant from a male donor in a futile attempt to treat her illness. Mezey's group exposed brain-tissue samples from the four females to a marker that attaches to a DNA sequence unique to a male's Y chromosome. The investigators also applied antibodies specific to nerve cells.

In each case, Mezey and her colleagues identified a small number of nerve cells with Y chromosomes. For example, one girl studied had received a bone-marrow transplant when she was 9 months old and died less than a year later. When researchers examined 182,000 of her brain cells, they found Y chromosomes in 519—and 19 of those males cells also displayed nerve cell markers.

Another research team's unpublished findings mirror Mezey's study. Last year, Martin Körbling of the University of Texas M.D. Anderson Cancer Center in Houston and his colleagues employed the same Y chromosome-based strategy to discover bone-marrow-derived skin, gut, and liver cells in a half-dozen women who had received marrow transplants before dying. Now, Körbling tells *Science News*, "we have data showing similar results in midbrain and cortex tissue."

Diane Krause of Yale University notes that her research team and many others are vigorously studying the mechanisms by which bone-marrow cells may transform into cells other than blood cells. Unless researchers can enhance the pace of this natural cellular makeover, the phenomenon is unlikely to be of much medical use, both she and Mezey caution.

21. State the basic research question about transplanted bone-marrow cells that was being discussed in this report.
22. Explain why brain tissue from two girls and two women was being analyzed by the researchers in Bethesda, Maryland.
23. State the conclusions made by the researchers about the transplanted bone-marrow cells in these four individuals.

21 Patterns of Inheritance

After you have finished reading this chapter, you should be able to:

Explain the importance of Mendel's work to our understanding of the patterns of heredity.

Describe Mendel's laws that govern heredity of traits.

Apply Mendel's work to the improvement of crops such as corn.

The past lives on in the present.

Sir Arthur J. Thompson

Introduction

In many ways, science is a product of its time. Throughout the ages, people have tried to understand and explain the natural world. Some of the explanations are, to our eyes, silly. But they served a purpose to the people who needed them to deal with a world that often was confusing. These explanations, no matter how far-fetched they may seem to us, mark people's attempts to understand their world. In this chapter, we will examine three explanations about living things that people once used. Then we will explore other, more scientifically accurate explanations that we use today.

EARLY ATTEMPTS TO EXPLAIN HEREDITY

The one-hump camel (more accurately called a dromedary) is a tall animal that is superbly adapted to desert life. Its huge feet spread its weight over a large area, allowing the camel to walk on shifting sand. (See Figure 21-1 on page 444.) Unless the animal kneels, it is not possible for a person to climb onto a camel's back without a ladder. The leopard is a very different animal in appearance and lifestyle. The leopard's fur has distinctive

Figure 21-1 The dromedary, a tall animal, is superbly adapted to desert life.

spotted markings. You may have heard the old expression, "A leopard can never change its spots." (See Figure 21-2.)

Now picture a giraffe, a tall animal with spots. (See Figure 21-3.) In the past, people offered an interesting explanation about where the strange-looking giraffe came from. They thought that the camel and the leopard occasionally mated, thereby producing a giraffe. According to this belief, the giraffe gets its height from the camel and its spots from the leopard. In fact, the scientific name of the giraffe, *Giraffa camelopardalis,* gives credit to this early idea about the animal's origin.

Figure 21-2 The leopard has distinctive spots.

Figure 21-3 In the past, people explained that the tall, spotted giraffe came from the mating of a camel and a leopard. Today, we know that this is not true.

Around 1850, gardeners were trying to explain how flowers got their characteristics. Most gardeners believed that the characteristics of one plant were blended together with the characteristics of another plant when they reproduced. In some ways, people thought it was like mixing paint. The offspring would have a mixture of the traits of each parent.

And then there was great confusion about the origins of humans. From as early as the 1600s, sperm cells and egg cells had been observed under microscopes. But where did our physical characteristics come from? Some people believed that, under the magnifying lens, they could observe a tiny person inside each sperm cell. (See Figure 21-4.) To them, all inherited traits came from the male parent. Others were sure that the egg contained a tiny future human being. To these people, the female determined the characteristics of the baby.

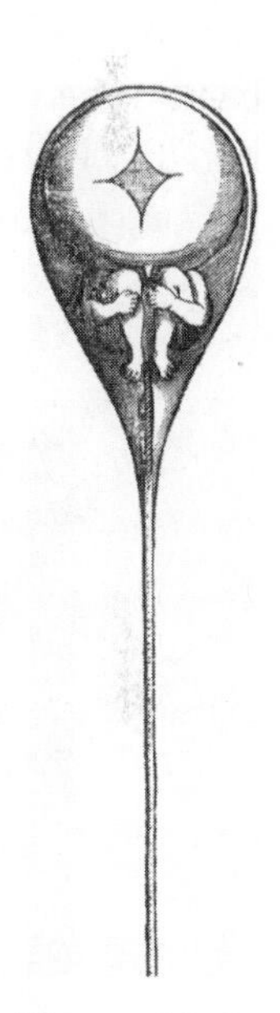

Figure 21-4 In the late 1600s, some people believed that, under a magnifying lens, they could see a tiny person inside each sperm cell.

As you know, these explanations are confused, confusing, and not very accurate from a scientific standpoint. How characteristics are passed from parents to offspring was a question that puzzled people for thousands of years. Whether it was about giraffes, garden plants, or humans, there were plenty of ideas offered, but no real answers.

A now-famous set of experiments completed more than 100 years ago by Gregor Mendel, an Austrian monk, changed our understanding of how characteristics are passed from one generation to the next.

GREGOR MENDEL, THE FATHER OF GENETICS

Mendel studied mathematics and science at a university but failed the tests to get his teaching certificate. So he entered a monastery, a place of quiet seclusion set apart from most of the cares of the world. However, Mendel was by nature a curious person. He was aware of the confusion about heredity and wanted to investigate how characteristics are passed from one generation to the next. After being denied permission to experiment with mice, he was given the use of a small garden to carry out his work. The common garden pea became the focus of his experiments. (See Figure 21-5.)

This was in 1857, and over the next 8 years, Mendel conducted hundreds of experiments with thousands and thousands of pea plants. He filled many notebooks with the results of his experiments. By applying careful mathematical analysis to his work—something that had rarely been done in biology before—Mendel discovered much about the ways heredity works. The science of genetics really began with Mendel. For this reason, he can rightly be called the "founder of the science of genetics."

Mendel discovered that hereditary information is passed from parents

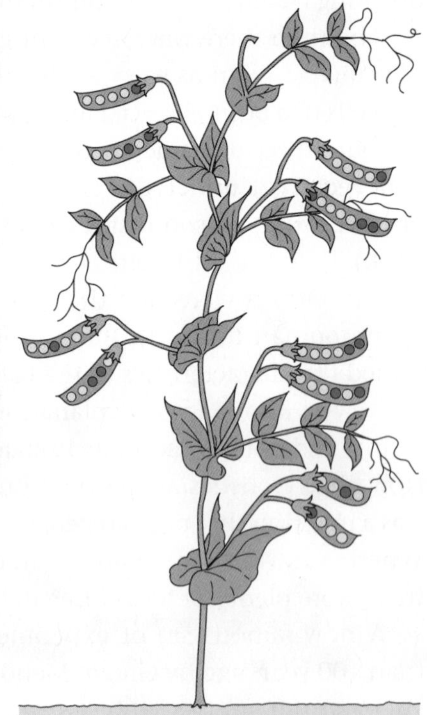

Figure 21-5 Gregor Mendel investigated the passing of characteristics from one generation to the next by studying the traits of the common garden pea.

to offspring in individual pieces he called *factors,* which we now call genes. He also discovered that the factors were passed on in specific, predictable patterns from both parents. These patterns of inheritance are described in Mendel's laws. You will now look more closely at Mendel's ideas and at his experiments.

THE GARDEN AS A LABORATORY

Mendel identified specific characteristics, or traits, in his pea plants. The characteristics he identified appeared in two different, contrasting versions. For example, some of the pea plants Mendel worked with had purple flowers; other plants had white flowers. The peas, actually the plant's seeds, develop within pods. Some of the plants produced yellow peas; others produced green peas. The shape of the peas varied, too. The peas were either smooth and round or wrinkled. In addition, the pea plants could be either tall or short. As you can see, the characteristics that Mendel studied occurred as contrasting traits that were easy to observe.

Mendel performed delicate surgery on the flowers when he did his experiments. You will recall from Chapter 16 that pollen grains—small structures that contain the male gametes—are produced in the stamens. Egg cells—the female gametes—are contained in the flower's ovary at the base of its pistil. Mendel recorded the characteristics of the plants he was working with. Let's say a particular plant produced green seeds. Mendel removed the stamens from this plant's flowers (in order to prevent self-pollination). Then, using a paintbrush, Mendel carefully placed pollen from another plant—perhaps one with yellow seeds—on the pistil of the first plant. By pollinating these flowers by hand, Mendel controlled the mating between the two parent plants. He controlled which pollen reached the egg cells in a particular flower.

The plant with the pollinated flowers continued to grow. Eventually this plant produced seeds, the peas in the pods. Each seed was a single offspring, the result of the fertilization of one egg by a single sperm. Mendel noted the characteristics of each seed—whether it was green or yellow, wrinkled or round. He kept accurate counts of the results of his crosses. It was these counts—the mathematics of genetics—that later allowed him to offer explanations for the way traits were inherited. In some cases, Mendel planted the seeds (peas) to see what kinds of plants were produced. In this way he could check to see how other traits besides the shape and color of the seeds were inherited. This type of experiment is called a **genetic cross**. (See Figure 21-6 on page 448.)

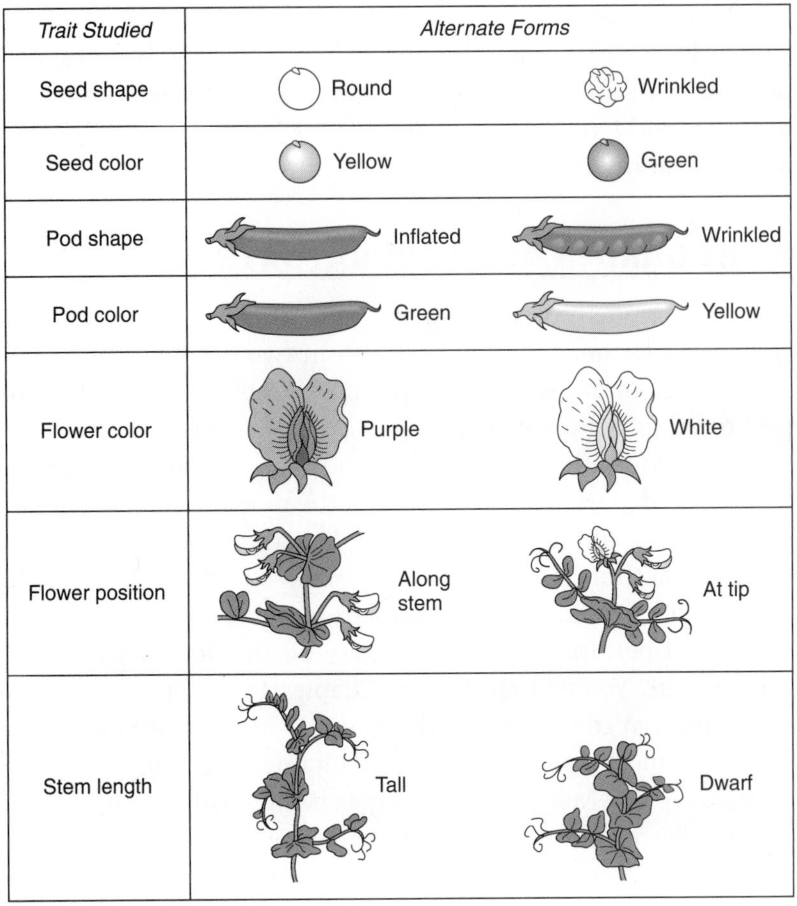

Trait Studied	Alternate Forms	
Seed shape	Round	Wrinkled
Seed color	Yellow	Green
Pod shape	Inflated	Wrinkled
Pod color	Green	Yellow
Flower color	Purple	White
Flower position	Along stem	At tip
Stem length	Tall	Dwarf

Figure 21-6 The characteristics that Mendel studied occurred as easy-to-observe, contrasting traits.

Mendel always began his experiments with plants that breed true. This means that if these plants are allowed to self-pollinate, they show characteristics that never vary from the parent plants. If a plant with purple flowers is true-breeding, then all of the seeds it produces will grow into plants that produce purple flowers.

The first true-breeding individuals used as the parents are called the P, or parental, generation. The offspring from the first mating of the parents are called the F_1 (first filial) generation. (The word *filial* comes from a Latin word that means "son.") Mendel then used individuals from the F_1 generation as the parents. He allowed these plants to self-pollinate. By producing a genetic cross of these individuals, he produced an F_2 (second filial) generation. (See Figure 21-7.)

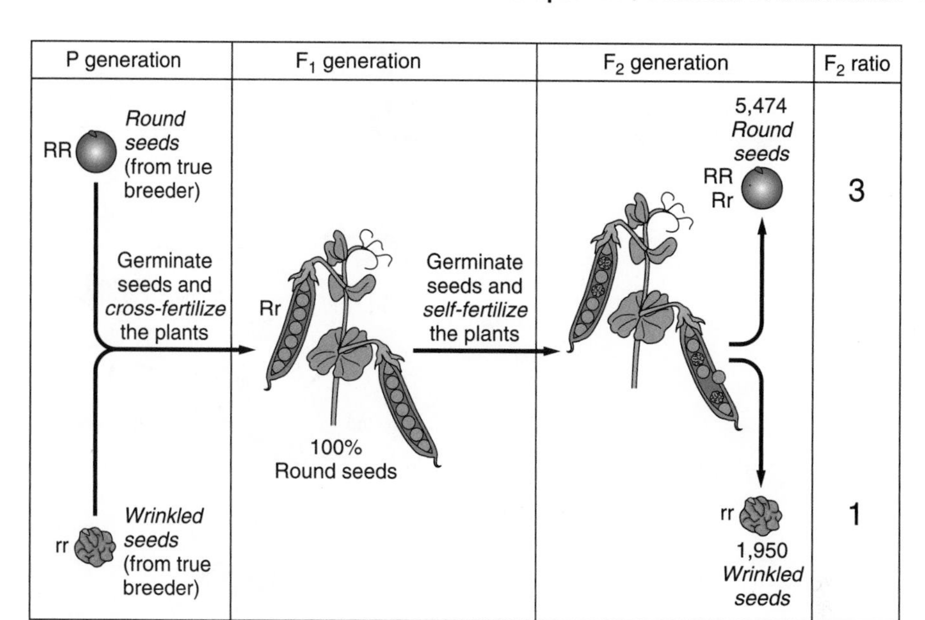

Figure 21-7 P is the parental generation, F_1 is the first filial generation, and F_2 is the second filial generation.

Mendel's work of growing pea plants, controlling the genetic crosses between the plants, observing the F_1 and F_2 generations, and counting the results lasted for eight years. Making accurate counts of the offspring was very important. By interpreting the numbers he accumulated, Mendel was able to show a beautiful, orderly pattern in the inheritance of an organism's traits.

A TYPICAL PEA PLANT EXPERIMENT

Mendel crossed true-breeding plants that had purple flowers with other plants that had white flowers. If inherited traits are blended, then the F_1 plants should have produced flowers that were a blend of purple and white, a kind of pink. Instead, Mendel found that all of the F_1 plants had purple flowers. The white flowers did not show up at all; they had completely disappeared in the offspring. (See Figure 21-8 on page 450.)

Had Mendel ended his experiment here, only a small amount of information would have been learned. Fortunately, he continued his work. By allowing the F_1 plants with the purple flowers to self-pollinate, he produced an F_2 generation. And in the F_2 plants, the white flowers reappeared! The actual count of flower color in this cross was 705 purple

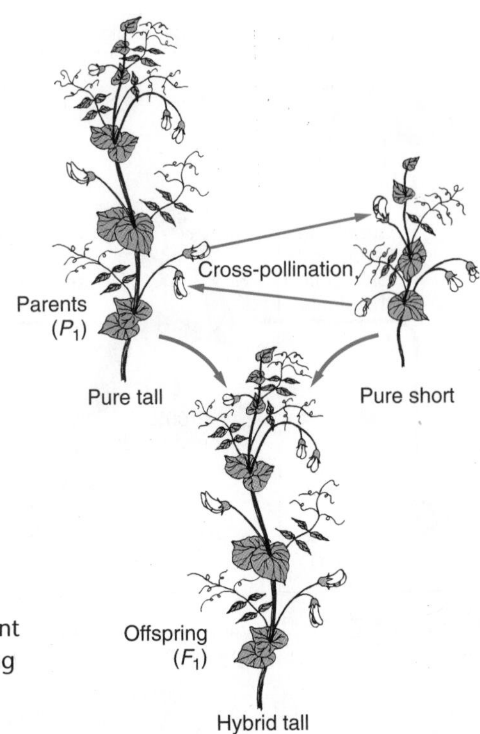

Figure 21-8 Mendel's first experiment was on the cross-pollinated contrasting pure types, such as flower color and stem length.

flowers, 224 white flowers. The ratio of purple flowers to white flowers was about 3 to 1. (See Figure 21-9.)

Mendel repeated the same experiment using the other true-breeding characteristics such as seed color, seed shape, and plant height. Each time, one of the traits disappeared in the F_1 generation and reappeared in the F_2. In addition, each time the ratio of the traits was about 3 to 1 in the F_2 generation. (See table below.) These were very meaningful results, and from these results, Mendel derived two of the main laws of genetics.

RESULTS OF MENDEL'S EXPERIMENTS WITH PEA PLANTS

	Original Crosses	Second Filial Generation (F_2)			
Trait	**Dominant × Recessive**	**Dominant**	**Recessive**	**Total**	**Ratio**
Seed form	Round × Wrinkled	5,474	1,850	7,324	2.96:1
Seed color	Yellow × Green	6,022	2,001	8,023	3.01:1
Flower position	Axial × Terminal	651	207	858	3.14:1
Flower color	Purple × White	705	224	929	3.15:1
Pod form	Inflated × Constricted	882	299	1,181	2.95:1
Pod color	Green × Yellow	428	152	580	2.82:1
Stem length	Tall × Dwarf	787	277	1,064	2.84:1

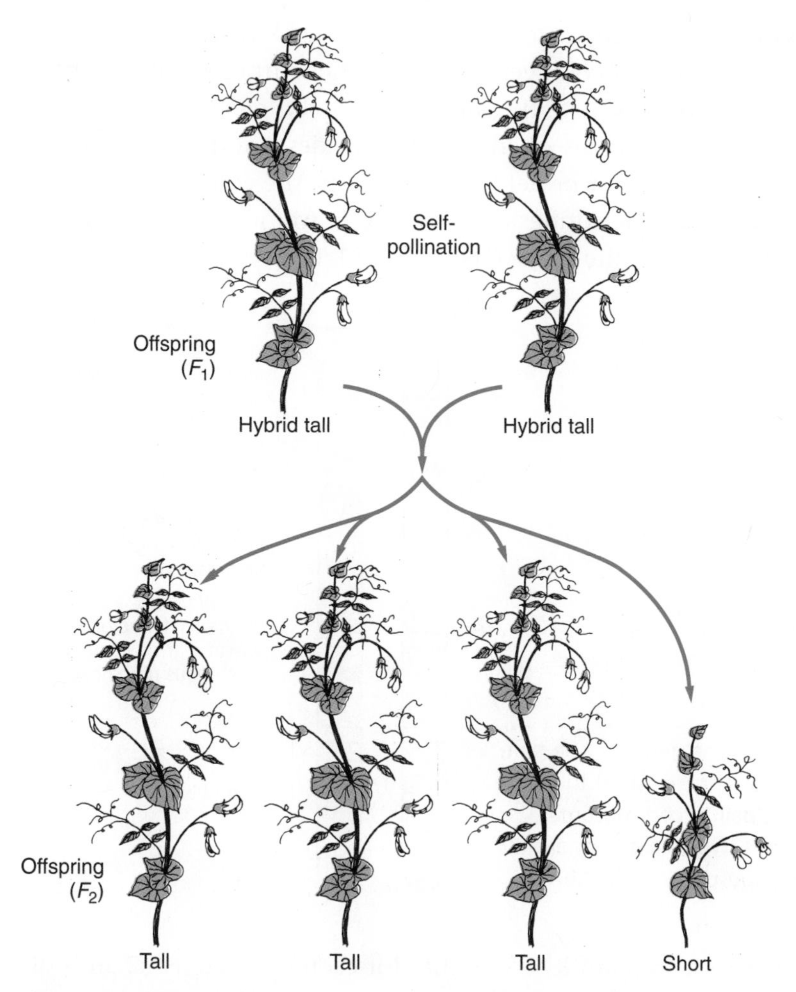

Figure 21-9 Mendel's second experiment was the self-pollination of the F_1 generation. The F_2 generation shows the 3-to-1 ratio.

THE LAW OF DOMINANCE

Mendel proposed several ideas that explained his results. What is amazing is that his ideas were correct, even though Mendel knew nothing about chromosomes, genes, or DNA. Scientists have been able to combine Mendel's ideas with what has been learned about genetics since his time. We have come a long way from those early days of confusion about heredity!

Mendel's first main idea was that each characteristic, or trait, exists in two versions. These two versions of a gene are called **alleles**. For example in pea plants, there are two alleles for the gene for flower color: one allele for purple flowers and one allele for white flowers. Genes exist at specific locations on chromosomes. One chromosome may have an allele for purple flowers at the flower color gene spot. Another chromosome may have an allele for white flowers at the same flower color gene spot. The DNA at these spots consists of a sequence of nucleotides. For the purple flower allele, the DNA codes for proteins that result in the color purple. A different sequence of nucleotides at the white flower allele codes for proteins that make the white color. (See Figure 21-10.)

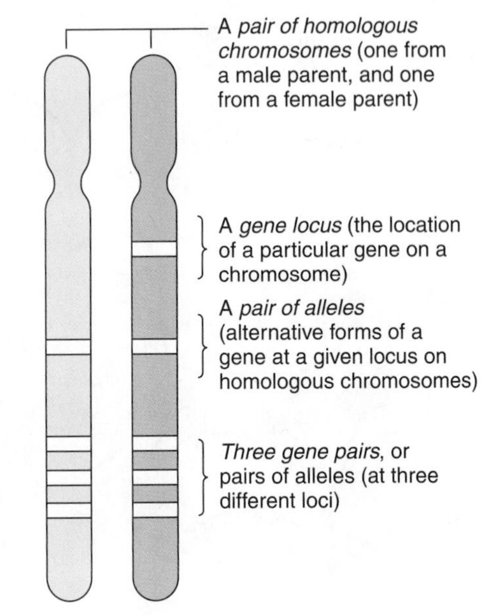

Figure 21-10 The gene for one trait, such as flower color, exists at the same place on the pair of chromosomes. Each gene has two versions, or alleles, for example, purple or white flowers.

Mendel's second main idea was that for each characteristic, an individual inherits two alleles, one from each parent. You learned in Chapter 16 that each of us has a diploid number of chromosomes in all our body's cells except for the gametes. Pea plants are also diploid. In fact, all offspring that result from sexual reproduction are diploid. In addition, the diploid set of chromosomes is made up of homologous pairs of chromosomes. Each homologous pair of chromosomes consists of one chromosome from one parent and one chromosome from the other parent. A particular gene occurs at the same place on each homologous chromosome. Therefore, a particular gene exists twice in each diploid cell—one from the mother and one from the father. The cell has two alleles for each gene.

The true-breeding parent pea plants in the P generation had two identical alleles. Both alleles in the true-breeding plants with purple flowers were for purple flowers. The true-breeding plants with white flowers had two alleles for white flowers. Individuals with identical alleles for a particular trait, or characteristic, are said to be pure or **homozygous**. Mendel also explained the results he found in the F_1 generation in terms of the alleles (although he didn't use this word). Because the F_1 plants obtained one allele from each parent, and because the parents differed from each other, the F_1 plants had two different alleles for flower color. The F_1 plants had one purple flower allele and one white flower allele. Organisms that have two different alleles for a particular trait, or characteristic, are said to be hybrid, or **heterozygous**.

Mendel next explained the results he found when observing the heterozygous plants. A heterozygous individual has two different alleles. The one that appears, or shows, in the offspring is the **dominant** allele. The one that does not appear in the offspring is the **recessive** allele. In the experiment we described, the allele for purple flower color is dominant, and the allele for white is recessive. Remember that in a cross between a pure plant with purple flowers and a pure plant with white flowers, only purple flowers showed up in the offspring. This is Mendel's Law of Dominance.

Why does one allele show up while the other allele remains hidden? How does dominance happen? Alleles are sections of DNA that code for a particular protein, usually an enzyme. A dominant allele usually codes for an enzyme that works. A recessive allele, being a slightly different DNA nucleotide sequence, usually codes for a form of the enzyme that does not do its job. The white flower allele codes for an enzyme that cannot make the purple color. The purple flower allele codes for an enzyme responsible for making the purple color. As long as one purple flower allele is present, as in a heterozygous plant, some of the enzyme will be produced to make the flower purple. Therefore, in this example, purple is dominant, and white is recessive.

THE LAW OF SEGREGATION

The final idea in Mendel's explanations of his experiments was that the two alleles for each trait, or characteristic, separate when the organism makes gametes. Remember that even though Mendel did not know about chromosomes or meiosis, he was certainly correct in his explanations of how traits are inherited. When an organism makes gametes, the chromosome number gets reduced by half. The homologous pairs of chromosomes get separated. Each gamete receives only one member of each pair. Therefore, each gamete gets only one of the two alleles. In this way, the two alleles for each characteristic get separated, or segregated. The gamete, with only one set of chromosomes, is haploid.

If the diploid parent cells were homozygous, then all the gametes

Breeding a Better Corn

Corn, originally a plant native to Mexico, is the most important agricultural crop grown in the United States today. People eat corn and it is also fed to livestock. Early in the twentieth century, researchers used the new science of genetics to improve the food value of corn grown on American farms. This is a story of how scientific understanding has made an important improvement in human life.

It is not good for most organisms to breed with close relatives, a process called *inbreeding*. Harmful traits are usually recessive and thus hidden in offspring if only one parent carried the gene. This is the case when offspring are heterozygous; with one normal gene present, the harmful traits are not expressed in the individual. If close relatives that carried the same flawed gene were to interbreed, the trait could become homozygous in the offspring. This occurs in corn plants, too. In corn, inbreeding causes the plant to become shorter, less strong, less productive, and more likely to develop diseases. The first person known to experiment with developing

produced would have the same alleles. However, if the parent was heterozygous, we can predict that 50 percent of the gametes would have the purple flower allele, and 50 percent of the gametes would have the white flower allele. This idea is Mendel's Law of Segregation.

TESTING MENDEL'S IDEAS

Let's take a closer look at Mendel's experiments. Since it is not possible to plant a garden, grow plants, and list results during this course of study, we will "cross" individuals by using a simple diagram.

We know from the F_1 generation that the purple flower trait is dominant, and the white flower is recessive. We can use the capital letter A for the dominant trait. A lowercase letter a can be used for the recessive trait. This is a convention used by many scientists when discussing the results of genetic crosses. The homozygous, purple flower parent plant could be symbolized as AA. We use two capital letters to represent the two dominant purple alleles carried on the chromosomes of the diploid pea plant. The combination of genes in an organism is called the **genotype**. The genotype determines what an organism looks like. The appearance of an organism is called its **phenotype**. Therefore, a pea plant with the genotype AA has the phenotype of purple flowers. Similarly, the plants with the genotype aa have the white flower phenotype. (See Figure 21-11.)

new strains of corn was the Pennsylvania farmer John Lorain. In 1812, he described experiments in which he crossed two different types of corn to make a hybrid that produced greater yields than either parent.

The American botanist and geneticist George H. Shull is well known for his work on hybrid corn. As a result of Shull's research, corn yields increased 25 to 50 percent. In 1917, at Harvard University, the chemist Edward East and his student Donald Jones successfully combined two single-cross hybrid corn varieties to produce the first new, highly productive corn variety that could be grown commercially. Since the 1930s, corn hybrids have been used in countries throughout the world. Now, molecular geneticists have produced even better breeds of corn by inserting genes to make the corn naturally resistant to pests. A controversy has arisen since some of this new "super corn" has apparently shown up uninvited in Mexico where the native corn still grows. Nevertheless, the study of genetics has helped provide more and better food for people all over the world.

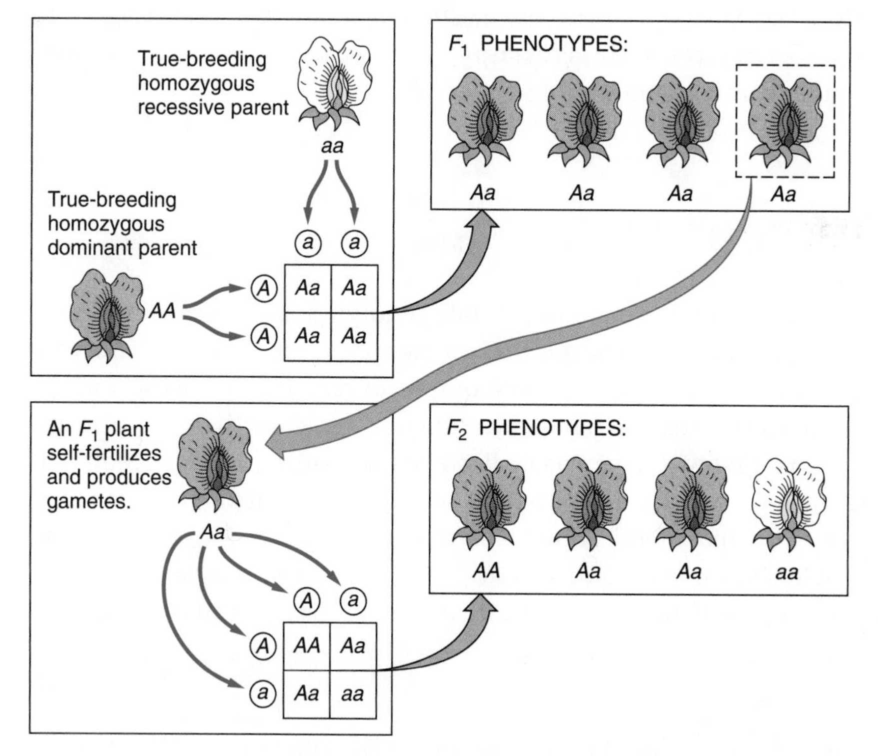

Figure 21-11 Flowers with the genotype AA or Aa have the purple phenotype. Those with the genotype aa have the white phenotype.

The first genetic cross in Mendel's experimental cross of a purple-flowered plant with a white-flowered plant can be represented as AA × aa. A useful way to examine the results of such a cross is by using a Punnett square. A Punnett square shows three things: the genotype of each parent, the segregation of the alleles that occurs when gametes are formed, and finally the possible genotypes of the offspring when the gametes come together during fertilization. Therefore, the Punnett square for the first cross in this experiment would look like this:

	A	A
a	Aa	Aa
a	Aa	Aa

The boxes in the Punnett square show the possible genotypes of the offspring. In this cross, we see that all offspring would have the heterozygous genotype, Aa. Therefore, all of the offspring would have purple flowers. Their phenotype is determined by the presence of a single, dominant gene for purple flower color. This is what Mendel observed. (See Figure 21-12.)

How can Mendel's next cross be shown? Using two heterozygous F_1 individuals as parents, this cross can be written as Aa × Aa. The Punnett square for this cross shows:

	A	a
A	AA	Aa
a	Aa	aa

In this case, one of the four possible genotypes (25 percent) in the offspring is AA, plants with purple flowers; two of the four possible genotypes (50 percent) are Aa, also plants with purple flowers; and one of the four (25 percent) is aa, plants that have white flowers. Three-fourths, or 75 percent, of the offspring will have purple flowers; one-fourth, or 25 percent, will have white flowers. Here is the remarkable 3-to-1 ratio in phenotypes of the F_2 generation that Mendel observed. Inheritance was no longer a mystery! We could show crosses on paper and demonstrate with actual crosses of organisms exactly how genetic traits are passed from parents to offspring.

Check Your Understanding

What is the difference between genotype and phenotype? How does an organism's genotype determine its phenotype?

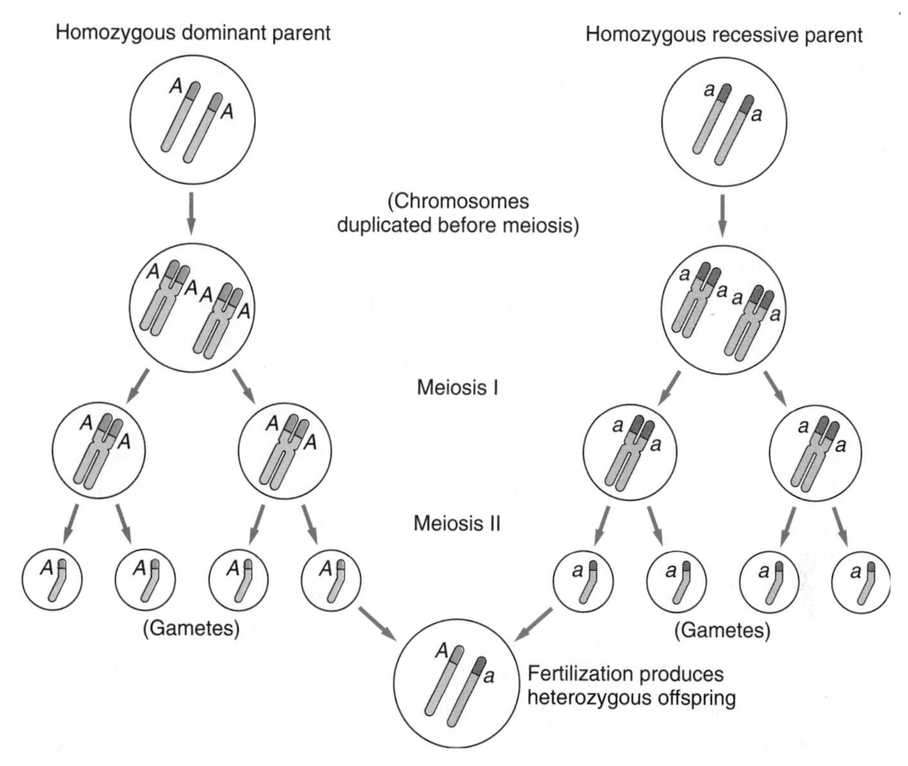

Figure 21-12 When one parent is homozygous dominant and the other is homozygous recessive, the offspring are all heterozygous and exhibit the dominant form (phenotype) of the trait.

THE LAW OF INDEPENDENT ASSORTMENT

Having demonstrated how a single characteristic is inherited, Mendel designed experiments to explain how two characteristics are inherited simultaneously. Again, Mendel used pea plants in his investigation. This time he examined how the traits for seed shape and seed color are inherited. Mendel used the same procedures he used in his earlier genetic crosses. Punnett squares explain his results. In making our Punnett square, let Y stand for yellow, y for green, R for round, and r for wrinkled. Mendel crossed a homozygous plant that had dominant genes with a homozygous plant that had recessive genes. In this case, the genotypes of the plants used in the cross were YYRR (yellow round seeds) × yyrr (green wrinkled seeds). All the plants in the F_1 generation were YyRr, heterozygous for both traits, and would produce round yellow seeds. Once again the F_2 generation showed different and important results. By crossing YyRr × YyRr, Mendel produced plants that had the following four phenotypes:

yellow round seeds	$\frac{9}{16}$
green round seeds	$\frac{3}{16}$
yellow wrinkled seeds	$\frac{3}{16}$
green wrinkled seeds	$\frac{1}{16}$

You can see that the ratio of the four different phenotypes in this cross is 9:3:3:1. This ratio is important. (See Figure 21-13.)

Looking at the results differently, Mendel saw that three-quarters of the seeds were round, and one-quarter were wrinkled. This is the familiar 3-to-1 ratio of dominant to recessive traits seen when two heterozygous (Rr) plants are crossed. For the other trait, the production of yellow or green seeds, the results also showed a 3-to-1 ratio. Again a familiar, and expected, result. Even though the two traits were being inherited together, they behaved as if they were inherited separately. In other words, the inheritance of one trait had no effect on the inheritance of the other. Mendel stated that the alleles (factors) for different traits segregate independently. Just because two alleles such as R and Y were together in the parent did not mean that they would stay together when gametes were formed. We now know why this is true. Chromosomes can line up in any order during meiosis. These traits would be found on different chromosomes and be inherited independently of each other. Mendel called this the Law of Independent Assortment. Once again, he was correct, even though he lacked knowledge of how chromosomes work.

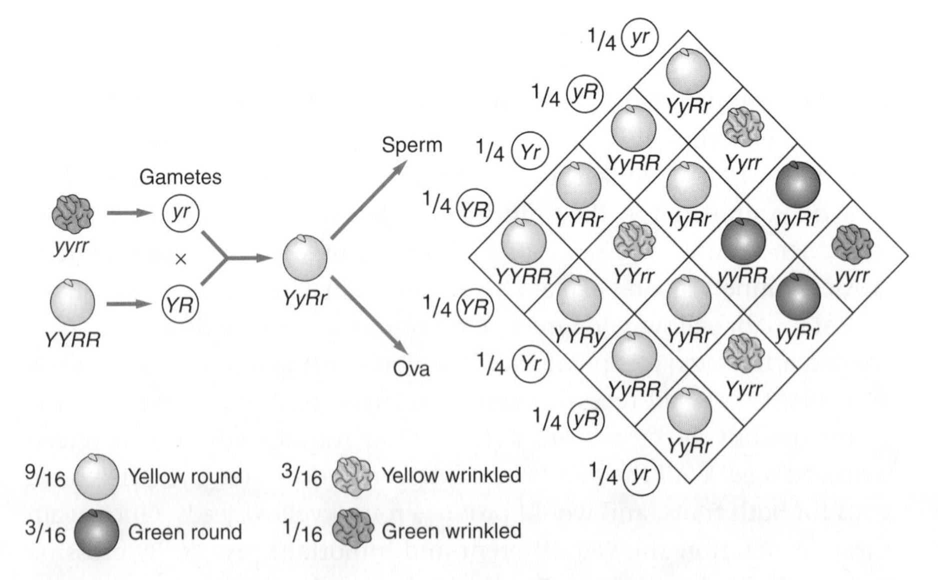

Figure 21-13 The Law of Independent Assortment is illustrated by crossing organisms that are hybrid for two traits; each trait shows a 3-to-1 ratio.

EXCEPTIONS TO THE RULES

Mendel's experiments opened the door to our understanding of the patterns of inheritance. However, as genetic research continued, it soon became clear that Mendel's laws did not provide sufficient explanations for all kinds of inheritance.

It seems that some dominant and recessive traits do mix, or seem to blend together, in some organisms. For example, crossing a red-flowered snapdragon with a white-flowered snapdragon produces plants that have pink flowers in the F_1 generation. However, blending of the red and white traits has not occurred. In the F_2 generation, red and white flowers reappear, along with pink flowers. If true blending had occurred, red or white flowers would never reappear. This process is known as **incomplete dominance**. In the heterozygous pink plants—with one allele for red and one allele for white—the dominant allele does not completely make up for what is missing in the recessive allele. However, once the dominant and recessive homozygous genotypes form again, the original dominant (red) and recessive (white) phenotypes reappear. (See Figure 21-14.)

Codominance is another related exception to Mendel's rules. In codominance, both of the alleles are expressed in heterozygotes. In one well-known example of codominance, a type of cattle seems to show three

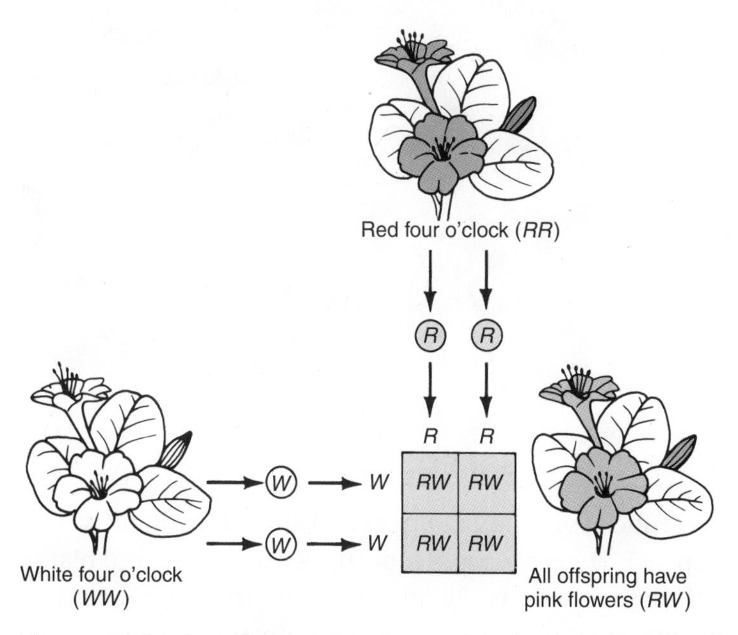

Figure 21-14 Snapdragons show incomplete dominance. When plants that have red flowers are crossed with plants that have white flowers, the offspring have pink flowers.

phenotypes. Some of these animals are brownish-red, some are white, and some are a color called roan. If you examine a roan animal, you can observe hairs that are brownish-red and other hairs that are all white. The roan cattle are heterozygous, with one allele for red and one allele for white. Both alleles are expressed. This is a case of codominance. The brownish-red cattle and white cattle are homozygous.

Siamese cats are another example of how genes are expressed. We know that the genotype of an organism determines its phenotype. The environment in which an organism lives can also affect the way genes are expressed. Siamese cats have a gene that codes for an enzyme that produces dark fur. However, this enzyme works only at cool temperatures. Most of a cat's body is too warm for this enzyme to work. If you look at a Siamese cat, you can easily identify the dark ears, paws, face, and tail—the cooler areas where the enzyme works. Even though all parts of the cat's body have the same combination of genes, the way that the genes are expressed differs from one part to another because of differing environmental influences. The environment frequently affects the final phenotype, or appearance, of an organism. (See Figure 21-15.)

Figure 21-15 The phenotype expression of the genotype for dark color points in Siamese cats depends on temperature.

PLANT AND ANIMAL BREEDING

The results of the artificial selection of plants and animals by humans was described in Chapter 1. For centuries, people have chosen to breed plants and animals with desirable traits. By allowing only these organisms to reproduce, traits such as size, shape, rate of development, and color have been altered.

The breeding of plants and animals has been greatly helped by the

Figure 21-16 As a result of plant breeding, the small kernels of wild wheat have been transformed into the large kernels of bread wheat.

discoveries of Mendel and other geneticists. Often, breeders can identify the exact traits in which they are interested. By determining the alleles for that genetic trait, they can prepare the best genetic crosses. For example, plant breeders have produced new crop plants that are resistant to disease, that can live in new environments, or that produce food that looks and tastes better. By changing the genetics of wheat and rice, scientists were able to create new varieties that produce much more food. As a result, more food can now be produced in countries that have experienced rapid population growth. When this transformation of crop plants began in the 1960s and 1970s, it was known as the green revolution. (See Figure 21-16.)

Animals also have been bred by people for many different purposes. For example, sheep have been bred to produce more wool of better quality, turkeys have been bred to produce more white meat, chickens to lay larger eggs, cows for more milk, and pigs to produce meat that contains less fat.

The next scientific step—the understanding of genetics on a molecular level—marks the beginning of an entirely new type of control of inheritance, as you will learn in Chapter 23.

LABORATORY INVESTIGATION 21
How Are Traits Passed From Parents to Offspring?

INTRODUCTION

Mendel has been called the founder of genetics because his research with pea plants led to an understanding of the laws that govern the inheritance of traits. The patterns that result as traits are passed from parents to offspring could now be understood and predicted. The laws Mendel discovered apply to plants and to all animals, including humans. This laboratory investigation will demonstrate the inheritance of a single characteristic that has a dominant and a recessive allele. You will study inheritance patterns that show when the parents are homozygous and heterozygous for this trait.

MATERIALS

Index card with a capital T on it and an index card with a lowercase t on it for each student, additional cards with a capital T

PROCEDURE

1. Take one card with a large T and one card with a small t. The T card represents the tall allele for the height of a pea plant. The t card represents the short allele for the height of a pea plant. The tall allele is dominant; the short allele is recessive.
2. Work in pairs. Each pair of students represents two parent pea plants. Because each student has a T card and a t card, each represents a heterozygous pea plant. Turn your two cards facedown and mix them up so you do not know which is which. Now select one card from your pair. Your card and your partner's card represent the genotype of the first offspring. Record the first offspring's genotype and phenotype.
3. Repeat step 2 nine more times. Keep recording the genotype and phenotype of each offspring.
4. Now summarize the results of the "crosses." Determine the number and percentages of TT, Tt, and tt genotypes and the number and percentages of tall and short plants that were produced.

5. Compare the results of all pairs and summarize them on the chalkboard. Calculate the total percentages of genotypes and phenotypes for these numbers also.
6. Repeat steps 1 to 5 with the following variation: One student has two T cards, and the partner has a T card and a t card.

INTERPRETIVE QUESTIONS

1. Does the sorting of alleles for one offspring affect the sorting of alleles for later offspring from the same two parents? Explain.
2. Determine the expected percentages of genotypes and phenotypes for each pattern of inheritance in this activity.
3. How do the percentages of genotypes and phenotypes from individual sets of parents of ten offspring compare to the expected results?
4. How do the total percentages of genotypes and phenotypes from the entire class compare to the expected results?
5. Compare your answers for questions 3 and 4 and explain what you discover.

CHAPTER 21 REVIEW

Answer these questions on a separate sheet of paper.

VOCABULARY

The following list contains all of the boldfaced terms in this chapter. Define each of these terms in your own words.

alleles, codominance, dominant, genetic cross, genotype, heterozygous, homozygous, incomplete dominance, phenotype, recessive

PART A—MULTIPLE CHOICE

Choose the response that best completes the sentence.

1. The allele for the trait that does not appear in a heterozygous individual is *a.* dominant *b.* codominant *c.* incompletely dominant *d.* recessive.
2. Gregor Mendel performed his genetics experiments on *a.* humans *b.* pea plants *c.* fruit flies *d.* mice.
3. If a pea that is homozygous for purple flowers is crossed with a pea that is homozygous for white flowers, the flowers in the offspring will be *a.* purple *b.* white *c.* pink *d.* purple or white.
4. The first generation of individuals used in a series of genetic crosses is designated by *a.* F_1 *b.* F_2 *c.* P *d.* A.
5. The phenotypic ratio 3:1 is seen in the offspring of *a.* two heterozygous individuals *b.* a homozygous dominant and a homozygous recessive *c.* two homozygous recessives *d.* a heterozygous individual and a homozygous recessive.
6. The dominant and recessive versions of a particular gene are known as *a.* alleles *b.* homozygotes *c.* heterozygotes *d.* codons.
7. In peaches and nectarines, fuzzy skin (peaches) is dominant, smooth skin (nectarines) is recessive, and yellow fruit is dominant over white fruit. If these genes are not linked, a cross between two heterozygous yellow peach trees will produce *a.* homozygous yellow nectarines and homozygous white peaches in a 1:1 ratio *b.* homozygous yellow peaches, heterozygous yellow peaches, and homozygous white nectarines in a 1:2:1 ratio *c.* yellow peaches and white nectarines in a 3:1 ratio *d.* yellow peaches, white peaches, yellow nectarines, and white nectarines in a 9:3:3:1 ratio.

8. In a cross between yellow peaches that are heterozygous for fruit color and skin fuzziness, the genotypic ratio of the offspring is *a.* 1:2:4:2:1 *b.* 1:1:1:1 *c.* 1:2:1 *d.* 9:3:3:1.
9. The offspring resulting from a cross between a homozygous dominant and a homozygous recessive are *a.* homozygous dominant *b.* homozygous recessive *c.* heterozygous dominant *d.* heterozygous recessive.
10. When red snapdragon flowers are crossed with white snapdragon flowers, the offspring have pink flowers. This is an example of *a.* simple dominance *b.* incomplete dominance *c.* codominance *d.* mutation.
11. According to the Law of Independent Assortment, *a.* the dominant gene will be expressed in heterozygous individuals *b.* if genes are not linked, the inheritance of one trait has no effect on the inheritance of another trait *c.* alleles are separated during gamete formation *d.* in organisms that reproduce sexually, genes are inherited from both parents.
12. A pea plant that is heterozygous for seed color would produce gametes that are *a.* 100 percent green *b.* 100 percent yellow *c.* 50 percent green, 50 percent yellow *d.* an unpredictable mix of green and yellow.
13. If two pea plants that are heterozygous for tallness are crossed, their offspring will *a.* all be tall *b.* all be medium height *c.* show a 1:2:1 phenotypic ratio *d.* have a 1:2:1 genotypic ratio.
14. In chickens, a cross between an individual that is homozygous for black feathers and an individual that is homozygous for white feathers produces offspring that are speckled black and white. This is an example of *a.* codominance *b.* incomplete dominance *c.* simple dominance *d.* incomplete penetrance.
15. In fruit flies, red eyes are dominant over white eyes. Suppose a fly that has red eyes is crossed with a fly that has white eyes. If the red-eyed fly is heterozygous, then the offspring will *a.* all have red eyes *b.* all have white eyes *c.* show red eyes and white eyes in a 3:1 ratio *d.* show red eyes and white eyes in a 1:1 ratio.

PART B—CONSTRUCTED RESPONSE

Use the information in the chapter to respond to these items. Refer to the following diagram to answer questions 16 and 17.

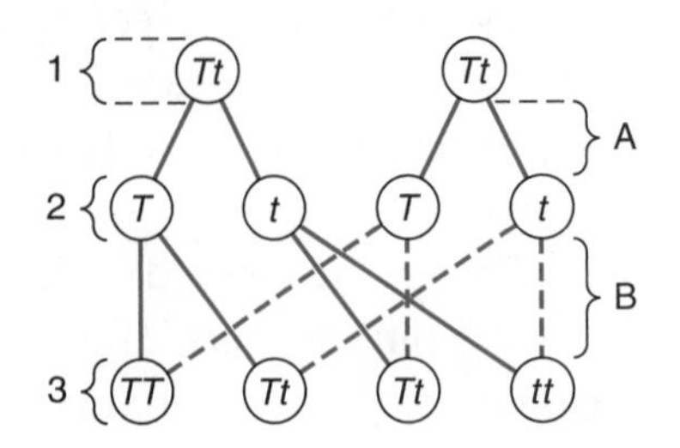

16. Identify the types of cells shown in rows 1, 2, and 3 in the diagram. Which cells are diploid? Which cells are haploid?
17. Identify the processes that take place in region A (between rows 1 and 2) and in region B (between rows 2 and 3). Which process illustrates one of Mendel's laws? Explain.
18. List Mendel's three laws and describe each in your own words.
19. Relate the Law of Segregation and the Law of Independent Assortment to the movement of chromosomes in meiosis.
20. In cartoon panthers, the color pink is dominant, and the color purple is recessive. Draw a Punnett square that shows the results of a cross between a heterozygous pink panther and a purple panther. What are the phenotypic and genotypic ratios in the offspring?

PART C—READING COMPREHENSION

Base your answers to questions 21 through 23 on the information below and on your knowledge of biology. Source: *Science News* (May 31, 2003): vol. 163, p. 350.

To Contain Gene-Altered Crops, Nip Them in the Seed

Canadian researchers have demonstrated that, in principle, they can engineer genetically modified (GM) crops to be incapable of breeding with conventional crops or wild relatives. The new approach could help contain the unintended spread of artificial traits, which is a major source of public concern about GM crops.

Although several alternative strategies for such containment exist, "there is no perfect solution," says Johann P. Schernthaner of Agriculture and Agri-Food Canada in Ottawa.

To engineer a crop that would theoretically require no intervention by farmers to keep it reproductively contained, Schernthaner and his colleagues inserted into some tobacco plants a genetic element called a seed-lethal trait. That element prevents the plants' seeds from germinating under any circumstances.

To enable the GM plants to reproduce amongst themselves, the researchers then inserted another artificial trait that represses the seed-lethal construct. GM plants with both traits develop and reproduce normally, the researchers report in an upcoming *Proceedings of the National Academy of Sciences.*

To create an inherently containable GM crop, the researchers suggest that a different seed-lethal construct be placed on each member of a pair of chromosomes, so the constructs wouldn't be inherited together. Each chromosome would also receive a repressor trait that inactivates the seed-lethal construct on the other chromosome. That way, unintended crossings of the GM crop with related plants should produce nonviable seeds because they'd contain only one of the two engineered chromosomes.

The new approach faces several potential problems, says Henry Daniell of the University of Central Florida in Orlando. For example, it's still possible that with certain chromosomal rearrangements, both the seed-lethal and the repressor traits could spread to a non-GM plant.

21. State the major concern that exists about growing genetically modified (GM) crops with artificial traits in open fields.
22. Explain the purpose of the first seed trait that was inserted into the tobacco plants' genes.
23. Describe the changes made to other chromosomes on the GM plants that allow them to reproduce among themselves.

22 Human Genetics

After you have finished reading this chapter, you should be able to:

Discuss ways human traits are inherited.

Distinguish between autosomes and sex chromosomes.

Discuss several sex-linked traits in humans.

Explain the work of a genetics counselor.

Good health and good sense are two of life's greatest blessings.

Publilius Syrus, *Maxim 827*

Introduction

"Is it a girl or a boy?" At the moment of birth, parents have asked this question in every language for as long as there have been languages. The next question, whether spoken out loud or whispered, is asked with a mixture of hope and fear: "Is it healthy?"

Almost always, the process of inheritance results in a wonderful newborn with a unique combination of genetic traits. With these traits and a supportive environment, the baby will lead a healthy life. But this is not the case for all newborns. (See Figure 22-1.)

Genetic disorders occur in some infants. These disorders are not caused by an infectious microorganism. Instead, genetic disorders result from inborn errors of metabolism that are caused by defects in genes. Can we predict how often genetic disorders will occur? Can we treat people born with genetic disorders? Is it possible to learn if a baby has a genetic disorder before birth? These and other important questions will be examined in this chapter on human genetics.

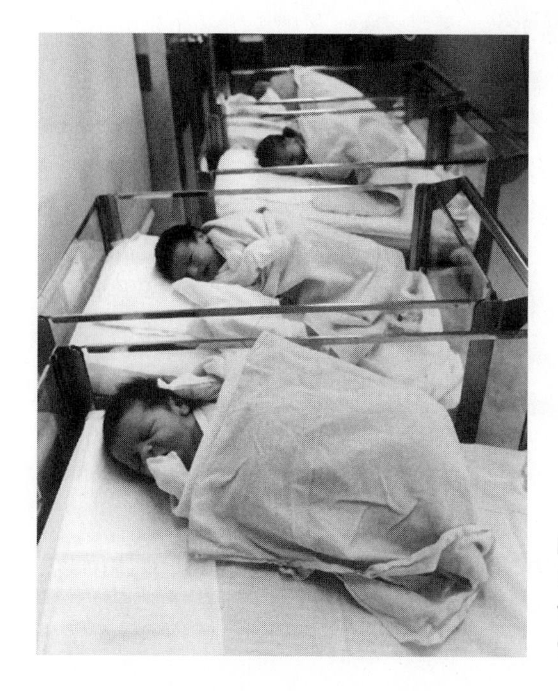

Figure 22-1 The process of inheritance results in a newborn with a unique combination of genetic traits.

GENETIC COMPLEXITIES: BEYOND THE PEA PLANTS

Much of Mendel's success was due to the traits he studied. The pea plants he worked with were either tall or short, and their seeds were either yellow or green, smooth or wrinkled. And so on. But look at your classmates. Are people either tall or short? Are there only two skin colors? Do people have only blond or black hair? Of course not. In humans, and in most organisms, almost all traits are not as clearly defined as the traits Mendel studied in pea plants. Was Mendel correct in his explanations? Was Mendel extraordinarily lucky in picking the traits he studied? The answer to both questions seems to be yes.

The traits Mendel picked to study were each determined by single genes. The height of the pea plants is due to a single gene that occurs in two different versions, or alleles. Human height is a different story. Height in humans is due to several genes. Pieces of DNA on different chromosomes code for the proteins that affect human height. A trait determined by several genes follows a pattern of **polygenic inheritance**. If a large group of people were arranged according to height, those with average height would be the most numerous. There would be fewer extremely short and fewer extremely tall people. Grouping individuals in this way produces a bell-shaped curve. (See Figure 22-2 on page 470.)

Figure 22-2 The bell-shaped curve produced when a group of men are arranged by height, which shows the variation within a population.

This is typical for polygenic inheritance. The many variations of human skin color are also due to polygenic inheritance.

LINKED TRAITS

Why do people with red hair also have freckles? Why do people with blond hair usually have blue eyes? The genetics needed to answer these kinds of questions was first studied in fruit flies. Shortly after Mendel's experiments with pea plants became widely known, geneticists turned to fruit flies for further research. Fruit flies are often seen flying around pieces of ripening fruit, where they lay their eggs. These tiny animals have made great contributions to our understanding of how traits are inherited.

In fruit flies, the inheritance of many different traits has been studied. These traits include eye color, wing size, body color, number of hairy bristles on the body, and much more. According to Mendel, traits are sorted and inherited independently. Just because two traits are together in a parent does not mean they will be inherited together in the offspring. But this is exactly what happens with some traits in fruit flies. In fact, in studying fruit flies, scientists were able to identify four main groups of traits. All traits in a group tend to be inherited together, not independently. In some ways, it appears that these traits are linked to each other. Having observed **linkage** in fruit flies, scientists realized that chromosomes suddenly made more sense. These linked traits are inherited together because the genes for these traits are located on the same chromosome. This knowledge explains the four groups of linked traits in fruit flies. These animals have four linkage groups because each cell of a fruit fly has four pairs of chromosomes. (See table on fruit fly traits.)

LINKAGE MAP OF FRUIT FLY TRAITS

Chromosome 1	Chromosome 2	Chromosome 3	Chromosome 4
0.0 yellow	6.1 Curly	26.0 sepia	0.0 + shaven
0.0 + scute	13.0 dumpy	40.7 Dichaete	0.0 + cubitus interruptus
0.8 prune	48.5 black	44.0 scarlet	0.0 + grooveless
1.5 white	54.5 purple	46.0 Wrinkled	0.0 + sparkling-polished
13.7 crossveinless	54.8 Bristle	47.0 radius incompletus	0.2 eyeless
20.0 cut	55.2 apterous	52.0 rosy	
21.0 singed	57.5 cinnabar	58.2 Stubble	
27.7 lozenge	67.0 vestigial	58.5 spineless	
33.0 vermilion	72.0 Lobe	64.0 kidney	
36.1 miniature	75.5 curved	69.5 Hairless	
51.5 scalloped	100.5 plexus	70.7 ebony	
56.7 forked	104.5 brown	79.1 bar-3	
57.0 Bar	107.0 speck	91.1 rough	
64.8 maroonlike		100.7 claret	

Many human traits are linked because they are found on the same chromosome. For example, the genes for red hair and freckles are on the same chromosome. The genes for blond hair and blue eyes are also located on another single chromosome. These genes are linked and tend to be inherited together. Now is the time to examine more closely the chromosomes found in human body cells.

IS IT A GIRL OR IS IT A BOY?

As described in Chapter 16, scientists can take a photograph of the chromosomes in a human cell by using a camera attached to a microscope. The chromosomes can be paired up and then numbered from largest to smallest. We can create a "picture" of the chromosomes. This picture is called a *karyotype*. If we examine a human karyotype, we find 22 perfectly matched pairs of chromosomes. These chromosomes, numbered from 1 to 22, are called **autosomes**. This accounts for 44 of the 46 chromosomes found in a diploid cell. However, the last two chromosomes do not always make a matched pair. It depends on whether the diploid cell came from a female or a male. If the cell came from a male, the last two chromosomes will not be an identical match. Cells from males contain an X and a much smaller Y chromosome. If the cell came from a female, the last two chromosomes will both be X chromosomes, a matched pair. Differences in the last pair of chromosomes—whethér a person has an X and a Y chromosome, or two X chromosomes—determine the person's sex. Therefore, these last two chromosomes are called the **sex chromosomes**.

Who determines a child's gender, the father or the mother? Let's ask the question another way. Is it the sperm cell from the father or the egg cell from the mother that determines the child's gender? (See Figure 22-3.)

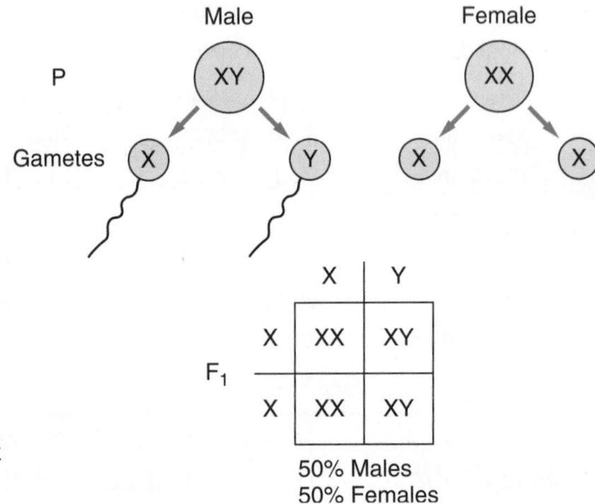

Figure 22-3 The X and the Y chromosomes are called the sex chromosomes.

Egg cells are produced in the mother by meiosis. Her cells have 22 pairs of autosomes and one pair of X's, her sex chromosomes, for a total of 46 chromosomes. Through meiosis, every egg cell gets one of each of the 22 autosomes and one X chromosome. Now consider sperm cells. Every cell in the father's body has 22 pairs of autosomes and an X and a Y chromosome. During meiosis, 50 percent of his sperm cells get 22 autosomes and an X chromosome, and 50 percent of his sperm cells get 22 autosomes and a Y chromosome. Egg cells have only X chromosomes in addition to the autosomes. So who determines the sex of the child? The father does. If a sperm with a Y chromosome fertilizes the egg, the zygote is XY, and a male develops. If a sperm with an X chromosome fertilizes the egg, the zygote is XX, and a female develops. (See Figure 22-4.)

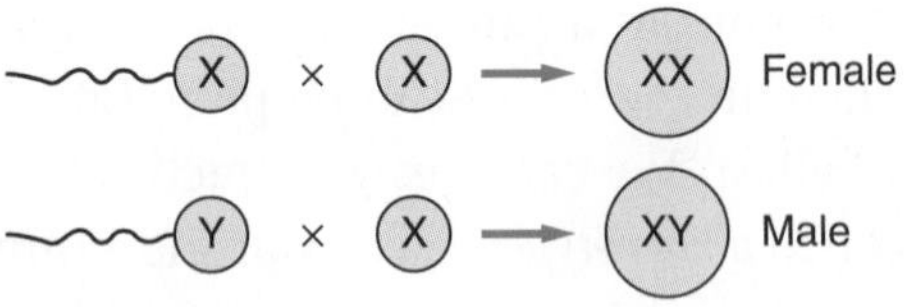

Figure 22-4 Sex determination in humans depends on the sperm cell.

SEX-LINKED TRAITS

Red-green color blindness, an inability to distinguish the colors red and green, occurs most often in males. Hemophilia, a disease in which blood does not clot properly, also occurs most often in males. Duchenne muscular dystrophy, a disease that slowly destroys muscles, occurs primarily in males, too. What accounts for this pattern? These traits are all **sex-linked**. They occur when a particular allele with a defective portion of DNA is present on the X sex chromosome. The much smaller Y chromosome lacks alleles for these three traits.

Why do these disorders occur rarely in females? Females have two X chromosomes. As long as a woman inherits one normal allele on one X chromosome, the presence of the defective allele on the other X chromosome will be hidden, since it is recessive. The normal allele is dominant. A male, on the other hand, has only one X chromosome. If the defective allele is on his single X chromosome, there is no dominant healthy allele on the Y chromosome. The boy or man will have the disease.

A female who has a recessive defective allele on one X chromosome but not on the other is called a **carrier**. These females carry the genetic disorder, but they do not have the condition in their phenotype. However, they can pass the condition on to offspring. Carriers are heterozygous for the gene that produces a genetic disorder. The only way a female can be red-green color blind or have hemophilia, for example, is if she is homozygous for the trait. Her father had the disorder and passed it on to her on his X chromosome. Also, her mother either had the disease or was a carrier and also passed on an X chromosome with a defective allele. (See Figure 22-5.)

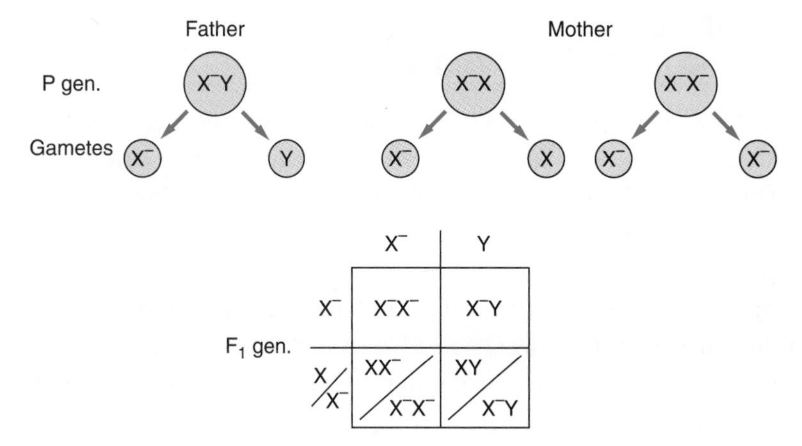

Figure 22-5 The only way females can get hemophilia is if the father was affected and the mother was a carrier or affected.

INHERITANCE PATTERNS FOR SEX-LINKED RECESSIVE GENETIC DISORDERS

Key

X	=	X chromosome with normal gene
X–	=	X chromosome with defective gene
Y	=	Y chromosome

X Y	→	Normal male
X X	→	Normal female
X– Y	→	Male with genetic disorder
X– X	→	Carrier female
X– X–	→	Female with genetic disorder

Finally, a sex-linked genetic disorder in a son is always inherited from his mother. Why is that? Remember that a son can inherit an X chromosome, where the defective allele is located, only from his mother. He receives a Y chromosome with no corresponding allele from his father. He therefore inherits the defective allele from his mother. (See the table above, which shows the inheritance patterns for sex-linked disorders.)

Check Your Understanding

Explain why there are more male hemophiliacs than female hemophiliacs. You might want to research the occurrence of hemophilia among the royal families of Europe who descended from Britain's Queen Victoria (a carrier of the flawed gene).

CYSTIC FIBROSIS: AN AUTOSOMAL RECESSIVE TRAIT

In the United States, cystic fibrosis (CF) is the most common genetic disorder in the Caucasian, or white, population. Approximately one in 25 people in this population carries one CF allele. CF is a recessive trait. To have the disease, a person must be homozygous for the trait. That is, both parents must be carriers. Unlike the three disorders already mentioned, the gene for cystic fibrosis is not located on the X or Y chromosomes. It is on an autosomal chromosome. Cystic fibrosis is therefore called an *autosomal recessive* trait. Approximately one in every 2000 Caucasian babies is born with this disease.

Children with cystic fibrosis have many digestive and respiratory prob-

Figure 22-6 A physiotherapist is giving percussion treatment (to loosen the mucus) to a 3-year-old cystic fibrosis patient.

lems. The most important health problem is that the lungs constantly fill with a thick mucus. It requires constant medical treatment to keep the lungs clear. People with cystic fibrosis also have frequent lung infections. These people can also have problems digesting food properly. Until recently, children with cystic fibrosis rarely lived beyond five years of age. (See Figure 22-6.)

In 1989, researchers announced with great excitement that they had found the gene that produces cystic fibrosis. In more than 70 percent of people with CF, three DNA nucleotides are missing in a single gene. These nucleotides code for the amino acid phenylalanine, at the 508th position in the polypeptide of that gene. This one missing amino acid in the polypeptide causes all the problems. This is a mutation. It is believed to have first occurred in a person who lived in northern Europe several hundred years ago. Since then, many people have inherited the recessive CF gene from this person's descendants.

New, powerful technologies have begun to assist researchers in finding the genes that cause human genetic disorders. Now that the first draft sequence of the human genome has been completed, a list of up to 300 possible genes per disorder is often available. A tool called "GeneSeeker," which helps narrow the list, was announced in 2003. It uses the Internet and links together information in an automated way from nine different databases around the world. The *European Journal of Human Genetics* called GeneSeeker a "web-based data mining tool for the identification of candidate genes for human genetic disorders." This type of information-processing tool on the World Wide Web is expected to become a useful addition to the actual experimentation done in the laboratory.

HUNTINGTON'S DISEASE: AN AUTOSOMAL DOMINANT TRAIT

Another genetic disorder is inherited in a different way. This disorder, called *Huntington's disease,* causes the nervous system to break down gradually over time. If one parent has the disease, it is very likely that 50 percent of his or her children also will develop the disease. Huntington's disease is caused by an autosomal dominant trait. Like cystic fibrosis, the gene is located on an autosome, not on a sex chromosome. However, unlike cystic fibrosis, the gene that causes Huntington's disease behaves as a dominant trait. As long as a person has one allele for this disease, even though he or she has another normal allele, he or she will develop the disease.

The symptoms of Huntington's disease do not begin to show until a person is about 40 years old. This late-developing disease creates tremendous worry for people in affected families. If a parent has the disease, it is quite possible that a son or daughter has inherited the allele. Even if the person is heterozygous for the trait, the disease will begin to show later in life.

KNOWING YOUR FAMILY HISTORY

As we begin to understand more about genetic disorders, it becomes clearer that there are not only wonderful possibilities but also risks in having children. Parents ask themselves if it is possible for their child to inherit a genetic disorder. To assess the risks, people must have information about their ancestors. It is necessary to learn about one's family history. A man and a woman who want to become parents can now go to a trained genetic counselor for help in determining their risks of having a child with a genetic disorder. One of the things a genetic counselor does is to help the couple prepare a chart that shows the occurrence of a disease in past generations of their families. This chart is called a **pedigree**.

A pedigree uses symbols to represent parents and children in several generations of a family. Squares represent males, and circles represent females. If the symbol is half shaded, the individual is heterozygous and a carrier. A symbol that is fully shaded shows that the individual is homozygous and has the disease. Sometimes a symbol is half shaded in a dark color, which means that the individual has the disease even though only a heterozygote. (See Figure 22-7.)

Why are pedigrees useful? A completed pedigree chart can show how a genetic disorder in a family is being inherited.

There are three main possibilities, which we have already examined:

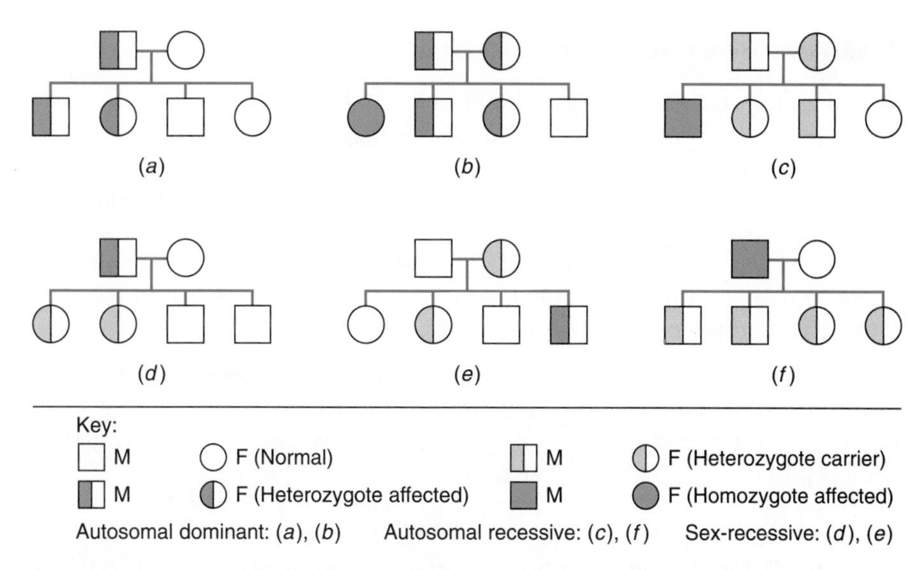

Figure 22-7 A completed pedigree chart shows the inheritance of a genetic disorder in a family.

- The disorder is a dominant condition carried on an autosome. In this case, roughly equal numbers of males and females are affected. The condition does not skip generations and then reappear. Also, whenever the gene is present, the disorder is expressed.
- The disorder is a recessive condition carried on an autosome. Again, roughly equal numbers of males and females are affected. However, the condition can skip generations when both parents are only carriers, to reappear later in homozygous individuals.
- The disorder is due to a recessive allele on the X chromosome. In this case, mostly males are affected. All daughters of an affected male and a normal female are carriers. Sons never inherit the disease from their father, only from their mother who is either a carrier or affected.

A pedigree that shows a person's family history for a particular trait can be analyzed, and patterns of inheritance can be determined. From this, predictions of the probabilities of the trait reappearing in future generations of males and females are made. Consulting with a genetics counselor enables prospective parents to make informed decisions about having children.

It is necessary to understand what probability means when making these kinds of important decisions. For example, a pedigree analysis might show that the chances of a man and a woman having a child with a particular genetic disorder is one in four. This same probability exists each

A Career in Genetics: Counseling

Joseph and his parents had been referred by their family doctor to the genetics clinic at a local hospital. The doctor had made this suggestion after testing the levels of sex hormones in Joseph's blood and ordering a chromosome analysis. Joseph, a healthy 16-year-old, was doing well in school. However, even though he was almost two meters tall, he showed no signs of sexual maturity. He had not developed additional body hair, and his voice retained the high pitch of a much younger person. A long talk took place in the genetics counselor's office.

The counselor explained that Joseph has one more X chromosome than is usually found in a male. This is due to an error that occurred during meiosis when the egg or sperm from which he developed formed. Joseph and his parents learned from the counselor that about one in every thousand males has this XXY chromosome makeup—a total of 47 chromosomes instead of the normal 46. The fact that this condition is called *Klinefelter syndrome* was not important to Joseph and his family. However, learning about the characteristics of the syndrome was very important. The counselor told them that the characteristics of men with this syndrome usually included tall stature, some minor birth defects, small

time the couple has a child. If one child has already been born with the disease, the next child still faces exactly the same one-in-four risk. If three children have been born without the disorder, the chance of the fourth child having the disorder is no greater; it is still one in four. Think of probability this way. If a coin is tossed once and lands heads up, the chance of tails the next time is still only 50 percent. If the same coin is tossed fifty times, and it comes up heads each time, the probability that it will come up tails on the next toss is still 50 percent. The probability of any event, whether it is tossing a coin or having a child, is not affected by what happened in previous events.

OTHER GENETIC DISORDERS

Genetic disorders can also result from an abnormal number of chromosomes. As described in Chapter 16, mistakes can happen during meiosis as the paired homologous chromosomes separate. This type of genetic disorder is very different from the single gene mutation that produces hemophilia, cystic fibrosis, or Huntington's disease. Abnormal chromosome numbers nevertheless cause disorders that can be detected prior to or at birth.

testicles, and sterility. Sometimes, but not in Joseph's case, mental retardation begins at birth.

Joseph's mother asked if the condition could occur in any other children she might have. It was explained that there are no known examples of more than one Klinefelter individual in a family. Furthermore, blood tests done on Joseph's relatives had not shown any chromosomal disorders.

As a result of the genetics counseling, Joseph realized that he could most likely look forward to a normal life. Indeed, in a few months, his beard began to grow. More important, Joseph learned about possible options for having a family. His parents also learned that amniocentesis might be a good idea if they decided to have another child, since some families have shown a tendency to produce other errors in meiosis.

A genetics counselor often has a master's degree, with special training in genetics counseling. Some are also physicians. Well-trained and experienced genetics counselors are in great demand. Many families, like Joseph's, rely on their expertise to answer urgent questions that can have profound effects on their lives.

Down syndrome is due to an extra chromosome number 21. This syndrome, which occurs in about one in 800 births, causes mild to severe mental retardation. (See Figure 22-8.) Another genetic disorder occurs in males who are born with an extra X chromosome (XXY). This extra X chromosome produces Klinefelter syndrome, which results in abnormal sexual development and infertility.

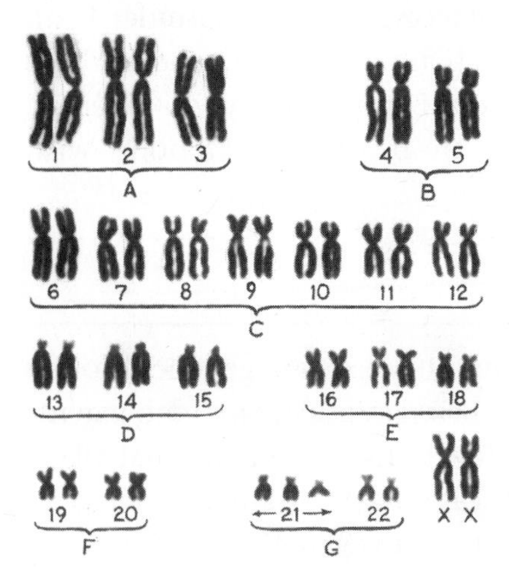

Figure 22-8 The karyotype of a person with Down syndrome. Notice that there are three copies of chromosome 21.

Sickle-cell anemia is a disorder due to a single gene defect, not an abnormal chromosome number. In Chapter 20, you learned that a mutation in the DNA nucleotide sequence of the gene for hemoglobin causes this disease. Sickle-cell anemia occurs most often in people whose ancestors came from West and Central Africa. It is inherited as an autosomal recessive trait, like cystic fibrosis. In the United States, people who are homozygous for sickle-cell anemia—approximately 0.2 percent of African Americans—have the disease. However, some red blood cells of people who are heterozygous—approximately 9 percent of African Americans—also become sickle-shaped under certain conditions. A simple blood test makes it easy to detect the sickle-cell trait in people. (See Figure 22-9.)

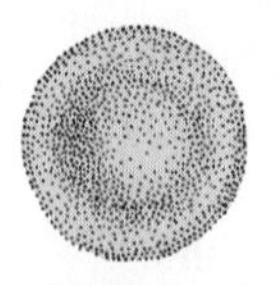

Normal red blood cell

Sickle-shaped red blood cell

Figure 22-9 The sickle-shaped red blood cell is characteristic of sickle-cell anemia—a genetic disorder.

Phenylketonuria, another autosomal recessive trait, is one of the most studied genetic disorders. About one in 15,000 infants born in the United States has the allele for this disease. This allele prevents the baby from producing an enzyme that breaks down the amino acid phenylalanine. As a result, phenylalanine builds up in the baby's blood. High levels of phenylalanine in the blood interfere with the development of the brain, causing mental retardation. Routine tests for this disease are now done on newborns. By detecting the disease early, the baby's diet can be changed to prevent the disease's effects from developing. This early detection can enable the baby to live a normal life.

Tay-Sachs disease causes a rapid breakdown of the nervous system during the first years of life. Children with Tay-Sachs disease rarely live more than five years. Tay-Sachs is an autosomal recessive genetic disorder. Until recently, this disease occurred most often among Jewish people whose ancestors came from Eastern and Central Europe. Today, however, an increased number of Tay-Sachs babies are being born to non-Jewish parents.

GENETIC SCREENING

Consider a husband and wife who know that a severe genetic disorder such as Tay-Sachs has affected people in their family. Perhaps both parents are heterozygotes, with one allele for Tay-Sachs and one normal allele. We know that a child will have the disease only if it has two defective alleles, that is, if the child is homozygous for the genetic disorder.

It has been explained to the couple that their chance of having a homozygous Tay-Sachs baby is 25 percent; that is, the child has a one in four chance of having Tay-Sachs disease. A Punnett square can show why this is so. The couple has a difficult decision to make. A child with Tay-Sachs will probably not survive beyond age five. The child's few years of life will be spent with an untreatable disease that destroys the brain. Many people, faced with this possibility, might choose not to have children. Since 1968, the presence of some genetic disorders can be determined before birth. This is called **prenatal diagnosis**.

The first prenatal diagnosis method that was developed is called **amniocentesis**. A needle is used to withdraw amniotic fluid from the uterus. In this fluid are some cells from the fetus. These cells are removed, grown in a laboratory dish, and tested. A karyotype of the fetal chromosomes is prepared. Biochemical tests that show the presence of a genetic disorder may also be done. More often today, the DNA of the fetus is studied directly to see if something is abnormal in the fetus's genes.

Other prenatal tests are also possible. Sound waves can make images of the fetus while it remains safely in its mother's uterus. The procedure that makes images with sound waves is called **ultrasonography**. The techniques for taking ultrasound pictures have improved a great deal. It is now possible to see some types of abnormalities in the fetus with these images. For example, defects in the kidneys, brain, and heart can be observed. Genetic disorders may also damage fetal muscles and nerves. When this happens, the fetus is not able to move normally. These abnormal movements can be recognized with ultrasonography. (See Figure 22-10.)

Chorionic villus sampling (CVS) is a technique that can be performed on an embryo two months earlier than amniocentesis. A CVS test removes cells from the chorion layer outside the amnion. Cells from this area are

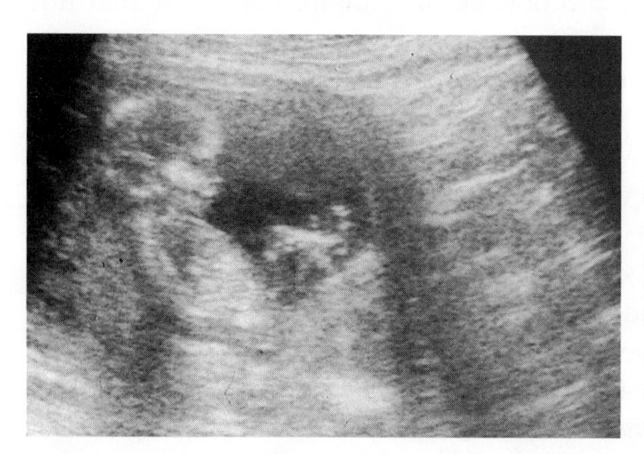

Figure 22-10 This ultrasonogram shows the outline of a developing fetus. (Note the head—which is facing to the right—and the chest and right arm, on the left side of the image.)

rapidly growing into the wall of the mother's uterus. These fetal cells can be analyzed for genetic abnormalities.

Another technique analyzes blood from the fetus. In this method, called **cordocentesis**, a needle is used to remove fetal blood from the umbilical cord. Remember that blood goes back and forth between the fetus and the placenta through the umbilical cord. Fetal blood cells can be studied for chromosomal abnormalities. Also, the blood itself can be checked with biochemical tests.

It is important to know that there is some risk involved in amniocentesis and CVS. Infections may occur, or the fetus itself may be harmed by the needle used to remove material to be studied. The chance of this happening is between one in 100 and one in 200. Newer, safer techniques are now being developed that will test the few fetal cells that are found in the mother's blood. These cells can be captured, and their DNA can then be studied for genetic disorders.

DIFFICULT DECISIONS

Most of the time, prenatal diagnosis determines that the developing fetus is normal. The couple can relax and prepare for the birth of a healthy child. However, when the results of a prenatal diagnosis indicate that the fetus is not genetically normal, difficult decisions have to be made. It is important for the couple to be able to discuss their course of action with their physician and with trained genetics counselors.

Often, the couple will use the information from the diagnosis to prepare for the birth of their child. Caring properly for the fetus during pregnancy is very important. Special plans can be made to have the appropriate medical personnel available at the time of birth. Specialists who can treat a particular disorder may be needed, too. Parents need to learn about the disorder and how best to care for the infant after it is born.

Amazingly, some inherited conditions can be treated and corrected in the fetus before birth. In February 1995, a four-month-old fetus was given a successful bone marrow transplant. Prenatal diagnosis had shown that the fetus had SCIDS, *s*evere *c*ombined *i*mmuno*d*eficiency *s*yndrome, as described in Chapter 14. Children with SCIDS rarely survive because they have no natural protection against disease. The first son born to this couple had the disorder and had died at seven months of age. With the second baby, normal bone marrow that produces white blood cells was transplanted from the father into the fetus during the fourth month of pregnancy to prevent the deadly condition from developing. By December 1996, when this procedure was reported, the couple had a healthy 18-month-old baby who was cured of the genetic disorder.

ETHICAL QUESTIONS CONCERNING PRENATAL DIAGNOSIS

Many birth defects still cannot be detected by prenatal diagnosis. The decision by a couple to proceed with prenatal diagnosis is made only after very careful evaluation by doctors and genetics counselors. From the results, difficult questions may arise. Individuals and society as a whole need to consider the ethics involved in these questions.

Sometimes, the couple faces the very difficult question of whether to continue the pregnancy. If the disorder is extremely serious, the couple may decide that an abortion should be performed. This is not an easy decision to make, and it is very much influenced by the couple's moral and religious values.

Should a woman be urged to have an abortion if testing shows that the fetus will be born with severe mental or physical handicaps? If a doctor is opposed to abortion, should he or she be allowed to withhold results of prenatal diagnosis that might cause a woman to choose an abortion? In addition to information about genetic disorders, prenatal testing can determine the child's gender. Should couples be informed of the fetus's gender? Should decisions to abort a pregnancy based only on the fetus's gender be made illegal? Should a couple be allowed to abort a pregnancy that involves twins if one of the fetuses is severely abnormal but the other one is normal?

As in many other areas of biology today, knowing the science of genetics is not enough. An important part of biology education is learning how to think and make decisions about difficult ethical issues.

LABORATORY INVESTIGATION 22
How Is a Human Karyotype Prepared?

INTRODUCTION

A karyotype is a picture or photograph that shows the chromosomes of an organism. It was not until 1956 that the human karyotype was determined. This karyotype established the human diploid number to be 46 chromosomes—44 autosomes and two sex chromosomes. Karyotyping has become a routine procedure for fetuses of couples that are at risk of having children with syndromes or conditions that are related to abnormal chromosome numbers.

To prepare a karyotype, a geneticist interrupts cell division at metaphase by the use of a special chemical. The treated cells are placed on a glass slide and broken open. Then they are stained. A group of chromosomes that has been freed from a cell is photographed through a microscope. The photograph is enlarged and the chromosomes are rearranged and counted. In this investigation, you will use a photograph of chromosomes to complete a human karyotype.

MATERIALS

Photocopy of "Set of Human Chromosomes" (from *Teacher's Manual*), scissors, glue or tape, paper

PROCEDURE

1. Carefully cut out the chromosomes from your photocopy of "Set of Human Chromosomes."
2. To aid in sorting, scientists have assigned human chromosomes to seven groups according to the size of a chromosome and the location of its centromere, the point where the two chromosomes are joined. Place the chromosomes in pairs; then decide in which of the seven groups the chromosomes belong.

Group	Chromosomes	Characteristics
A	1–3	Very long, with centromeres midway from the ends
B	4 and 5	Long, with centromeres somewhat closer to one end
C	6–12 and X	Medium length, with centromeres somewhat closer to one end
D	13–15	Medium length, with centromeres at or very near the end
E	16–18	Somewhat short, with centromeres somewhat closer to one end
F	19 and 20	Short, with centromeres midway from the ends
G	21 and 22 and Y	Very short, with centromeres at or very near the end

3. Arrange the pairs of chromosomes in order from the largest to the smallest, and in their groups. Arrange the chromosomes in a line with the centromeres on the line. Identify each chromosome with a number or letter. Glue or tape the chromosomes in place on your sheet of paper to complete your karyotype.

INTERPRETIVE QUESTIONS

1. From your karyotype, determine if the individual these chromosomes came from was male or female. How would the chromosomes differ if they came from a member of the opposite sex?
2. Explain how the karyotype would be different if its chromosomes came from a person with Down syndrome.
3. Describe how a genetics counselor might use the information from a karyotype to guide prospective parents.

CHAPTER 22 REVIEW

Answer these questions on a separate sheet of paper.

VOCABULARY

The following list contains all of the boldfaced terms in this chapter. Define each of these terms in your own words.

amniocentesis, autosomes, carrier, chorionic villus, cordocentesis, linkage, pedigree, polygenic inheritance, prenatal diagnosis, sampling, sex chromosomes, sex-linked, ultrasonography

MULTIPLE CHOICE

Choose the response that best completes the sentence or answers the question.

1. Red-green color blindness is an example of a genetic disorder that is *a.* sex-linked dominant *b.* sex-linked recessive *c.* autosomal dominant *d.* autosomal recessive.
2. What is the probability of a person with Huntington's disease transmitting the disease to his or her child? *a.* 0 percent *b.* 25 percent *c.* 50 percent *d.* 100 percent
3. Genes that are found on the X and Y chromosomes are said to be *a.* autosomal *b.* sex-linked *c.* polygenic *d.* teratogenic.
4. A chart that shows the occurrence of a genetic disorder in a family is called a *a.* pedigree *b.* Punnett square *c.* karyotype *d.* sonogram.
5. Genetic disorders are caused by *a.* bacteria *b.* viruses *c.* prenatal exposure to certain chemicals *d.* defects in genes.
6. Which of the following is an example of polygenic inheritance? *a.* height in humans *b.* flower color in peas *c.* hemophilia *d.* sickle-cell anemia
7. An example of a disorder caused by a dominant allele is *a.* cystic fibrosis *b.* Duchenne muscular dystrophy *c.* sickle-cell anemia *d.* Huntington's disease.
8. Linked traits *a.* are located on the same chromosome *b.* do not follow the law of independent assortment *c.* are inherited together *d.* all of these.
9. How many pairs of autosomes are found in humans? *a.* 1 *b.* 2 *c.* 22 *d.* 46
10. In humans, the gender of the offspring is determined by *a.* the mother's egg *b.* the father's sperm *c.* the time of day fertilization takes place *d.* whether the egg is fertilized.

11. A person who is heterozygous for a recessive allele that causes a genetic disorder *a.* usually shows moderate symptoms of the disorder *b.* is called a carrier *c.* will not transmit the disorder to his or her offspring *d.* all of these.
12. A sample of the fluid surrounding the fetus is removed in the diagnostic procedure called *a.* amniocentesis *b.* chorionic villus sampling *c.* cordocentesis *d.* ultrasonography.
13. A female with Down syndrome would have the chromosomal makeup *a.* 45XY *b.* 44XX *c.* 45XX *d.* 44XXY.
14. What must be true of the parents of a female hemophiliac? *a.* mother is a carrier, and father is normal *b.* mother is normal, and father has hemophilia *c.* mother has hemophilia, and father is normal *d.* mother is a carrier, and father has hemophilia.
15. Which of these disorders is caused by an autosomal recessive allele? *a.* cystic fibrosis *b.* red-green color blindness *c.* hemophilia *d.* Duchenne muscular dystrophy

PART B—CONSTRUCTED RESPONSE

Use the information in the chapter to respond to these items.

This pedigree shows the incidence of polydactyly (extra fingers or toes) in a family.

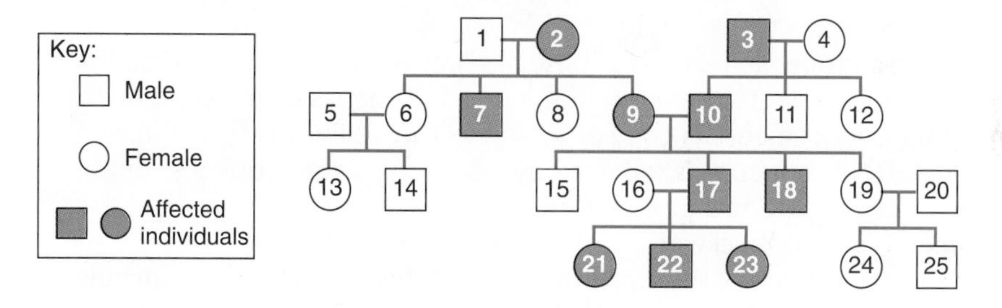

16. Based on the findings in this pedigree, polydactyly is *a.* sex-linked recessive *b.* sex-linked dominant *c.* autosomal recessive *d.* autosomal dominant. How can you tell?
17. Describe the genotypes of as many individuals in the pedigree as possible. What can you conclude about individual 17's genotype if he and his wife have a fourth child who is not polydactyl?
18. How is PKU passed from parents to children? Why is early detection of this disease important?
19. Why is prenatal diagnosis important for a couple with a family history of genetic disorders? Briefly describe four techniques used in prenatal diagnosis.

20. A grandmother fears that her school-age grandchildren may catch cystic fibrosis or Tay-Sachs disease from their classmates. What would you say to reassure her?

PART C—READING COMPREHENSION

Base your answers to questions 21 through 23 on the information below and on your knowledge of biology. Source: *Science News* (October 19, 2003): vol. 162, p. 254.

Long Live the Y?

Rumors of the human Y chromosome's eventual death may be exaggerated. David Page of the Whitehead Institute for Biomedical Research in Cambridge, Mass., and his colleagues have identified a means by which the Y chromosome may forestall, or at least delay, the gradual degradation that some biologists argue will ultimately delete it from the human genome.

In the Feb. 28 *Nature*, two Australian scientists summarized recent research showing that the human Y chromosome gradually accumulates mutations that deactivate its few genes. Non-sex chromosome pairs have a chance to replace mutation-ridden sections by swapping DNA with each other. But most of the Y is unable to exchange DNA with its partner, the X chromosome. "At the present rate of decay, the Y chromosome will self-destruct in around 10 million years," the two researchers said.

Page's group isn't so sure. These researchers have sequenced all the DNA on the Y and discovered that many of its genes have neighboring doubles. A gene and its copy often form DNA palindromes, in which a gene's DNA sequence is followed by nearly the same sequence, but in reverse. (Palindromes in literature are words or phrases that read the same forward or backward, such as, "Was it a rat I saw?").

Further scrutiny of the Y palindromes led the researchers to conclude that each gene and its backward copy are so similar in sequence that they must regularly exchange DNA in a novel form of the recombination observed in non-sex chromosomes.

"There is, in fact, intense recombination within the arms of the palindrome," says Page. "It really changes the way we think about the Y chromosome." He speculates that the palindromes enable the human Y chromosome to constantly repair mutations, staving off its own degradation.

"It's a great story," comments Eric Green of the National Human Genome Research Institute in Bethesda, Md. He suggests that researchers should look for such palindromes in sex chromosomes of other animals.

21. Explain the conclusion that was reached by two Australian scientists about the human Y chromosome and the deactivation of its genes.
22. State what Page and his research team have discovered next to many of the genes on the human Y chromosome.
23. Explain the importance of recombination between sections of DNA on the human Y chromosome.

23 Biotechnology

After you have finished reading this chapter, you should be able to:

Discuss some of the basic tools of recombinant DNA technology.

Explain ways scientists identify a particular gene.

Discuss some of the ethical questions posed by biotechnology.

Genetically engineered cells are the largest nonunion workforce in the world.

Anonymous biotechnologist

Introduction

Seventy million years ago, a mosquito lands on a dinosaur. The mosquito withdraws a few drops of blood from the huge animal. Its hunger satisfied and its gut filled with blood, it flies off. Moments later it lands on a tree and gets trapped in the sticky sap oozing from the bark. The sap covers the mosquito and hardens around it. Eventually the tree dies, falls to the ground, and rolls into a lake. Through a combination of chemical and physical changes, the sap changes into amber—a hard, clear, rocklike substance. The amber remains completely unchanged until it gets discovered by a fossil hunter. (See Figure 23-1.)

Figure 23-1
Mosquitoes and other insects have been preserved in amber.

Figure 23-2 This dinosaur egg is a fossil. It will never hatch and it will not provide DNA for genetic research.

Today, the piece of amber that contains the mosquito has ended up in a laboratory. Within the clear golden amber, the long-dead mosquito can still be seen. Carefully, a scientist drills into the amber and enters the mosquito's body. Blood cells, the remains of the mosquito's last meal, remain inside the insect. Dinosaur DNA from these blood cells is captured and removed. Genetic engineers analyze and reassemble the dinosaur's DNA. They turn the DNA on, instructing it to begin using the information stored in it. Time passes, soon an egg hatches, and once again a baby dinosaur lives on Earth. (See Figure 23-2.)

Is this story science fact or science fiction? Will we ever be able to recreate extinct animals from ancient DNA? Probably not. Is all of the story fictional? Most of it is. A few scientists have claimed to find DNA—insect DNA, not dinosaur DNA—from insects trapped in amber. Other scientists have been unable to confirm these results. Nevertheless, we have made great advances with the DNA of living organisms. Today it is possible to take pieces of DNA and move them from one place to another. This DNA can then be made to express the information it stores in new places. This is called **recombinant DNA** technology. It is now an important part of our lives, exerting influences on us in many different ways.

RECOMBINANT DNA TECHNOLOGY: A BRIEF DESCRIPTION

A revolution in biology occurred in the second half of the twentieth century. This revolution began with the discovery of the structure and function of DNA, the molecule of life. The revolution has increased in importance through recombinant DNA technology.

We know that genes are made of DNA and that they determine the characteristics of every organism on Earth. Now we have learned how to identify and find individual genes. Once found, these pieces of DNA can be removed and put together, or recombined, with other pieces. The genes can then be moved into new places. The methods for doing this make up recombinant DNA technology. The power of this technology is truly amazing. It is possible to modify the genetic makeup of an organism. This has been done for thousands of years as people have selected certain characteristics in plants and animals. Types of crops, breeds of dogs, even varieties of pigeons have been developed in this way. What is truly new is that genetic engineering through recombinant DNA technology makes it possible to put "new" genes into organisms.

For example, through recombinant DNA technology, also called **biotechnology**, and genetic engineering, human genes can be inserted into bacteria. These altered bacteria become factories that produce human proteins. Many other types of genes can be inserted. Today, industry uses these techniques to create organisms that produce a kind of oil. Agriculturists are improving crops by giving them new genes that enable the plants to resist diseases. Animals are being genetically engineered to grow bigger faster or to produce more milk. A new type of evidence called DNA fingerprinting is now used in legal cases. And perhaps most important, through biotechnology, human gene therapy is possible in some cases. There have been many setbacks and progress is very slow. Still, some day it may be possible for defective genes—those that cause genetic disorders—to be removed. In their place, new healthy genes may be inserted. These inserted genes will prevent an individual from suffering from the genetic disorder.

BASIC TOOLS OF RECOMBINANT DNA TECHNOLOGY

It is now important to look at some of the basic techniques that make this revolution in biotechnology possible.

Restriction Enzymes. New questions arose after the structure and function of DNA had been discovered. Scientists knew that, even though the DNA nucleotides occurred in a seemingly random order, the sequence of these nucleotides was undoubtedly not random and in fact held the "key to life." Scientists also knew that there was an incredibly long sequence of letters—about three billion of them—in each of our cells. This sequence contains the instructions for making the proteins needed to direct the chemical activity in us.

It was estimated that we have between 30,000 and 50,000 different types of proteins. As stated earlier, the lengthy stream of nucleotide letters in our genetic material is enough to fill 2000 books, each the size of

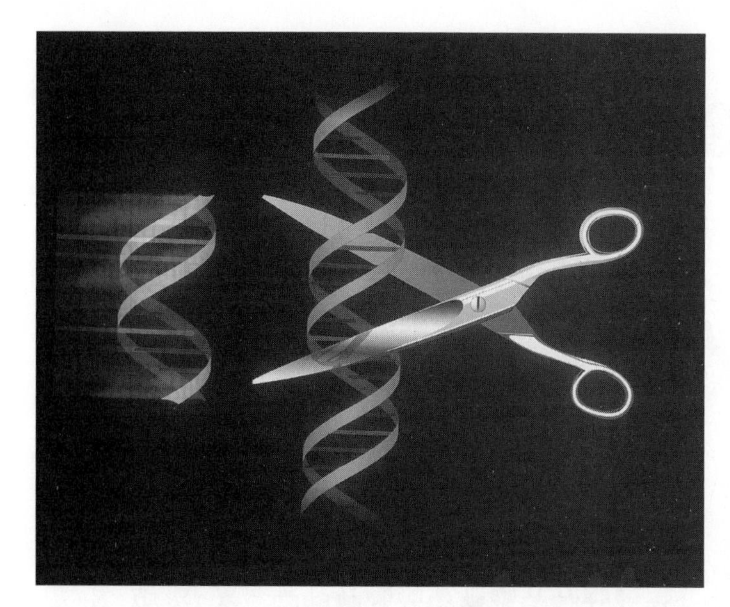

Figure 23-3 In this piece of computer artwork, the scissors represent the restriction enzyme used to cut a piece of DNA. Another piece, perhaps the gene to produce human insulin, will be inserted.

this one. In those books, there are between 30,000 and 50,000 genes, each one for a specific amino acid sequence, or polypeptide, that forms a specific protein.

The puzzle that faced scientists was how to make sense of all this DNA. Scientists wanted to know where genes for specific proteins were located in the DNA. They dreamed of being able to remove the genes and move them from one organism to another. But with the overwhelming quantity of DNA in every living cell, the task seemed hopelessly complex.

However, an unexpected series of discoveries occurred around 1970 that changed everything. Scientists working with bacteria discovered enzymes that are able to cut a strand of DNA in very specific places. Different types of bacteria each have their own kind of enzyme. The bacteria use the enzyme to destroy, or restrict, any DNA that comes into them from other sources, such as viruses. Therefore, these enzymes are called **restriction enzymes.** (See Figure 23-3.)

A restriction enzyme recognizes a sequence of between four and six nucleotide base pairs. Whenever it finds this sequence it cuts the DNA. The place where a cut occurs is called a **restriction site**. A large number of restriction enzymes cut the two strands of a DNA molecule at slightly different places, that is, off center. Short, single-stranded leftover pieces of DNA remain at the cut ends. These DNA fragments are said to have "sticky ends" because of their unpaired bases. A restriction enzyme is a kind of molecular scissors. You will see that the sticky ends can act as a kind of molecular glue. (See Figure 23-4 on page 494.)

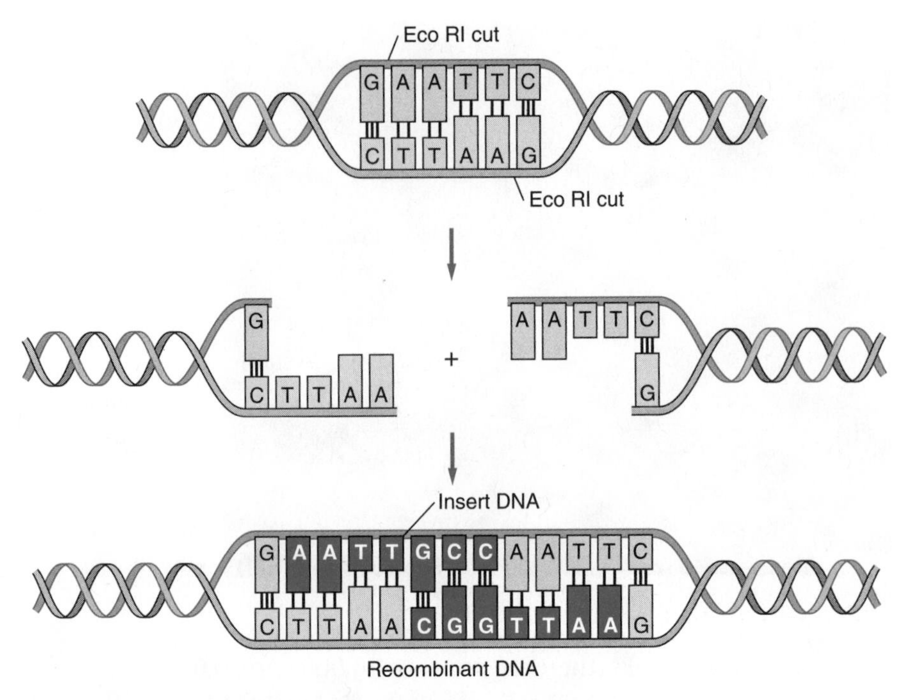

Figure 23-4 The action of restriction enzyme Eco RI.

Gene Splicing. For a scientist, restriction enzymes are very powerful tools. With them, DNA can now be cut at precise locations. In addition, the same restriction enzyme can be used to cut DNA from two completely different organisms, perhaps from a frog cell and a bacterium. Because the same sticky ends are produced in both sets of DNA, pieces of DNA from one organism can now be glued, or spliced, into the DNA of another organism. This process is called **gene splicing**. This technique of genetic manipulation makes use of DNA ligase, an enzyme that joins the DNA pieces together. Gene splicing was done in 1973 for the first time. In fact, a gene from frog DNA was joined to the DNA in a bacterial cell.

Vectors. Restriction enzymes that act as molecular scissors are not enough. Scientists also needed to find a way to carry pieces of DNA from one location to another. They wanted to be able to move it out of one cell and into another cell, usually in a different organism. Once in the new location, scientists wanted the foreign DNA to replicate. Who are the movers, the carriers of DNA? They are called **vectors**.

There are two main types of vectors: bacterial plasmids and viruses. In addition to a single large chromosome, many types of bacteria contain small circular pieces of DNA called **plasmids**. A plasmid can replicate even though it is separate from the chromosome. Plasmids also have the abil-

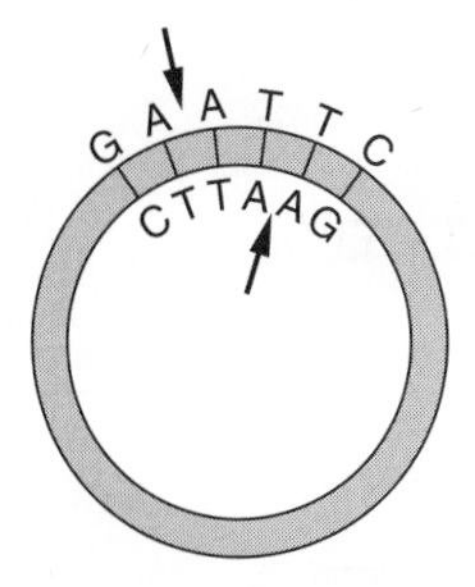

Intact Plasmid

1. Restriction enzymes make cuts on both DNA strands of a plasmid.

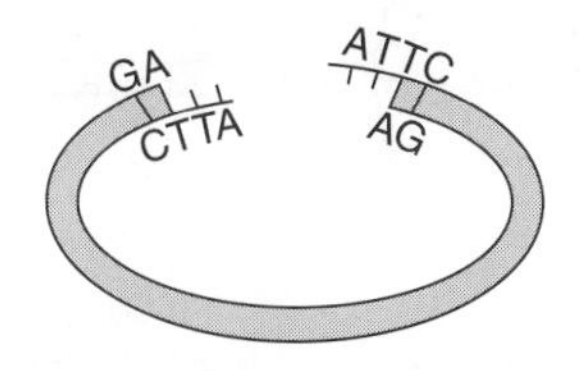

Broken Plasmid

2. The free ends of the plasmid have bases that will pair with complementary exposed bases on any other DNA stand.

DNA fragment

3. A DNA fragment containing useful genes is chosen to be inserted into the plasmid.

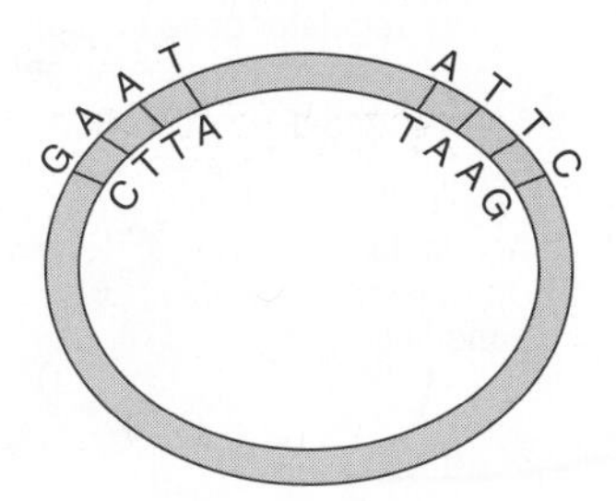

Recombinant DNA

4. The two different DNA molecules can base pair at their sticky end; they then can be sealed together by DNA ligase.

Figure 23-5 Inserting a DNA fragment into a plasmid.

ity to replicate in a new location, such as in another cell. By splicing a foreign gene into a plasmid, scientists can transport the gene to a new bacterial cell and replicate it there. (See Figure 23-5.)

Viruses can act in a similar manner. A foreign gene that is joined to the viral genome will move along with it. After a virus infects a cell, it begins to reproduce inside the infected cell. The foreign gene taken in with the virus's genetic material is also replicated in the new location.

Host Cells. Bacterial cells are used most often to receive the piece of DNA from a vector. Bacteria reproduce quickly, and soon there are thousands of bacterial cells—each containing the newly transplanted gene. Now that the recombinant DNA is in the reproducing bacteria, the amount of modified DNA is also increasing. This is one way to make much larger quantities of the recombinant DNA.

There is another reason why the bacteria are encouraged to reproduce.

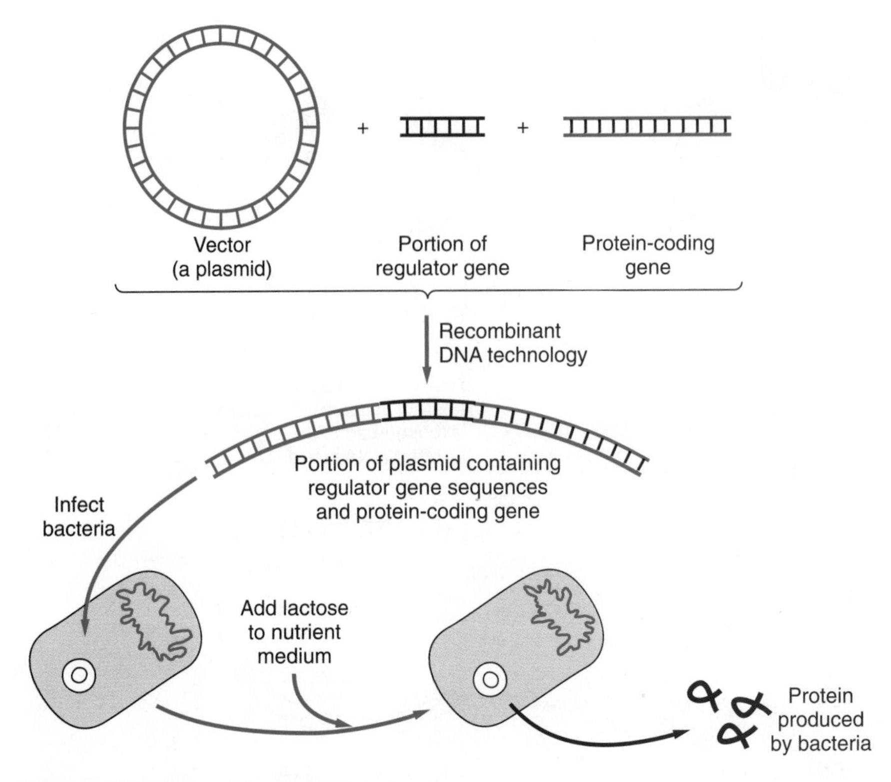

Figure 23-6 Recombinant DNA technology.

The modified DNA in the bacteria is coding for the production of a protein of interest to scientists. In effect, each bacterial cell becomes a mini-factory. The more cells, the more factories, and the more protein produced. (See Figure 23-6.)

Check Your Understanding

Why are microorganisms so important to many techniques of biotechnology?

HOW SCIENTISTS IDENTIFY GENES OF INTEREST

You might wonder how scientists identify a particular gene in the first place. One way is to collect cells that use the gene of interest to make a particular protein. For example, if we wanted to find the gene that codes for the protein insulin, we would use cells from that part of the pancreas that makes insulin. We would collect mRNA molecules from these cells. Because the pancreatic cells are insulin-making "factories," we would

The Horse Genome Project

The horse has played a major role in people's lives throughout history. Horses are important to people as a means of transportation, sport, recreation, and even companionship. People have fought wars, chased cattle, entered competitions, and won races—all while riding on the backs of horses. And headlines that concerned horses abounded in 2003, with the attempt at the Triple Crown by the up-and-coming New York racehorse Funny Cide, and with the release of both a documentary and a Hollywood film (based on the book) about Seabiscuit, the unlikely champion racehorse of the 1930s. Indeed, the question of what makes a great racehorse has intrigued people for decades.

In 1990, the U.S. Department of Agriculture invited veterinary scientists to begin mapping the genes of domestic animals, just as researchers were doing for human genes. Work began quickly on the DNA of cattle, sheep, pigs, and chickens. But not horses. In spite of people's interest in horses, there did not seem to be enough money for that work at the time.

All that changed in 1995 when geneticists working with horses decided to begin to share information. By 2001, the horse project involved 25 laboratories in 15 countries. Already, genes have been found on horse chromosomes for a number of bone and muscle diseases. Breeders of champion racehorses are paying close attention to the progress of the horse genome project. Certainly, it would be important to know in advance if a racehorse had a gene for a particular disease. However, will genes for exceptional racing performance ever be found? In other words, will DNA technology help build a better racehorse? For now, the outcome is uncertain. But it's probably a safe bet that it is just a matter of time!

probably be correct in guessing that the most abundant mRNA molecules from these cells are insulin mRNAs. Remember that the mRNA carries genetic information to the ribosomes from genes located in the DNA in a cell's nucleus. For the next step in the research, the enzyme reverse transcriptase from RNA viruses is used to convert this mRNA back into DNA, called complementary DNA or cDNA. We now have DNA for the gene that is of interest to us. If the mRNA was for insulin, we now have DNA that contains the gene for insulin.

Bacterial cells are used to express genes from all sorts of plants and animals. This is another dramatic piece of evidence that supports the theory that all life evolved from common ancestral cells. Cells show close relationships with other cells because genetic activities can be moved back and forth between cells of very different types.

For some purposes, bacterial cells are not useful host cells. Instead, eukaryotic cells are used. These hosts may be yeast cells, plant cells, or animal cells—including human cells—that are grown or cultured in the laboratory.

Gel Electrophoresis. In addition to cutting, pasting, and moving DNA, genetic engineers need to be able to sort DNA. A restriction enzyme cuts wherever its particular restriction site occurs. DNA is said to be "digested," or broken into smaller pieces, when the restriction enzyme cuts it. Therefore, the DNA from a cell is cut with a restriction enzyme into thousands of small pieces. The pieces are of many different lengths, depending on the location of the restriction sites. These pieces are called **restriction fragments.** Scientists can sort the restriction fragments according to their lengths by using a technique called **gel electrophoresis**. Agarose gel, a material somewhat like a gelatin dessert, is used. The digested DNA is moved along in a liquid solution through the gel by an electric current. The smaller the restriction fragments, the easier and faster they can move through the gel. As a result, smaller fragments of DNA move farther along in a given amount of time than do larger fragments. Running a DNA gel of restriction fragments produces a pattern of bands. Each band consists of DNA fragments of a particular size.

Polymerase Chain Reaction (PCR). A technique called **polymerase chain reaction** (PCR) is frequently used as a kind of molecular photocopy machine. Using a special group of enzymes, PCR can take an extremely small amount of DNA and make large quantities of it. This increase in the amount of DNA is called **amplification**. Once the DNA is amplified, it can be analyzed for specific purposes. (See Figure 23-7.)

PCR is a technique that is proving useful at trials. For example, blood or hair is sometimes found at a crime scene. Using PCR, the DNA from a drop of blood or a single hair can be amplified. Then it can be analyzed. Through this technique, the identity of a criminal can be determined and, in some cases, an innocent person can be cleared of involvement in a crime.

Short Tandem Repeats (STRs). A knife is found at the scene of a murder. The knife is placed in a plastic bag without being touched by bare hands. In the crime lab, the knife is examined for fingerprints. Through FBI computers, a fingerprint match is made. A person is arrested and eventually convicted of the murder based on the fingerprints left on the weapon.

What does DNA have to do with identifying a criminal? In the past, the answer would have been "nothing." Only fingerprints were used. However, in the last 25 years, it has been discovered that the DNA of every individual—other than identical twins—is unique. One technique that

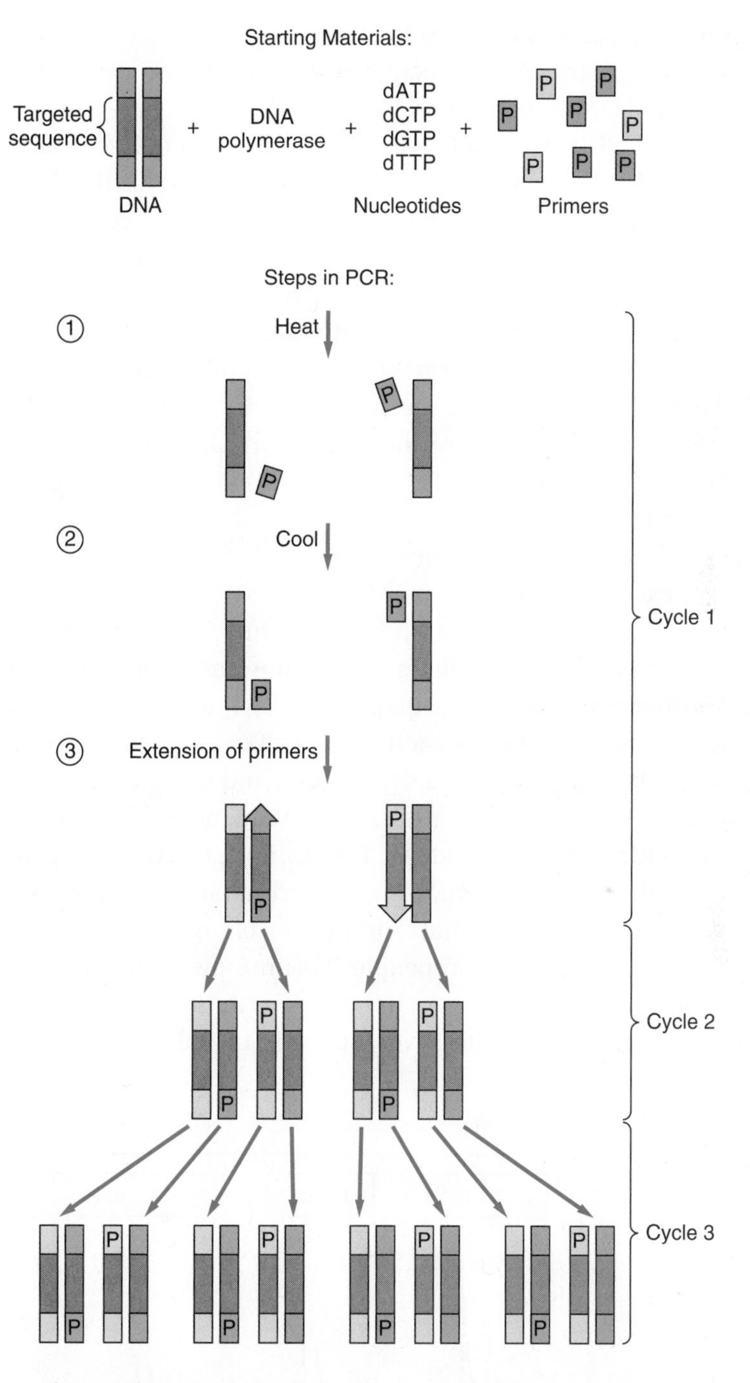

Figure 23-7 The polymerase chain reaction (PCR) is used to increase the amount of DNA in a sample.

shows the uniqueness of an individual's DNA makes use of restriction enzymes. As a result of such techniques, every person can be shown to have a unique DNA "fingerprint."

In 1997, the U.S. Federal Bureau of Investigation (FBI) established the Combined DNA Index System (CODIS). Through CODIS, the DNA analysis from a sample found at a crime scene can now be compared in the system to help identify a criminal. The information in CODIS is based upon regions in human DNA that can vary from person to person. Variations in DNA sequences between individuals are called *polymorphisms*. DNA sequences that show the greatest amount of polymorphism, or variation, between individuals are the most useful in forensics. The DNA polymorphisms used for CODIS are known as **short tandem repeats**, or **STRs.** (*Note:* The word *tandem* means "a group of two or more parts arranged one behind the other.")

STRs are very short sequences of DNA that are repeated numerous times. For example, GATA is a four-base-pair STR sequence. It is repeated from 6 to 15 times in people on chromosome 7. (*Note*: This sequence would be repeated in both alleles of chromosome 7, one from each parent.) Another four-base-pair sequence is TCAT, and it too is repeated a specific number of times in each person's DNA. By analyzing enough STRs, researchers can develop a very precise and unique identification of a person. The CODIS system uses 13 STR sequences. The system is very powerful in identifying individuals. For example, by combining the information on all 13 STRs for one suspected criminal, it was possible to calculate that the chance of that one person being identical to another person is 1 in 7.7 quadrillion people. This number is one million times larger than the total human population of Earth; so the CODIS system in this example was able to positively identify the suspect with great certainty. (See Figure 23-8.)

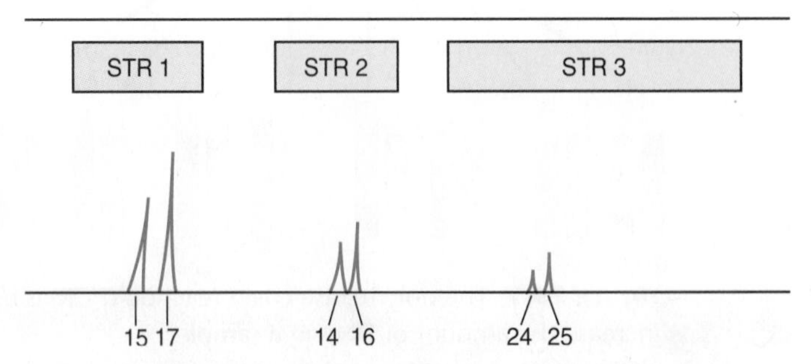

Figure 23-8 PCR analysis of three STRs for an individual shows that his genotype (with one allele from each parent) is 15 and 17 repeats for one STR, 14 and 16 repeats for a second STR, and 24 and 25 repeats for a third STR.

USES OF RECOMBINANT DNA TECHNOLOGY

Medicine is benefiting from the use of biotechnology. More than 200 human genetic disorders can now be diagnosed with recombinant DNA technology. Often individuals can be diagnosed with a disease even before the disease shows any symptoms. This is because the gene that causes the disease can be identified in their DNA. This identification can even be performed on a fetus in the mother's womb.

One hundred years from now, people may consider the medicine practiced today to treat inherited diseases very strange. In the future, physicians may assume that the only logical thing to do is to change the DNA in a person so that the disease does not occur. This has already been done in a few cases. By the year 2003 this treatment, called *gene therapy*, had been successfully used in France on a number of children with severe combined immunodeficiency disease (SCIDS). Without treatment these children, having no natural protection from diseases, would likely have died. However, some new disease problems, such as leukemia, have occurred in a few of the gene-therapy patients. One immunodeficiency disease results from a lack of the enzyme adenosine deaminase (ADA). Cells from some ADA-deficient patients have been given the missing ADA gene. Then these cells have been returned to the patients' bodies to cure the disease. Although treating diseases on the genetic level will probably occur more frequently sometime in the future, the immediate future of gene therapy is not at all certain.

Some medicines are now being produced in bacteria through biotechnology. For example, genetically engineered bacteria use a human gene to produce insulin. It is, therefore, pure human insulin. Previously, insulin was collected from the pancreas of pigs and cattle. These pancreatic tissues were left over when the animals were killed for food. In this case, the insulin from these animals was not completely identical in structure to human insulin therefore caused medical problems in some people. Human insulin (commercial name *Humulin*), synthesized in bacteria, eliminates these problems. (See Figures 23-9a and 23-9b on page 502.)

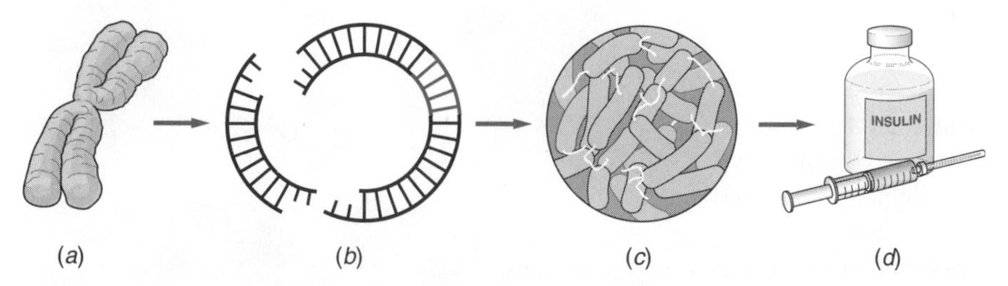

Figure 23-9a The insulin gene is inserted into bacterial plasmids. The bacteria that contain this gene produce insulin, which is used by people with diabetes.

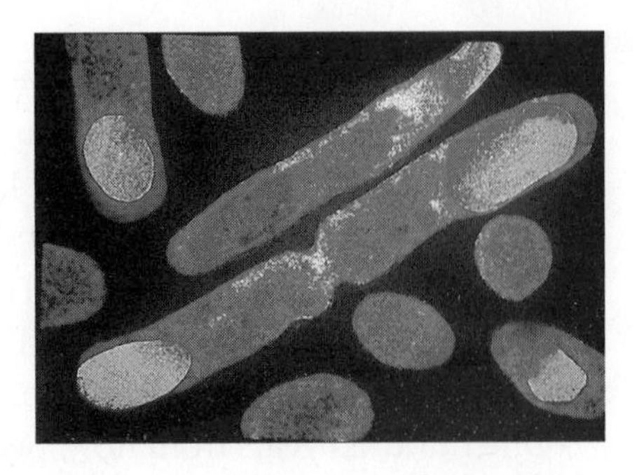

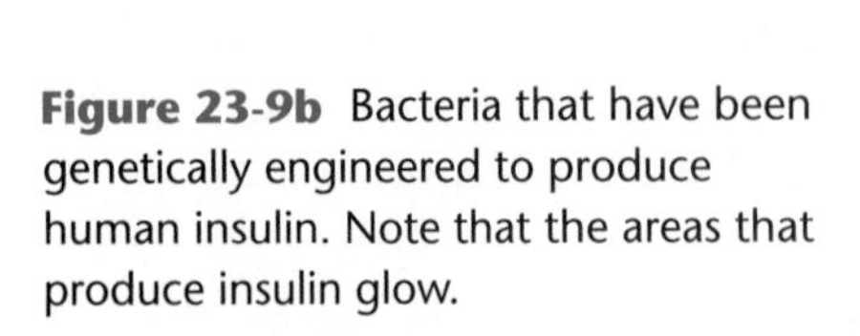

Figure 23-9b Bacteria that have been genetically engineered to produce human insulin. Note that the areas that produce insulin glow.

As has been mentioned, DNA fingerprinting is being used to determine the guilt or innocence of individuals in criminal trials. In addition to collecting DNA from skin, hair, or blood samples, investigators can collect DNA from semen samples. Through PCR and STR analysis, the guilt or innocence of individuals in rape trials is also being argued. (See Figure 23-10.)

Biotechnology has been enlisted to help clean up the environment, too. Genes have been inserted into bacteria to give them the ability to remove hazardous materials from the environment.

Animals that are raised for food are receiving genes from other animals. Trout and salmon are being farmed to provide food for people. DNA for a growth hormone has been injected into the fertilized eggs of these fish. As a result, they grow faster, producing more food in a shorter time.

In plants, an early example of biotechnology allowed scientists to develop tomatoes that do not become overripe. A gene inserted into these tomatoes prevents a gene that causes early ripening from working. Today, the biggest challenge plant scientists are working on involves genes for nitrogen fixation. These genes allow plants to take nitrogen gas from the air and change it into a form they can use. All plants need nitrogen, but they cannot directly use atmospheric nitrogen. Farmers often have to use fertilizers that contain usable forms of nitrogen to provide this important

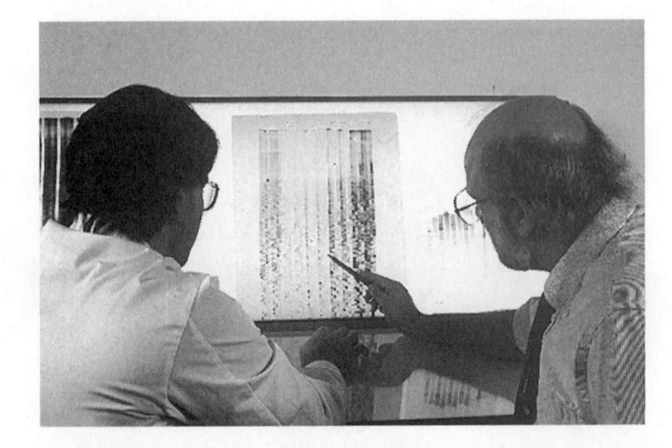

Figure 23-10 Results of the analysis of DNA fingerprints can be used to argue the guilt or innocence of individuals in criminal proceedings.

nutrient to their crops. If the gene for "fixing" this nutrient could be inserted into crop plants, smaller amounts of expensive chemical fertilizers would be needed. It is no wonder that many scientists are working on this problem.

ETHICAL QUESTIONS ABOUT BIOTECHNOLOGY

It is more than 50 years now since Watson and Crick first discovered the structure of DNA in 1953. Recombinant DNA technology has given humans the ability to alter living organisms, including ourselves. The potential benefits of this are enormous. However, according to many people, there are dangers associated with the ability to alter the genetic makeup of organisms.

For example, genes that make agricultural plants resistant to disease can be added to a crop plant's genes. In some cases, genes may move naturally between plants. What might happen if these disease-resistant genes were to enter a weed's genes? Weeds already cause a lot of problems for farmers. The new "superweeds" with a genetic immunity to certain diseases might begin to increase their numbers greatly. These superweeds would have a genetic advantage over other plants.

Some scientists fear that inserting foreign genes into plants and animals that humans eat may produce severe allergic reactions in some people. Will government agencies require sufficient testing to make certain this does not occur with genetically altered plants?

Human gene therapy also raises tremendous ethical questions. Few people argue the benefit of giving normal genes to a person in order to treat a genetic disorder. But should these genes also be added to a person's eggs or sperm, for the benefit of future children? If this is done, then the genetic change occurs in all future generations of humans that are descended from individuals with those altered gametes. The genetic makeup of the future human population would change. Do we know enough about the consequences of these actions to allow ourselves to have the power to do this? Do we know all the potential consequences of these changes? At this time, the answers are clearly no.

Scientists are thrilled that the first draft sequence of the human genome has been completed. But then who should be given the right to examine another person's genes? In what ways should that information be used? Should people be able to use information about an individual's genetic makeup to decide whether to give that person a job or to provide health or life insurance? Clearly the future of biotechnology holds many challenges, scientific as well as ethical.

EXTENDED LABORATORY INVESTIGATION 23

What Does the Future Hold for Biotechnology?

INTRODUCTION

Rapid advances in technology produce explosions of knowledge in the fields of genetics and molecular biology. The cloning of mammals is no longer impossible. The entire human genome is being deciphered and recorded. This major research project identifies every gene that determines who we are and how we function. The field of biotechnology is providing an understanding, and a degree of control, of life processes that is unprecedented. It is vitally important that every person understands the information revealed by biotechnology research. Decisions and choices that require this understanding will have to be made by individuals. For this activity, you will prepare a report on an aspect of biotechnology.

POSSIBLE TOPICS

genetic engineering

cloning of mammals

biotechnology and agriculture

designer drugs

human gene therapy

Human Genome Project

DNA fingerprinting

POSSIBLE INFORMATION SOURCES

local public or school library

newspapers and magazines

the Internet

computer databases and encyclopedias

research experts

federal and local government agencies

PROCEDURE

1. Choose one of the suggested topics or another topic that interests you. Have your teacher approve your choice.
2. Decide if you want to work by yourself, with a partner, or with a group. If you work with a group, you may want to assign jobs to make sure that everyone participates and that no part of the project gets overlooked.
3. As you research and write your report, keep these tips in mind:
   - Choose a purpose and state it at the beginning of your report. For example, you might want to describe current progress in gene therapy, or you might want to try to persuade people that cloning organisms is either a good or a bad idea.
   - Use different kinds of sources. For example, read newspaper articles and encyclopedia entries, check out scientific Web sites, or interview an expert.
   - Include facts, details, and statistics. You can also include your opinions, as long as you can give reasons to support them.
   - Include visuals—pictures, graphs, maps, and anything else that illustrates your topic and makes it more interesting.
   - Organize your report. Begin by stating your purpose. Then give the results of your research in the body of the report. Finally, write a concluding paragraph.
   - Proofread!
   - Be sure to include a bibliography that lists all your sources.
4. Present your report to your class.

CHAPTER 23 REVIEW

Answer these questions on a separate sheet of paper.

VOCABULARY

The following list contains all of the boldfaced terms in this chapter. Define each of these terms in your own words.

amplification, biotechnology, gel electrophoresis, gene splicing, plasmids, polymerase chain reaction (PCR), recombinant DNA, restriction enzymes, restriction fragments, restriction site, short tandem repeats (STRs), vectors

PART A—MULTIPLE CHOICE

Choose the response that best completes the sentence or answers the question.

1. What is the best explanation for why a restriction enzyme does not cut the DNA of the cell that produces it? *a.* The cell's DNA is chemically shielded against restriction enzymes. *b.* The restriction enzyme can distinguish between "self" and "foreign" DNA. *c.* The cell's DNA does not contain the restriction site. *d.* The cell uses RNA rather than DNA for its genetic material.
2. Which term does *not* mean basically the same thing as the other three? *a.* biotechnology *b.* recombinant DNA technology *c.* genetic engineering *d.* electrophoresis
3. Restriction enzymes were first identified in *a.* bacteria *b.* viruses *c.* humans *d.* mosquitoes.
4. "Sticky ends" result from treating DNA with *a.* gel electrophoresis *b.* restriction enzymes *c.* polymerase *d.* plasmids.
5. Inserting a piece of DNA with "sticky ends" from one organism into the DNA of another organism is called *a.* gene splicing *b.* gene sequencing *c.* electrophoresis *d.* PCR analysis.
6. DNA "fingerprints" are based on banding patterns resulting from the analysis of *a.* plasmids *b.* restriction enzymes *c.* STRs *d.* PCRs.
7. In addition to a large circular chromosome, bacteria may contain small rings of DNA known as *a.* STRs *b.* centromeres *c.* vectors *d.* plasmids.
8. Which of these is *not* needed to produce complementary DNA (cDNA)? *a.* mRNA *b.* reverse transcriptase *c.* target gene (DNA) *d.* sugar-phosphate-nitrogenous base groups
9. A tiny drop of blood is found in a murder suspect's car. Which technique would be the first one performed on the sample?

a. gel electrophoresis *b.* restriction enzyme cutting *c.* DNA fingerprinting *d.* PCR analysis

10. A genetically engineered virus is used to deliver the gene for normal hemoglobin into the bone marrow of a person with sickle-cell anemia. In this example, the vector is the *a.* virus *b.* gene for normal hemoglobin *c.* bone marrow *d.* patient.
11. Gel electrophoresis is used to *a.* sort restriction fragments by size and to sequence DNA *b.* amplify the DNA in a small sample *c.* cut DNA into pieces *d.* synthesize DNA from RNA.
12. In genetics, PCR stands for *a.* protein-carbon reaction *b.* polymerase chain reaction *c.* polymorphic cell reduction *d.* polymorphic chain replication.
13. The place where a restriction enzyme cuts is called a (an) *a.* STR *b.* chimera *c.* recombinant *d.* restriction site.
14. Which organisms are most commonly genetically engineered to create "protein factories"? *a.* cows *b.* bacteria *c.* viruses *d.* tobacco plants

PART B—CONSTRUCTED RESPONSE

Use the information in the chapter to respond to these items.

Enzyme	Source Organism	Restriction Site
*Eco*RI	*Escherichia coli*	–C–T–T–A–A┼G –G┼A–A–T–T–C
*Hind*III	*Haemophilus influenzae*	–T–T–C–G–A┼A– –A┼A–G–C–T–T–
*Hae*III	*Haemophilus aegyptius*	–C–C┼G–G– –G–G┼C–C–
*Hpa*II	*Haemophilus parainfluenzae*	–G–G–C┼C– –C┼C–G–G–
*Sma*I	*Serratia marcescens*	–G–G–G┼C–C–C– –C–C–C┼G–G–G–

15. According to the enzyme table, which enzymes would be most useful for gene splicing? Explain.
16. How might the enzymes that are not useful for gene splicing be used?
17. List three reasons why recombinant gene technology should be encouraged and three reasons why it should be banned. Then give your own opinion about this issue and explain your reasoning.
18. How has recombinant gene technology helped improve the quality of medication for patients with diabetes?
19. Describe how techniques of recombinant gene technology are being used to solve crimes.

PART C—READING COMPREHENSION

Base your answers to questions 20 through 22 on the information below and on your knowledge of biology. Source: *Science News* (January 11, 2003): vol. 163, pp. 29-30.

Researchers Target Sickle-Cell Cure

Stem cell transplants have long been a therapy option for children with life-threatening cases of sickle-cell disease, but the procedure itself can be deadly. Researchers in France now report that transplants have cured 30 consecutive patients over several years, thanks largely to an immunity-suppressing drug that has shown only mixed effectiveness in the past. The study of 69 children and young adults began in 1988.

The results document "an unprecedented cure rate for children" with this ailment, says Robert I. Handin of Brigham and Women's Hospital in Boston.

Most of the transplants were of bone marrow. Seven consisted of umbilical cord blood. Like marrow, cord blood contains stem cells, which can develop into various types of blood cells. All the transplants came from siblings of the patients.

In the transplant procedure, doctors first destroy the bone marrow of a patient, clearing the way for a sibling's stem cells to take over and repopulate the person's body with healthy blood cells. A major risk is that the recipient's immune system will attack and kill the implanted cells.

In the study's earlier years, the first 12 children with sickle-cell disease received a bone marrow transplant plus busulfan and cyclophosphamide, two drugs known to facilitate transplants. However, four of these children rejected the cells.

The next 57 patients received these two drugs with their cell transplants plus the immune suppressant, a protein called antithymocyte globulin. These patients fared better, says study coauthor Françoise Bernaudin of Saint Louis Hospital in Paris. Not one patient rejected the cells.

Overall, six patients—all among the first 39 to enter the study—died from complications of stem cell transplants.

Bernaudin presented the research last month in Philadelphia at a meeting of the American Society of Hematology.

20. State the procedure that is used to take stem cells from one person and give them to patients with sickle-cell disease.
21. Explain why the stem cell transplant procedure is very risky to the patient.
22. How have researchers in France helped to reduce the risks associated with stem cell transplants?

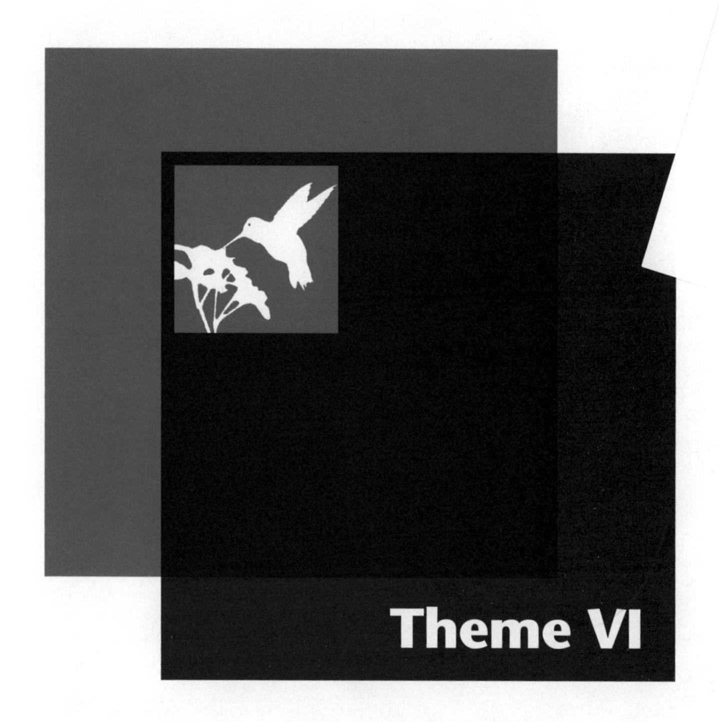

Theme VI

Interaction and Interdependence

An Introduction to Ecology

After you have finished reading this chapter, you should be able to:

Define ecology and ecosystems.

Identify some biotic and abiotic factors and explain their effects on organisms.

Discuss the ways in which limiting factors affect organisms.

If the Earth were only a few feet in diameter, floating a few feet above a field somewhere, people would come from everywhere to marvel at it.

Anonymous

Introduction

An aquarium is a complete small world in itself. Plants live together with animals in one place. By looking closely, you can see a snail crawling along the leaf of an underwater plant. Elsewhere on the plant, hidden between some leaves, are masses of snail eggs surrounded by small globs of clear jelly. Within the eggs are the individual dark spots that are the developing snail embryos. A goldfish swims slowly by. The mouth of the fish opens and closes again and again. Water enters its mouth, bringing the oxygen the fish needs. The fish takes a bite out of a plant leaf for food, perhaps eating some snail eggs, too. (See Figure 24-1 on page 512.)

The aquarium is near a window. Light pours into the tank, and the plants grow toward the sun. The plants are anchored by their roots in a mixture of crushed stones and gravel. In the corner of this small aquatic world is a plastic tube that releases bubbles. The fish swim in and out of the bubbles that form a curtain of air.

Why do some people spend long periods of time gazing quietly into an aquarium? Some answers become obvious upon reflection.

An aquarium is a self-contained miniature world of life. The basis of

Figure 24-1 The aquarium fulfills the life-support needs of all its inhabitants.

life on Earth can be found in an aquarium. Like the living things in an aquarium, every organism on Earth lives in its surroundings, its environment. No living thing exists alone. All living things interact, or affect other living things and their environment. All living things depend on each other and their environment—they are interdependent. These relationships of interaction and interdependence between living organisms and their environment are studied in the branch of biology known as ecology.

ECOLOGY AND ECOSYSTEMS

Every plant and animal lives somewhere. The environment in which an organism lives is molded by many different factors. These include availability of food and water, amount of sunlight, temperature, and type of soil in the area. If these conditions involve other living organisms, they are known as **biotic factors**. For a bird, biotic factors include, but are not limited to, trees, other birds, insects, and worms. Other conditions involve nonliving or **abiotic factors**. For example, some abiotic factors for a bird are water, wind, temperature, and light.

Earth is a living planet. It is far too complex to try to understand all at once. However, the interaction and interdependence of organisms and their environment can be understood by examining specific areas. Ecologists call a specific place an **ecosystem** when all the living organisms and nonliving factors in that one place are considered together. A pond is an ecosystem. A forest is an ecosystem. So, too, is a rocky seashore. Our

Figure 24-2 Earth is the ultimate ecosystem—it fulfills the needs of all its inhabitants.

aquarium, as small as it is, is also an ecosystem and can therefore be used to study ecology. (See Figure 24-2.)

ADAPTATIONS AND EVOLUTION

It is truly amazing how so many organisms seem to fit so well with their environment. Camels have extremely wide two-toed feet with skin that stretches between the toes. With these feet, camels can avoid sinking into the desert sand. (See Figure 24-3 on page 514.) In fast-moving tropical streams, some species of catfish have suckers on the bottom of their bodies. The suckers hold onto a rock and thus prevent the catfish from being swept downstream. The larval stage of the tiger swallowtail butterfly resembles a bird's dropping. With this disguise, it is not the first choice of a hungry bird searching for a meal!

In a tropical rain forest, the fertile soil is extremely shallow. Most of the nutrients available for plants are located in this thin layer. Many rain forest trees have shallow root systems. To support themselves upright, these trees have wide supports called buttresses. The tree's buttresses, located at

Figure 24-3 The wide, two-toed feet of camels enable them to walk in sand without sinking.

the base of their trunks, support the tree in much the same way as the stone buttresses of a Gothic cathedral support its walls. (See Figure 24-4.)

The giant saguaro cactus is able to take in enormous amounts of water during the rare desert rainstorms. The stem of this cactus, once it is swollen with water, can keep the plant alive throughout the long dry months ahead. (See Figure 24-5 on page 516.) The tiny pond skater insect has a waxy substance on its six feet. The wax keeps the insect from breaking through the top layer of water molecules, and it can easily walk across the surface of a pond. Wherever we look on this living planet, we see more examples of adaptations that enable organisms to survive Earth's vastly different living conditions. How did all this happen?

First, we must be clear about what did *not* happen. No individual organism *intentionally* changed to survive in a particular environment. No individual rosebush purposely developed brighter flowers to attract more insects. No individual arctic hare intentionally became whiter to hide in the snow.

What *did* happen is that individuals in a population have always been somewhat different from one another. Sometimes the differences are obvious. Two dogs from the same litter, for example, might be very different sizes or colors. Sometimes the differences are not obvious to an observer. A slight difference in the biological makeup of an organism, for example, the ability to make a particular enzyme, might provide that organism

Figure 24-4 Some rain-forest trees with shallow roots have woody buttresses for support.

with a slight survival advantage over other organisms. In the examples given, not all individuals in the hare population or in the rose population are identical. These variations in individuals are due to inherited genetic differences. Because of the differences, some individuals are better suited to certain environmental factors. In an ancestral population, some hares had somewhat lighter-colored fur and could hide better. Those hares would have been more likely to be missed by a predator and survive to pass on their genetic traits to future generations. It is through this process of natural selection that a species, not individual organisms, evolves. This is the basis of evolution. Over time, a species' traits make a remarkable fit with its environment. If they do not, the species will probably not survive in that particular environment. The characteristics of an organism that give it this fit are called **adaptations**. Adaptations can be in physical traits, such as the size, shape, or color of an organism. They can also be in behavioral traits, such as the building of nests by birds or the yearly release of seeds by plants. Humans, too, have many physical and behavioral adaptations.

Every organism is therefore adapted through evolution to live in a particular place. Each species of organism is adapted to a specific set of

Figure 24-5 The giant saguaro cactus, which lives in the desert, stores water in its stem.

conditions. The place where an organism lives is its **habitat**. The habitat of a bullfrog is a pond. The habitat of a giant anteater is open grassland. An organism's habitat is its "address."

To understand an organism's relationship to its environment, we must know more than its address, or habitat. We must know what it does, that is, its "occupation." The occupation of an organism is called its **niche**. The niche of an organism includes how it gets its food, reproduces, and avoids predators, among other things. The behavioral adaptations of an organism make up its niche. These adaptations are the result of evolution just as are the organism's physical adaptations. For example, woodpeckers that were best able to find insects in the bark of trees survived. Their offspring inherited a whole series of structural and behavioral adaptations. The niche of an organism determines its habitat. In other words, the ways that an organism has evolved to survive will also determine where it can live. A woodpecker cannot live in the grasslands. Its niche involves finding its insect food in the trunks of trees. Woodpeckers need trees, and the insects that live in them, to survive.

Just how important niches are has been made clearer with new research on fossils reported in 2002 by scientists at the University of California–Berkeley. Their investigations on mass extinctions of long ago showed that it took much longer than had been thought for the diversity of species to return to Earth. The report stated, "One possible explanation for why diversification takes so long after an extinction is that extinction eliminates not merely species or groups of species, but takes away ecological niches. It eliminates both organisms and the roles those organisms played in the ecosystem. Recovery thus becomes more complicated."

ENVIRONMENTAL FACTORS

Evolution occurs by natural selection. Natural selection is the process by which the organisms that are best adapted survive. Now we can ask another question: Adapted to what? Every organism is adapted to the conditions present in its environment. To understand why a plant grows the way it does, or why an animal behaves the way it does, we must look at the factors present in the environment of each organism.

If you have a garden, you already know that certain plants grow best under particular conditions. (See Figure 24-6.) One type of plant grows best in full sun, while other plants grow best in shade. Some plants grow best under dry conditions, while others grow best in high humidity. Animals also thrive under certain conditions. Organisms often have

Figure 24-6 The conditions present in a garden determine which plants will thrive there. Some plants prefer shade; others prefer more direct sunlight.

preferences and thrive best in environments that approximate these conditions. Environmental factors can therefore limit where organisms live. Each species is adapted to a specific set of conditions. These factors are called **limiting factors**.

For example, the availability of sunlight is an important limiting factor for plants. Tall trees living in a tropical rain forest near the equator receive a great deal of sunlight. Areas far to the north and south of the equator receive less intense sunlight. The amount of sunlight in the oceans and lakes also varies a great deal. Light penetrates water only to a depth of about 200 meters. Below that depth, it is too dark for plants to grow.

Temperature, of course, is another major limiting factor. Most organisms live in the range that is above 0°C and below 42°C. A few species of bacteria can survive in conditions well outside of this temperature range. Some live in hot springs, for example, where the water temperature is nearly at the boiling point. Each type of plant or animal or microorganism has a fairly narrow temperature range that it prefers. In other words, the organism has a tolerance only for temperatures within this range. Fish show this tolerance range very clearly. If you measure water temperatures at different depths in a lake during the summer, you will find that the temperature drops as the depth increases. Each type of fish in the lake has a different tolerance range for temperature. A species that prefers warmer water lives in the relatively shallow regions or near the water's surface. A fish species that prefers colder water lives at greater depths. This can also help explain why fish generally move to deeper water during hot summer months. The water near the lake's surface becomes too warm for many fish to survive in, since warmer water contains less dissolved oxygen. (See Figure 24-7.)

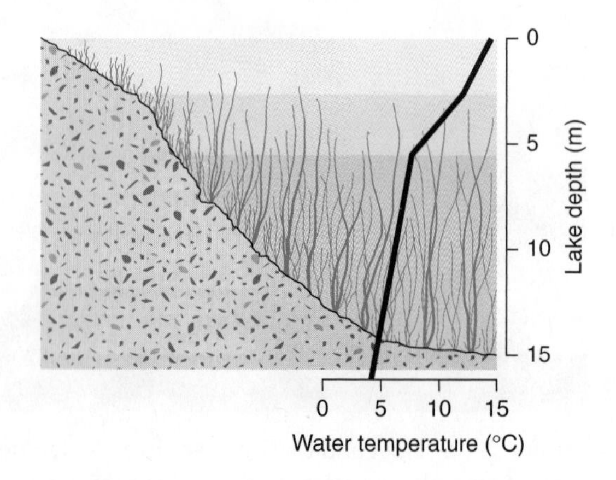

Figure 24-7 A lake's temperature decreases with increasing depth.

Figure 24-8 The maple is a deciduous tree.

All life depends on water. However, organisms have evolved adaptations that enable them to survive in very specific ranges of available moisture. **Deciduous** trees, such as maples, have large, flat leaves. Because water is constantly moving out of the openings on the leaf surfaces, trees with large, flat leaves live in areas with abundant rainfall. Such trees can live only where rainfall exceeds 100 centimeters per year. (See Figure 24-8.) Other species of trees have evolved ways to save water. **Coniferous** trees, such as pines, have narrow leaves called *needles*. Little moisture is lost from this type of leaf; thus, coniferous trees can survive in areas that have less water available. (See Figure 24-9.)

Other environmental factors are much less obvious, but they are still very important for a species' survival. For example, chemical nutrients in the environment may determine which organisms survive in a particular place. Two such nutrients are the elements nitrogen and phosphorus. Not only do these substances have to be present in adequate amounts for many

Figure 24-9 The pine is a coniferous tree.

plants, they also have to be present in the right forms. Farmers have long been aware of this. In many cases, soil contains less than the minimum amount of nutrients needed by plants to grow well. Many crops cannot be grown unless fertilizers are added to the soil. Fertilizers increase the amounts of nitrogen and phosphorus to levels needed by the crops. Almost every gardener has experienced another important limitation. While too little of a limiting factor may prevent the growth of an organism, too much may also limit the organism's growth. Some gardeners think that the more fertilizer they give their plants, the more flowers and vegetables they will get. To their surprise, they find that their plants often die when greater amounts of fertilizer are added to the soil. Organisms have a tolerance range, sometimes a very narrow one. The amount of an environmental factor should be neither too low nor too high. The tolerance range determines the best conditions for a specific type of organism in a specific location.

Check Your Understanding

Discuss two ways biotic factors and two ways abiotic factors can affect the life of a tree.

SALTWATER ECOSYSTEMS

Because we live on land, we tend to think of Earth as a land planet. However, 75 percent of Earth's surface is covered by water. One of the main reasons that life exists on Earth, and probably not on other planets in our solar system, is that only the temperatures present on Earth allow for the existence of liquid water. On Earth, life evolved first in water, not on land. Today, the climates of all areas, including land areas, depend heavily on the parts of Earth's surface covered by water. Because water is so important to life, we will begin our study of ecosystems by looking at those in the water.

A map of the world shows individual oceans. A closer look, however, shows that all the world's oceans are connected. There is really one world ocean. Some ecologists consider this ocean to be one tremendously large ecosystem. Scientists call a very large area in which a common group of organisms lives a **biome**. Because the oceans share many of the same types of organisms, we will call this environment the ocean, or marine, biome.

Two of the main limiting factors in the ocean biome are the saltiness, or salinity, and the temperature of the water. We drink freshwater—water that contains little or no salt. However, water in the ocean contains about

3.5 percent salt. Anyone who swims in the ocean knows the taste of salt water and probably also knows that it is not a good idea to drink it.

The actual amount of salt in ocean water varies. As the amount of salt varies, the density of the water also changes. The more salt, the denser the water. This is why a person can float easily in very salty water, while it is difficult for many people to float at all in freshwater.

The temperature of ocean water also differs from place to place. Temperature affects water density, just as the level of salt does. Cold water is more dense than warm water. Because of density differences, water in the ocean is constantly moving. Denser water sinks, while less dense water rises. These water movements, along with winds on the surface, cause currents to occur in the ocean. Some ocean currents are so large that they actually form "rivers" in the ocean. The Gulf Stream is an example of a major ocean current. It originates in the Gulf of Mexico, travels north along the east coast of the United States, and moves across the Atlantic Ocean to bring warm water toward Great Britain. In fact, some of the islands off the coast of Britain are so warm that semitropical plants can survive there. (See Figure 24-10.)

Light is another important limiting factor in oceans. Different parts of the ocean have different sets of conditions. Because of this, the ocean is considered to be made up of different zones. The top zone in the ocean down to about 200 meters is where sunlight penetrates. Below this depth, it is always completely dark. Because plants need light to make food

Figure 24-10 The Gulf Stream is one of Earth's ocean currents.

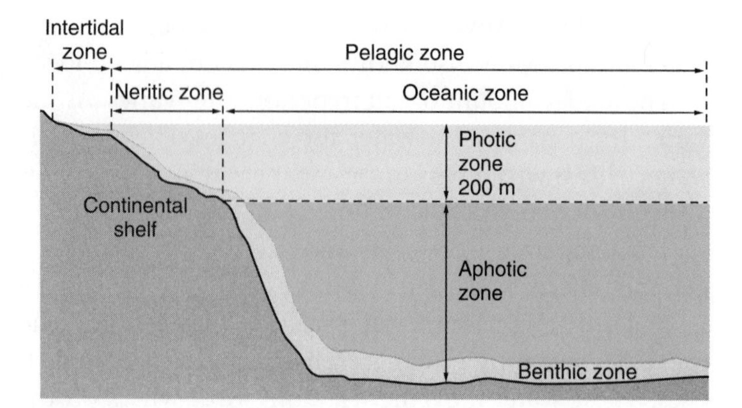

Figure 24-11 Sunlight penetrates about 200 meters below the ocean's surface.

through photosynthesis, all plant life in the ocean lives only in the top zone. (See Figure 24-11.)

Another zone occurs at the very edge of the ocean. This zone experiences very unusual environmental conditions—it varies, at different times of the day, between being either wet or dry. Why do alternating wet and dry conditions occur here?

The force of gravity from the moon and sun pulls on Earth. The land does not move, but the water does. The water in the world's oceans therefore rises and falls twice each day. These movements of the ocean water are called **tides**. At high tide, the shoreline of the ocean is covered by water. At low tide, the same place is left exposed to the air. The area that is alternately covered with water and left exposed to the air is called the intertidal zone. Only a very special group of plants and animals can live under the unique environmental conditions present in the intertidal zone. (See Figure 24-12.)

Moving into the ocean away from the shore, the water is quite shallow. Because light penetrates the water in this near-shore zone, a great many organisms live here. Much of the abundant sea life caught for food by humans inhabits the near-shore zone. Unfortunately, this important part of the ocean is often mismanaged by humans. The near-shore zone is where some of the greatest harm is being done to the ocean and the life it supports. Pollution, mostly from human causes, and overfishing are two major problems in this part of the ocean.

Finally, the great open areas of the ocean—where the water is tremendously deep—make up another zone. This zone has its own conditions. Because the bottom is far beyond the reach of sunlight, photosynthesis occurs only in organisms that float near the surface of deep water. These tiny photosynthetic floating microorganisms are called **phytoplankton**.

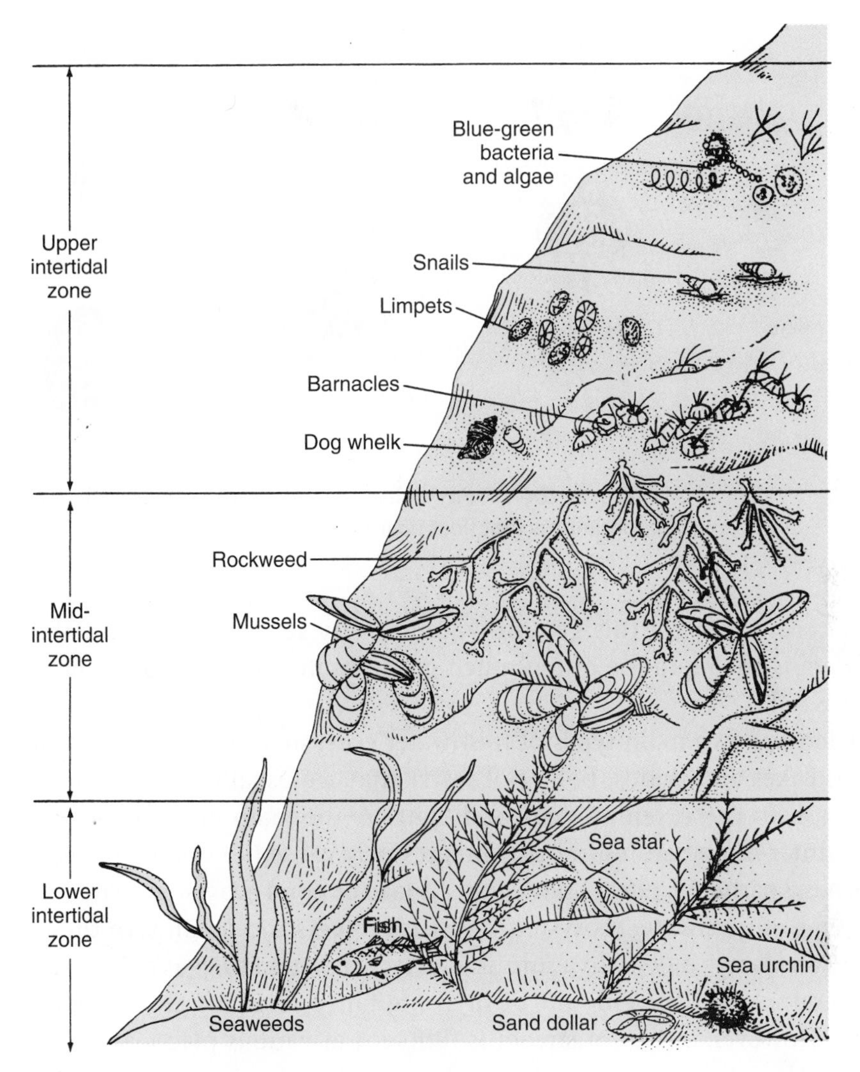

Figure 24-12 The intertidal zone is home to a special group of plants and animals.

Phytoplankton are an extremely important source of food for much of the life in the ocean. (See Figure 24-13 on page 524.)

FRESHWATER ECOSYSTEMS

Less than three percent of Earth's water is freshwater. Of this, two-thirds is frozen near the North and South Poles and at the tops of mountains. Of the liquid freshwater, most of it is stored in the ground. We must take

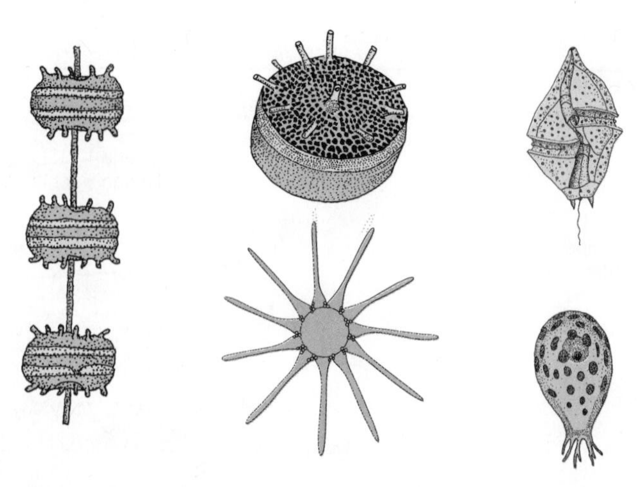

Figure 24-13 Phytoplankton are photosynthetic microorganisms that float near the surface of the ocean.

great care to protect this **groundwater**. A very serious environmental problem is the rapid use of too much groundwater and the polluting of what is left.

There are two main types of freshwater ecosystems on the surface of Earth. Lakes and ponds, bodies of water that are usually still, make up one type. In these, temperature and light are the main limiting factors. The winter temperatures in cold climates may cause the water to freeze. This has a particular effect on the life of a lake or pond ecosystem. The second type of freshwater ecosystem consists of rivers and streams. In these, the water always moves. In a running-water ecosystem such as a river, the temperature and the amount of light are fairly constant at any given point. However, a river or stream is different at various places along its length. Near its beginning high in the mountains, the river is young; it is cold, flowing swiftly, and usually clear. Near its end at the ocean, the river is old; the water has become warmer, moves more slowly, and is often muddy. Trout are powerful swimmers and are well adapted to the swift-moving water of young rivers and streams. Freshwater bass, on the other hand, live well among the thick growth of plants in an older, slower river.

LAND ECOSYSTEMS

For an interesting reason, life on land is much more diverse than life in water. Conditions in water are fairly constant over wide areas. Conditions in water also change little over time. When conditions do change in water, they change slowly.

Conditions on land are very different. From season to season, and from one part of the day to another, temperatures on land can vary widely. The temperature in a desert can be above 50°C during the day and drop to well below freezing at night. Conditions on land can change rapidly over relatively short distances, particularly when changes in altitude are involved. Life in a valley differs from life on either side of the valley as you move away from the valley floor.

Variations in temperature, amount of moisture, soil type, length of days and nights, and seasonal variations, along with altitude, operate together to produce different land ecosystems. Sometimes humans have found out the hard way just how very diverse land ecosystems can be. For example, people who have moved to new lands have tried to grow crops in the same way in their new home as they did in the place they came from. The failure of the crops to grow well made it very clear that too little was understood about the new environment. The study of ecology has many practical applications.

Ecologists use the idea of biomes to describe and study life in various land ecosystems. A land biome is a large area in which one main type of

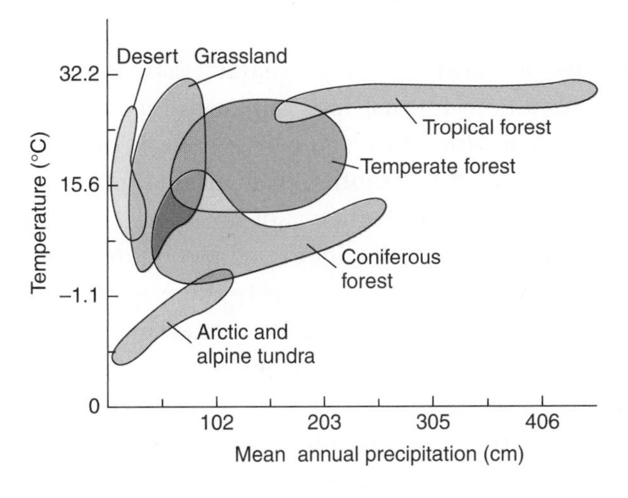

Figure 24-14 The temperature and moisture in a biome determine which plants will grow there.

plant lives. It is mainly a biome's climate—that is, its temperature and moisture—that determines which plants will grow. (See Figure 24-14.) Therefore, a map of land biomes looks approximately the same as a map of the main climate zones on Earth. Because animals depend on plants as a food source, a biome over one part of the world usually has the same or similar types of animals as that biome in another part of the world. Let's tour the main land biomes of Earth. (See Figure 24-15 on page 526.)

Trees are obviously an important type of plant on land. In fact, there

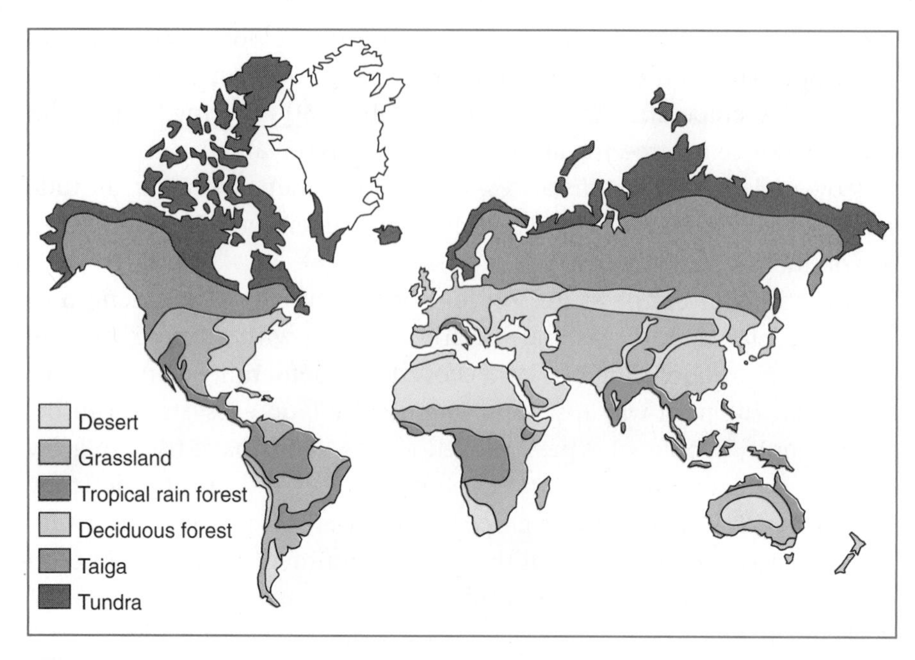

Figure 24-15 Land biomes of the world.

are three main forest biomes on Earth, each with characteristic types of trees: tropical rain forests, deciduous forests, and coniferous forests.

Tropical rain forests exist in a wide band between the Tropic of Cancer and the Tropic of Capricorn, north and south of Earth's equator. In the tropical area, there are more than 250 centimeters of rainfall each year. The warm temperatures and the hours of daylight remain roughly equal throughout the year. Abundant life of all kinds exists in the tropical rain forest biome. A small area in a tropical rain forest may contain more than 100 tree species, 300 orchid species, and thousands of species of animals, most of them insects. Monkeys, bats, squirrels, parrots, snakes, frogs, and lizards are just a few of the kinds of animals that inhabit this biome. Conditions in a tropical rain forest vary. At ground level, it is dark because little light penetrates the leaves of trees that grow overhead. Plants that live on the floor of a tropical rain forest are adapted to life under conditions of low light. Many houseplants, also adapted to low-light conditions, are originally from the tropical rain-forest floor. It is also moist on the floor. Higher up in the trees, it is cooler, windier, and brighter. (See Figure 24-16.)

Dead plants and animals decay very quickly in the warm, moist conditions on the rain-forest floor. However, rain quickly washes away most soil nutrients. The nutrients that remain are confined to the few top cen-

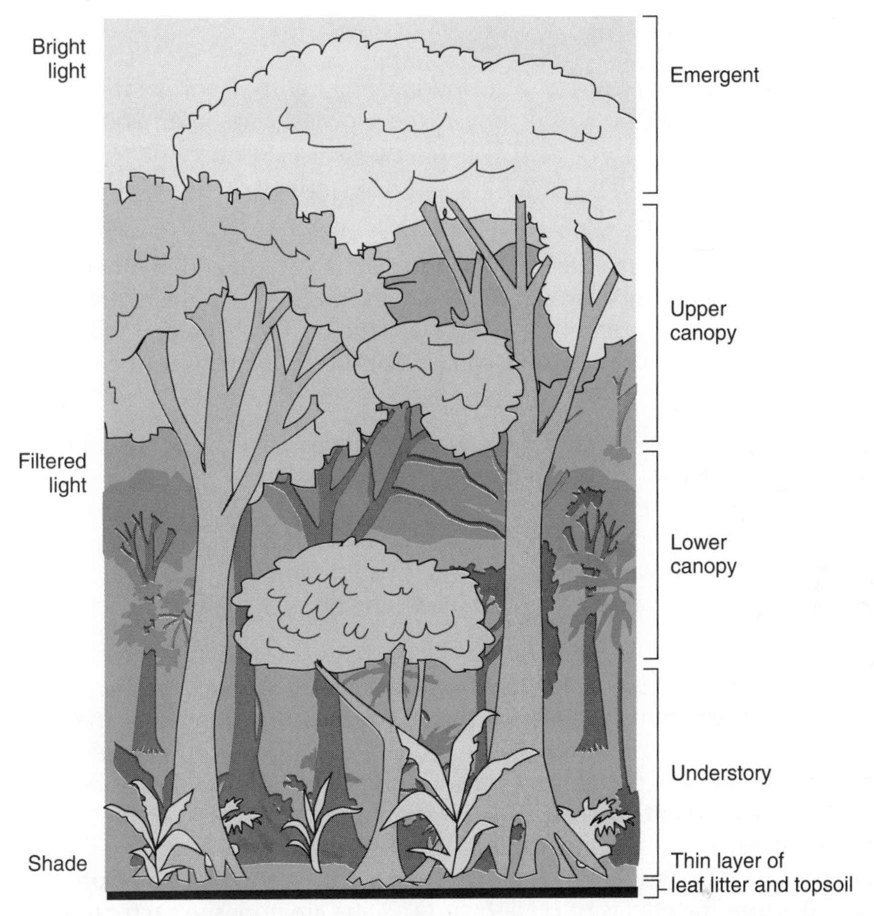

Figure 24-16 Conditions on the rain-forest floor are warm, moist, and dark; higher up in the trees, it is cooler, windier, and brighter.

timeters of soil. As a result, tropical rain-forest soils are poor soils. Farmland created by destroying tropical rain forests quickly becomes useless as the nutrients are used up. As a result, more tropical rain forests are being destroyed to make more farmland. This ever-increasing need for farmland to grow crops is the main reason tropical rain forests—along with the remarkable diversity of life they contain—are destroyed.

Farther north and south of the equator, climate patterns change. Definite seasons, from hot wet summers to cold drier winters, occur. The seasonal nature of these areas creates the deciduous forest biome. Here, fewer tree species are found, although there are many individual members of the same tree species. More light reaches the forest floor where a wide variety of plants live. Deciduous trees, such as maple, beech, hickory, and oak, lose their leaves each autumn. This seasonal dropping of leaves is

El Niño and La Niña

Will the risk of a major hurricane in Florida be higher than normal this summer? Will Buffalo, New York, one of the snowiest cities in America, have another winter with no snowfall? Will intense Pacific storms cause destructive mudslides that bring soil and rocks crashing down hillsides along the California coast?

Questions like these have been asked in the United States and in other countries for years. The answers to these questions focus on the effects produced by two atmospheric events: the "little boy" El Niño and the "little girl" La Niña. Both El Niño and La Niña events occur regularly in the Pacific Ocean.

El Niño is a periodic warming of the water in the eastern Pacific Ocean near the equator. La Niña—sometimes called the cooler sister—is marked by cool waters near the equator in the Pacific. Both of these events change the atmosphere above the ocean. A change in the wind causes a change in the movement of the ocean's surface waters. Changes in water temperature change the amount of moisture that leaves the ocean's surface and enters the atmosphere.

El Niño events tend to transfer heat from the ocean to the atmosphere. This increases the strength and speed of the jet stream—the high winds that move from west to east across the United States—and contributes to relatively predictable climatic patterns. During La Niña, the colder Pacific water near the equator weakens the jet stream that helps to keep our weather constant. As a result, La Niña events cause unusual weather patterns in the United States. El Niño events occur every three to seven years. Because the effects of El Niño and La Niña are opposite each other, we are always living in one event and heading toward the other. Without a doubt, we will continue to learn more about how changes in water temperatures in the Pacific Ocean are related to changing weather patterns that occur in distant parts of the world and right at home.

typical of trees in these areas. Dropping their leaves prevents trees from losing water during cold weather. Deciduous forest animals include wolves, gray foxes, deer, raccoons, squirrels, birds, and insects. The soil in temperate deciduous forests is much richer than the soil in a tropical rain forest.

Farther north of the equator are the great northern coniferous forests of North America, Asia, and Europe. The main plants in these forests are the needle-leaved evergreen trees, such as spruce, pine, fir, hemlock, and cedar. These are called conifers, or coniferous trees, because they produce their seeds in cones. In a coniferous forest, the winter is long; the warmer growing season is short. Because of cold temperatures over much of the year, little decay occurs. These forest floors develop a thick layer of dead needles.

Farthest from the equator, the temperatures are too low, and the amount of precipitation too small, for trees to grow. Only one or two months of the year are warm enough to support plant growth. The biome here is the tundra. Only the upper layer of the soil warms during the summer. Below this layer, the soil remains permanently frozen all year. A thick layer of plants, such as mosses, lichens, grasses, and low shrubs, covers the surface. Although life in the tundra is harsh, more animals live there than you might expect. These include caribou, arctic foxes, snowy owls, great numbers of insects, and migrating geese and ducks during the summer.

So far, in our tour of land biomes, we have been following decreasing temperatures as we moved away from the equator. If we explored climates where precipitation—but not temperature—decreases, we would first find ourselves in grasslands. Grasslands are usually found in the interiors of continents, where limited moisture prevents the growth of trees. With less than 100 centimeters of rain per year, only grasses can grow. In different places in the world, grasslands have different names. Grasslands are called prairies or plains in North America, steppes in Russia, savanna in parts of Africa, veldt in South Africa, and pampas in South America. The rich soil that has slowly developed in grasslands has created some of the best agricultural land on Earth. Grasslands produce most of the world's grain crops. Because of this, grasslands are often called the "world's breadbasket." (See Figure 24-17.)

Figure 24-17 Wildebeest migrate across the grasslands, or savanna, of Africa in search of food.

Large animals that eat grasses are native to grasslands. These include bison and antelope. In many areas, domestic cattle and sheep have replaced the native grazing animals. Grasshoppers, prairie dogs and other small rodents, hawks, and snakes are also typical grassland animals.

Finally, our tour brings us to the driest places on Earth, which produce the desert biomes. In a typical desert, temperatures may be very high during the day. However, because of limited water vapor in the air, the heat accumulated during the day is quickly lost at night. As a result, temperatures fall dramatically at night. Plants include water-storing cactuses, shrubs with roots that grow deep underground to reach water, and wildflowers and grasses that grow and mature in the short period of time after the rare rain. Animals also show adaptations in this area of limited water availability. The kangaroo rat lives underground for much of the day. It gets its water from the tissues of green plants and by breaking down the fat in dry seeds that it eats. Predators such as coyotes and foxes are often active at night to avoid and survive the harsh desert conditions.

The table below lists the six major land biomes and their main features.

THE LAND BIOMES OF EARTH

Biome	Features/Conditions
Tropical rain forests	Found just north and south of equator; have more than 250 cm rain per year; warm temperatures and stable day length all year; abundant and diverse forms of plant of animal life.
Deciduous (temperate) forests	Farther north and south of equator; definite seasons, from hot wet summers to cold drier winters; compared to tropical rain forests, have fewer tree and animal species, but richer soil.
Coniferous forests	Farther north than deciduous forest; mostly needle-leaved evergreen trees; winters are longer; growing season shorter; very little decay of leaf litter (due to cold).
Arctic & alpine tundra	Farthest north; temperatures and rainfall too low for trees; brief growing season for mosses, lichens, grasses, shrubs; have resident and migratory bird and mammal species.
Grasslands	In interior of continents; warm temperatures, but less than 100 cm rain per year; have rich soil; support large grazing animals (antelope); produce most of world's grain crops.
Deserts	Driest places (lowest rainfall); temperatures high in day, low at night; have plants that store water (cactus), and animals that stay underground during day and are active at night.

EXTENDED LABORATORY INVESTIGATION 24
What Can We Learn by "Adopting" a Tree?

INTRODUCTION

Observations of a single organism in its environment over an extended time period can provide important information. You can learn about the relationship of an organism to the physical conditions in its surroundings. Common organisms in our environment are trees. Trees have an amazing variety of ways to carry out such life functions as producing food and producing more trees. In many parts of the United States, the best time to observe trees is during the spring. At this time, deciduous trees come out of their dormant winter state and begin active growth and reproduction. In this investigation, you will "adopt" a deciduous tree and make observations during the spring growing season. You can write your observations in your science journal. (No laboratory materials required.)

PROCEDURE

1. During February or March, select a deciduous tree that you will be able to observe daily. No leaves should be present at this time. Your tree should have at least one branch close to the ground so that you can observe the ends of the branches.
2. Begin a journal for your periodic observations. Describe the location of your tree, including specific geographic references so that another person reading your journal can locate your tree.
3. Describe the appearance of your tree as fully as possible. You can add drawings or photographs to your description. In your written description include:
    - overall appearance and shape
    - the approximate size of your tree: height, spread of branches
    - bark color and texture
    - trunk diameter: make this measurement a standard distance from the ground; 1.5 meters is a good height at which to measure the diameter
    - kind of branching
    - appearance of buds and twigs

4. Begin close-up observations of the buds and twigs on your tree on a regular basis. Keep daily journal entries when you first observe changes in the buds and twigs.
5. Throughout your observation period, keep records of the daily high and low temperatures as well as the sunrise and sunset times. Graph these data.
6. Write a complete description of your tree as it appears at the end of May.

INTERPRETIVE QUESTIONS

1. In the library, study the sexual reproduction of trees. Explain how your observations relate to this process.
2. Research the species of trees that commonly grow in your area. What features are usually used to identify a species? To what genus and species does your tree belong?
3. Relate the changes you observed on your tree to the changes in temperature and daylight.
4. What have you learned about the process of observation by keeping your tree journal?

CHAPTER 24 REVIEW

Answer these questions on a separate sheet of paper.

VOCABULARY

The following list contains all of the boldfaced terms in this chapter. Define each of these terms in your own words.

abiotic factors, adaptations, biome, biotic factors, coniferous, deciduous, ecosystem, groundwater, habitat, limiting factors, niche, phytoplankton, tides

PART A—MULTIPLE CHOICE

Choose the response that best completes the sentence or answers the question.

1. The interactions among living things and between living things and their nonliving environment are the focus of the area of biology known as *a.* system dynamics *b.* cryptozoology *c.* herpetology *d.* ecology.
2. An example of an abiotic factor in a pond ecosystem is *a.* floating duckweed *b.* a dragonfly *c.* mud at the bottom *d.* a school of fish.
3. An adaptation for life in the desert might be *a.* large, flat leaves with many stomates *b.* a fleshy, water-storing stem *c.* long, sharp thorns on branches *d.* shedding leaves in winter.
4. Together, the living and nonliving factors in a particular place, such as a pond or field, are called *a.* a biome *b.* abiotic factors *c.* an ecosystem *d.* a population.
5. The land biome that is characterized by a warm, wet climate and the greatest number of species of plants and animals is the *a.* tropical rain forest *b.* deciduous forest *c.* coniferous forest *d.* tundra.
6. The insects called walking sticks look exactly as their name suggests. Factors in the walking stick's environment might include *a.* insect eaters that use sight to find their prey *b.* cactuses that have a single large, barrel-shaped stem *c.* brightly colored flowers *d.* plants that are poisonous to plant eaters.
7. An organism's "address," or where it lives, is best described as its *a.* biome *b.* habitat *c.* ecosystem *d.* niche.
8. An organism's "occupation," or what it does, is best described as its *a.* habitat *b.* ecosystem *c.* niche *d.* biome.

9. In a desert, the limiting factor for most plants is probably *a.* temperature *b.* sunlight *c.* nutrients in the soil *d.* water.
10. Trees that shed their leaves in winter are *a.* coniferous *b.* crepuscular *c.* benthic *d.* deciduous.
11. Approximately how much of Earth's surface is covered by water? *a.* 25 percent *b.* 50 percent *c.* 75 percent *d.* 99 percent
12. A very large area characterized by a particular climate and set of plants and animals is *a.* an ecosystem *b.* a niche *c.* a habitat *d.* a biome.
13. In the marine biome, photosynthesis takes place primarily in *a.* deciduous plants *b.* phytoplankton *c.* coniferous trees *d.* grasses.
14. Most of the world's grain crops are produced in *a.* deciduous forests *b.* grasslands *c.* coniferous forests *d.* tropical rain forests.
15. The biome closest to Earth's poles is *a.* deciduous forest *b.* grassland *c.* tundra *d.* coniferous forest.

PART B—CONSTRUCTED RESPONSE

Use the information in the chapter to respond to these items.

16. For each of the following descriptions, identify the area of the map that most closely fits that description.
    a. "Moose, goose, spruce"
    b. Fields of wheat where bison once roamed.
    c. Snowy owls hunt mice among the lichens on frozen subsoil.
    d. Monkeys and toucans move in a warm, humid canopy of leaves.

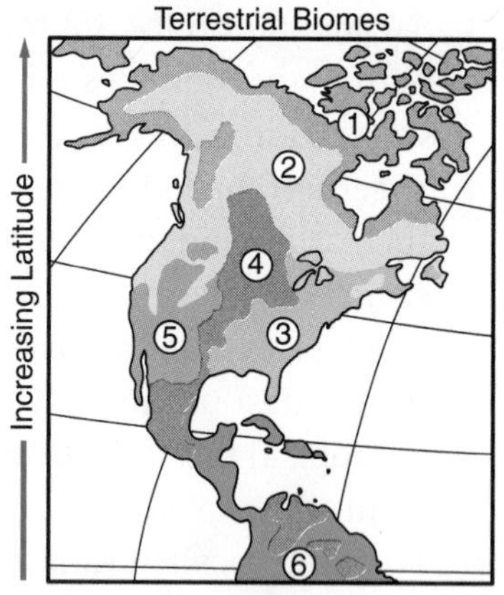

17. Identify the biomes in areas 3, 4, and 5. Explain why these areas are so different from one another.
18. Distinguish between the two types of freshwater ecosystems. Explain what effect damming rivers has on these ecosystems.
19. Describe the three major areas of the marine biome. Which area is most strongly affected by tides?
20. Select an organism that lives in your environment. Make a list of at least five biotic and five abiotic factors that affect the organism.

PART C—READING COMPREHENSION

Base your answers to questions 21 through 23 on the information below and on your knowledge of biology. Source: *Science News* (October 19, 2002): vol. 162, p. 253.

Tests Revise Image of Kangaroo Rats

The textbook case of how to survive in a desert may have important details wrong, according to new studies of kangaroo rats.

Species in the genus *Dipodomys,* nocturnal rodents that scurry through North America's deserts, have epitomized toughness in punishing climates, says Randall Tracy of the University of Connecticut in Storrs. Earlier researchers, he says, marveled at how the creatures apparently got water by metabolizing seeds and avoided overheating by staying in cool burrows until late at night.

In an upcoming issue of *Oecologia,* Tracy and Glenn E. Walsberg from Arizona State University in Tempe challenge those views. They made their observations near Yuma, Ariz., in the Sonoran desert.

The animals' burrows get hotter than expected, the researchers found. For more than 100 days of the year, soil temperatures rose to over 30°C at depths of 2 meters. Yet during most of the summer, the kangaroo rats remained less than a meter deep, where it's about 35°C. Nor did the animals emerge only in the cool part of night; they ventured above ground right after sundown. Also, forget the seeds-only menu. The rodents ate a considerable amount of green plant tissue, presumably a substantial water source during tough times.

21. Explain the two methods that scientists had first thought kangaroo rats used to survive in the desert.
22. State the research findings that challenge the traditional explanation of how kangaroo rats stay cool.
23. Explain why scientists now doubt that kangaroo rats get their water from the metabolism of seeds.

Populations and Communities

After you have finished reading this chapter, you should be able to:

Differentiate between a population and a community.

Identify several types of symbiotic relationships.

Define succession and give an example of one kind of succession.

There is no exception to the rule that every organic being naturally increases at so high a rate, that, if not destroyed, the Earth would soon be covered by the progeny [offspring] of a single pair.

Charles Darwin, On the Origin of Species

Introduction

Article I, Section 2, of the Constitution states that every person in the United States is to be counted every 10 years. The Constitution indicates that this counting, or census, determines how many representatives will be sent from each state to Congress. "The number of representatives shall not exceed one for every thirty thousand."

A census is a way of determining the size of the human population in a country at a particular moment in time. In many ways, a census has a lot to do with ecology. Much of the work done by scientists who study ecology is about numbers. Ecologists are more interested in groups of organisms than in individuals. All the organisms of one species that live in one place at a particular time make up a **population**. No population ever lives alone. Other organisms—plants, animals, fungi, and microorganisms—are also present. For example, a field has a population of mice, but populations of wildflowers, insects, mushrooms, birds, and snakes also live in the field. All of the populations that interact with each other in a particular place make up a **community**. In large part, the study of ecology is about populations and communities. (See Figure 25-1.)

Figure 25-1 The bison population and the pronghorn antelope population interact with each other and with populations of plants and other organisms in the area to make up a community.

To begin our study of populations, it is helpful to think about the census again. The data from Census 2000 are being used to answer questions concerning how our government will work. Some of the questions that the census information is answering are:

- How many people live here now?
- How is the size of the population changing?
- Why are people moving from one area of the country to another?

As you will see, similar questions and answers are related to our study of ecology.

PROPERTIES OF A POPULATION

As you study populations, it is necessary to pay attention to some different ideas from when you studied the structure and function of individual organisms. These ideas are called properties of populations, and they apply only to populations.

Density is one important property of a population. If we want to determine the density of a population, we must know the size of the population in a given area. For example, if 100 rabbits live within a 1-square-kilometer area, the density of rabbits here can be considered low. However, if 100 rabbits live in a small suburban backyard, the density of rabbits is considered high. The census conducted in 2000 counted 281,400,000 people in the United States. If all the people counted were evenly spaced across the

Figure 25-2 You can visualize the population density of New York City from this photograph taken at night by the crew of a space shuttle. The brighter areas are those where population density is higher.

United States, this would work out to a population density of 30.7 persons per square kilometer, a relatively low density. In this case, however, population density is more meaningful if we examine the density in a specific area. For example, in Alaska, the population density is less than 0.4 person per square kilometer, while in Manhattan, New York, population density is greater than 25,800 people per square kilometer. In nature, many environmental factors affect population density. Some of these factors are climate, availability of food, the number of predators, and disease. (See Figure 25-2.)

Ecologists also study the **distribution** of individuals in a population throughout the habitat. One pattern of distribution is called clumped. Elephants are usually clumped; they live in herds. Daffodils grow in clumps, as do palm trees. Members of many other populations are spread out evenly throughout a habitat. This is called uniform distribution. Dandelions grow in a uniform pattern in a field, as do oak trees. Individual plants of a particular species in these populations grow roughly equidistant from each other. Tigers do not live in herds. Their distribution is uniform. Uniform distribution occurs when individuals in a population interfere in some ways with other individuals of the same species. For example, they may compete for food. This kind of competition keeps them from living too close together. (See Figure 25-3.) Finally, individu-

Figure 25-3 The distribution of tigers in their habitat is uniform; competition keeps them from living too close together.

als in a population may appear to be distributed randomly in a habitat. Sometimes they live in groups, sometimes they live by themselves. This happens when environmental factors do not favor one part of the habitat and when individuals have little or no effect on each other. Obviously, random distribution hardly ever occurs in nature.

Another property of a population is **growth rate**. This is the increase or decrease in the density of a population over a specific period of time. Population densities increase when the number of births and individuals moving into an area (immigration) is greater than the number of deaths and individuals leaving the area.

FACTORS THAT AFFECT POPULATION GROWTH

Most organisms are able to reproduce rapidly. Even if a pair of individuals produces only two offspring each year, the growth rate is enormous, especially if all offspring survive. Suppose a population was founded by two individuals. By doubling each year, after only ten years the two original individuals would have created a population of 1024 individuals. (See Figure 25-4 on page 540.) Many organisms produce even greater numbers of individuals. Some fish are able to produce thousands of eggs per year. If these eggs were all fertilized and survived to reproduce, the number of resulting fish would be astronomical. Trees produce many thousands of seeds per year. Female rats and mice produce about 24 offspring each year. As you can see, with these growth rates, Earth would quickly be overcrowded with organisms. Of course, this does not occur. An important area in ecology is the study of what controls population growth.

When a population's density is low, there is usually sufficient food and space for existing organisms. The birth rate increases, while the death rate drops. The density of the population begins to increase at a faster rate. Can this rate of increase last forever? We know that the answer is *no*. At some point, the population gets crowded.

Data:

Year	1	2	3	4	5	6	7	8	9	10
Number of Individuals	2	4	8	16	32	64	128	256	512	1024

Graph:

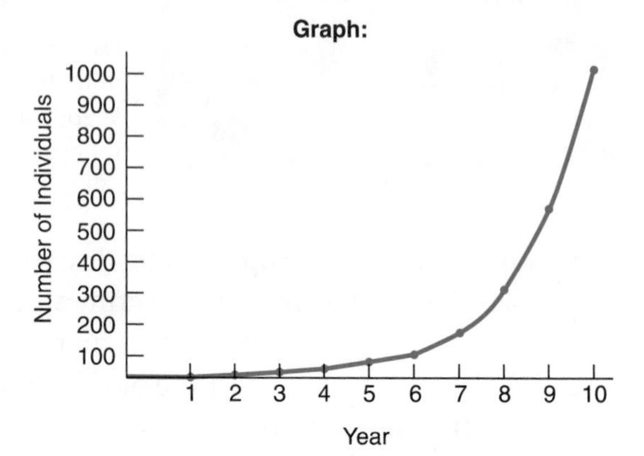

Figure 25-4 The data table and graph show population growth when doubling occurs every year.

The most basic needs of organisms from their habitats are food and space. However, every habitat has limits. When population density increases too much, the available food and space decrease. More organisms die, and the birth rate drops. The population density has reached a maximum for the particular habitat. The size of a population that can be supported by an ecosystem is called the **carrying capacity**. Population growth slows and may reach a stage of zero growth as the population density comes close to an area's carrying capacity. Zero growth means that the size of the population is no longer increasing—the birth rate and death rate are about equal. (See Figure 25-5.)

The rate at which a population grows can be shown in a graph. At the beginning, the population is small. Because there are few individuals present, the growth rate of the population is slow. As the population increases, the rate at which it grows begins to speed up as more individuals reproduce. After some time, however, the growth slows and finally reaches zero growth as the population approaches the area's carrying capacity. This population growth curve is known as an S-shaped curve.

Factors that limit the size of a population were described in Chapter 24. Some of these factors become more important as population density increases. These are called **density-dependent** factors. Obviously, competition for food and space increases as population density increases. For example, owls usually build nests in tall dead trees. As the number of owls in an area increases, the availability of tall nesting trees decreases. Some owls will have to nest in shorter trees. In a shorter tree, a predator may more easily find the owl's nest.

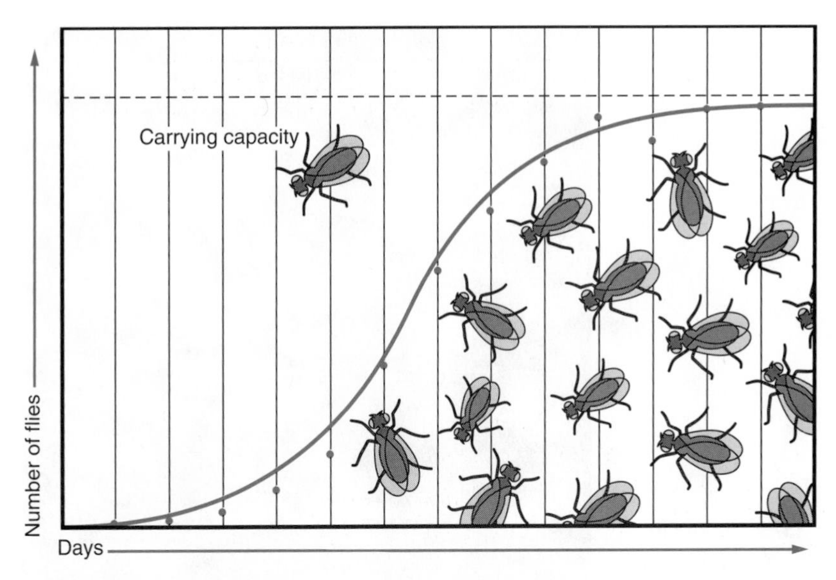

Figure 25-5 A new population increases rapidly at first, then levels off to zero growth when it reaches the carrying capacity.

The more crowded a population is, the easier it is for predators to find prey. In a crowded population, there is also a greater chance for disease to spread among individuals. In addition, ecologists determined that greater stress occurs in individuals in an overcrowded population. For example, when rats are kept in crowded conditions, they attack each other, develop irregular mating behaviors, and males have lower sperm counts. Along with disease, competition, and predation, stress has been included as a density-dependent limiting factor. Some scientists have suggested that stress may cause problems in the human population as the density of our own species increases. (See Figure 25-6 on page 542.)

Natural disasters can decrease population size. Earthquakes, hurricanes, extremely cold winters, and periods of drought can quickly kill large numbers of individuals. It does not matter if the population density was high or low before the disaster occurred. These types of events are known as **density-independent** factors. The growth curve for a population experiencing density-independent limiting factors is not S-shaped. Instead, the population density falls rapidly after a disaster and then may grow rapidly again. These kinds of sudden population changes may occur after each disastrous event.

In addition to individual events, it is also possible that a series of density-independent events may affect a population. For example, an event such as a drought might be followed by an extremely cold winter. Perhaps a number of deer had starved after the drought. Then, without

Figure 25-6 People in densely populated areas may be subject to stressful conditions, such as traffic jams.

sufficient time for the population of deer to increase, the cold winter killed the remaining deer. Now the population size is zero. No more deer exist in the habitat. If this was the last remaining population of those deer, the species would no longer exist. Extinction would have occurred. The deer would never be seen alive again. (See Figure 25-7.)

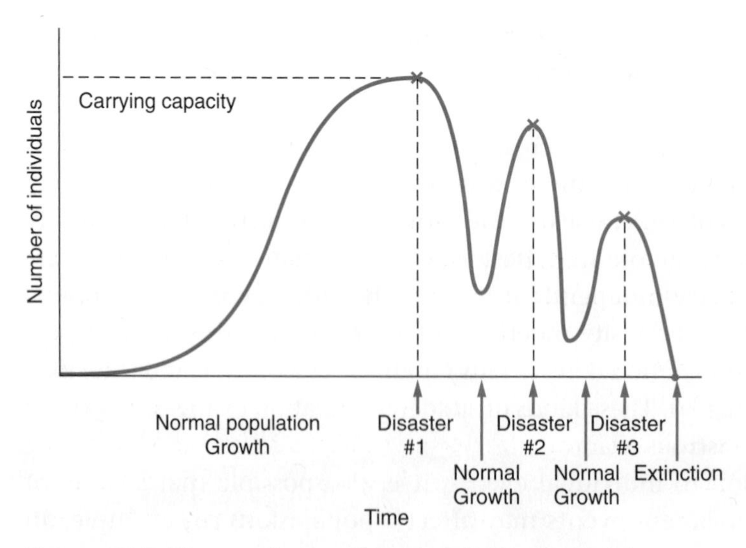

Figure 25-7 The growth curve for a population experiencing density-independent limiting factors is not S-shaped.

COMMUNITY INTERACTIONS

In 1624, the poet John Donne wrote, "No man is an island, entire of itself; every man is a piece of the continent, a part of the main." The poet's words are actually one of the main ideas of ecology. No organism lives alone. No population lives alone. Organisms that live in the same environment are always interacting with each other. The survival of many plants and animals often depends on the relationships they have with other organisms. We will examine some of the main types of these relationships now. (See Figure 25-8.)

Competition is one of the main interactions between organisms. Suppose a population of small brown birds lived in a forest. These birds eat yellow butterflies found at the tops of trees. If a population of small blue

Figure 25-8 The survival of many plants and animals depends on the relationships they have with other organisms in their environment.

birds lived in the same forest and ate the same yellow butterflies from the treetops, there would be competition between the two types of birds. This competition between two species is **interspecific competition**, and it would be intense, in this case, because of the overlap in the birds' niches. However, if the two bird species fed in the same treetops on different insects, there would be less overlap in their niches and less competition between the two species.

The greatest competition of all occurs between members of the same species. This is called **intraspecific competition**. Intraspecific competition occurs because individuals of the same species most likely share identical niches. In other words, they live in exactly the same way. Therefore, they compete most intensely for the limited resources in the environment. Both intraspecific and interspecific competition result in natural

Figure 25-9 This female lion—a predator—competes with other lions for food.

selection. Only the most fit individuals survive and pass on their genes to their offspring. Therefore, competition is an important force in the process of evolution. (See Figure 25-9.)

Predation is another of the most basic relationships that occur in nature. In many cases, as an individual member of a population, you either "eat or are eaten." That is, you are either a predator or the prey. In predation, members of one population are the food source for members of another population. It may seem a strange way to think about it, but grass seeds are prey when a mouse—a predator—eats them. Of course the mouse can become the prey when a snake, another predator, eats it. (See Figures 25-10a and 25-10b.)

Figure 25-10a The mouse preys on plants and seeds.

Figure 25-10b The rattlesnake preys on the mouse.

In a predator-prey relationship, one individual in a population is benefited while another individual is harmed. The density of the prey population often determines the density of the predator population. The lynx, a large wild cat, preys on snowshoe hares, a relative of the rabbit. A study that lasted for more than 90 years counted the population of lynx and the population of hares in a particular area. The scientists noticed that each time the number of hares increased, after a short time, the number of lynx increased also. More hares meant more food that could support a greater population of lynx. However, when the number of predators increased, the number of hares then decreased. Because the hares were the food source for the lynx, the predator population size then decreased. A pattern occurred again and again as the numbers of prey and predator alternately rose and fell together. (See Figure 25-11.)

Some of the most wonderful adaptations have evolved as variations have been selected for by nature to help members of a species avoid becoming prey. The woodcock, a brownish-red bird, uses camouflage to avoid being seen. If you look at a pile of brownish-red leaves on a forest

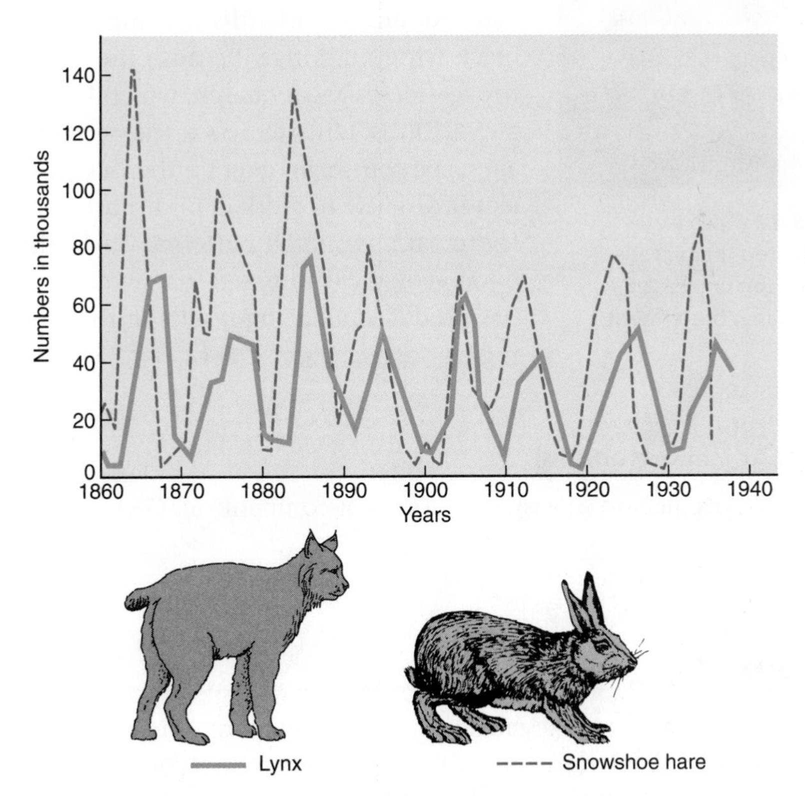

Figure 25-11 The number of hares determines the number of lynx in a cyclical pattern.

floor, you might not be able to see the woodcock lying on top of the leaves. By keeping a low profile, and with good camouflage, the woodcock might not be seen by a passing predator.

Other organisms have different types of adaptations that help them avoid becoming prey. For example, the bombardier beetle uses a chemical-attack method. It produces an irritating chemical that it sprays into the mouth of any animal that tries to eat it. This annoying chemical confuses the predator and gives the beetle time to escape.

Another defense adaptation makes use of disguises that allow an organism to pretend to be something it is not. The brilliantly colored orange-and-black monarch butterfly has a bitter taste. A bird that tries to eat a monarch butterfly will spit it out in disgust. The viceroy, another type of butterfly, tastes just fine, and birds would be glad to eat it. However, they leave this tasty butterfly alone. Why? The viceroy resembles a monarch. The viceroy uses a technique called **mimicry**. As a mimic of the monarch butterfly, it has evolved almost identical orange-and-black wing markings. Because the viceroy resembles the monarch, which tastes awful, birds tend to leave the viceroy alone. Of course, in order for the viceroy's clever disguise to work, a bird must first try to eat a monarch butterfly! Through basic ecological interactions, evolution has produced amazing results in the living world. (See Figure 25-12.)

Figure 25-12 The viceroy butterfly, at top, mimics the markings of the unpleasant-tasting monarch butterfly, at bottom.

Check Your Understanding

Define interspecific and intraspecific forms of competition. Give an example of each type.

SYMBIOSIS

As we have said, almost no organism lives entirely alone. In fact, most organisms have close relationships with at least one other type of organism. This close relationship is called **symbiosis**. In this relationship, one type of organism can live near, on, or even in another organism. Each partner in the relationship can help the other, harm the other, or have no effect on the other partner.

TYPES OF SYMBIOTIC RELATIONSHIPS

Parasitism. A **parasite** lives on or in another organism that it uses for food and, sometimes, for shelter. The organism the parasite uses is called the host. In this type of interaction, one organism is helped while the other is harmed. This relationship is different from the predator-prey relationship. In that relationship, the prey is usually killed. In **parasitism**, the host organism usually continues to live, but it is harmed. Parasites usually evolve together with their host. Parasites have characteristics that make them specifically adapted to a particular host. For example, the human tapeworm is a parasite that can live inside the human intestines. It has no eyes, no muscles, and no digestive system. It absorbs nutrients through its outer surface. The outer surface is tough enough to withstand the enzymes in the intestine. The tapeworm has hooks on its head, which it uses to attach itself to the inside of the intestine. A single tapeworm has both male and female reproductive organs. It can reproduce by itself without even having to find another tapeworm.

The tapeworm is a parasite that is large enough to be easily noticed. However, every organism seems to host some parasites, many too small to be seen. All plants and animals have other organisms that live in them and on them. These parasites include viruses, bacteria, fungi, worms, fleas, and lice. A single bird may be the host of up to 20 different kinds of parasites. (See Figure 25-13 on page 548.) Even parasites have parasites. For example, a leech that attaches to and sucks blood from its host animal may have leeches that live inside of it.

Parasitism is a natural result of the process of evolution. Over time, parasites have developed adaptations that allow them to take advantage of other organisms as both habitats and food sources.

Commensalism. In some symbiotic relationships, one organism benefits while the other organism remains unaffected. This is called **commensalism**. Many examples of this type of relationship exist. When the Cape buffalo—a wild African herbivore—walks through grass, it disturbs insects. Cattle egrets, beautiful white birds, gather around the buffalo and feed on the insects. The birds are helped, and the Cape buffalo is unaffected.

In the ocean, the remora—a small fish—attaches itself near the mouth

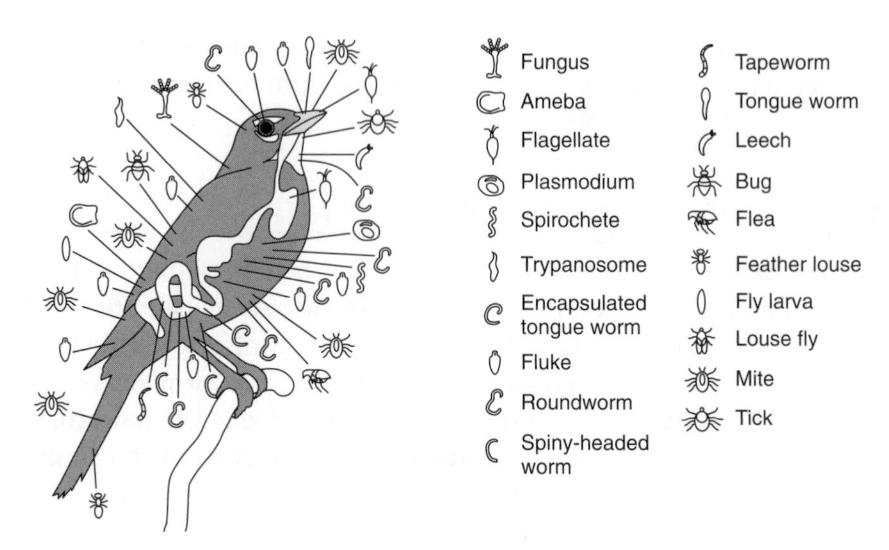

Figure 25-13 One bird may be host to many different kinds of parasites. Some of those parasites may have parasites of their own as well.

of a shark. It feeds on the leftovers from the shark's meals. The remora benefits; the shark, as far as we know, is unaffected. In the rain forest, epiphytic plants, such as some bromeliads, attach themselves to the bark of trees far above the ground. Living up high, these plants are able to capture more of the sun's light while not affecting the tree they are growing on. (See Figure 25-14.)

Mutualism. Finally, there are symbiotic relationships where both parties benefit. These kinds of interactions are called **mutualism**. A fascinating example of mutualism is the relationship between the South

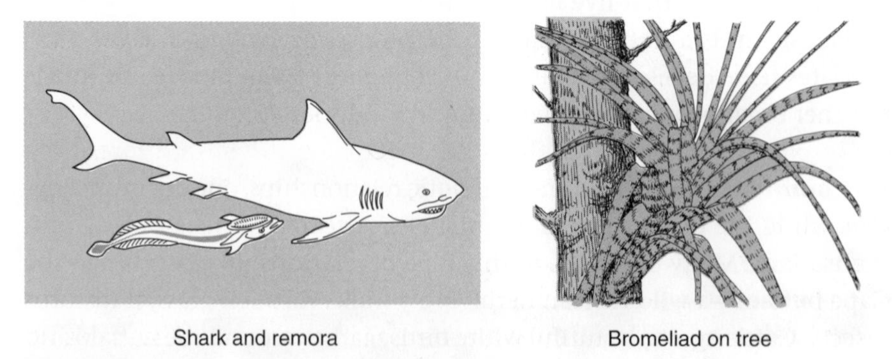

Figure 25-14 The relationship between the shark and remora, as well as that of the bromeliad and tree, are examples of commensalism.

American acacia tree and a species of stinging ants. The trees produce hollow thorns. The ants make their nest in these thorns and feed on sugars produced by the plant. If any other insect lands on the acacia, the ants quickly surround and kill it. The ants also cut down competing plants that grow near the tree. In one experiment, scientists removed all the ants that lived on several trees. Without the ants, after a period of time, the trees died.

The living world contains many examples of mutualism: the bacteria that live in our own intestines and make some of the vitamins we use; the microorganisms that live in a cow's intestines and help the cow digest the cellulose in the grass it eats; insects that collect food from flowers and unknowingly pick up pollen and pollinate the flowers as they move among them. (See Figure 25-15.)

Figure 25-15 The relationship of the cow and the bacteria in its intestines is an example of mutualism.

Remember that these relationships are the results of evolution. Species have been extremely successful at finding ways to gain an advantage over others. Often this advantage comes from developing a close relationship with another species, which also benefits.

CHANGING COMMUNITIES

A complex variety of interactions exist within an ecosystem. Many different populations live together and affect each other. In a forest, for example, there are grasses, shrubs, trees, fungi, insects, worms, bacteria, birds, reptiles, and mammals. All these populations make up the community in the forest. Will this community remain the same over time? It's not very likely. Some populations of organisms may disappear entirely while other populations may move in from somewhere else. Existing populations may increase or decrease in number. Often, the community changes with the changing seasons. Despite these kinds of changes,

The Mystery of the Baby Elephants

Once, the relatives of elephants roamed the grasslands of the United States. Today, Asia and Africa are the only two continents where elephants freely roam. Asian and African elephants are separate breeding populations and differ enough from each other to be considered two distinct species. African elephants are generally larger in size, have larger ears, and have a different tip on their trunk. If history is correct, African elephants were domesticated in Roman times, when they were used to carry materials and to act as a kind of living battle "tank." Today, however, African elephants are considered to be truculent and are rarely domesticated. Occasionally, an African elephant is part of a circus act and is taught tricks for the amusement of patrons. Asian elephants are considered to be gentler and easier to train than African elephants. In fact, Asian elephants often form close bonds with their trainers. In Asia, the elephants are trained to carry lumber and other heavy materials. Obviously, in nature, Asian and African elephants never meet. However, in zoos and in circuses, these animals are brought together. Sometimes they even share the same enclosures, with—it is now known—often tragic results.

In the early 1980s, a baby Asian elephant was born in the Bronx Zoo. It was heralded as the first elephant born in New York in more than 10,000 years, and its charming antics made it a crowd favorite. But a year and a half later, it died of unknown causes. (See Figure 25-16.)

In the 1990s, another baby Asian elephant was born at the National Zoo in Washington, DC. Like the New York baby, this elephant also died at about the same age. Several years later, scientists learned that the Washington baby elephant died after being infected with a herpesvirus. Later, tests of its preserved tissues revealed that the New York baby elephant had also died of a herpesvirus infection.

however, the forest—especially a very old forest—may remain essentially the same over many years. It has become a stable community.

And then a sudden, profound change occurs. For example, a fire destroys much of the existing plants. Some animals die, while others are able to flee the flames. What happens in the forest now? Scientists have learned a lot by observing areas that fires have destroyed. One interesting example that has been well studied occurred in Yellowstone Park in Montana. The fires that burned there from June through September 1988 affected more than 1.6 million acres in Yellowstone Park and in surrounding forests. The fires cost $120 million and required 10,000 firefighters to battle them.

The virus that killed the baby Asian elephants is commonly found in African elephants. In these animals, the virus produces only a mild skin infection and sores. However, when the virus infects Asian elephants, especially young Asian elephants, it can produce deadly results. The reverse can also happen. It is now thought that a similar virus is found in Asian elephants, but it does not kill them. However, when this virus enters an African elephant, it may cause death. By 2002, more than 20 baby elephants in the United States had died due to infection by a virus.

Figure 25-16 A young Asian elephant and its mother in a zoo.

Yet there is hope for infected baby elephants. One of the earliest symptoms of infection is a purple tongue. If this symptom is noted, treatment with the human antiviral drug famciclovir can be started. This drug can save animals' lives if it is administered early enough. Scientists also hope to develop tests to identify elephants that carry the virus and do not become ill. It is thought that these elephants could pass the virus on to other, healthy elephants. In time, scientists hope to develop a vaccine to prevent infection, but that goal is still far in the future.

Did these tremendous fires destroy life in the most famous national park in the United States? Definitely not. What scientists observed is the process of **succession**. Usually, succession is a series of slow changes that occur in an area until a stable community is reached. It takes a long time. Most naturally occurring successions in an area take longer than a person's lifetime. However, a stable community after a sudden disturbance quickly begins to go through a series of changes. These changes often follow similar patterns in different places whenever the same kind of disturbance has occurred. (See Figure 25-17 on page 552.)

As a result of the 1988 fires, the forests in Yellowstone became open fields filled with blackened tree stumps. Soon the fields were covered with

Figure 25-17 This picture, taken in 1998, shows that regrowth and succession had begun to take place in the fire-ravaged forests of Yellowstone Park.

a rich carpet of green grasses. In the years that followed, more plants began to grow, small animals returned, and after a while young trees began to grow. One community after another came to these once-burned forests.

Succession also occurs when a new environment appears for the first time. This happens, for example, when a volcano produces new rock, or when a glacier melts, exposing bare, lifeless soil. Scientists are very interested in studying the gradual succession of communities that appear in these places. This tells them a great deal about how communities on Earth evolved during ancient times to create the diverse ecosystems we have today.

EXTENDED LABORATORY INVESTIGATION 25

How Do Populations Change Over Time?

INTRODUCTION

Ecologists spend much of their time studying populations of organisms and the interactions between populations. Ecologists also study how populations change over time. In these studies, the growth of the population is investigated, as well as the ways in which population growth is dependent on available resources. In this investigation you will study a population of fruit flies as it changes over time.

MATERIALS

Sets of culture vials of fruit flies of various ages (each vial should have been started in the same manner, with approximately six male and six female flies; the vials should be one day, one week, two weeks, three weeks, and four weeks old), hand lens, chart or handouts of the life cycle of the fruit fly, petri dishes that contain fruit flies in each life-cycle stage, dissecting microscope

PROCEDURE

1. Observe a starter vial. Note your observations in your log book. Record the environmental conditions in the vial and any organisms in the vial.
2. Continue your observations of each of the other vials. Make notes of the environmental conditions present in each vial. In a data table, record the number of organisms in each vial. Also record the number of individuals in each of the stages in a fruit fly's life cycle. Use the hand lens when necessary.
3. If you have a dissecting microscope available, use it to observe the fruit flies in the vials, and make drawings of the organisms you observed.

INTERPRETIVE QUESTIONS

1. What changes did you observe in the number of organisms in each vial?
2. What evidence did you observe in the vials of the various stages of fruit fly development?
3. How did the conditions within the culture vials change over time?
4. Prepare a graph that shows changes in population size within the vials over time. Use the information in the graph to predict the size of the fruit fly population after five weeks, six weeks, and two months.
5. What is the relationship between resources available for a population and the continued growth of that population?
6. From your observations, describe what is needed in an environment for a population to continue to grow or to remain stable in size.

CHAPTER 25 REVIEW

Answer these questions on a separate sheet of paper.

VOCABULARY

The following list contains all of the boldfaced terms in this chapter. Define each of these terms in your own words.

carrying capacity, commensalism, community, competition, density, density-dependent, density-independent, distribution, growth rate, interspecific competition, intraspecific competition, mimicry, mutualism, parasite, parasitism, population, predation, succession, symbiosis

PART A—MULTIPLE CHOICE

Choose the response that best completes the sentence or answers the question.

1. All the populations of organisms living together in a given area form a *a.* species *b.* biome *c.* community *d.* commensalism.
2. Population density is *a.* the number of individuals of a species in a given area *b.* the total number of individuals in a population *c.* the rate at which the population increases or decreases *d.* the way individuals live in groups or are scattered through an area.
3. Density-independent limiting factors include *a.* interspecific competition *b.* parasitism *c.* predation *d.* forest fires.
4. The size of a population that can be supported by an ecosystem is called the *a.* population density *b.* carrying capacity *c.* census *d.* steady state.
5. A graph that shows the rate at which a population grows until it reaches a steady state will generally look like the letter *a.* J *b.* M *c.* S *d.* Z.
6. The relationship between foxes and mice is best described as *a.* predation *b.* interspecific competition *c.* parasitism *d.* commensalism.
7. Creosote bushes release chemicals into the soil that prevent other creosote bushes from growing nearby. As a result, creosote bushes have a *a.* clumped distribution *b.* uniform distribution *c.* high population density *d.* high growth rate.
8. All the organisms of one species that live in one place make up a *a.* symbiosis *b.* community *c.* ecosystem *d.* population.

9. Certain moths have spots on their wings that look like owl eyes. Flashing these spots at an insect-eating bird usually confuses the bird into thinking it has discovered a bird-eating owl, so the bird leaves the moth alone. This is an example of *a.* symbiosis *b.* mimicry *c.* commensalism *d.* competition.
10. Which of these would tend to limit population growth? *a.* crowding *b.* increase in food supply *c.* decrease in predators *d.* decrease in competitors
11. Fleas live on a dog and drink its blood, causing itchy bites that make the dog miserable. This is an example of *a.* commensalism *b.* mutualism *c.* mimicry *d.* parasitism.
12. Over a long period of time, a pond fills up and becomes first a field and then a forest. This is an example of *a.* succession *b.* commensalism *c.* interspecific competition *d.* predation.
13. Population density increases with an increase in *a.* emigration *b.* birth rate *c.* death rate *d.* density-dependent limiting factors.
14. Which of the following are competitors to wild grazers such as zebras and antelope? *a.* tick-eating birds *b.* vultures *c.* lions *d.* cattle
15. Many corals have algae that live inside their tissues. The algae are protected by the coral, and the coral needs the algae to survive. This is an example of *a.* parasitism *b.* mutualism *c.* intraspecific competition *d.* commensalism.

PART B—CONSTRUCTED RESPONSE

Use the information in the chapter to respond to these items.

This graph shows the population growth curves of two different species of aquatic organisms.

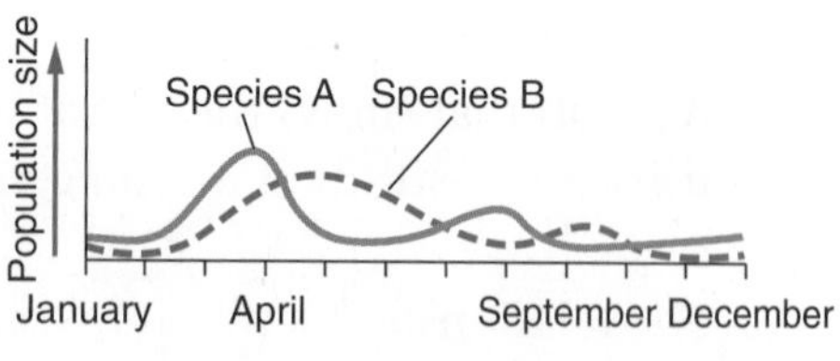

16. At which time of year is the population density of species B the lowest?
17. What is a possible relationship between species A and species B? Explain. What appear to be the limiting factors on the two populations?
18. What is a valid prediction based on this graph? *a.* Species A will not be present in the water during the winter months. *b.* Species A will eliminate species B from the water after one year. *c.* Species B will attain maximum population size each autumn due

to a decrease in water temperature. *d.* Species B will decrease in population size approximately one month after a decrease in the population size of species A.

19. Explain the relationship between niche overlap and interspecific competition.
20. Why does a disaster such as a forest fire provide an opportunity to observe the process of succession?

PART C—READING COMPREHENSION

Base your answers to questions 21 through 23 on the information below and on your knowledge of biology. Source: *Science News* (November 2, 2002): vol. 162, p. 285.

Ants Cheat Plant; Plant Cheats Back

A tropical tree acts as a fickle but pragmatic landlord, report researchers in Brazil.

This small tree of the central Amazon, *Hirtella myrmecophila*, grows a pair of little pouches at the base of young leaves. Two-millimeter-long ants of the species *Allomerus octoarticulatus* move into the pouches.

Plenty of plants offer ants special shelters in exchange for protection from other insects, says Heraldo L. Vasconcelos of Universidade Federal de Uberlândia in Brazil. Sometimes, however, ants exploit the trees' good will by not doing their part. In the case of *H. myrmecophila*, the trees take an unusual countermeasure, report Vasconcelos and Thiago J. Izzo of the Instituto Nacional de Pesquisas da Amazônia in Manaus, Brazil. In a preserve north of Manaus, they found that as leaves age, they shed their little ant pockets.

To figure out why, the researchers excluded ants from some branches. Those branches lost about half their new growth to pests. But the ant-free branches also kept about eight times as many flowers as ant-laden counterparts did. In the October *Oecologia*, the researchers propose that the tree benefits from ant security guards when foliage is young but then drops its cozy pouches from branches of mature leaves, preventing the ants from damaging too many flowers.

21. How does a small Brazilian tree provide a home for a particular species of tiny *Allomerus* ants?
22. Explain why researchers allowed some of the branches of these trees to grow without any ants in them.
23. Explain how the tiny ants are both beneficial and harmful to the tree branches, and state the action taken by the trees to minimize the ant damage.

26 Ecosystems

After you have finished reading this chapter, you should be able to:

Relate energy flow to trophic levels of organisms.

Define biological magnification and explain why it is important.

Explain the importance of biodiversity to human life.

We must protect the forests for our children, grandchildren, and children not yet born. We must protect the forests for those who can't speak for themselves such as the birds, animals, fish and trees.

Qwatsinas (Hereditary Chief Edward Moody), Nuxalk Nation

Introduction

Populations interact with and depend on each other. They also interact with the physical environment around them. Interaction and interdependence are two important relationships. Because of these relationships, living organisms and their physical surroundings are often studied together, as parts of a single unit called an **ecosystem**. The flow of energy and matter through ecosystems will now be the focus of our attention.

To begin, let's imagine ourselves in one of the most impressive of Earth's ecosystems—a tropical rain forest. (See Figure 26-1.) Charles Darwin described his feelings as he first set foot in a South American rain forest in 1832:

> The day has passed delightfully. Delight itself, however, is a weak term to express the feelings of a naturalist who for the first time, has wandered by himself in a Brazilian forest. The elegance of the grasses, the novelty of the parasitical plants, the beauty of the flowers, the glossy green of the foliage, but above all the general luxuriance of the vegetation, filled me with admiration. A most paradoxical mixture

Figure 26-1 A tropical rain forest in Guatemala.

of sound and silence pervades the shady parts of the wood. The noise from the insects is so loud that it may be heard even in a vessel anchored several hundred yards from the shore; yet within the recesses of the forest a universal silence appears to reign. To a person fond of natural history, such a day as this brings with it a deeper pleasure than he can ever hope to experience again.

THE BASIC CHARACTERISTICS OF ECOSYSTEMS

An ecosystem is made up of living (biotic) and nonliving (abiotic) factors. In other words, biotic factors, such as living organisms, and abiotic factors, such as water, air, light, or temperature, function together in an ecosystem. For living organisms to survive, there must be a source of energy. The energy flowing between organisms and their environment is a basic characteristic of an ecosystem. Organisms are made up of matter. The flow of matter between organisms and their environment is another main characteristic of an ecosystem.

What is the source of energy for a tropical rain forest? As with almost all ecosystems on Earth, it is the sun. While energy is constantly reaching Earth from the sun, matter is not. The amount of matter on Earth remains constant. However, matter moves back and forth between organisms and the environment. As with energy, the cycling of matter can be seen in all ecosystems.

ENERGY FLOW THROUGH ECOSYSTEMS

In most ecosystems, energy arrives as sunlight. Some organisms are able to use the energy from sunlight directly. Other organisms use this energy indirectly. They get their energy by eating other organisms. Scientists describe and group organisms in an ecosystem based on whether an organism can make its own food or must eat food made by other organisms.

Organisms are grouped in a system of **trophic levels.** *Trophic* means "feeding." On the first level are organisms that use energy directly from the environment, such as the energy in sunlight. These first-level organisms are called **producers**. Plants are producers because they use the process of photosynthesis to make their own food with water, carbon dioxide, and the energy from sunlight. Both the grass in a field and the mighty oak trees in a forest are producers. Organisms that feed on producers are in the next trophic level. These organisms are called **primary consumers.** A caterpillar that eats oak leaves is a primary consumer. In the next trophic level are the organisms that eat primary consumers. These organisms are called **secondary consumers**. A bird that eats the caterpillar is a secondary consumer. A large hawk or a cat that eats the bird is a **tertiary consumer.** Each of these steps is called a trophic level because it describes the source of the organisms' food. We can describe the flow of energy in an ecosystem by using trophic levels. (See Figure 26-2.)

Energy enters an ecosystem at the producer level. Energy flows through an ecosystem by being passed along from an organism in one trophic level to an organism in a higher trophic level. This transfer of food energy from one organism to the next is called a **food chain**. Oak tree leaf to caterpillar to small bird to hawk is a food chain. In a real ecosystem, a single, simple food chain like the one described is never found. Obviously, caterpillars are not the only animals that eat oak leaves. Other insects, deer, and perhaps rabbits may also eat oak leaves. Caterpillars are eaten not only by birds but also by frogs and other small animals. Food chains are interconnected. In reality, food chains make up a complex pattern in which food energy may be passed in many different directions and to many different organisms. This complex pattern

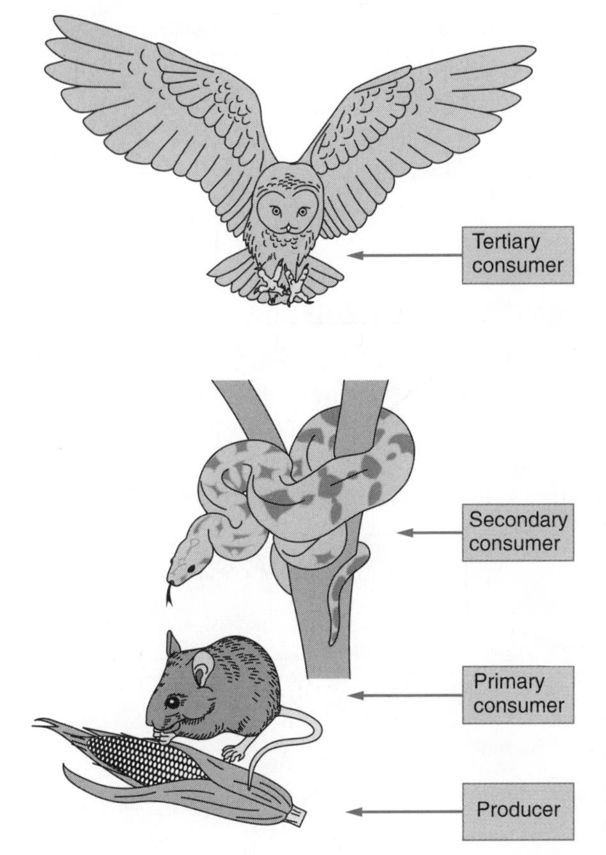

Figure 26-2 The trophic levels in an ecosystem describe the energy flow.

formed by food chains in an ecosystem is known as a **food web**. (See Figure 26-3 on page 562.)

However, no matter how complex a food web is, energy always moves in one direction—from a lower to a higher trophic level. Energy does not get recycled. As energy moves through each trophic level, some of it is used and some of it is lost when it is changed to heat during energy transfer and conversion. The greatest amount of energy is present at the lowest trophic level (the producers); the least is present at the highest level (upper-level consumers). For this reason, additional energy must constantly enter an ecosystem. In other words, for life on Earth to continue, the sun must continue to shine.

Ecologists use a pyramid to describe the flow of energy through an ecosystem. The wide base of the pyramid represents the amount of usable energy in all of the producers. The energy gathered from the sun and stored in all of the plants in the forest is represented by this level. The next step up in the pyramid shows the energy that primary consumers get

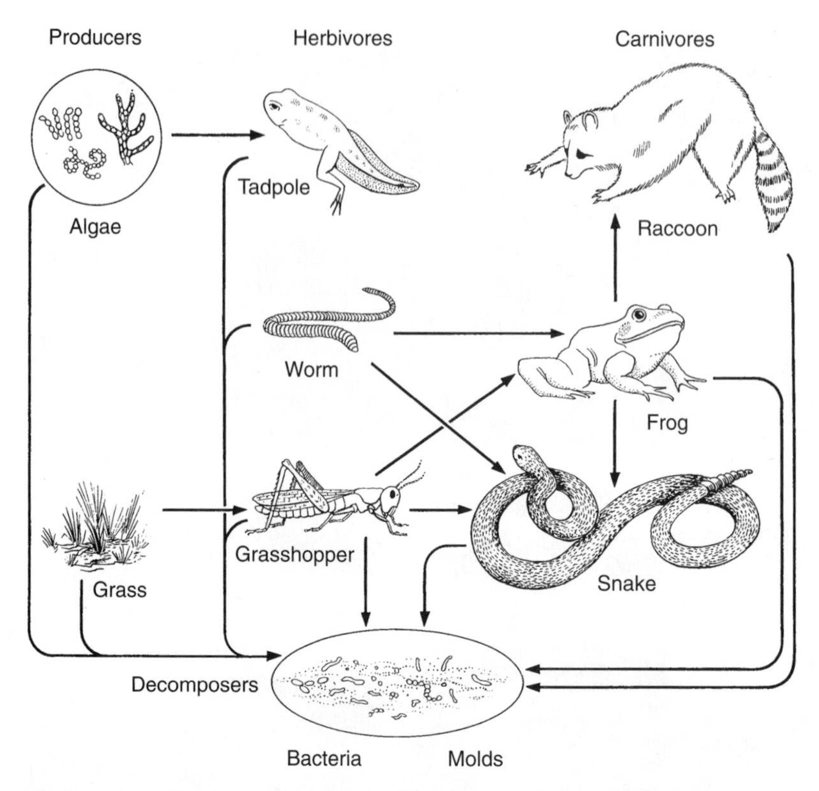

Figure 26-3 The interconnection of food chains forms a food web.

from producers. This layer is smaller than the layer that represented the energy present in the producers. The passing of energy from one level to the next is actually not a very efficient process. Only about 10 percent of all energy gets passed from one level to the next. This is true as we move up the pyramid from primary consumers to secondary consumers and then to the top level. (See Figure 26-4.)

A pyramid of energy can provide an important lesson in how to feed the ever-increasing human population. Leaves such as spinach, and seeds such as rice and beans, come from plants, which are producers. Cattle, a source of meat for human consumption, are consumers. Throughout the world, much more food energy is present at the producer level than at the consumer level. Which type of food is more abundant and available for everyone? Which type of food makes a more efficient use of energy sources—spinach with rice and beans or hamburgers?

Scientists also show the flow of energy through an ecosystem in two other types of pyramids. Usually there are more individual organisms at the producer level. At each higher trophic level, the number of organisms decreases. Therefore, a pyramid of numbers can be made. A pyramid of

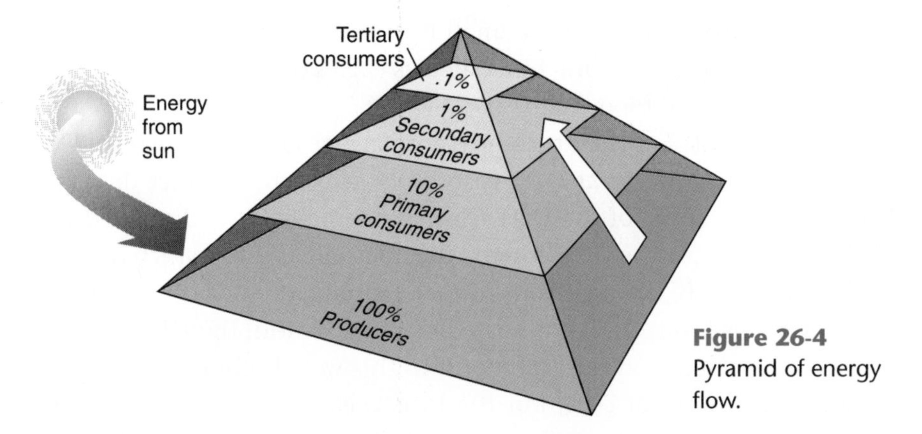

Figure 26-4 Pyramid of energy flow.

numbers shows that fewer organisms are supported at each higher trophic level in an ecosystem. Also, if the total mass of all organisms at each trophic level is measured, a pyramid of biomass can be drawn. Once again, the lowest trophic levels usually have the largest biomass; the highest levels have the smallest biomass. (See Figure 26-5.)

A HIDDEN DANGER IN A FOOD CHAIN

It is a wonderful thing to go for a walk in the country and perhaps discover a small stream. The water sparkles like diamonds as sunlight plays on its surface. The stream makes gurgling sounds as it flows over and around rocks. In water as clear as glass you can even see some small fish as they dart along the stream's rocky bottom. The following food chain might be found in this stream: The stream's microorganisms are eaten by

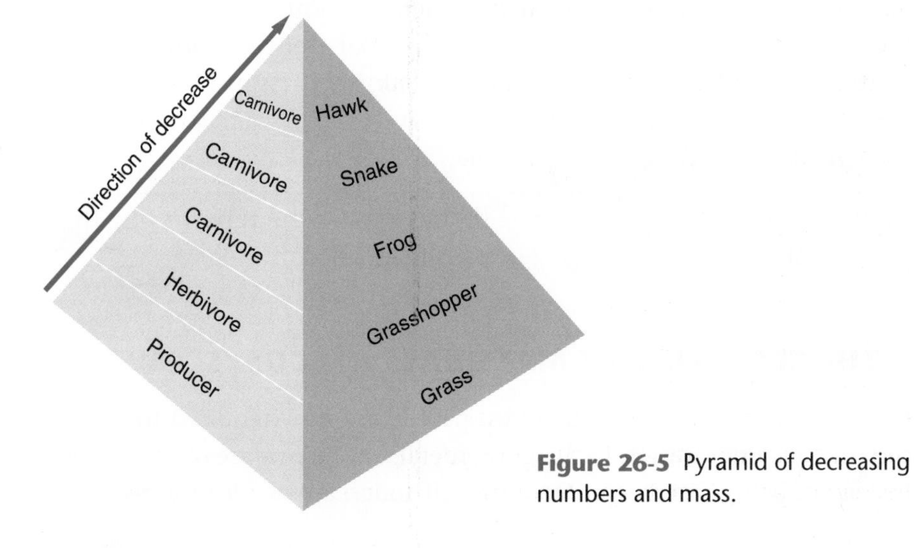

Figure 26-5 Pyramid of decreasing numbers and mass.

insects. The insects are eaten by small fish; these small fish are eaten by larger fish. Finally, fish-eating birds such as eagles or osprey eat the larger fish. This is a typical food chain.

Now suppose this stream runs next to a farmer's field. In the past, U.S. farmers used the chemical DDT to protect crops from insect damage. When it rained, some of the DDT washed off the fields and entered the stream. In the water, the DDT entered the microorganisms, which became food for the insects. As organisms in each trophic level fed on organisms in the previous level, DDT was passed on. In addition, the DDT became more concentrated. You can think of it in this way. Each microorganism contained a tiny bit of DDT, but the insects eat a great many microorganisms, so the tiny bit of DDT in each became a more concentrated level of DDT when it was stored in the insects. (See Figure 26-6.)

This is true for organisms at each trophic level. The level of DDT in each organism increases as the DDT is moved along the food chain. The little fish contain more DDT, the larger fish even more, and finally, the birds the most. This process is known as **biological magnification**. Years ago, when this happened to eagles and osprey, the high levels of DDT interfered with the proper buildup of calcium in the shells of their eggs. Egg shells contaminated with DDT are very fragile. The eggs usually broke before the developing birds hatched. Although they still laid eggs, few birds were able successfully to produce young. This was especially true of bald eagles and osprey, birds whose diet consists mostly of fish. In parts of the United States, populations of these birds began to diminish quickly. It is only since the use of DDT has been banned in this country that these magnificent fish-eating birds have reestablished their populations. The bald eagle has recently been removed from the endangered species list—a great success story in wildlife management. Although no longer used in the United States, DDT is still used in some parts of the world, where it continues to enter various natural food chains.

Check Your Understanding

Why is a food web more accurate than a food chain in portraying the relationships that exist among organisms in an ecosystem?

THE RECYCLING OF MATERIALS IN ECOSYSTEMS

In many parts of the United States, people are now required to recycle certain consumer wastes. Paper, glass, metal, and plastic are often recycled instead of being discarded. Recycling, although a new idea for people, is

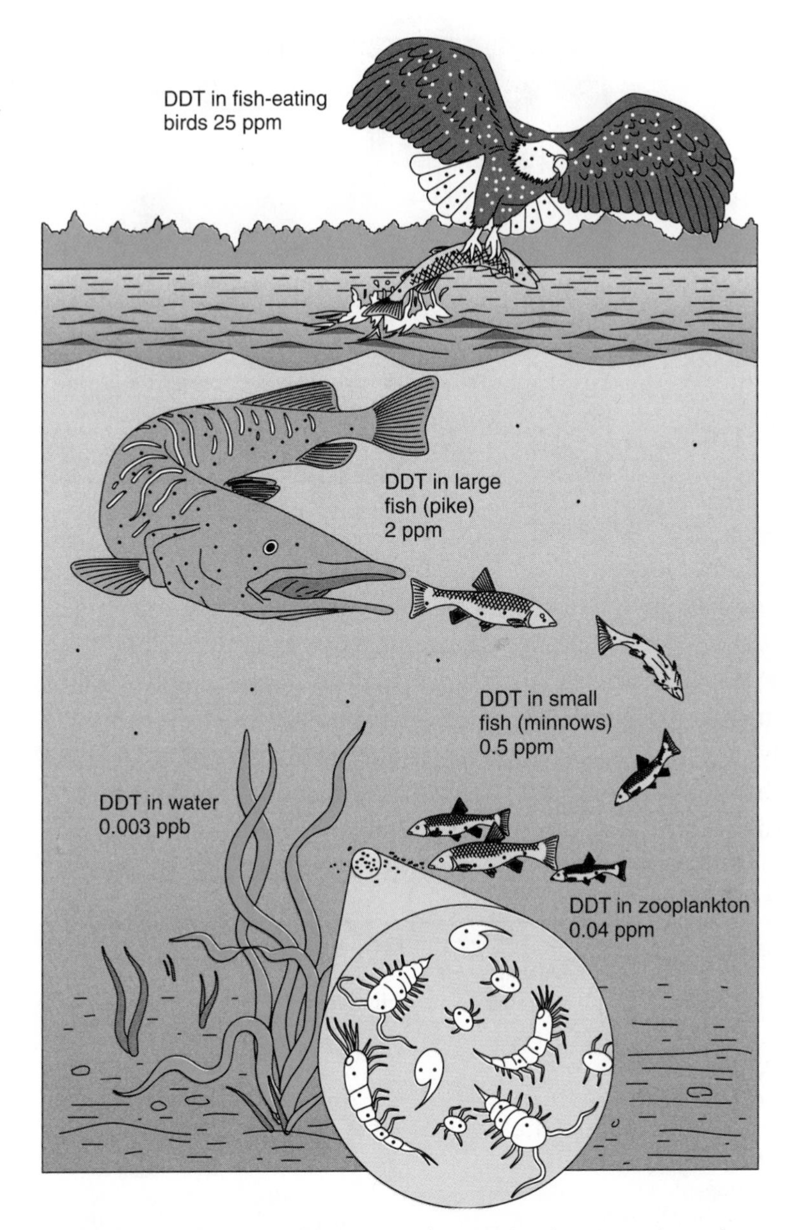

Figure 26-6 The level of DDT in each organism increases through biomagnification as the chemical moves along the food chain.

not a new idea in nature. (See Figure 26-7 on page 566.) Natural ecosystems have recycled materials since life began on Earth. Life would not continue without this recycling of materials. Why is this so?

All substances are made up of chemical elements. There are about 90 chemical elements that occur in nature. Of these, only a relatively small number of elements are found in significant amounts in organisms. These

Figure 26-7 These teenagers are helping to recycle newspapers.

include carbon, hydrogen, oxygen, phosphorus, sulfur, and nitrogen. The amount of these elements on Earth today is approximately the same as when the planet formed. Because they are needed by living things, and their supply does not increase, these elements need to be reused, or recycled, again and again. The atoms of these elements that are present today in your body may have been present in other organisms before. You may have atoms in you that were once part of a tree that grew in an ancient forest, or in a dinosaur that walked through that forest. (See table.)

SOME ELEMENTS IN LIVING MATTER

Symbol	Element	Atomic Number	Percentage of Earth's Crust by Weight	Percentage of Human Body by Weight
Ca	Calcium	20	3.6	1.5
C	Carbon	6	0.03	18.5
Cl	Chlorine	17	0.01	0.2
H	Hydrogen	1	0.14	9.5
Mg	Magnesium	12	2.1	0.1
N	Nitrogen	7	Trace	3.3
O	Oxygen	8	46.6	65.0
P	Phosphorus	15	0.07	1.0
K	Potassium	19	2.6	0.4
Na	Sodium	11	2.8	0.2
S	Sulfur	16	0.03	0.3

The Carbon Cycle. How do these elements get recycled in the natural world? Let's look at the element carbon. All organisms are made of molecules that contain carbon. This carbon is obtained from carbon dioxide in the air. Producers such as grasses, trees, and other plants take in carbon dioxide from the air during photosynthesis. They use the carbon from the CO_2 gas to build carbohydrates—sugars and starches. Consumers, including humans, obtain carbon from producers and sometimes from other consumers that serve as food. To complete carbon's recycling, plants and animals return carbon to the atmosphere. This occurs through respiration as we breathe out carbon dioxide.

Recycling of carbon also occurs after an animal or plant dies. Another very important part of the recycling process occurs through the actions of **decomposers**. Decomposers are heterotrophs—organisms that are unable to make their own food. They get their food by feeding on dead

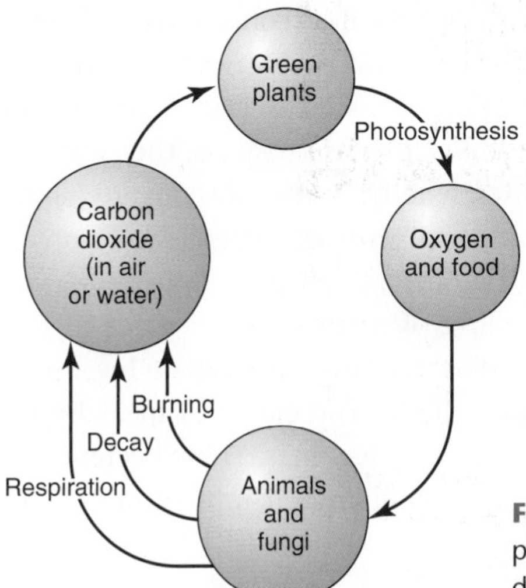

Figure 26-8 Carbon is recycled via photosynthesis, respiration, decay (or decomposition), and burning.

organisms. The most common decomposers are bacteria and fungi. As they carry out their life processes, they too release carbon as CO_2 into the atmosphere. (See Figure 26-8.)

The Oxygen Cycle. Oxygen, another element, also moves between living organisms. All animals need oxygen for respiration. Respiration is the process that releases the chemical energy stored in food. Land animals obtain oxygen for respiration from the air they breathe. Fully aquatic animals like fish get the oxygen they need from the water they live in. (Oxygen can dissolve in water, and this is the oxygen that fish use.)

Almost all the oxygen in Earth's atmosphere originally came from the metabolic activities of plants. During the process of photosynthesis, plants give off oxygen as a waste product. So we are breathing in a waste gas given off by plants. This is natural recycling.

The Nitrogen Cycle. Nitrogen is an element that is used by organisms when they make proteins. Nitrogen is the most abundant gas in the air; approximately 78 percent of the atmosphere is nitrogen. So it would seem that there is plenty of it around. The problem is that most living things cannot use free nitrogen, the nitrogen present in air.

How do plants obtain the nitrogen they need? Plants can take in and use nitrogen in the form of nitrates. Nitrate compounds are combinations of nitrogen, oxygen, and some other element. One way that nitrates are formed naturally is during lightning storms. These nitrates fall to the ground and enter the soil. Plants get their nitrates from the soil by absorbing them through their roots. The nitrogen in plants is passed on to primary, secondary, and tertiary consumers when these consumers eat plants or organisms that eat plants. Each of these organisms releases nitrogen back to the environment in the form of nitrogenous wastes. Decomposers also recycle nitrogen when they digest proteins into amino acids, then ammonia, and even into another form, nitrites. Finally, nitrites are converted into nitrates, which can be used by plants. Special kinds of bacteria are involved in each of the steps of this conversion process. (See Figure 26-9.)

One remarkable step in the recycling of nitrogen takes place in the roots of a special group of plants. These plants are called **legumes**. Legumes include peas, beans, peanuts, alfalfa, and clover. A special type of bacteria lives inside nodules on the roots of legumes. These bacteria are able to take free nitrogen from the air and change it into nitrates. The

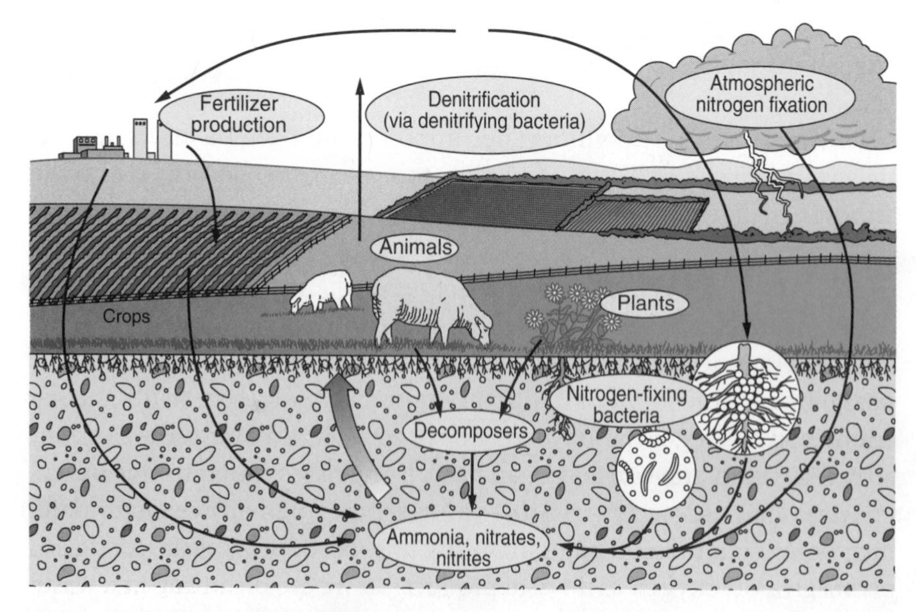

Figure 26-9 The nitrogen cycle.

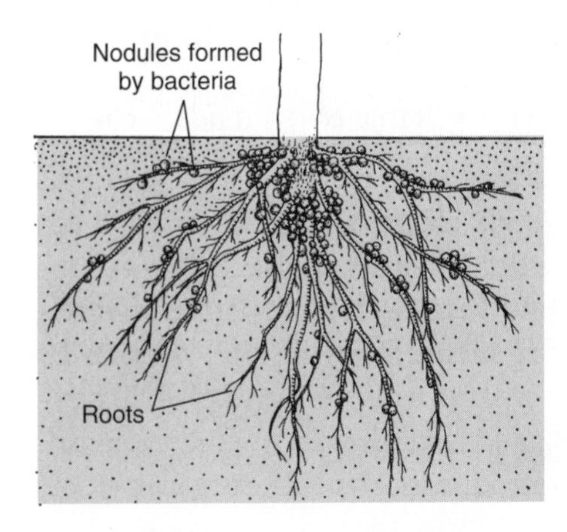

Figure 26-10 Nitrogen fixation in the nodules on the roots of legumes is a good example of mutualism.

process is called **nitrogen fixation**. The nitrates are used by plants to grow, and the plants provide nutrients for the bacteria. This is a wonderful example of mutualism. It is also very important to life on land. Soil in which legumes live becomes richer because the bacteria add more nitrates to it. In turn, this natural fertilizer helps other plants grow. (See Figure 26-10.)

Long ago, farmers realized that by sometimes growing legumes in a field, they could improve the growth of other crops in other years. For example, corn is an important food crop. However, because corn needs a great deal of nutrients, it can be grown in a field for only a few years. By planting clover one year, farmers made the soil ready to grow more corn the following year. Farmers were allowing the nitrogen-fixing bacteria to recycle necessary nitrates into the soil from the air. This kind of crop rotation allows nature to replace soil nutrients and is much less expensive than the continuous application of fertilizers that would be needed to produce corn year after year in the same field.

CHANGE AND STABILITY IN POPULATIONS AND COMMUNITIES: THE IMPORTANCE OF BIODIVERSITY

Biologists have long wondered what causes populations to change rather than remain the same. For example, why do rabbits in England get larger over a long period of time, while rabbits in the United States remain the same size? Remember that all the characteristics that are passed from one generation to the next in a population are determined by genes. Biologists wondered under what conditions the genes in a population would remain the same from generation to generation. A mathematician, Godfrey Harold Hardy, and a physician, Wilhelm Weinberg, answered this

Restoring the River of Grass

The Everglades is a vast, wide freshwater marsh that covers much of the southern part of Florida. The Everglades begins at the northern edge of Lake Okeechobee, with the overflow of rainwater out of the lake, and extends all the way to the southern tip of the state just before the Florida Keys. Although some cypresses, mangroves, palms, and oaks grow in scattered clumps, the vast majority of the Everglades is covered by a dense growth of saw grass. A very slow, steady flow of water moves through the saw grass from north to south. To your eyes, the Everglades looks like a huge swampy area of tall, sharp-edged grass; but in many ways it is really a wide river because of the water that flows through it. The Everglades is therefore called a "River of Grass."

One hundred years ago, the Everglades was considered a wasteland. In 1906, construction projects built drainage canals that altered the flow of this water that had moved unchallenged for centuries. After the land was drained, huge areas were turned over to agriculture. By the 1980s, continued growth of sugarcane farms, housing developments, and highway construction had reduced the Everglades ecosystem of grass marshes to about one-half its original size.

Voices of alarm were raised for many years. The survival of many plant and animal species was threatened. Finally, it was noticed that the quantity and quality of the entire underground freshwater supply on which humans in southern Florida depend were being placed at risk.

In 1996, the federal government endorsed the Everglades restoration project. The project will be one of the largest ecological restoration efforts

question. According to the Hardy-Weinberg Law, five conditions must occur for a population *not* to change:

- There must be no mutations.
- There must be no arriving (immigration) or leaving (emigration) of individuals to or from the population.
- The population must be large.
- All individuals must have the same chance of surviving.
- The matings of individual organisms must be random; no mates can have preferences.

Once these conditions were described by Hardy and Weinberg, evolution made more sense. It was clear that most of the time, for most populations, some of these conditions are *not* met. As a result, populations change. Size of individuals, shape, structure, coloring, behavior, and any other inherited characteristic may change from generation to generation.

anywhere in the world. Hundreds of millions of dollars will be spent on protecting the fragile Everglades ecosystem. Included in the plan is the removal from sugarcane production of 100,000 acres of farmland in ecologically sensitive areas.

Much of the water that flows through the Everglades has become contaminated by pesticides and fertilizers that are used to increase crop yields on farms in the area. One of the main goals of the restoration project is to let large areas of land act as natural water filters to remove some of the waterborne contaminants. Today, there are too many farms and too few natural areas in the Everglades. The restoration project will help restore the balance. Six large wetlands are being constructed between Lake Okeechobee and the Everglades. One of these projects, for example, was begun in 2001 and is scheduled for completion in 2009. These areas, called storm-water treatment areas, will use naturally occurring biological processes to reduce the levels of phosphorus carried by the water that moves through the Everglades.

Another important part of the Everglades project will restore the natural north-south flow of water. The natural pattern of water flow through the Everglades was disrupted by the canals, pumping stations, and water-control structures that were built to create flood-control and water-supply systems for southern Florida. In fact, these unnatural attempts to control the Everglades' water flow have been harmful to the entire ecosystem. Today, planning is under way to find alternatives that can meet flood-control and water-supply needs while ensuring the long-term health of the Everglades.

This is evolution in action. The Hardy-Weinberg Law shows, by looking at the genes of a total population, how the environment interacts with a population to produce evolutionary change.

Do entire communities in an ecosystem stay the same? What causes a particular community to change? Is the number of species that make up a community critical? These important questions are now being studied by ecologists. The amount of variety in a community is called species diversity or **biodiversity**. A community with only a few species of plants and animals has low biodiversity. The Great Salt Lake, because of its high salinity, has few species that are adapted to living there. A community with many species has great biodiversity. A tropical rain forest community may have the greatest biodiversity of any community on Earth.

Biodiversity is one major concern of ecologists today. In fact, at the 1992 Earth Summit held in Rio de Janeiro, Brazil, it was hoped that all countries attending the conference would sign The Biodiversity Treaty. Although 167 nations have signed the treaty since 1992, some large nations, including the United States, have not. Why is biodiversity such a great concern to scientists? One in five known species present on Earth when you were born is already extinct. For the most part, human actions are the cause of these extinctions. As species disappear, biodiversity decreases. Scientists are concerned about the effects of diminished biodiversity on ecosystems. They have learned that there is a great deal of interaction and interdependence in ecosystems. Does a community in an ecosystem need a certain number of species interacting with each other to remain viable? How many species can a community afford to lose without being harmed? The ability of an ecosystem to continue and to remain healthy is called its stability. If all the insects in a forest died, would the forest survive? Would the plants that the insects ate grow too quickly? Would these plants interfere with the growth of trees? Would bird populations suffer with no insects to eat?

Many studies that investigate biodiversity and stability in specific communities are currently being conducted. In the 1960s, the ecologist Robert Paine studied the animals that live along a stretch of rocky seashore. In this community, there were 15 species of small animals, including barnacles, clams, and one large predator species, a sea star. In one experiment, Paine removed all the sea stars. After a time, the biodiversity decreased greatly. Instead of the original 15 species present before the sea star was removed, only eight species of smaller animals remained. Seven species had disappeared. The population of one type of mussel had increased dramatically. The community had changed a great deal. Stability no longer existed. Paine realized that by removing the sea star—a predator—the interactions among the smaller animals had changed. The sea star had kept the density of other populations low. Without the predators, competition among the other animals increased for the limited space. Only a few species survived and increased their numbers.

How much loss of biodiversity can occur before Earth's ecosystems stop functioning properly? This is a very serious concern of many people. It is also an important concern for all species that depend on Earth's ecosystems. We are one of those species.

HABITAT DESTRUCTION

There is one main reason why biodiversity is decreasing. Many species are disappearing because of habitat loss. Humans are using and changing many places where organisms formerly lived. For example, in the Midwest, many fields contained low-lying areas. These low places remained filled with water all year long. Many birds, such as ducks and geese, found food in these bodies of water. Although the birds did not live in the ponds all year, they visited these same places during their migrations each spring and fall. The birds were flying between their winter feeding grounds and summer breeding grounds and were able to rest and find food here on their long journeys.

However, the farmers could not grow crops in these wet places. Wheat and corn need drier land. As a result, most of these wet places were filled in. This was a critical loss of habitat for the migrating water birds. Over time, the populations of geese and ducks decreased. Some species even became extinct. Biodiversity was reduced. This is only one example of the loss of a habitat affecting biodiversity. In many other places, habitat loss has also occurred. Most of the forests in the eastern United States are gone, replaced by farms and cities. Habitat loss occurs on a river when a dam is built. Fish that can survive only in moving water die in the still water of the lake that forms behind a dam. Today, the greatest habitat destruction is occurring in the world's tropical rain forests. It is estimated that 70 to 90 percent of Earth's biodiversity will be lost if the rain forests are destroyed. Sadly, this is happening while scientists are trying to identify and classify the many organisms still being discovered in these forests. In addition, many fear that species containing substances that could prove to be extremely valuable medicines are being lost forever before even being discovered. (See Figure 26-11.)

Figure 26-11 This rain-forest habitat has been destroyed to make room for a banana plantation.

LABORATORY INVESTIGATION 26
What Role Do Mushrooms Play in Ecosystems?

INTRODUCTION

Mushrooms and molds are members of the fungus kingdom. Organisms in this kingdom lack chloroplasts and so are unable to carry on photosynthesis. Fungi, like animals, are heterotrophs—organisms that cannot make their own food. They are thus dependent on other organisms for their energy. Unlike animals, which digest their food by using enzymes they produce in their digestive system, fungi secrete enzymes into the food on which they grow. Fungi then absorb the nutrients.

In ecosystems, fungi play a very important role. As decomposers, fungi return nutrients from dead plants and animals to the soil. This investigation focuses on a particular decomposer available in a grocery store, the edible mushroom.

MATERIALS

Edible mushrooms, cardboard, scalpels, spore prints prepared in advance, compound and dissecting microscopes, glass slides, water, coverslips

PROCEDURE

1. Observe a mushroom. Draw the structures of the mushroom that you can see with your unaided eyes. Label as many of the structures as you can.
2. Place the mushroom on a piece of cardboard so that it rests on its flat side. Use a scalpel to slice the mushroom in half. Make another drawing of the internal structures that are now exposed. Label the structures you observe.
3. Observe a spore print. Compare this print to the structures of the mushroom you have just observed.
4. Place the mushroom under the dissecting microscope and record any new structures that become visible under magnification.

5. Use the scalpel to prepare thin sections of the mushroom from various parts. Try to make slices from the stem, cap, and gills. Place a thin slice on a glass slide. Add a drop of water and a coverslip. Place the slide on your compound microscope and examine it using the low-power objective. Draw what you observe. Make notes of any similarities and differences you observe in the various structures.
6. Prepare a wet mount of spores from the gills. Observe the spores under the high-power objective. Draw what you observe.

INTERPRETIVE QUESTIONS

1. Study the structures and life cycle of a mushroom from reference materials. Describe the function of the structures of the mushroom you observed. List the parts of the mushroom you found in the reference materials that you were not able to observe in your specimen.
2. What parts of the mushroom are involved in obtaining nutrients? How do these parts obtain nutrients?
3. What parts of the mushroom are involved with reproduction? How does a mushroom reproduce?
4. Why is reproduction by spores considered to be an example of asexual reproduction?
5. Why are the actions of fungi, and other decomposers, essential for the continuation of life on Earth?

CHAPTER 26 REVIEW

Answer these questions on a separate sheet of paper.

VOCABULARY

The following list contains all of the boldfaced terms in this chapter. Define each of these terms in your own words.

biodiversity, biological magnification, decomposers, ecosystem, food chain, food web, legumes, nitrogen fixation, primary consumers, producers, secondary consumers, tertiary consumers, trophic levels

PART A—MULTIPLE CHOICE

Choose the response that best completes the sentence or answers the question.

1. The source of energy for almost all ecosystems on Earth is *a.* the sun *b.* geothermal forces *c.* photosynthesis *d.* cellular respiration.
2. The organisms on the first trophic level are *a.* primary consumers *b.* tertiary consumers *c.* heterotrophs *d.* producers.
3. During the process of nitrogen fixation, *a.* animals release nitrogen back to the environment in the form of nitrogenous wastes *b.* the nitrogen in nitrates is used to make plant proteins *c.* bacteria take free nitrogen from the air and change it to nitrates *d.* lightning breaks down nitrates to form nitrogen gas.
4. A clam that feeds on the phytoplankton it filters from the water is eaten by a walrus, which is in turn eaten by a polar bear. In this example, the primary consumer is the *a.* clam *b.* photoplankton *c.* walrus *d.* polar bear.
5. Which of these communities has the highest biodiversity? *a.* the Great Salt Lake in Utah *b.* a deciduous forest in upstate New York *c.* the open ocean off the shore of California *d.* a rain forest in Costa Rica.
6. The series of events in which food energy is transferred from an organism at one trophic level to an organism at a higher trophic level, and from that organism to the next, forms *a.* a feeding pyramid *b.* a food web *c.* a food chain *d.* an ecological succession.
7. The interacting biotic and abiotic factors in an area make up *a.* a trophic level *b.* an ecosystem *c.* biodiversity *d.* a food web.

8. During the oxygen cycle, oxygen is released into the atmosphere by
   a. producers during the process of photosynthesis
   b. decomposers during the process of fermentation
   c. consumers during the process of respiration
   d. bacteria during the process of nitrogen fixation.
9. Which condition is necessary for a population *not* to change?
   a. mutations *b.* movement of individuals into and out of the population *c.* individuals having different chances of surviving *d.* random matings.
10. In an ecological pyramid, the most food energy is found
    a. at the top of the pyramid *b.* in the middle of the pyramid
    c. at the bottom of the pyramid *d.* none of these.
11. Scientists have found very few fossils of *Tyrannosaurus rex* and comparatively large numbers of the dinosaurs on which *T. rex* fed. The best explanation for this fact is that *a.* energy is lost with each successive link in the food chain *b.* there are fewer individual organisms at each successive level of an ecological pyramid *c.* consumers do not form fossils as readily as producers *d.* scientists have not been searching in the best places to find *T. rex* fossils.
12. The complex pattern of interlocking food chains in an ecosystem forms a *a.* biomass pyramid *b.* food chain *c.* biogeochemical cycle *d.* food web.
13. The main reason for loss of biodiversity is *a.* disease *b.* the limited number of crop and livestock varieties used in modern agriculture *c.* habitat loss *d.* hunting and fishing.
14. Biological magnification involves *a.* an increase in the number of individuals at successive levels of an ecological pyramid *b.* the increasing concentration of harmful chemicals at each successive step of a food chain *c.* the viewing of microscopic life-forms with special equipment *d.* the increasing amount of energy represented by a unit of biomass at each successive level of an ecological pyramid.
15. The organisms that break down dead plants and animals, thereby releasing CO_2 into the atmosphere, are *a.* decomposers *b.* autotrophs *c.* producers *d.* consumers.

PART B—CONSTRUCTED RESPONSE

Use the information in the chapter to respond to these items.

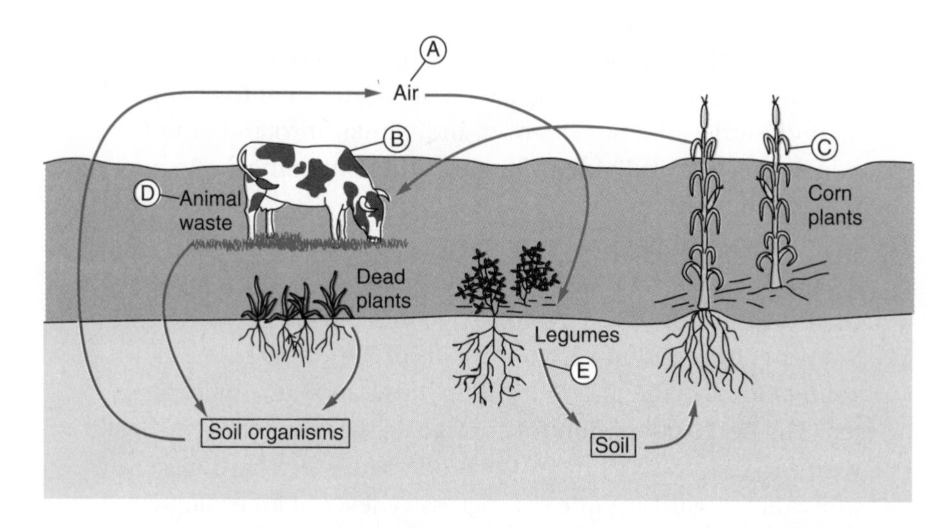

16. The events in the diagram show that materials are cycled between *a.* living things only *b.* heterotrophs only *c.* the living and nonliving parts of the environment *d.* the nonliving parts of the environment through evaporation, condensation, and precipitation.
17. What process does this diagram illustrate? Identify the type of nitrogen compounds found at the points lettered A, B, C, D, and E.
18. Explain why a farmer may plant a field with alfalfa rather than corn every few years.
19. Why is it ecologically wise to eat "lower on the food chain"?
20. What is the Hardy-Weinberg Law? How does it tie ecology to evolution?

PART C—READING COMPREHENSION

Base your answers to questions 21 through 23 on the information below and on your knowledge of biology. Source: *Science News* (October 26, 2002): vol. 162, p. 269.

Insects, Pollen, [and] Seeds Travel Wildlife Corridors

In an unusual test of a conservation strategy called wildlife corridors, strips of habitat boosted insect movement, plant pollination, and seed dispersal among patches of the same ecosystem.

Theory predicts that adding such corridors enhances the benefits of otherwise isolated preserves, says Joshua Tewksbury of the University of Washington in Seattle. He and his colleagues tested that strategy in South Carolina pine forests.

At eight locations, the researchers cleared mature vegetation and created open habitat on five 1-hectare plots—arranged as a central plot with four satellites. In each case, a 150-meter-long corridor connected the central plot to one outlier, while the others remained isolated. The unlinked patches had dead-end corridors or additional area so they matched the habitat area of another patch and its connecting corridor. Thus, scientists could distinguish between effects of biological entities' ease of movement and of extra habitat.

Butterflies, pollen, and seeds all moved most often between the corridor-connected patches, the researchers report in an upcoming *Proceedings of the National Academy of Sciences.* Variegated fritillary and common buckeye butterflies that the researchers captured, marked, and released in the central patch proved two to four times as likely to show up in connected patches as in unconnected ones.

When researchers placed male holly plants in the center patches, females in connected patches showed an average increase in seed production of nearly 70 percent, compared with that of female hollies in unconnected patches. Also, bird droppings in connected patches harbored more berries from shrubs in the center patches than did droppings in patches not connected to the central patch.

This is the first test of a corridor's effect on plant-animal interactions, says Tewksbury.

21. Explain how (at all eight locations) three of the test plots were made different from the central and fourth plots.
22. State the difference that was observed in the number of butterflies (released from the central plot) that showed up in the connected plots as opposed to those that showed up in the isolated plots. How would you explain this difference?
23. Why did researchers study the droppings of birds in these test plots, and what was their conclusion from this study?

27 People and the Environment

After you have finished reading this chapter, you should be able to:

Identify several ways humans have affected ecosystems.

Relate the importance of topsoil to farm productivity.

Discuss the importance of several types of air and water pollution.

Explain how the "greenhouse effect" can alter Earth's climate.

To see the world in a grain of sand
And heaven in a wild flower,
To hold infinity in your hand
And eternity in an hour.

William Blake, *Auguries of Innocence*

Introduction

In 1962, the American biologist Rachel Carson wrote *Silent Spring*. Her book begins with a description of a lovely town filled with beautiful trees and flowers. The fields around the town are full of deer and foxes. Countless birds sing in the trees. The clear, cold streams are full of fish. And then things begin to change. The farmers' chickens get sick and their sheep die. Physicians diagnose new illnesses that they have never previously seen in their patients. And perhaps even more ominous, people in the town notice that there are far fewer birds.

The "voices of spring" were silenced, Carson said. What had caused this? The answer was stated for the first time in this book. It was no magic spell, no foreign enemy that had produced a spring without birdsong. The people themselves had done it by using chemicals to kill organisms considered pests, such as insects and weeds. *Silent Spring* was one of the first and most important books that made people aware of the fragility of the environment and the interdependence of all living things. Rachel

Carson asked us to consider this all-important question: What effects do humans have on the environment?

A HISTORY OF PEOPLE CHANGING THE ENVIRONMENT

A black bear gathers insects, honey, and berries for food. It spends much of its day in forests and fields searching for things to eat. A hawk soaring high in the sky hunts for mice. Swooping down to catch a mouse, it gets a meal. Almost all animals obtain their food through hunting and gathering. Humans, however, are different. (See Figure 27-1.)

We have not always been different. Up until about 10,000 years ago, humans gathered and hunted food just as other animals do. However, because of our intelligence, we learned which plants made the best food. Then we collected and planted seeds from these plants to ensure a steady supply. Animals such as goats, chickens, and cattle were domesticated and raised for food. These new activities—planting crops and domesticating animals—marked the beginnings of agriculture.

As a result of agriculture, people use the land differently from other animals. When people cut down and burn trees to make room to plant crops, wild animals are often forced to leave the area. Usually, people plant only one or two crops, such as corn and beans. In some cases, instead of growing a crop, people fence in an area where grass is grown for sheep or cattle to graze. Either way, wild animals lose their source of food and a place to live. (See Figure 27-2 on page 582.)

Since the development of agriculture, humans have been extremely successful in increasing their population. This is because the steady supply of food enables people to settle in one area, live there for a long time,

Figure 27-1 This Alaskan brown bear and her cubs are looking for something to eat.

Figure 27-2 Early agriculture: This painting of people picking grapes is from the Tombs of the Nobles in Thebes, Egypt.

and raise larger families. Food crops such as corn and wheat can be grown for large numbers of people. Domesticated animals provide a continuous supply of leather, wool, meat, milk, and eggs. Cotton and flax are grown to supply material for clothing. However, through agriculture, humans cause great changes to the environment. The natural habitat for many different species is destroyed, while only those few plant and animal species that suit our purposes are raised in their place.

Advances in science and technology have produced even greater changes in the environment. Science has taught us much about the world around us. Through technology, we use what we have learned in science to make our lives more comfortable and convenient. For example, technology can be applied to agriculture. Chemical fertilizers increase crop yields. Modern farm machines are applications of technology that increase the productivity of people.

During the eighteenth and nineteenth centuries, developments in science and technology led to the Industrial Revolution. One effect of the Industrial Revolution was that many products were made by machines instead of by hand. Factories were built to rapidly make enormous quantities of manufactured goods that ranged from clothes to medicines. However, industrialization also greatly increased the ways that humans affected the environment. (See Figure 27-3.)

Figure 27-3 In addition to producing enormous quantities of manufactured goods, factories produce air and water pollution.

CHANGES TO THE LAND—ADDING WASTES

All organisms produce wastes as a normal by-product of their life processes. However, since the time of the Industrial Revolution, the amount of wastes produced by humans has increased greatly. Also, since that time, the kinds of wastes are very different. Many of the wastes produced by our modern society do not break down and disappear quickly in the environment. Styrofoam, other plastics, and synthetic fabrics, for example, do not decompose. In addition, wastes often contain harmful chemicals. These waste materials, called *solid wastes*, must be put somewhere. One type of place solid wastes are often deposited in is called a *landfill*—an area in which garbage is buried. Most communities now deposit their solid wastes in sanitary landfills. In a sanitary landfill, a layer of wastes is covered by a layer of soil, and then by another layer of wastes, and then more soil. This layering process is continued until a mountain of garbage forms. Often, sanitary landfills are located near cities, where the amount of waste produced by the large population is enormous. In New York City, LaGuardia Airport, one of the busiest airports in the country, is located partly on a sanitary landfill. (See Figure 27-4.)

Sanitary landfills are scattered through the environment. In a sanitary landfill, attempts are made to limit the effects of the wastes on the surrounding environment. Other kinds of landfills are much more injurious to the environment. One of the most harmful is a toxic waste dump. Very dangerous wastes, usually including poisonous chemicals, are often buried in these sites. The most dangerous of all toxic wastes are radioactive

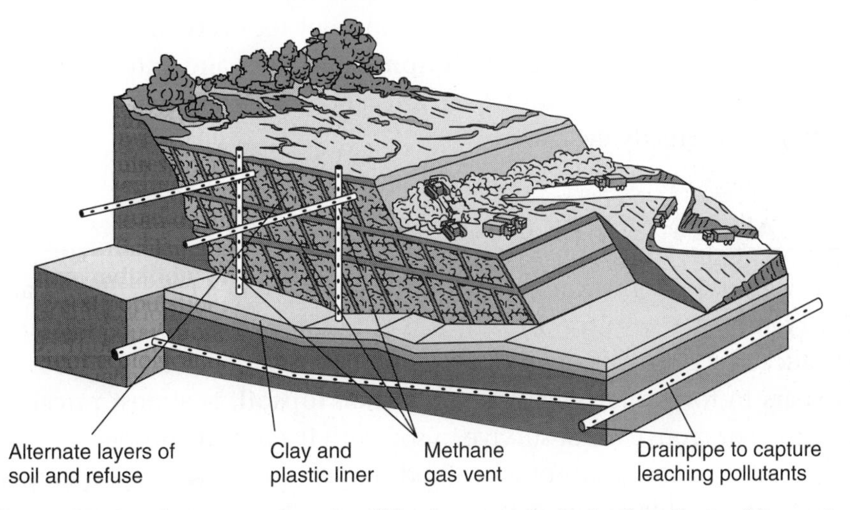

Figure 27-4 By using a sanitary landfill, a community limits the effects of its wastes on the surrounding environment.

Figure 27-5 This area is considered the worst chemical dump site in Michigan.

substances. Some of these wastes have the potential to remain in the environment for a long time. Radioactive wastes are produced by nuclear power plants, in some industrial processes, and as a result of certain medical procedures. Many radioactive wastes continue to give off harmful radiation for thousands or even millions of years. As a consequence, radioactive wastes are highly regulated. At this time, they do not represent nearly the hazard that dumps and landfills do.

Sometimes, houses are built near a toxic waste dump, or even on top of one that has been filled and covered over. (See Figure 27-5.) Often, the materials in the dump pose so strong a risk to human health that people living there become ill. The site must be cleaned and the materials removed and relocated to a safe place where they can be buried or destroyed. In 1980, the U.S. government passed the Superfund Act. Under the guidance of the Environmental Protection Agency, this act provides billions of dollars to clean up toxic waste dumps. This is important not only for the people who live near dumps, but for everyone—toxic substances have been known to leach from dump sites and enter the freshwater supply. Thus, toxic wastes can affect areas far from the site at which they were originally deposited.

CHANGES TO THE LAND—LOSING SOIL

Although soil is sometimes called dirt, it is not "dirt cheap." It is actually a very valuable resource. In fact, this combination of organic material produced by organisms and inorganic mineral particles takes hundreds of years to form. Nutrient-rich soil, called **topsoil**, is almost priceless. Because our crops cannot survive in soil with inadequate nutrients, without good topsoil we cannot grow food. Terrestrial ecosystems depend on this most valuable resource. (See Figure 27-6.)

Soil is now being lost because of human activities. For example, when

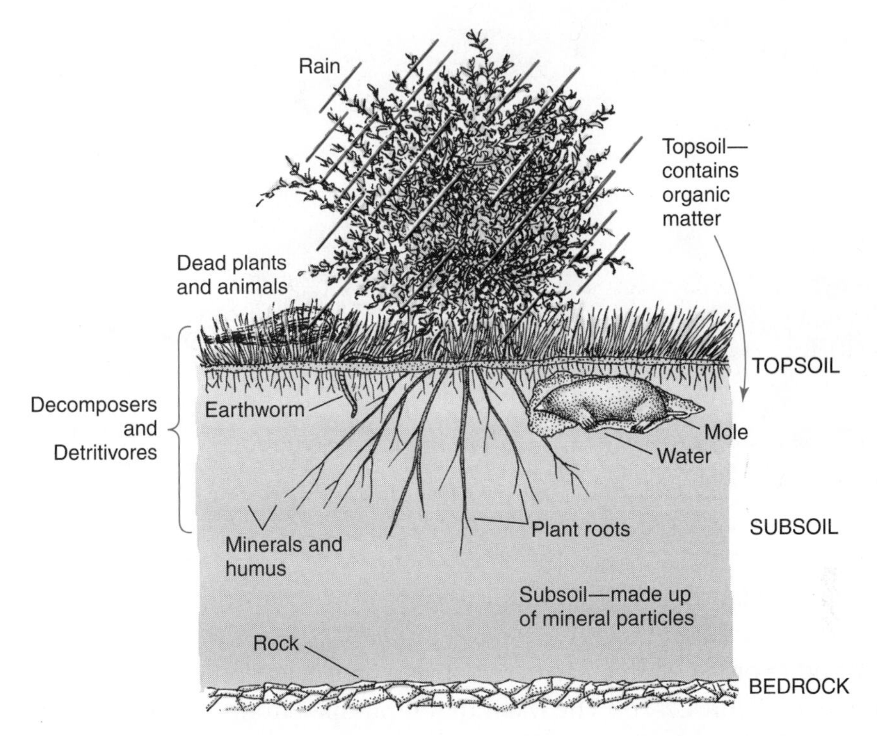

Figure 27-6 Crops cannot grow in soil that does not have adequate nutrients, which makes topsoil just about priceless.

toxic chemicals enter the ground, the soil becomes unusable. Some toxic chemicals interfere with plant growth. Others can be taken in by plants and incorporated into their tissues. Then, these plants become toxic to the animals that eat them. Also, some soil is lost when roads and buildings are constructed. These areas covered over by human construction become permanently unavailable to other life-forms.

A severe threat to our soil supply is the increasing problem called **desertification**. In desertification, arable soil is lost due to drought or poor farming practices. The land becomes dry and nutrient-poor. For example, land used to graze cattle and sheep is often stripped of all vegetation by these animals' large populations. The animals eat all the grass down to its roots and unwittingly compress the ground into a hard surface with their feet. (See Figure 27-7a on page 586.) The land becomes bare. At that point, few plants can grow. The normal plant growth that holds and enriches the soil is no longer present. Then, when weather patterns change in these areas—for example, if less rain falls—the land becomes a desert. This process is especially noticeable in areas that lie on the edge of normal desert areas. The Sahara in Africa is increasing in size

Figure 27-7a
This government-owned land in Utah has been overgrazed by cattle.

as adjoining lands dry up. Desertification is now occurring on every continent except Antarctica.

Another problem occurs in desert areas where irrigation is used to grow crops. (See Figure 27-7b.) The water that is pumped from the ground contains minerals, including salt. As the water evaporates in the dry air, the minerals remain in the soil. This process of salinization eventually makes the land unsuitable for agriculture.

As humans cut forests or remove the plants growing naturally in an

Figure 27-7b
Through irrigation, crops can be grown in deserts.

Figure 27-8 A dust storm in Cimarron County, Oklahoma, in April 1936. Poor agricultural practices, wind, and drought led to the Dust Bowl.

area, there is an increase in soil **erosion**. Both wind and water can cause erosion. Rain washes away loose soil; strong winds blow it away. In the 1930s, the Dust Bowl developed in the Great Plains when the wind, in addition to human activities, caused severe erosion. Great dust storms blew away the rich topsoil. Thousands of families left the area to begin new lives in other places. Farmers have since learned, through experiences like the Dust Bowl, how to grow crops in ways that reduce soil erosion. (See Figure 27-8.)

CHANGES TO THE WATER

Two-thirds of Earth is covered by water, but it is salt water. Only about one percent of all water on this planet is available as flowing freshwater. Yet we have taken this one percent for granted. Because flowing water carries away wastes, a stream or river seems the perfect place to dump garbage. Clear, clean water flows past the area in which the garbage is dumped, and the wastes are carried away. (See Figure 27-9 on page 588.)

In the past, this was often true. Towns were usually built next to rivers. Most human organic wastes, or **sewage**, were dumped into these rivers. Microorganisms—bacteria and protozoa—fed on the wastes. Their actions broke down the sewage into simpler substances and made the water clean again. To carry out their life processes, the microorganisms took oxygen from the water. As the water moved and splashed along, more oxygen was dissolved in it, naturally replacing what was taken out by respiration.

Today, as the human population increases in size, the situation is very different. Towns have grown into cities, and the increasing size of cities means that they now lie closer to other cities. In some metropolitan areas,

Figure 27-9 When clean, clear water such as this flows past an area that dumps its garbage in the stream, people living downstream must deal with the problem.

it is difficult to know just where one city ends and the next one begins. Many more wastes are placed into streams and rivers. The cleansing action of the water still occurs; but the helpful microorganisms are not sufficient to deal effectively with the wastes. There are simply more wastes introduced than the natural ecosystem can handle. Sometimes, harmful bacteria such as *E. coli,* which are naturally found in human sewage, are also introduced into the water. More oxygen is removed from the water by all these microorganisms than can be added back by natural processes. The river or stream, once full of life, begins to lose its ability to support the same species of organisms.

In addition, the types of wastes deposited have changed. Industries sometimes dump toxic chemicals into rivers. Industrial wastes such as mercury and PCBs cannot be broken down by natural systems into safer substances. In time, these toxic wastes build up and become more concentrated in the environment and in the life-forms that inhabit it. When the river, or lake it flows into, becomes polluted, its ecosystem changes.

Fish that need clean, well-oxygenated water are replaced by fish that can live in water with lower levels of oxygen. If the levels of pollutants keep increasing, these fish will die, too. Only scavenger fish that can survive in fetid water—with high levels of pollution and low levels of oxygen—will survive. One more point to consider: Can you predict which fish are preferred by people? Most fish that are eaten by people are species that survive only in clean water, for example, trout. Few fish species that survive in dirty water are eaten by people. And those fish are often unfit for consumption due to their high levels of contaminants.

During the past century, a major advance in dealing with the problem of water pollution has been the development of sewage treatment plants. In these treatment plants, water from reservoirs and organic wastes are treated in large tanks. Wastes in the water are chemically digested by bacteria. The remaining solids, including dead bacteria, then settle to the bottom and are removed. Chlorine is added to the water to kill bacteria. Finally, the purified water is released into a river or stream. In some areas, the solids removed from the water are sold as fertilizers after they have been treated.

Heat that is added to natural waterways is another type of water pollution. Water is often taken from rivers and used in many industrial processes. Sometimes it is used to cool machinery or to generate electricity. As a result, the water is hotter when it leaves the factory or generating station than when it entered. Because the water being returned to the river is now warmer, **thermal pollution** has occurred. It might seem that heat should not be much of a problem when added to a body of water. However, the added warm water changes the ecosystem it enters. Often, aquatic plants and animals can live in a very narrow temperature range. Higher temperatures may kill some organisms outright or interfere with their ability to feed or reproduce. Then, other organisms that are better adapted to the warmer water can increase in number. These new communities of plants and animals may not be as desirable for that habitat as the organisms they replaced. Today, power-generating plants are required to build large cooling towers. Heated water is kept in these towers to cool before it is released. Of course, the towers add heat to the air that surrounds them.

Many cities obtain their water from underground wells and aquifers. In fact, aquifers supply more than 50 percent of the drinking water in this country. Water used to irrigate crops also comes from underground sources. This water, called **groundwater**, accumulates over time and is stored naturally between layers of rock. The Ogallala Aquifer, which extends (underground, of course!) from South Dakota to Texas, is actually

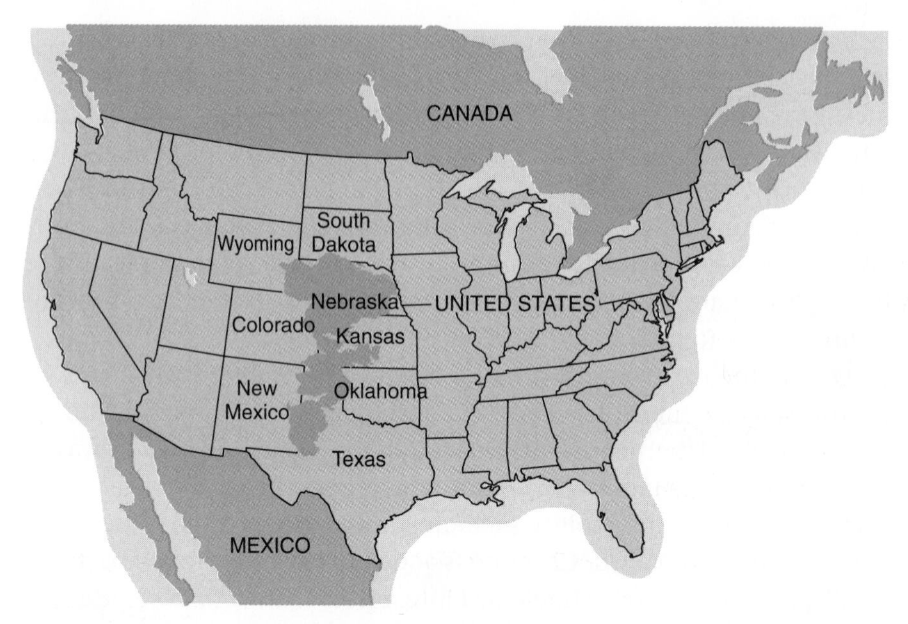

Figure 27-10 The Ogallala Aquifer, which extends from South Dakota to Texas, is an important source of fresh groundwater.

the largest body of freshwater in the world. (See Figure 27-10.) However, groundwater in the United States is being pumped out and used at a faster rate than it can be replenished by natural processes. Groundwater supplies are also vulnerable to pollution from human activities. Some scientists have predicted that U.S. groundwater will be used up within the next 50 years.

There are aboveground reservoirs of freshwater, too. (See Figure 27-11.) Much of the growth of New York City has been attributed to the water-supply system that was begun in the nineteenth century. This system transports water from upstate reservoirs through a series of huge underground pipes, or water tunnels. Construction of new water tunnels continues to this day. In spite of this seemingly endless supply of fresh water in New York, there are times when the water supply reaches low levels. In years when natural rainfall and runoff are not sufficient to replace the water being used, people are asked to use less water in their daily activities. At times, even more stringent restrictions on water use are put into place. Can new toilets help us use less water? Yes! The toilet uses up more water in the house than anything else. Now, low-flow toilets are being sold. In some cities, such as New York, these are the only kinds of toilets that can be sold. Several gallons of water are saved with each flush.

In other areas of the United States, such as the Southwest, water has

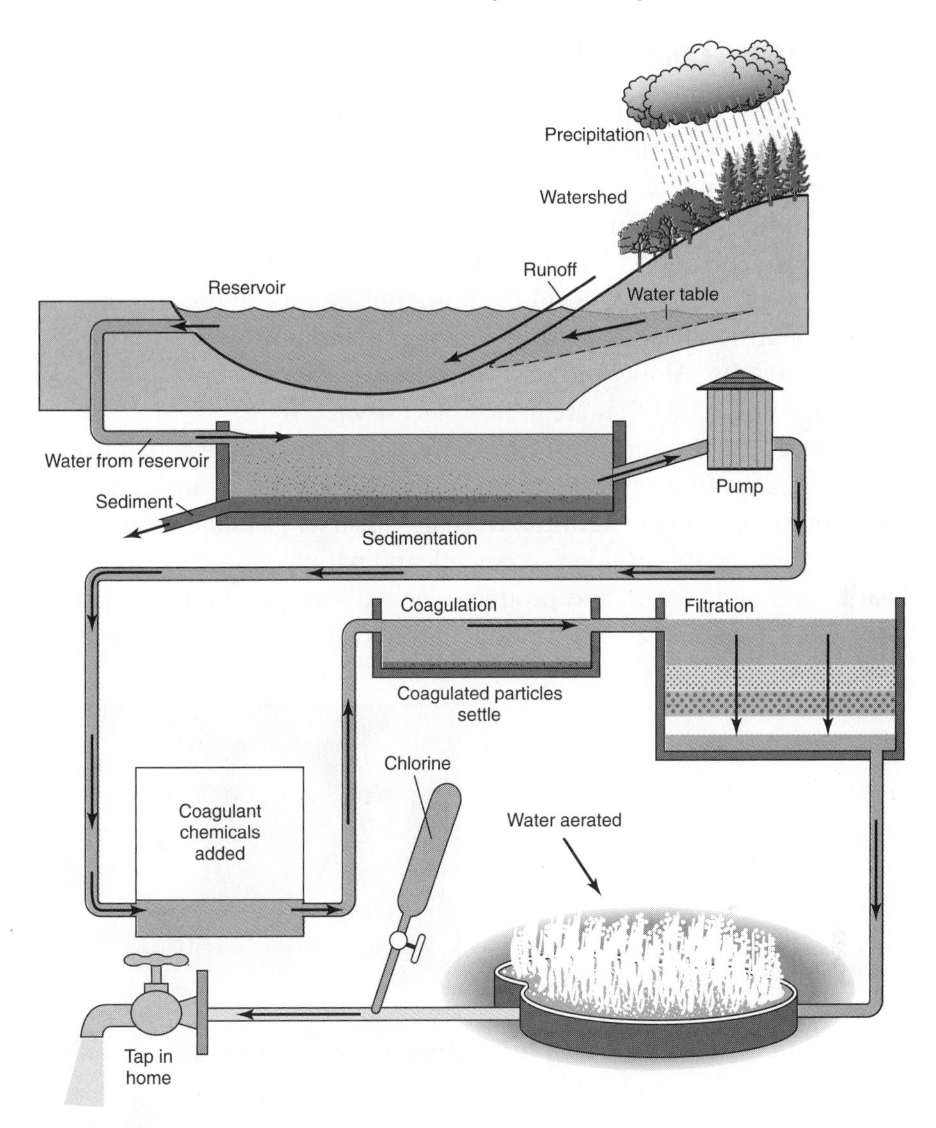

Figure 27-11 Water treatment for freshwater from an aboveground reservoir.

to be piped in from greater distances. In these places, water use is strictly controlled either by law or by diminished supply. Remember that, even if you have a plentiful water supply, there are many places that do not share this bounty.

Check Your Understanding

How can pollutants added to water at one point affect a community's drinking water far from the pollutant's point of origin?

CHANGES TO THE AIR

Air is different from land and water in that no one owns the air. Land can be owned as private or government property. Lakes can be owned by a person or a group of people. Even the coastal waters of the ocean belong to the countries they surround. But we all share the air, which forms a continuous blanket over Earth. If the air becomes polluted in one place, that pollution can easily spread to another place. Part of the continuous blanket of air that covers Earth is moving in and out of our lungs each time we breathe. Along with the air that we need to survive, we also inhale any pollutants that are in it.

What is air pollution? Gases and tiny solid particles are added to the air by human activities all the time. If these substances are not normally found in the air and are harmful, they are called *air pollutants*. For example, the burning of fossil fuels—coal, oil, and natural gas—to power cars, heat homes and offices, and produce electricity creates much air pollution. (See Figure 27-12.) Many industries release pollutants into the air

Figure 27-12 The exhaust from cars is a form of air pollution caused by human activities.

from huge smokestacks. Natural events such as forest fires and dust storms can also cause pollutants to enter the air. The worst natural sources of air pollution are erupting volcanoes, which can release great quantities of ash particles and toxic fumes into the air. (See Figure 27-13.)

You learned that when pollutants are added to water, the natural environment has the ability to cleanse itself to some degree. However, this

Figure 27-13 Volcanoes are a natural source of air pollution. The 1980 eruption of Mount St. Helens, in the state of Washington, ejected one cubic kilometer of ash and rock.

natural cleansing can occur only if the quantity of pollutants is low. This is also true about air pollutants. Pollutants can spread out in the atmosphere until the amount in any one spot is low, and they are no longer concentrated at harmful levels. Air pollutants can be absorbed by the land and water, too. But, again, this absorption can happen only if the amount of pollution is low. Unfortunately, the level of air pollution is so high in some cities that it cannot be sufficiently reduced by natural processes.

The multitude of vehicles, homes, offices, and factories that are crowded together in cities has worsened the air pollution problem. Los Angeles, London, Mexico City, Milan, and New York City are cities where large amounts of air pollutants are produced. In Tokyo, the situation is so severe that bottled oxygen is often supplied to policemen at busy traffic intersections. The worst air pollution disasters have occurred in cities when weather conditions trap pollutants and hold them near the ground. Then, the natural processes that would have dispersed, diluted, and blown away the pollutants to other places do not occur. Sometimes a cold layer of air gets trapped under a warm layer of air. This weather condition is called a **temperature inversion**. When a temperature inversion occurs, pollutants get trapped in the cold air and are held in place until the weather changes. In 1952, before there were better air pollution controls, a temperature inversion produced conditions that resulted in more than 3000 deaths in London. (See Figure 27-14 on page 594.)

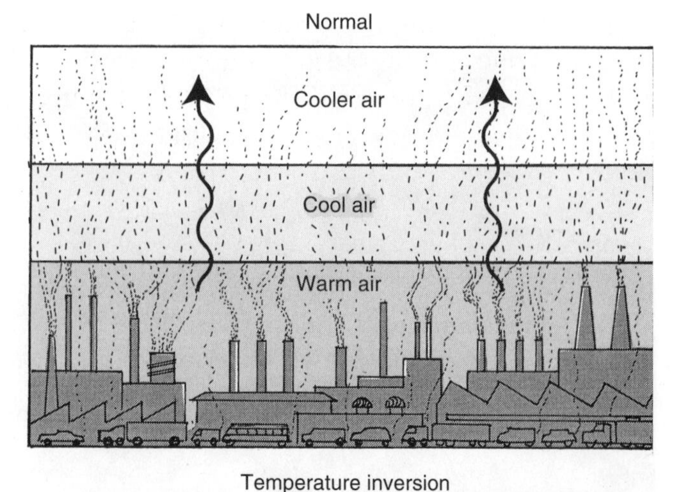

Figure 27-14 When a temperature inversion occurs, pollutants build up in the cool layer of air trapped near the ground.

One type of air pollution first noticed in cities appeared to be a mixture of smoke and fog. Thus, it came to be known as "smog." We now know that sunlight helps produce a special kind of smog, called *photochemical smog*. The pollutants from cars include carbon monoxide, complex chemicals called hydrocarbons, and gases made up of nitrogen and oxygen. When sunlight shines on these air pollutants, chemical reactions that produce brown photochemical smog occur. Two main pollutants that result from this condition are nitrogen dioxide (NO_2) and ozone (O_3). Ozone is poisonous to plants and animals. High levels of ozone in photochemical smog can be harmful to human health. Photochemical smog often can be seen in the sky over Los Angeles and New York on warm days. Weather reports may refer to these as "ozone alert" days.

Major improvements have been made in efforts to reduce air pollution. For example, car engines have built-in pollution control devices.

These devices reduce the amount of pollutants (such as carbon monoxide and hydrocarbons) added to the air when fossil fuels are burned. In addition, gasoline for cars no longer contains lead, a harmful additive that also entered the air in auto exhaust fumes.

Today, laws require factories to reduce and/or prevent the release of pollutants from smokestacks. Devices called "scrubbers" are installed, which reduce the emission of harmful compounds. New technologies for producing energy also have been developed. Solar collectors and photovoltaic cells provide us with heat or electricity without polluting the air. Thus, they help reduce overall pollutant levels in the air. In the United States, passage of the Clean Air acts of 1970 and 1977 were responsible for beginning many of these changes. As a result, the air is now cleaner than it was just a few decades ago.

GLOBAL AIR POLLUTION PROBLEMS

The atmosphere completely surrounds Earth. It protects us from certain forms of radiation from space, absorbs gases that we produce, and supplies the oxygen we need to live. The atmosphere, or air, is a mixture of several gases. The proportion of gases in the air remains fairly constant around the globe. For example, the percentage of nitrogen in the air over Europe is about 78 percent, the same as it is over South America and Africa. But mixtures can vary, so the composition of air varies. You can imagine that there is more carbon monoxide in the air over a large city than over a cornfield. However, since there is free movement of air molecules, the higher number of carbon monoxide molecules in the city will spread out to other areas of Earth. As a result of prevailing wind patterns, some types of air pollution move far from their point of origin. In fact, some types of air pollution now seem to be affecting the entire planet.

Acid rain is a form of pollution that produces effects far from where the pollutants entered the air. Sulfur dioxide and nitrogen oxides are produced when coal, oil, and gasoline in power plants and vehicles are burned as a source of energy. Winds carry these gases high into the atmosphere and over long distances. At some point, these gases combine with water droplets in the air. The water becomes acidic and falls back to the ground as acid rain. Actually, acid rain is more correctly called *acid precipitation,* since snow and ice crystals can be acidic, too. Many forests in North America and Europe have been damaged by acid rain. Fish in thousands of lakes have died because acidification of the water makes it unsuitable to support life. (See Figures 27-15a and 27-15b on page 596.)

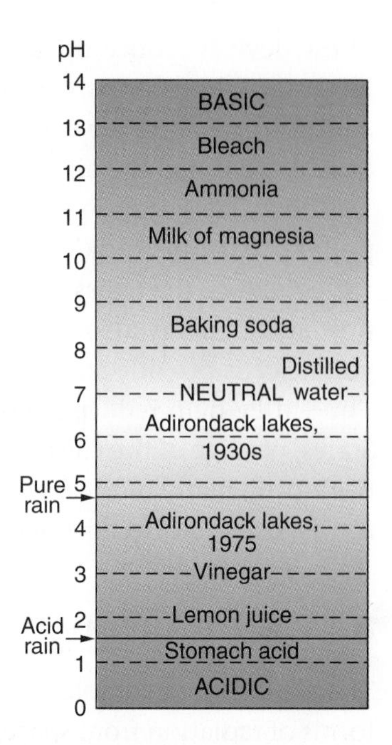

Figure 27-15a Acid rain has caused lakes in the Adirondack Mountains to become more acidic, causing harm to wildlife such as fish.

Perhaps even more important to the future of Earth are the effects of **global warming** and ozone depletion. Carbon dioxide (CO_2) in the atmosphere helps keep Earth warm. It prevents infrared wavelengths of energy (which warm the planet) from escaping back out to space. This is called the "greenhouse effect." The glass of a gardener's greenhouse does the same thing, trapping in the heat. (See Figure 27-16.) Because of all the fossil fuels we have been burning, the amount of CO_2 has been

Figure 27-15b Acid rain also hurts trees, and has damaged many forests in North America and Europe.

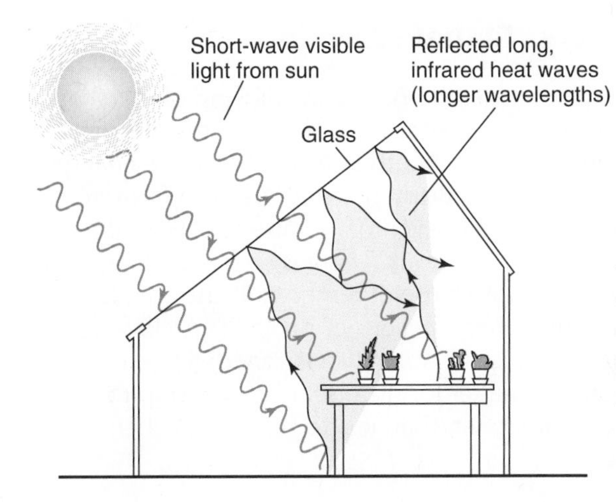

Figure 27-16 The greenhouse effect—carbon dioxide in the air traps infrared energy, warming the atmosphere. This is similar to the way the glass in a greenhouse traps heat, keeping the plants warm.

increasing in the atmosphere, particularly so in the last 200 years, since the start of the Industrial Revolution. In addition, the burning of rain forests to clear the land has added more CO_2 to the atmosphere, and destroyed countless trees that used to absorb CO_2. (See Figure 27-17.)

With more CO_2 in the atmosphere, more heat is trapped. Many people are concerned that the climate of Earth is getting warmer for this reason. The average global temperature has increased in the past 100 years. In fact, the 1990s was the warmest decade on record. Such a change in Earth's climate could have major effects on habitats and organisms everywhere, as warm-temperature plant and animal species move toward the poles.

The atmosphere protects life on Earth in another very important way. Ultraviolet (UV) rays are part of the energy that reaches Earth from the sun. UV rays can damage the DNA in cells, including the DNA in our cells. Skin cancer is a serious disease that can be caused by overexposure to UV radiation. There is a layer of ozone (O_3) gas that surrounds Earth high in the atmosphere and blocks out much of this UV radiation. Scientists have

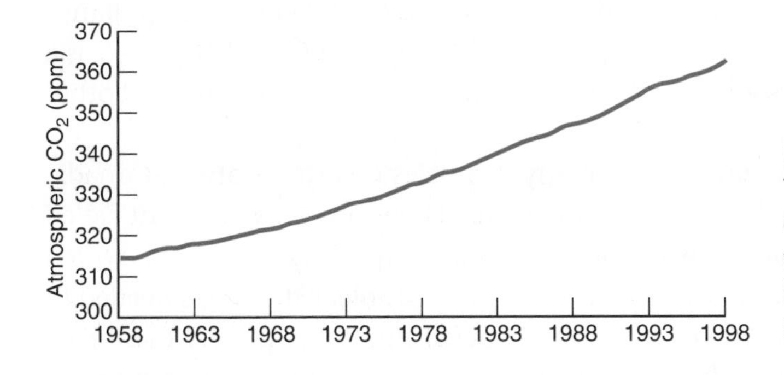

Figure 27-17 The amount of carbon dioxide in the atmosphere has been increasing.

Antarctic Ice Shelves and Global Warming

Dr. David Vaughan is a British scientist who has been studying glaciers for a long time. As a glaciologist, he is very interested in the ice that covers and surrounds the great landmass at the South Pole. If the climate change that is occurring on Earth causes global warming, then the ice of Antarctica will start to melt. The melting ice would, eventually, raise sea levels around the world and the results would be disastrous. There is enough ice in just the western part of Antarctica to cause a rise of 5 meters in the sea level. This would flood many coastlines where millions of people live. However, knowing what is happening to the ice of Antarctica is very difficult. Dr. Vaughan has said that it would be easier for us to send a spaceship to study Europa, the ice-covered moon of Jupiter, than it is to send researchers to the West Antarctic Ice Sheet.

Much attention has been given in recent years to ice shelves, floating masses of ice that surround much of Antarctica. While their melting will not directly affect sea levels—that ice is already in the sea—the loss of the ice shelves would make it much easier for the huge masses of land ice to melt. So, Dr. Vaughan and other glaciologists have been closely monitoring a series of collapses of ice shelves that began in 1995. In January 1995, a 770-square-mile section of an ice shelf along the Antarctic Peninsula broke apart suddenly. This area of ice was 35 times larger than all of Manhattan. Another even larger ice shelf that was at least 400 years old broke apart in 1998. And, in just 35 days beginning on January 31, 2002, the largest

suggested that without the protection of the ozone layer, early life-forms might never have moved out of the oceans and continued to evolve on land. Now scientists are concerned about the effects of air pollutants in the atmosphere because certain pollutants harm the ozone layer. This effect is known as **ozone depletion**. Chlorofluorocarbons (CFCs) are suspected of causing the most ozone depletion. CFCs are found in air conditioners and refrigerators. It is the compression and expansion of these compounds that actually cools the air or the food. When the appliance leaks, the CFCs get into the air. Until the 1970s, CFCs were also used as a propellant in aerosol cans of hair spray, deodorant, spray paint, and other consumer products.

In 1987, representatives of many countries met in Montreal, Canada, to discuss the problems caused by ozone depletion. An agreement called the Montreal Protocol was made at that meeting. This agreement, which was far tougher than anyone had thought possible, listed steps needed to protect the ozone layer by limiting or eliminating the use of ozone-depleting chemicals by 1996. The agreement came into effect on time collapse to be seen in 30 years occurred. About this ice shelf, Dr. Vaughan said, "We knew what was left would collapse, but the speed of it is staggering." The area of ice that disappeared was larger than the state of Rhode Island, 220 meters thick, and contained 720 billion tons of ice—enough to fill 12 trillion 19-kilogram bags!

These ice-shelf collapses are not entirely unexpected. The temperatures in the area of the Antarctic Peninsula, a long sliver of land pointing toward South America from Antarctica, have been rising steadily since the 1940s. The average temperature is now 2.5 degrees Celsius higher than it was in 1945. This is the fastest rate of warming seen any place on Earth. With the higher temperatures comes the melting of ice. The really big question remains: Is this change occurring only in this one place on Earth, or is this an early warning sign of global warming everywhere?

Scientists cannot agree on the answer to this big question. However, they are determined to study ice around the world—especially in the Antarctic—even more closely to get an answer. But how can this study be done, knowing, as Dr. Vaughan has said, how difficult it is to get to Antarctica? Go into outer space! Which is exactly what has been done. A satellite called *ICESat* (Ice Cloud & Land Elevation Satellite), launched in 2003, is now orbiting Earth to track precise changes in ice sheets around the world.

when the required number of countries, including the United States, ratified it. Much progress has been made since the agreement was signed.

HUMAN POPULATION GROWTH

The most serious problem that now affects all life on Earth is a problem of numbers: the rate at which the human population is increasing.

Over most of the past 2 million years, the human population has increased slowly. However, about 600,000 years ago, when people first began to use tools, a more rapid population increase occurred. Since then, sudden population increases have occurred several times. The beginnings of agriculture, about 10,000 years ago, caused a rapid increase as people settled down with a more secure source of food. And again, about 200 years ago, the Industrial Revolution began the most recent and rapid population increase. This last event, combined with scientific advances in farming and medicine, has caused an explosion in the size of the human population.

Let's look at some amazing numbers. In 1800, the human population reached 1 billion. It had taken about 2 million years to reach that number. In 1930, after only 130 more years, the human population reached 2 billion. In 1960, after only 30 more years, the human population reached 3 billion. Only 15 years later, in 1975, the number was up to 4 billion. A dozen years later, in 1987, it was 5 billion. By the year 2000, the human population had reached 6 billion. The United States Bureau of the Census predicts a world population of 7 billion by 2012, 8 billion by 2027, and 9 billion by 2047. (See Figure 27-18.)

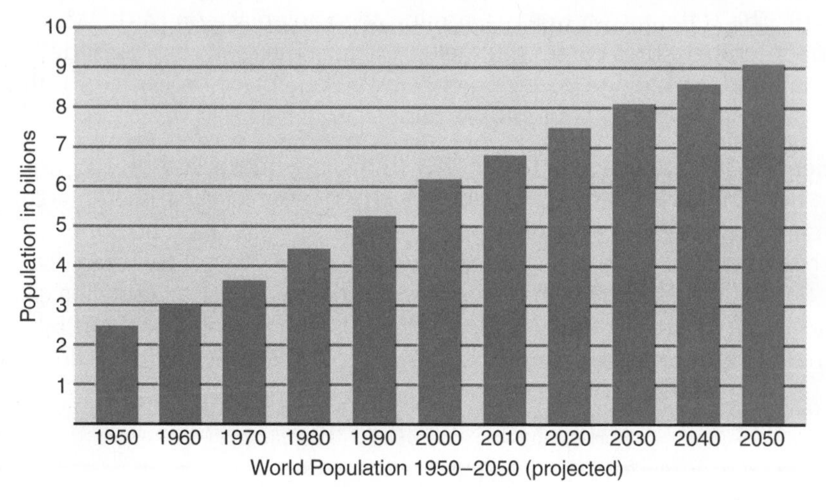

Figure 27-18 The world's human population is rapidly increasing.

Human population growth rates are not the same everywhere. In highly industrialized countries such as Japan, Canada, the United States, and most European nations, population growth rates are low. In developing countries such as India, Brazil, Mexico, and Kenya, the human population grows at a much faster rate. Even in China, where population control through family planning has been quite successful, the growth rate is still much higher than in developed countries. How will the growth rates of the different countries in the world affect Earth's total human population in the future?

What is now known is that the average annual growth rate for the world human population reached a peak around 1990 and has been declining ever since. In other words, the human population is still growing but more slowly now. This is because—in almost every country, including the poorer developing nations—standards of living are rising, the status of women is improving, and there is an increased use of birth control. While the death rates are declining, so are the birth rates. People are having fewer babies, partly because parents are more certain that the children they are having will survive and grow to maturity. As a result, a general agreement is developing among scientists that the world population will continue to grow until the middle of this century, reaching a peak of about nine billion, and then may start to decline in the second half of the century.

Regardless of future growth patterns, however, the big question now is: What is Earth's carrying capacity for humans? Some scientists think that the population is already past the planet's carrying capacity and that very serious environmental problems are already beginning.

Today's human population of six billion may lead to the end of many life-forms on Earth. Habitat destruction, such as that of tropical rain forests, will cause the rapid extinction of vast numbers of species. Many organisms are already being pushed toward extinction due to loss of habitat and other human factors. Industrialization in developed and developing countries will create more air, water, and land pollution. Acid rain, global warming, and ozone depletion are worldwide concerns. Earth—as one large, complex, intact ecosystem—cannot go on forever under the pressures exerted by an ever-increasing human population.

In 1994, in Cairo, Egypt, 160 countries met at the International Conference on Population and Development. The participants at the conference overwhelmingly agreed that Earth's population cannot continue to grow at its current rate—that is, not without terrible consequences. Yet there was much disagreement about how to lower the growth rate. What is clear is that our population, just like the population of any other organism, cannot increase forever. Either we will find a way to control the size of the human population, or nature will do it for us. Polluted land, water, and air; lack of food and space; and widespread disease may ultimately limit human population size. On the other hand, individual choices and government planning could also limit human population size. Let us be wise enough to make the right decisions, while we still can.

LABORATORY INVESTIGATION 27
How Do Human Activities Affect Natural Ecosystems?

INTRODUCTION

In this investigation, you will study how the ecosystem of a hypothetical island, Key Mangrove, has been changed by people over a 30-year period. You will probably guess from its name that the most important characteristic of this island is its mangrove swamps—a kind of wetland ecosystem found in the tropics. Mangrove trees are one of the few trees that grow in brackish water. These trees are able to absorb minerals and salt from the ocean. Mangrove trees provide food and shelter for many species of animals that breed and grow along warm ocean coasts.

People are attracted to beautiful natural environments. However, they often make alterations to these environments without being aware of the consequences of such changes. The effect of human development on Key Mangrove is the subject of this activity.

MATERIALS

Maps of Key Mangrove that cover a period of more than 30 years (from the *Teacher's Manual*)

PROCEDURE

1. Examine the first map of Key Mangrove. About what percent of this island was originally covered by mangrove swamps?
2. The other three maps were made at 10-year intervals. Prepare a chart to describe the changes that have occurred during each 10-year period. Keep the following questions in mind as you examine the different maps.
   - How has transportation to and around the island changed?
   - What industries have been added?
   - How have these industries changed over time?
   - How has residential use changed over time?
   - What recreational facilities have been added?
   - What municipal services have been added to meet the needs of increased residential and manufacturing developments?

3. Compare map 1 with map 4. What percentage of the mangrove swamps in map 1 remain on map 4? Approximately what percentage of mangrove swamps was lost during each 10-year time period?
4. What kind of development was built first? How did this development change the pattern of later development?
5. In what ways could people use the native mangrove swamps for recreation?
6. In what ways has the recreational use of the mangrove swamps been altered by increased development?
7. How has the recreational use of the island changed over time?

INTERPRETIVE QUESTIONS

1. From library or Internet sources, research and describe the benefits that mangrove swamps provide for coastal areas.
2. What did the mangrove swamps provide to the first human settlers of Key Mangrove?
3. Only one large area of mangrove swamp remains on Key Mangrove after 30 years of human settlements. What interests could the following people be expected to have in this area? Prepare a short statement of each person's intentions for this area: vacation homeowner, member of the local Chamber of Commerce, National Wildlife Federation member, shrimp-packing plant owner, member of the shrimp-packers' union, permanent homeowner, oil refinery manager, bank president.

CHAPTER 27 REVIEW

Answer these questions on a separate sheet of paper.

VOCABULARY

The following list contains all of the boldfaced terms in this chapter. Define each of these terms in your own words.

acid rain, desertification, erosion, global warming, groundwater, ozone depletion, sewage, temperature inversion, thermal pollution, topsoil

PART A—MULTIPLE CHOICE

Choose the response that best completes the sentence or answers the question.

1. Toxic waste dumps *a.* become harmless after a few years *b.* contain alternating layers of soil and garbage *c.* affect only the immediate area *d.* are the target of the Superfund Act.
2. Which of these is *not* a result of excess sewage being dumped into water? *a.* increase in harmful bacteria in water *b.* increase in oxygen levels *c.* different types of fish found in the water *d.* buildup of toxic chemicals in aquatic organisms
3. The rich, uppermost layer of soil is *a.* bedrock *b.* topsoil *c.* subsoil *d.* ozone.
4. Arable soil is lost due to drought or poor farming practices in a process called *a.* desertification *b.* thermal pollution *c.* ozone depletion *d.* desalination.
5. Which of these specifically addresses the problem of thermal pollution? *a.* cooling towers *b.* Superfund Act *c.* Endangered Species Act *d.* sanitary landfills
6. Nitrogen oxides and sulfur dioxide are the main causes of *a.* thermal pollution *b.* acid rain *c.* the greenhouse effect *d.* ozone depletion.
7. In *Silent Spring,* the town's loss of birds and other problems turn out to be the result of *a.* terrorist activities by a foreign enemy *b.* nuclear radiation *c.* a viral epidemic *d.* the people's own actions.
8. In the United States, groundwater *a.* is found in aboveground reservoirs *b.* is unaffected by toxic waste dumps *c.* is being used at a faster rate than it can be replenished *d.* supplies less than 50 percent of the drinking water.
9. Humans have a greater impact on the environment than most species because they *a.* are more numerous *b.* continually

improve on methods to grow food and to make things *c.* hunt other animals for food *d.* use tools.

10. The human population *a.* increases at a fairly uniform rate throughout the world *b.* is starting to decrease *c.* will continue to increase rapidly even if the rate of growth decreases *d.* does not compete for space and resources with other living things.
11. The wearing away of rock or soil by wind or water is called *a.* desertification *b.* erosion *c.* ozone depletion *d.* siltation.
12. Which of these statements about air pollution is false? *a.* It does not occur naturally. *b.* It is involved in global warming. *c.* It causes acid rain. *d.* It has decreased in the United States.
13. The world's total human population reached six billion individuals by the year *a.* 1800 *b.* 1900 *c.* 1950 *d.* 2000.
14. The Dust Bowl of the 1930s was the result of *a.* soil erosion *b.* overpopulation *c.* air pollution *d.* ozone depletion.
15. In the atmosphere, carbon dioxide *a.* reacts with water to form acid rain *b.* filters out harmful UV rays *c.* prevents heat from escaping into space *d.* is an important component of smog.

PART B—CONSTRUCTED RESPONSE

Use the information in the chapter to respond to these items.

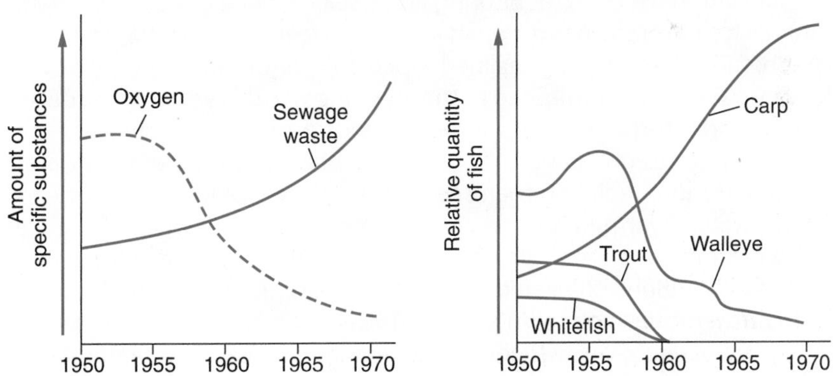

16. According to these graphs, what is the relationship between sewage and oxygen content?
17. What impact have these changes had on the fish populations? What might be done to restore fish populations to the relative quantities they had in 1950?
18. Explain why the growth of the human population creates an ecological problem.

19. Why has solid waste disposal become an increasingly difficult problem since the time of the Industrial Revolution? Name and describe two types of landfills.
20. List and describe three ways of reducing air pollution.

PART C—READING COMPREHENSION

Base your answers to questions 21 through 23 on the information below and on your knowledge of biology. Source: *Science News* (December 21 & 28, 2002): vol. 162, p. 400.

Warm Arctic Summer Melted Much Ice

Satellite observations of the Arctic Ocean show that the amount of sea ice there this year was the lowest it's been in more than 20 years.

In September, the extent of the sea ice—defined as the area in which ice covers at least 15 percent of the ocean's surface—was 5.27 million square kilometers, says Julienne C. Stroeve, a climatologic at the National Snow and Ice Data Center in Boulder, Colo. Of that area, sea ice actually covered about 3.6 million square kilometers, a figure 17 percent lower than normal for that time of year and 9 percent below the previous minimum for a September. The earlier record low was set in 1998, during the late stages of the strongest El Niño ever seen and when the global average temperature had been much higher than normal for several months.

Satellites have been monitoring arctic sea ice since 1978. Since then, annual ice coverage has dropped about 3 percent per decade, and September ice coverage has declined 8 percent per decade.

Several factors contributed to the low ice cover this year, says Stroeve. From March through May, southerly winds pushed ice away from the northern shores of Eurasia and North America. Because the open water absorbed more radiation than snow-covered ice would have, the near-shore waters warmed and accelerated melting at the edges of the ice packs. From June through August, unusually warm and persistently stormy conditions blanketed the Arctic Ocean, fracturing the [sea ice] and further fostering melting.

Unlike earlier years with low ice coverage, this year the sea northeast of Greenland was relatively free of ice.

21. State the method that is used to conduct observations of the amount of sea ice on the Arctic Ocean.
22. Explain what was unusual about the amount of sea ice observed in the Arctic in September of 2002.
23. State the two reasons why the sea ice coverage may have been so low this past year.

Saving the Biosphere

After you have finished reading this chapter, you should be able to:

Discuss the Gaia hypothesis that Earth is a living organism.

Relate the importance of the 3 Rs—reduce, reuse, and recycle—to the health of the biosphere.

Compare renewable and nonrenewable resources.

Discuss reasons why protecting the environment must be a worldwide effort.

Plant a new Truffula. Treat it with care.
Give it clean water. And feed it fresh air.
Grow a forest. Protect it from axes that hack.
Then the Lorax
and all of his friends
may come back.

Dr. Seuss, *The Lorax*

Introduction

April 22, 1970, was a beautiful spring day across the United States. More people joined demonstrations on that day for a single purpose than ever before in history. That was the first Earth Day, and its purpose was to rally support to save the environment. (See Figure 28-1 on page 608.) Speeches, parades, marches, and rallies occurred everywhere. Fifth Avenue in New York City was closed to traffic so that people could walk in safety down the street. Even the United States Congress closed so that politicians could return home to attend local events.

On that first Earth Day, it was as if everyone awoke at the same time. Everywhere, people realized that they had shared concerns. The environment, on which all living things depend—our land, our air, our water—was

Figure 28-1 The environment includes cities and wilderness areas, such as the Grand Teton Mountains and Snake River in Wyoming.

becoming increasingly polluted. This first Earth Day marked the beginning of the modern environmental movement. It also marked the beginning of the realization that something had to be done. Many environmental organizations were founded at this time. The United States government passed the Clean Air Act and the Clean Water Act. Other laws that created the Environmental Protection Agency were also passed.

WHAT NEEDS TO BE SAVED?

All environmental problems are local, regional, and, in the end, global. Some examples may illustrate these points. Some streams that flow through towns are badly polluted. The water has a vile odor, the fish in the stream are dying, and garbage is present everywhere. This is a local problem. Most of the lakes in the northeastern United States have become acidic from acid rain. Today, the pH in these lakes is lower than it used to be, and certain species of fish cannot reproduce. Acid rain creates regional problems. Finally, the average temperature of Earth may be rising. This warming of the planet because of human and natural changes in the atmosphere is a global problem. (See Figure 28-2.)

The biosphere is the total area of the land, water, and air on Earth's surface where life is found. Saving the biosphere means paying attention to local, regional, and global problems. Can we find ways to live that allow us to protect the environment instead of hurting it? Can we protect the environment not only for our own species but for all species? What must we do to protect the environment for organisms that will be alive 50, 100, or 200 years from now?

Figure 28-2 Although environmental problems start out local, they become regional, and in the end have a global impact.

A CHANGE IN ATTITUDE

We will look at some ways we can protect the biosphere. Sometimes the most important changes are the most difficult to make. It is often good to examine some accepted attitudes in our society. For example, what if we thought it was most important that our lifestyle not harm our relationship with the environment and other living species? Would we be willing to make the necessary changes in our lifestyle to accomplish this?

Attitudes would also have to change in how we view ourselves in relation to the whole Earth. Often, an industrialized society views Earth only as a source of valuable resources for its use. In contrast, ecology teaches us that humans are just one of many species. All species on Earth, including ours, interact and are interdependent. From our study of ecology, we learn that in order for our species to survive, we must make sure that these important relationships continue. (See Figure 28-3 on page 610.)

THE GAIA HYPOTHESIS

In 1972, a powerful idea was introduced by British scientist James Lovelock. Through his study of Earth and its atmosphere, Lovelock came to a surprising conclusion. The whole Earth, he said, including its atmosphere, should be considered to be a single living organism. As a living organism, Earth is constantly adjusting its own conditions to keep life going. He named his idea the Gaia hypothesis. (*Gaia* means "Earth" in Greek.) To support his hypothesis, Lovelock and other scientists have shown that the mixture of gases in the atmosphere is continually changing. Earth

Figure 28-3 People, such as this young shepherd, who work closely with animals and the land often are more aware of our relationship with Earth and other species.

itself and the living things on Earth make these adjustments. The purpose, he says, is to make sure that life can continue. Lovelock said that these adjustments have been occurring for billions of years. Because it continually changes itself, Earth as a single, huge organism is able to keep itself alive. Most scientists are not sure if the Gaia hypothesis is correct. But the idea that Earth is a living organism has made a big difference in the way many people think about the planet. The Gaia hypothesis forces us to regard Earth as one mighty being made up of many, many interrelated components. We, as a single species, cannot interfere with these relationships. If we do, other changes will occur on Earth that could be very harmful to ourselves and other species.

We need to find a way to balance how we live with how the environment functions. A change in attitude may be one of the most difficult things to do. However, it may also be one of the most important in order to maintain life as we know it.

THINK GLOBALLY, ACT LOCALLY

Environmental problems are often large and seemingly resistant to solutions. For example, if the air that covers your city is polluted, what can you—just one person—do about it?

You can do what a baby does when it learns to walk. It makes many little steps while gaining confidence as its small body learns to better coordinate its movements. One teacher offered the following advice to her class. Begin by making a list. On the list, write down the things you can do as an individual to protect the environment. Many of the really large problems facing us today are the result of an accumulation of a number of small problems. Let's follow this procedure with an example. Sanitary

Figure 28-4
The Staten Island sanitary landfill in New York City was closed in 2001. Although recycling has reduced the amount of garbage, the city must make plans for the future.

landfill sites are filling up quickly. (See Figure 28-4.) Everyone produces too much garbage. The list could include a number of ways a person can reduce the amount of waste he or she produces. A sample list may look like the following:

- Use a mug or a glass instead of paper cups for drinks.
- Make double-sided photocopies instead of single-sided.
- Bring your own shopping bags to the grocery store.

Try making your own list of ways you can reduce the amount of wastes you produce. Everyone should be able to think of at least ten ways that they could change their behavior to protect the environment.

The health of the environment will depend on people's attitudes and behaviors. In a consumer society, people are often encouraged to buy and use more and more. In fact, the health of the economy often depends on people's spending habits. Many people purchase more things than they need, things that enhance their self-esteem or impress others. Some environmentalists have suggested that people must learn to appreciate the "hidden costs" of many consumer goods. The environment pays a price for the perceived needs of people who live in a modern industrial society.

Some environmentalists have encouraged people to live by the "3 Rs": reduce, reuse, recycle. *Reduce* means "use less," that is, reduce consumption. Five paper towels to clean a spill on a counter may not be necessary; two may do the job just as well. (A reusable cloth may be better still.)

Figure 28-5 Recycling is everybody's concern. As shown here, students can help in the recycling effort.

A well-designed low-flow toilet uses only a fraction of the water of a regular toilet. Using fewer towels and less water aids the environment by protecting resources. Another advantage is that fewer materials are discarded in landfills, and sewage treatment facilities have to deal with less wastewater.

Many products can be *reused.* Paper or plastic grocery bags that you bring food home in can be taken back the next week and reused. In this case, less material is used to make more bags and fewer items are added to sanitary landfills.

Finally, many products can and should be *recycled* to make other products. Most cities now have programs to recycle plastic, glass, metal, and paper. (See Figure 28-5.) Recycling requires that individuals separate recyclable materials that have some value. These recyclable materials must be collected separately. Most important, ways must be found to use these materials again. Newspapers are now required to have at least ten percent of their paper made from recycled material. (See Figure 28-6.) In New York City, the black asphalt on many streets contains crushed recycled glass instead of gravel. Building materials are now being made that look like pieces of wood. The "wood" is actually a form of plastic made from recycled materials.

Figure 28-6 These newspapers are tied and ready to be recycled.

RENEWABLE VERSUS NONRENEWABLE RESOURCES

Air, water, and sunlight are some of the important resources that are basic to life on Earth. The world economy since the Industrial Revolution requires other resources. These resources include coal, oil, and raw materials such as metal ores.

On Earth, resources can be considered renewable or nonrenewable. A **renewable resource** can be replaced within a generation. Enough of the resource is being made to replace what is being used. For example, oxygen in the air used by organisms for respiration is renewed by the life processes of green plants. Wood used to build houses can be replaced as trees are replanted and grow. Foods derived from plants and solar energy from the sun are renewable resources. Even the wind can be considered a renewable resource. (See Figure 28-7.)

Nonrenewable resources cannot be replaced. A limited amount of nonrenewable resources exists on Earth. Nonrenewable resources include such energy sources as coal, oil, and natural gas. In addition, metals such as gold, silver, iron, copper, and aluminum and nonmetals like sand, gravel, and limestone are also nonrenewable resources.

An important step toward protecting the biosphere is to begin to use renewable energy sources. Electricity can be made from burning coal—a nonrenewable resource—or renewable resources can be used. For example, electricity can be made in a dam from the power of falling water, and wind power can turn the blades on giant windmills to generate

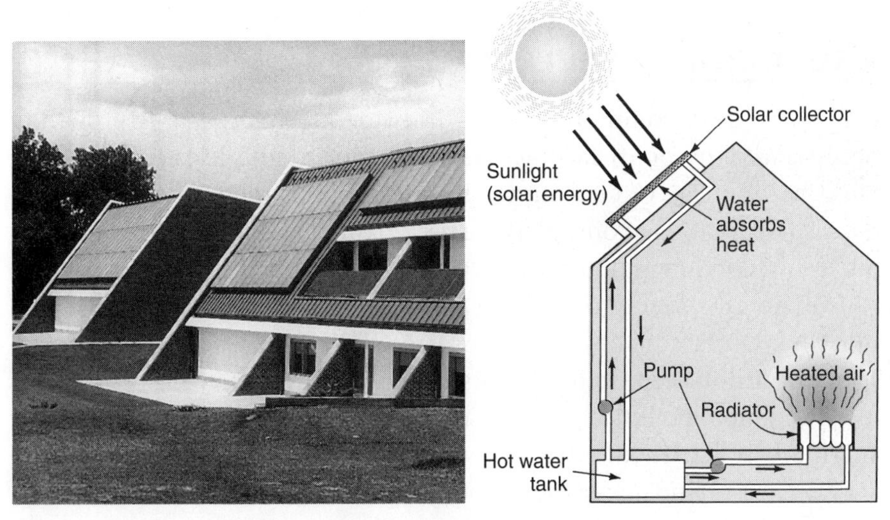

Figure 28-7 The solar panels on this school in Denver, Colorado, provide 70 percent of the building's heat and hot water by using a renewable resource.

Figure 28-8 Windfarms in the Altamont Pass area of California provide electricity by using a renewable resource.

electricity. (See Figure 28-8.) Sunlight can be used to heat water and buildings, and to produce electricity. Finally, hot water from deep beneath Earth's surface can be used to heat buildings and to make electricity. This is geothermal energy. All of these—water power, wind power, solar energy, and geothermal energy—are renewable energy sources.

SUSTAINABLE DEVELOPMENT

One of the most common beliefs worldwide is that economic growth is necessary. Many politicians get elected by promising to keep the economy growing. It is often assumed that more of everything is better. This means larger cities, more roads, faster airplanes, taller buildings, and so on. If the economy is not growing, people think that something is wrong.

We are now seeing that there are many problems with this way of thinking. On this planet, unlimited growth is not possible. The human species cannot keep increasing its demands on Earth. We cannot add more and more pollution to our water, air, and land. We cannot use more and more of Earth's resources forever.

We need to find a way to live on Earth that is sustainable. Sustainable means living in a way that does not interfere with Earth's ability to support life in the future. Changing and improving the way we live without harming the environment for people in the future is called **sustainable development**. Taking the idea of sustainable development seriously means making big, real changes in how we do things today. These changes may mean the difference between a healthy future and a future with very big problems.

Check Your Understanding

What is the importance of sustainable development in maintaining the health of life on Earth?

SUSTAINABLE DEVELOPMENT OF FORESTS

An example of sustainable development involves the forestry industry. Some of the greatest damage to the environment is being done as we cut down trees for our use. Trees are a source of lumber for the construction industry. They are also used to make wood pulp, which is used to make paper. In many parts of the world, wood from trees is used as a source of fuel to heat homes and cook food. (See Figure 28-9.)

The most economical way to harvest trees is to cut down an entire forest. Trees are felled with huge chain saws. Logs are trimmed of branches and removed from a forest area by cranes and large trucks. Roads are made through the forest to accommodate the large trucks moving back and forth with their heavy loads of wood. Many places where old forests once stood are now just vast wide-open areas. This method of forestry is called *clear-cutting*. As a result of this practice, forest animals leave or die. Tree roots make the soil act like a sponge to soak up rainwater. Without the trees, rain runs right into the streams and rivers, carrying the topsoil with it. Plants can no longer grow. Floods occur more often. Soil erosion affects

Figure 28-9 The wood from forests has many uses: lumber, wood pulp for paper, and (in some areas) fuel.

Figure 28-10 This hillside has been clear-cut. Although cutting all the trees in one area is economical, it takes a severe toll on the environment.

rivers and streams. Sand and gravel flow into them, changing and destroying the habitat for fish. (See Figure 28-10.)

Sometimes only some trees are cut. This can be less damaging to a forested area than clear-cutting. However, to get to the trees, logging roads must be built. These roads cause tremendous damage to the forest environment. There are now hundreds of thousands of kilometers of logging roads in the remaining forests of the United States.

Sustainable development in forestry would mean using the trees in a way that does not harm or destroy the forest. This would mean replacing every tree that was cut with a seedling and making sure that the seedling survives. Only a few logging roads could be constructed, and these roads would have to be built carefully with environmental concerns in mind. In addition, sustainable development in the forest industry would also have to consider the matter of numbers. Suppose it takes 60 years for a tree to grow large enough to be useful as lumber. In a forest of 300 trees, removing and replacing five trees every year would be sustainable. At the end of 60 years, all the trees would have been removed once ($5 \times 60 = 300$), and a forest would still remain. However, it would be a forest of trees of all sizes and different ages, continually regrowing. Such a forest would provide a healthy habitat for other woodland species. Soil erosion would not occur. This method of using a forest makes sense and is entirely possible. It simply means not taking more than Earth can replace.

Another interesting method for protecting forests involves using different plants as a replacement for tree-derived products. The kenaf plant has been grown in Egypt for thousands of years. It grows extremely quickly, often reaching 3.5–5.5 meters in only five months. Kenaf looks like a tall bamboo or sugarcane plant. When kenaf plants are cut and ground up, they make excellent pulp to produce paper. In fact, kenaf paper is actually of a better quality than paper made from wood pulp. Kenaf plants grow very well in the southern United States. By using kenaf plants as a source of paper pulp, we would be able to cut fewer forests. Instead, we could grow kenaf on farms like any other crop. This would help us live in a sustainable manner with the forests.

Rubber trees provide another example of sustainable forest development. Wild rubber trees that grow in Brazil's tropical rain forests produce the milky liquid from which natural rubber is made. People make cuts in the tree's bark and collect the liquid that oozes out. The trees are not harmed by the process. The native people understand that what is good for the trees is good for them.

Other rain-forest trees produce the valuable Brazil nuts. These nuts, high in fat and protein, are sold throughout the world. Collectors of Brazil nuts, like the people who collect the rubber, know how important it is to keep the forest healthy. These uses of the forests are sustainable. How different is the use of tropical rain-forest land for cattle ranching, farming, and logging! In order to use the land in these ways, the rain forests are cut, burned, and destroyed at an alarming rate. Sustainable development is not just a fancy idea. It may be the difference between a good or a very bad future for life on Earth.

ENVIRONMENTAL PROTECTION IN A DEVELOPED COUNTRY

The United States is a highly industrialized, developed country. Like other developed countries, the United States uses up more than its share of energy and resources. Also, as in other developed nations, the environment has been seriously affected by human activities. As a result, awareness of the environment has increased in the last 30 years.

The Environmental Protection Agency (EPA) is one important result of an increased public awareness about environmental problems. As an independent agency of the United States government, the EPA is responsible

The Invasion of Alien Species

To an ecologist, an **alien species** is not an invader from outer space. Rather, it is an organism that has been introduced into a new community by humans, either by accident or on purpose. Sometimes the alien species has no competition in the new community. It begins to reproduce without natural controls. The results of a new population of organisms can be disastrous. Government and private efforts to control and prevent the spread of alien species are intensifying.

Some examples of the effects of alien species include the following:

- During the 1980s, the zebra mussel—a shellfish—arrived in the St. Lawrence Seaway that separates eastern Canada and the United States. The mussel traveled in ballast water inside a cargo ship. Ballast water keeps a ship, especially an empty ship, stable in the water. When a ship is ready to take on cargo, the ballast water is pumped out. The weight of the cargo keeps the ship stable in the water. When the ballast water was released in the St. Lawrence Seaway, the zebra mussels, native to the Caspian Sea in Russia, were suddenly introduced into a new home. With no local predators and weak local competitors, their numbers quickly began to increase. The mussels attached themselves to many surfaces. They crowded together inside the pipes through which water flows to city water supplies. They also made it difficult for native shellfish species to survive. By the year 2002, the zebra mussels had moved throughout the Mississippi River system as far south as New Orleans and as far west as Oklahoma, and they were costing the United States about $5 billion each year in economic losses and control efforts.
- In 1877, the mongoose—a small, carnivorous mammal—was introduced intentionally into Puerto Rico to control the rat population that was damaging the sugarcane crop. For a while, the number of rats decreased. In time, the rats learned to avoid the mongoose predators by climbing trees. Then the mongooses turned to other sources of food.

for safeguarding the environment. Created in 1970, the EPA is in charge of protecting the environment for future generations. Its job is to control and limit air, land, and water pollution, noise pollution, and pollution by radiation, chemical pesticides, and toxic substances. The EPA works with local and state governments to do this. It also conducts scientific research that determines the environmental effects and impacts of various substances and activities. Since our environment is interconnected with other environments on Earth, in the late 1980s, the EPA began investigating worldwide environmental problems. For example, it has created a Climate Change Division that conducts research on global warming.

Figure 28-11 The mongoose was introduced to Puerto Rico to control the rat population. However, its introduction produced unexpected, harmful results—an increase in the June beetle larvae population.

They began to eat poultry, small birds, and lizards. The decrease in the number of lizards led to an explosion in the population of June beetle larvae. Soon the larvae became a more serious pest to sugarcane than the rats. (See Figure 28-11.)

- Purple loosestrife grows in vast numbers in many wet fields in the United States. Its tall stalks of purple flowers can be seen in large areas, with few other plant species visible. Purple loosestrife was introduced from Europe as an ornamental plant in the early 1800s. Today, it has invaded wetlands in 48 states, crowding out 44 native plants and endangering the wildlife that depends on those native plants. Approximately $45 million a year is spent to try to control the spread of this alien species.
- The English sparrow was introduced into the United States in 1853 to control canker worms—a garden pest. Instead, the hardy little sparrow has itself become a pest by eating crops, displacing some native birds, harassing others, and carrying 29 diseases that affect humans and domestic animals. Also, in spite of the good intentions of the people who introduced the English sparrow to this country, the canker worms are still pests in American gardens!

The Endangered Species Act is another response in the United States to a worldwide environmental crisis. The Endangered Species Act was passed in 1973. This act regulates a wide range of activities that affect threatened or endangered plant and animal species. For example, a turtle species may have a very limited range; its population may be confined to a particular swamp and live nowhere else. Thanks to the Endangered Species Act, the swamp itself is protected as a home for this turtle. The swamp cannot be drained, filled, and/or developed for any commercial purposes. Because of the Endangered Species Act, logging can no longer be done in large forest areas in the northwestern United States. These

Figure 28-12 The spotted owl of the Pacific Northwest is an endangered species. Protecting the owl also protects the forests it lives in.

forests are home to the spotted owl, an endangered species, and so the forests themselves are protected. (See Figure 28-12.) Without this protection, habitats would likely be altered or destroyed. Without habitat protection, most endangered species—especially those with a limited home range—would likely cease to exist.

DEVELOPED AND DEVELOPING COUNTRIES

People's lives are very different in different parts of the world. In the industrialized, developed countries of Europe and North America, and in Japan, the average standard of living is high. Most citizens of these countries eat well, have decent housing, and get good medical care. These people can expect to live for 70 or more years. On the other hand, in many countries of Asia, Africa, and Latin America, most people are poor. They do not have adequate food, housing, or medical care. In these countries, people often have a much shorter life span. Saving the biosphere means different things in rich and poor nations. Environmental leaders are now learning about these differences.

For example, some of the plant and animal species that are most in need of habitat protection live in developing countries. All of the world's tropical rain forests are in Africa, Asia, and Latin America. These are also the areas with the greatest population growth and the most need for more land to provide food and places for people to live. People also need jobs to improve their standard of living. It is not fair to tell people that they must protect the environment and endangered species if it means that they cannot also improve their lives. (See Figure 28-13.)

Environmentalists realize that it is very difficult to set aside parkland to save endangered tigers if doing this interferes with people getting

Figure 28-13
China's giant panda is endangered because more and more of its habitat is being used for human activities.

enough food and housing. One solution that has worked in some countries is to involve people directly in protecting the environment. This idea has sometimes been called "parks for people." For example, in Kenya, villagers work as guides for tourists who come to see the wildlife. (See Figure 28-14 on page 622.) In Peru, native Amuesha Indians harvest trees in the tropical forests on a sustainable basis. In Nepal, a park has been created to protect the few remaining native tigers. People need grasses to make thatch roofs for their houses. The tiger park, surrounded by farmland, is the only place where this grass now grows. These villagers are permitted to collect the grass in the park twice a year. As a result, the people protect the tiger park so they can continue to collect grass for their houses. Both tigers and people are now better off because of the park.

SAVING THE BIOSPHERE: A WORLDWIDE EFFORT

Imagine 30,000 people from all over the world meeting for 11 days to seek ways to protect the environment. From June 3 to 14, 1992, the United Nations Conference on Environment and Development (UNCED) took place in Rio de Janeiro, Brazil. The largest meeting of this type ever held, it was known as the Earth Summit. Representatives from 178 countries attended. The leaders of 118 of these countries were present. Some 1400 nongovernmental organizations were present. There were 8000 journalists. UNCED was planned in order to see what progress had been made in the 20 years since the 1972 United Nations Conference on the Human Environment was held in Stockholm, Sweden. The central theme of UNCED was sustainable development.

Figure 28-14 In Africa, near Kenya's Ewasu River, local guides take guests on camel safaris. This is one way in which people can earn a living from their natural environment without doing it harm.

Five major agreements were discussed during the conference. These were as follows: (1) *Rio Declaration on Environment and Development*: 27 principles building on the Stockholm meeting of 1972; (2) *Agenda 21*: a plan of action for sustainable development; (3) *Statement of Forest Principles*: the rights of countries to use their forests, but also the responsibilities of these countries to preserve remaining forests; (4) *Biodiversity Treaty*: a proposed set of laws to protect species and ecosystems; and (5) *Framework Convention on Climate Change*: a proposed set of laws to control the global-warming gases, mainly carbon dioxide and methane.

Countries from around the world are attempting to work together through the United Nations on these agreements. By 2002, however, it was widely recognized that progress in implementing sustainable development had been extremely disappointing since the 1992 Earth Summit. With poverty deepening and environmental degradation worsening, the United Nations convened a World Summit on Sustainable Development in Johannesburg, South Africa, in September 2002. With the leaders of 104 countries present, decisions were made on specific issues such as providing people with clean water, proper sanitation, and modern energy services; reducing waterborne diseases; fighting HIV/AIDS; restoring fisheries; and improving agriculture in drylands to combat desertification. In addition, after the 1992 Earth Summit, a special conference on the Convention on Climate Change took place in December 1997 in Kyoto, Japan. The Kyoto Protocol, as it came to be known, was the plan for how countries around the world would actually begin to reduce carbon dioxide emissions in order to limit global warming. The plan will enter into effect when enough countries have signed it to account for at least 55 percent of the total carbon dioxide emissions for 1990. This goal had not yet been reached by 2003, with major countries such as the United States not yet agreeing to the plan. Although some progress is occurring on a global scale, political and economic differences among countries interfere with more progress.

While local and regional efforts to save the environment are important, the global effort matters most. The biosphere takes in all of Planet Earth. To protect it for the future requires the efforts of all people. It is the responsibility of all of us to understand, support, and demand these efforts.

Being a part of life on Earth is a gift. We must save the biosphere for the future in order to protect this precious gift for all those who will come after us.

LABORATORY INVESTIGATION 28
What Can We Learn by Observing a Lawn Ecosystem?

INTRODUCTION

For ecologists, an ecosystem is a unit of study. Ecosystems have boundaries. Scientists often study populations and communities, and the physical conditions that affect these organisms, within an ecosystem. In addition, the relationships between organisms and the matter and energy that flow through an ecosystem can be studied. In this laboratory investigation, you will analyze a one-meter-square plot of lawn—a small ecosystem. The information you gather will enrich your understanding of the study of ecology.

MATERIALS

Meterstick, 4-meter length of string, hand lens, thermometer, small resealable plastic bags or covered containers, spoon or trowel

PROCEDURE

1. Select an area of lawn outside your school. Use the meterstick to mark off a one-meter-square area. Use the string to make the boundary of this area.
2. In your notebook, make notes of the physical conditions that surround your sample area. Write about the slope of the land, the proximity (closeness) of buildings and other physical structures, trees and other large plants, total size of the lawn, amount of sun or shade, and so on.
3. Describe your sample area. Note the color of the lawn, the kinds of plants, insects, and other animals that are present, areas where the soil is bare, and other observations you think may contribute to your understanding of this square meter.
4. Now make observations while kneeling on the lawn. Write down the same kinds of observations you made while standing. Often, closeness allows you to observe an area in more detail.
5. Observe even more closely. Peer down between the blades of grass. Use the hand lens to help you observe the ground between the grass plants. Observe as many places as you can. Look for any organisms living on the plants. Make notes of your observations.

6. Develop a method to estimate the number of grass plants in your square meter. Scientists use the method called sampling to estimate the size of a population, especially one that is too numerous to count.
7. Use the thermometer to take the temperature of the soil surface. If possible, carefully measure the soil temperature at several depths below the soil surface. Use a pencil or other wooden object to make a hole in the soil. *Do not use the thermometer to poke a hole in the soil.*
8. Collect some organisms from the ecosystem for further study back in your classroom. Use plastic bags or containers to collect and carry samples.

INTERPRETIVE QUESTIONS

1. How did your observations of your sample area compare with the observations made by classmates? How can you explain any differences in observations?
2. How could your method of estimating the population size of the grass plants in your study area be used to estimate the size of a population of animals? What problems would you have if you were asked to estimate the size of an animal population in an area?
3. Why would a more thorough study of the lawn ecosystem need to be conducted over an extended time period?

CHAPTER 28 REVIEW

Answer these questions on a separate sheet of paper.

VOCABULARY

The following list contains all of the boldfaced terms in this chapter. Define each of these terms in your own words.

alien species, nonrenewable resources, renewable resources, sustainable development

PART A—MULTIPLE CHOICE

Choose the response that best completes the sentence or answers the question.

1. Which of these attitudes is harmful to the overall health of the biosphere? *a.* Greed is good. *b.* Look out for number one. *c.* Humans are the masters of nature. *d.* All of these.
2. An example of a renewable resource is *a.* gold *b.* coal *c.* solar energy *d.* aluminum.
3. Which of these will *not* make forestry more sustainable? *a.* replanting *b.* selective harvesting *c.* building a more extensive network of logging roads *d.* finding replacements for forest-derived products
4. According to the Gaia hypothesis, *a.* Earth can be thought of as a single organism *b.* recycling helps to conserve natural resources *c.* overpopulation leads to war, famine, and disease *d.* ecosystems are infinitely adjustable.
5. To environmentalists, the "3 Rs" stand for *a.* reading, writing, 'rithmetic *b.* reduce, reuse, recycle *c.* reuse, recycle, replant *d.* rat race and retire.
6. Which of these will help protect tropical rain forests? *a.* increased demand for beef *b.* higher wholesale coffee prices *c.* increased demand for rubber and Brazil nuts *d.* subsidies to encourage tropical farmers to expand the size of their fields
7. A dam cannot be built on a river because it is the only home to a fish that exists nowhere else in the world. The dam was stopped by the *a.* Endangered Species Act *b.* Environmental Protection Agency *c.* Clean Water Act *d.* Tennessee Valley Authority.
8. Nonrenewable resources *a.* exist in limitless supply *b.* include substances like wood *c.* include fossil fuels *d.* all of these.
9. Developing economies can be helped and the environment preserved through *a.* the creation of large-scale plantations

b. "parks for people" *c.* boycott of goods from noncompliant countries *d.* the building of luxury resorts.

10. An example of a way to "reduce" is *a.* installing a low-flow toilet *b.* buying clothes at a discount store *c.* using paper plates at a party *d.* melting down old bottles to make new ones.
11. A paper company claims it does no harm to the environment because its workers plant an area with white pine seedlings after they have finished removing all the trees that were there before. However, an ecologist says that this is not good enough because *a.* natural forests do not consist of just one type of plant *b.* natural forests contain trees of different ages *c.* wildlife habitat is not restored by the replanting *d.* all of these.
12. Sustainable development involves *a.* achieving zero unemployment *b.* creating an economy that grows by at least five percent per year *c.* supporting a consumer society *d.* improving life now without harming life in the future.
13. Clear-cutting *a.* involves cutting down trees selectively *b.* increases erosion *c.* restores wildlife habitats *d.* improves the ability of the soil to hold water.
14. International agreements and treaties address *a.* biodiversity *b.* sustainable development *c.* global warming *d.* all of these.
15. An example of recycling is *a.* using a mug instead of a paper cup *b.* turning off the lights when you leave a room *c.* using plastic from soda bottles to make polyester fleece clothing *d.* using two paper towels rather than five to clean up a spill.

PART B—CONSTRUCTED RESPONSE

Use the information in the chapter to respond to these items.

16. What basic conclusions can you draw from the table on page 628?
17. Prepare a bar graph based on the table that compares the size and change in forests in tropical and nontropical areas.
18. Describe the steps that were taken by the United States government in the 1970s to protect the environment.
19. According to some sources, the average American consumes 55 kilograms of natural resources every day. Explain why this number is likely to be true.
20. "Whoever Dies with the Most Stuff Wins" reads a bumper sticker on a deluxe-model SUV. What might an environmentally conscious person have to say in response to that philosophy?

GLOBAL FOREST STATUS, 1991–1995 (in millions of hectares)

	All Countries		Developing Countries	
Region	Forest Area, 1995	Annual Change, 1991–95	Natural Forest Area, 1995	Annual Change, 1991–95
Tropical Regions				
Africa	504.90	– 3.69	502.74	– 3.70
Asia/Oceania	321.67	– 3.21	297.50	– 3.51
Latin America/Caribbean	950.04	– 5.69	901.34	– 5.69
Total Tropical	1,776.61	–12.59	1,701.59	–12.90
Nontropical Regions				
Africa	15.34	– 0.05	12.71	– 0.05
Asia/Oceania	243.20	– 0.21	123.16	– 0.66
Latin America/Caribbean	42.65	– 0.12	40.93	– 0.12
Europe	145.99	+ 0.39	—	—
Former Soviet Union	816.17	+ 0.56	—	—
North America	457.09	+ 0.76	—	—
Total Nontropical	1,720.42	+ 1.32	176.80	– 0.83
Total Forest Area	**3,454.38**	**–11.27**	**1,878.39**	**–13.73**

Source: U.N. Food and Agriculture Organization, *State of the World's Forests 1997* (Rome: 1997).

PART C—READING COMPREHENSION

Base your answers to questions 21 through 23 on the information below and on your knowledge of biology. Source: *Science News* (October 12, 2002): vol. 162, p. 237.

Rare Animals Get U.N. Protection

The great white shark, river dolphins, several types of whales, and an unusual two-humped camel are among animals that were designated on Sept. 24 to receive new or heightened protection under the Convention on Migratory Species, a U.N. treaty.

Meeting in Bonn, Germany, delegates placed the great white shark in the treaty's Appendix I. Such a listing bans the 80 nations that have ratified the treaty from catching or harming the species inside their boundaries. It also requires the nations to protect the species' habitat.

Three whale species—fin, sperm, and sei—also made it into Appendix I, along with the nearly extinct Ganges and Indus River dolphins (*Platanista gangetica gangetica*) in India, Pakistan, and Bangladesh. These virtually blind freshwater dolphins, which navigate by sonar, are threatened by pollution, hunting, and entanglements in fishnets.

Perhaps the most unusual animals added to Appendix I are the 300 hairy-kneed camels, found in three isolated pockets of China and Mongolia, including a former nuclear test range. They survive on salt water, and scientists suspect they're a new species.

Even nations that haven't ratified the Convention on Migratory Species, such as the United States, Japan, Russia, and China, sometimes adopt its policies.

21. State two ways in which animals get protection when they are listed in Appendix I of the U.N. treaty.
22. Explain why the river dolphins of India, Pakistan, and Bangladesh need special Appendix I protection.
23. State two facts about the hairy-kneed camels that support the decision to protect them under the U.N. treaty.

Glossary

- **abiotic factors** the nonliving parts of an organism's environment

absolute dating determining the age of a fossil by radioactive dating techniques; more precise than relative dating

absorption process by which food molecules enter cells after digestion

absorption spectrum shows the amount of light at each wavelength that is absorbed by a substance

abstinence the avoidance of sexual intercourse

acetylcholine a neurotransmitter; it carries nerve impulses from a neuron to a muscle

acid rain precipitation that has a low pH because of dissolved acids

activation energy the energy required to start a chemical reaction

active immunity occurs when the immune system remembers previously encountered microorganisms and is better prepared to fight them later

active site the part of an enzyme where substrates fit and are acted upon

active transport movement of substances across a membrane from an area of lower to higher concentration; requires energy

adaptations special characteristics that make an organism well suited for a particular environment

adaptive radiation process by which several populations evolve from a parent population, adapted to different ecological niches

adolescence a period of rapid growth and sexual maturation between the ages of 13 and 19

adrenal gland a gland on top of each kidney that regulates the removal of water from the body and the response to stress

advanced trait a trait that is modified by evolution to be different from the primitive trait it evolved from

aerobic respiration energy-releasing process that uses oxygen to produce ATP in eukaryotic cells

afterbirth the placenta after it is expelled from the vagina

aldosterone a hormone secreted by the adrenal glands which regulates the amount of salt in the blood and urine

alien species a new organism that has been introduced to an ecosystem, either intentionally or by accident

allantois membrane in a terrestrial vertebrate embryo that contains waste

allele a version of a particular gene

allergic reaction a condition caused by an overreaction of the body's immune system

alveoli tiny saclike structures in the lungs at which gas exchange takes place

amino acid organic compounds that are the building blocks of polypeptides and proteins

ammonia a nitrogenous waste that is toxic to the body

amniocentesis a test during pregnancy that checks chromosomes in fetal cells from the amniotic fluid for genetic diseases

amnion a watertight membrane that surrounds and protects terrestrial vertebrate embryos

amniotic fluid the fluid inside the amnion that surrounds and protects terrestrial vertebrate embryos

amplification increasing the number of copies of DNA manyfold

analogous structures features in different species that have different evolutionary origins but serve similar functions; for example, the wings on a fly, a bird, and a bat

anaphase the third stage in mitosis; sister chromatids separate and become single-stranded chromosomes

anther the portion of the stamen that produces and releases pollen

antibiotic a chemical that kills organisms of selected species; frequently used to combat infectious disease

antibodies molecules that individuals produce as a defense against foreign objects in the body; antibodies bind to specific antigens

anticodon a three-nucleotide sequence on tRNA that complements and binds to a specific codon

antidiuretic hormone (ADH) a hormone secreted by the pituitary gland to increase the reabsorption of water from the urine to the bloodstream

antihistamine a drug that counteracts histamine to reduce inflammation

antigen a protein on a foreign object that stimulates the immune system to produce antibodies

arboreal describes organisms that live in trees

artery a blood vessel that carries blood away from the heart

artificial selection process by which humans encourage the development of specific traits by increasing the reproductive success of individuals with those traits

asexual reproduction reproduction that involves only one parent

atom the smallest unit of an element that can combine with other elements

ATP **(adenosine triphosphate)** a substance used by cells as an immediate source of chemical energy for the cell

atrium (*plural,* **atria**) the thin-walled, upper chambers that receive blood into the heart

autoimmune disease condition that occurs when the immune system attacks the body's healthy tissues

autosome a chromosome other than the sex chromosome

autotroph a self-feeding organism that obtains its energy from inorganic sources

auxin a plant hormone that encourages growth

axon the part of a neuron that transmits signals away from the cell body

■ **behavior** every action that an animal takes, either learned or instinctive; usually to aid survival

benign causing no bodily damage

bile a fluid produced by the liver and released into the small intestine to help digest fats

binary fission division of a parent cell into two daughter cells

binomial nomenclature a system of naming organisms that uses two names to denote genus and species

biochemistry the chemistry of living things

biodiversity the variety of different species in an ecosystem

biogeography the study of the geographic distribution of organisms

biological magnification the process by which concentrations of pollutants increase at higher trophic levels in a food chain

biome a very large area characterized by a certain climate and types of plants and animals

biotechnology new procedures and devices that utilize discoveries in biology

biotic factors the living parts of an organism's environment

bipedalism the ability to walk upright on two feet

blastocyst in mammals, the blastula, which consists of a hollow ball of cells

blastula an early embryo that consists of a hollow ball of cells

bottleneck a process by which the genetic makeup of a population changes because all but a few individuals in the population are eliminated

Bowman's capsule a cuplike structure at the start of the nephron that funnels fluid into the nephron tubule

bronchiole one of several small tubes that branch off from the bronchi and lead into the alveoli of the respiratory system

bronchus (*plural,* **bronchi**) one of two tubes that branch off from the trachea and enter the lungs

budding a form of asexual reproduction where an offspring grows out of the side of the parent

■ **capillary** the smallest blood vessel in a closed circulatory system; where materials are exchanged with surrounding tissues

carbon fixation the addition of carbon dioxide to sugar molecules during the light-independent part of photosynthesis

carcinogen a chemical that increases the chance of developing cancer

carrier an individual with one copy of a defective allele but no symptoms

carrying capacity the size of a population that an ecosystem can support

catalyst something that increases the rate of a chemical reaction and is not changed during the reaction

cell the smallest living unit of an organism; all organisms are made of at least one cell

cell body the main part of a cell, such as a neuron, that contains the nucleus

cell cycle the series of events that occurs in the lifetime of a cell

cell differentiation the creation of specialized cells from less specialized parent cells through controlled gene expression

cell membrane a selectively permeable plasma membrane that separates, and regulates substances that pass between, the inside and outside of a cell

cell plate a structure that divides a plant cell at the end of mitosis and becomes part of the new cell walls

cellular respiration the process that uses oxygen to create ATP for energy use

central nervous system (CNS) the brain and nerve cords (invertebrates); the brain and spinal cord (vertebrates); controls receptors, glands, and muscles

centrioles the organelles that help guide cell division

centromere the part of a chromosome that joins the two chromatids together

cerebellum the part of the brain below the cerebrum that helps process sensory perception and muscular control

cerebral cortex the outer surface of the cerebrum; responsible for most of the processing in the cerebrum

cerebrum the largest part of the brain, responsible for learning, sensory perception, voluntary action, and motor control

cervix the opening of the uterus

chemosynthesis the production of organic (food) molecules using inorganic chemicals as an energy source

chlorophyll the green pigment in photosynthetic plants that traps the energy in sunlight

chloroplast the organelle in a plant that contains chlorophyll and carries out photosynthesis

chorion a membrane in a terrestrial vertebrate embryo that interacts with the environment

chorionic villus sampling (CVS) a test that checks cells from the chorion to detect genetic diseases in the embryo

chromatid one strand of DNA; after replication, a chromosome is made up of two identical chromatids attached by a centromere

chromosomal abnormality a deficiency or excess of one or more chromosomes

chromosome a structure, composed of DNA, that contains the genetic material

cilia small, hairlike structures on the surface of cells that aid life processes

circulation the movement of blood through the blood vessels of an animal

cirrhosis the scarring of an organ, particularly the liver

cleavage the first few cell divisions at the start of embryonic development

cloning the creation of a new individual from one cell of another individual

closed circulatory system a transport system in which the blood does not leave the blood vessels
codominance when the heterozygous individual expresses both of its alleles
codon a three-nucleotide sequence that codes for the production of a specific amino acid
coenzyme a substance that helps an enzyme catalyze a reaction
coevolution the process by which two or more species evolve in response to each other
cofactors substances that help an enzyme function
commensalism a symbiotic relationship in which one organism benefits and the other is unaffected
community populations of different species that interact within an area
competition the struggle between organisms for limited resources
concentration gradient a difference in the concentration of a substance at different points in space
condensation (dehydration) synthesis process by which two molecules are joined, while one molecule of water is removed
conditioned response an organism's response to a conditioned stimulus
conditioned stimulus a stimulus that an organism learns to associate with another stimulus
coniferous possessing narrow needle like leaves
contraception a method of reducing the chance of pregnancy
contractile vacuole an organelle that eliminates excess water from protists
convergent evolution the process by which unrelated species evolve similar characteristics that help them survive in similar environments
cordocentesis a blood test done on a fetus to check for disease
corpus luteum the follicle after it has released its egg
crossing-over the process in meiosis whereby homologous genes are exchanged between homologous chromosomes
cross section a view of an organism or organ after it has been bisected
cytokinesis the division of the cytoplasm and organelles of one parent cell into two daughter cells
cytoplasm the watery fluid that fills a cell, surrounding its organelles
cytotoxic T cells white blood cells that recognize and kill cells that are invaded by microorganisms

■ **decay product** the atoms that result when a radioactive element decays
deciduous possessing broad, flat leaves, which are dropped seasonally
decomposer a heterotrophic organism that obtains its food from dead organisms
dendrite the portion of a neuron that receives signals and transmits them to the cell body
density the number of individuals in a population per a given size area
density-dependent factors affected by the density of a population
density-independent factors not affected by the density of a population
deoxyribonucleic acid (DNA) the hereditary material of all organisms, which contains the instructions for all cellular activities
dermis the inner layer of the skin
desertification the process that turns land into desert, unsuitable for farming
detoxification the conversion of toxic chemicals to harmless chemicals by the liver
development the changes in an organism that occur from fertilization until death
dialysis the process of using a machine to remove wastes from the blood
diastolic pressure the pressure of the blood between heart contractions, at its lowest point
diffusion the movement of molecules from an area of high concentration to an area of low concentration
digestion the process of breaking down food particles into molecules small enough to be absorbed by cells
digestive cavity a sac inside a planaria in which digestion takes place

diploid having two sets (the normal species number) of homologous chromosomes

distribution the way individuals in a population are scattered across their environment

divergent evolution the process by which two or more populations evolve differently from a common parent population

DNA see *deoxyribonucleic acid*

DNA hybridization mixing together single-stranded DNA from different individuals to form double-stranded DNA, one strand from each individual

DNA replication the process of making two copies of DNA from one template

dominant the allele of a gene expressed in the heterozygous individual

dynamic equilibrium a state in which conditions change but always stay near an equilibrium point

■ **ecosystem** an area that contains living and nonliving parts that interact

ectoderm the outer layer of cells in a gastrula

ejaculation the expulsion of semen from the body

electron transport chain a complex of molecules that uses energy to create a concentration gradient across the inner membrane of the mitochondria

element a pure substance with set chemical properties, which cannot be separated into smaller substances by ordinary means

elimination the removal of indigestible materials from the digestive system

embryo an organism in an early stage of development before it is hatched, born, or germinated

embryonic development the sequence of events that changes a zygote into a functional organism

endocrine system the glands that communicate with each other and affect other parts of the body by secreting hormones into the bloodstream

endoderm the inner layer of cells in a gastrula

enzyme a protein that acts as a catalyst for a biological reaction

enzyme-substrate complex a very short-lived molecule that is formed when an enzyme and one or more substrates combine at the start of a reaction

epidermis the outer layer of the skin

epididymis organ in male reproductive system in which spermatids mature into sperm

erosion the slow, steady removal of soil from an area; caused by wind and rain

esophagus tube in digestive system that connects the pharynx and the stomach

estrogen a steroid hormone; along with progesterone, a major female hormone

estrus period of time when a female mammal is fertile and receptive to mating

excretion the removal of metabolic wastes from the body

exhalation the movement of air out of the respiratory system

extinct refers to a species that no longer has any living members

extinction the death of all individuals of a species

extracellular digestion the digestion of food outside the body cells

■ **fate map** a map of an early embryo that shows what each region will become

fermentation anaerobic process that converts chemical energy from organic chemicals into usable forms, such as ATP

fertilization process by which an egg cell and a sperm cell unite to form a zygote

fetus a developing embryo after the first trimester

filament the portion of the stamen that physically supports the anther

filtrate the liquid that has been filtered

fitness the ability to carry out normal activities and still have enough energy and strength to overcome unusual challenges

fixed action pattern a behavior that does not change with repetition
fluid mosaic model describes the cell membrane as a phospholipid bilayer that contains freely moving molecules such as proteins
follicle a bundle of cells in an ovary that contains an egg
follicle-stimulating hormone (FSH) a hormone secreted by the anterior pituitary
food chain the linear transfer of energy from one organism to the next
food web the complex interconnecting food chains in a community
fossil traces or remains of dead organisms, preserved by natural processes
founder effect process by which a population acquires a very different gene pool from its parent population because it descends from a few ancestors

■ **gametes** the haploid cells that combine to form a zygote during fertilization
gametogenesis the production of gametes in the sex organs
gastric juice a fluid secreted by the stomach wall that chemically digests food
gastrulation the development of the blastula (early animal embryo) into a gastrula that has three germ layers
gel electrophoresis a technique for sorting pieces of DNA by size
gene the segment of DNA that contains the genetic information for a given trait or protein
gene expression the use of the genetic information in a gene to affect the cell or its environment
gene flow the movement of alleles into and out of a population as individuals move into and out of the population
gene splicing combining two pieces of DNA lengthwise to create one molecule
genetic code the code in all living things that makes every codon correspond to a specific amino acid
genetic cross an experiment in which two organisms are bred to better understand their genetic makeup or how their traits are inherited
genetic drift the process by which allele frequencies change over time due to chance
genetic variation the differences among offspring in their genetic makeup
genotype the combination of alleles in an individual
geographic isolation reproductive isolation that is caused by a geographical barrier
gills the respiratory organs of water-breathing vertebrates
global warming an increase in the average atmospheric temperature of Earth
glomerulus a bundle of capillaries in the Bowman's capsule of a nephron
glycolysis anaerobic breakdown of glucose that produces molecules of ATP
gradualism process by which species evolve at a continuous, slow, steady pace
groundwater freshwater that is found below the surface of the earth
growth factor a substance that directs the growth of cells or organs
growth rate the change in population density divided by the time in which the change takes place

■ **habit** a behavior that becomes automatic after it is repeated many times
habitat the place in which an organism lives
habituation learning to ignore a repeated stimulus that is unimportant
haploid having one-half the normal, or diploid, number of chromosomes
heterotroph an organism that obtains its energy from other living things
heterozygous containing two different alleles for a given gene
hibernation a state of rest in which an organism maintains a low rate of metabolism
histamine chemical released by the body in response to an allergic reaction
homeobox sequences stretches of DNA, called homeotic genes, which

act like switches to turn genes on and off

homeostasis the ability to maintain a constant internal environment

hominid the family of hominoids that includes humans and their humanlike ancestors

hominoid the group of primates that includes humans, their ancestors, and apes

homologous chromosomes a pair of chromosomes that contains the same genes; each homologous chromosome may contain different alleles

homologous structures features in different species that have the same evolutionary origin, but may perform different functions at present

homozygous containing two identical alleles for a given gene

Human Genome Project a collaborative effort to determine the DNA sequence of the entire human genome

hydrolysis reaction in which one complex molecule is broken down into two simpler molecules by the addition of a molecule of water

hypertension abnormally high blood pressure

hypothalamus region in the brain that controls the release of hormones from the pituitary and helps maintain homeostasis in the neuroendocrine system

■ **imitative learning** when an animal learns to model its behavior after an observed behavior

imprinting behavior that occurs when a newly born animal responds to the first moving object it sees as its parent and follows it automatically

immune system the organ system that protects the body against infection

immunodeficiency disease a condition that weakens the immune system

implantation the attachment of a blastocyst to the uterus

incomplete dominance when the heterozygous offspring has a phenotype that is a mixture of the homozygous parents' phenotypes

independent assortment idea that each character for a trait is inherited independently of any other character for a trait

industrial melanism process by which populations near polluted areas evolve to have darker (more protective) coloring than those in more pristine areas

infectious disease an illness that can be passed from one individual to another

infertility the inability to produce offspring

inflammation a reddening or swelling in the body in response to an infection

inhalation the movement of air into the respiratory system

ingestion the process of taking food into an organism

interneuron a neuron that connects the sensory neuron to the motor neuron

interphase period before cell division during which DNA replication occurs

interspecies competition competition between two or more species

intracellular digestion the digestion of food inside a cell

intraspecies competition competition between members of the same species

in vitro fertilization fertilization of an egg within a laboratory dish

■ **karyotype** picture of a full set of chromosomes arranged by size and shape

kidneys the pair of organs in terrestrial vertebrates that regulates blood chemistry and removes water and metabolic wastes from the plasma

killer T cells see *cytotoxic T cells*

Krebs cycle in cellular respiration, aerobic reactions inside the mitochondria, in which energy is released and stored in molecules of ATP

■ **larva** an immature animal that is morphologically different from the adult

larynx the part of the respiratory system that contains the voice box; it connects the pharynx with the trachea

legume a plant that has a symbiotic relationship with nitrogen-fixing bacteria in its roots

level of organization a scale for looking at the structure of a system; from atomic to worldwide

limiting factor a part of an organism's environment that determines where that organism can live

linkage tendency of two genes to be inherited together on same chromosome

lipids a group of organic compounds that includes fats and oils

luteinizing hormone (LH) a hormone secreted by the anterior pituitary

lysosome an organelle that contains enzymes for digesting food inside cells

■ **macroevolution** a change in species over a long time and on a large scale

macrophage a white blood cell that engulfs invading microorganisms

malignant causing bodily damage; often refers to a cancerous tumor

mammals a class of vertebrates characterized by being warm-blooded, having hair, and producing milk to feed offspring

mechanical digestion the process of breaking food into smaller pieces by physical means such as chewing

medulla oblongata the part of the brain stem, connecting brain to spinal cord, that controls involuntary activities such as breathing and heartbeat

meiosis the division of one parent cell into four daughter cells; reduces the diploid number of chromosomes to the haploid number (half)

menopause stage in mammalian life cycle when a female is no longer capable of producing offspring

menstrual cycle a cycle in human females that involves ovulation and the preparation of the uterus to support an embryo

menstruation the monthly shedding of the inner lining of the uterus

mesoderm the middle layer of cells in a gastrula

metabolism the chemical reactions (building up and breaking down) in an organism

metaphase the second stage in mitosis; chromosomes line up along the center of the cell

metastasis the spread of cancer when parts of a tumor break off and move to new sites in the body

microevolution change in species over a short time and on a small scale

migration the periodic movement of animals over long distances, usually for breeding and feeding

mimicry when one species resembles another species, to enhance its survival

mitochondria organelles in eukaryotic cells in which the Krebs cycle occurs

mitosis the division of one cell's nucleus into two identical daughter cell nuclei

morphology the shape and structure of an organism

motor neuron a neuron that transmits nerve impulses from the central nervous system (brain and spinal cord) to the effectors (muscles)

mutagen a substance that increases the chance that a genetic mutation will occur

mutation an error in the linear sequence (gene) of a DNA molecule

mutualism a symbiotic relationship in which both species' organisms benefit

■ **nasal cavity** the start of the respiratory system at which the air is filtered, odors are detected, and humidity and temperature are adjusted for respiration

natural selection the process by which organisms having the most adaptive traits for an environment are more likely to survive and reproduce

negative feedback system in which movement away from an equilibrium point is balanced by movement in the opposite direction

nephridium excretory organ of earthworms that removes metabolic wastes

nephron a microscopic structure in the kidney that removes metabolic wastes from the blood and forms urine
nerve impulse a signal that is transmitted along and between neurons
neural tube part of vertebrate embryo that becomes the central nervous system
neuroendocrine system the combined nervous and endocrine systems; regulates homeostasis in the body and response to the environment
neuron in animals, the cell that transmits nerve impulses to other neurons and to other types of cells
neurotransmitter a chemical that transmits nerve impulses between a neuron and another cell, usually another neuron
niche an organism's role in (or interaction with) its ecosystem
nitrogen fixation converting free nitrogen in the air into usable nitrates
nitrogenous waste metabolic waste formed when amino acids are broken down
nonrenewable resource a resource that is depleted with use and cannot be replaced
nucleotide the building block of DNA; consists of a five-carbon sugar, a phosphate, and a nitrogenous base
nucleus dense region of a eukaryotic cell that contains the genetic material

■ **open circulatory system** system in which blood circulates around the organs
opposable capable of touching something opposite; an opposable thumb can touch all the other digits on the same hand
organ a structure made up of similar tissues that work together to perform a task; for example, the stomach
organelles structures within a cell that perform a particular task, for example, the mitochondrion
organic relating to compounds that contain carbon and hydrogen; living things
organ system a group of organs that works together to perform a major task; for example, the digestive system
ornithologist a scientist who studies birds
osmoregulation maintaining the proper amount of water in a cell or body
osmosis diffusion of water molecules across a membrane, from a region of higher to lower concentration
ovary the female reproductive organ that produces the egg cells
oviducts (fallopian tubes) pair of tubes that connects the ovaries to the uterus
ovulation the monthly release of a mature egg from its follicle
ozone depletion a reduction in the amount of ozone in the ozone layer

■ **parallel evolution** the process by which two or more species evolve separately, but in similar ways, from a common parent population
parasite an organism that lives on or in another organism and causes it harm
parasitism a symbiotic relationship in which one organism benefits while the other (the host) is harmed
passive immunity when antibodies are injected into an individual to make his or her immune system better able to fight an infectious disease
passive transport movement of substances across a membrane; requires no use of energy
pathogens microscopic organisms, such as bacteria and viruses, that cause diseases
pecking order in animal groups, a behavior based on which individuals are dominant or subordinate to any other individual
pedigree a chart that shows members of a family, their genetic relationships to each other, and whether they carry or exhibit a heritable disease
penis the organ in the male reproductive system that delivers sperm into the female

peptide bond a strong covalent bond between adjacent amino acids in a polypeptide

peripheral nervous system (PNS) the cranial nerves and spinal nerves; controls reflexes and involuntary responses, and helps maintain homeostasis

peristalsis the wave of muscle contractions that helps move food down the digestive tract

pH a measure of the concentration of hydrogen ions in a solution

phagocytosis the process by which a cell engulfs matter

phenotype the physical expression of an inherited trait in an individual

pheromone a chemical released by one individual that affects the behavior of another individual

phloem tubes that carry carbohydrates from the leaves to the rest of a plant

phospholipids lipids composed of two fatty acids and a phosphate group; they make up a cell's plasma membrane

photolysis the splitting of water during the light-dependent phase of photosynthesis

photon the smallest possible unit of light energy

photosynthesis the process that, in the presence of light energy, produces chemical energy (glucose) and water

phylogeny the evolutionary history and relationships of a species

physiology the way an organism and its internal parts function

phytoplankton photosynthetic microorganisms that live near the ocean surface

pistil the female part of a flower; contains the ovary, style, and stigma

pituitary gland tiny gland in the skull that regulates the functioning of other endocrine glands by releasing hormones

placenta the organ that develops in the uterus of mammals to nourish an embryo and remove its waste

plasma the liquid portion of blood; contains water and dissolved substances

plasmid a circular segment of DNA in some types of bacteria

plasmolysis in plant cells, movement of the cell membrane away from the cell wall due to water loss

platelets fragments of blood cells that help the blood clot normally

pollination the delivery of pollen grains from an anther to a stigma

polygenetic inheritance occurs when a trait is determined by several genes

polymer a large molecule, such as the blood protein hemoglobin, made by linking smaller molecules together

polymerase chain reaction (PCR) a technique for amplifying a strand of DNA by letting enzymes replicate it many times in a test tube

polypeptide a chain of amino acids; long chains make up protein molecules

population all of the individuals of the same species that live in the same area

postnatal the period after birth

predation when one organism uses another living organism as a source of food

predator an organism that feeds on another living organism (the prey)

prenatal the period before birth

prenatal diagnosis process of detecting a genetic disorder before birth

primary consumer an organism that obtains its energy by eating producers

primitive trait a trait that evolved in an early, common ancestor

producer an organism (autotroph) on the first trophic level; obtains its energy from inorganic sources (via photosynthesis or chemosynthesis)

progesterone a steroid hormone; along with estrogen, a major female hormone

prophase the first stage of mitosis; the nuclear membrane breaks down

proteins a group of organic compounds made up of chains of amino acids

protein hormones a group of hormones that are made up of protein

puberty a stage in the life cycle of vertebrates when sexual maturation occurs

pulmonary pertaining to structure and/or function of the lungs

punctuated equilibrium process by which species undergo little or no change for long periods of time and then evolve via sudden, substantial changes

■ **radioactive isotopes** certain atoms of an element that decay at a uniform rate

recessive the allele of a gene not expressed in the heterozygous individual

recombinant DNA pieces of DNA from two or more different species that are joined into one DNA molecule

recombination the creation of new combinations or groups

reduction division process of cell division in meiosis that reduces the chromosome number in half (from diploid to haploid)

reflex an inborn, automatic response to a stimulus, which aids survival

regeneration the process of regrowing organs or a whole organism from a part of the organism

relative dating determining the age of a fossil by the layer in which it is found

renewable resource a resource that can be replaced through growth and regeneration

replication the duplication of DNA during cell division

reproduction the production of offspring, by either sexual or asexual means

reproductive isolation occurs when one population of a species is prevented from breeding with members of another population of that species

respiratory surface the moist area of an animal at which gas exchange occurs

response an organism's reaction to a stimulus; can be inborn or learned

restriction enzyme a bacterial enzyme that cuts DNA at points where specific nucleotide sequences (codons) occur

restriction fragments pieces of DNA that have been cut by restriction enzymes

restriction site the specific nucleotide sequence where a restriction enzyme cuts

retrovirus a virus that injects RNA into a host cell and then makes a DNA copy

ribonucleic acid (RNA) the single-stranded nucleotide molecule that helps in protein synthesis; contains the base uracil instead of thymine

ribosome the small organelle at which protein synthesis occurs; contains RNA

RNA see *ribonucleic acid*

robust refers to the more heavy-boned australopithecine hominid species

root hair an extension of the cells that line roots, which increases the surface area for absorption

■ **saliva** fluid in the mouth that is produced by salivary glands and starts the chemical digestion of carbohydrates

schooling describes behavior of animals that travel in groups, such as fish

scrotum the external sac of skin in mammals that contains the testes

sebaceous glands glands in the skin that secrete oils to keep skin and hair soft

secondary consumer an organism on the third trophic level; obtains its energy by eating primary consumers

sedimentary rock formed by the compression of sediments such as sand

selective permeability the ability of a membrane to let some substances pass through it and to block the passage of others

semen the fluid containing the sperm that male animals release during mating

sensory neuron a neuron that transmits nerve impulses from receptors to the central nervous system (brain and spinal cord)

sensory perception the ability to receive information about the environment

sensory receptor a specialized structure that receives nerve impulses and then transmits messages to the neurons

sewage organic waste produced by humans
sex chromosomes the chromosomes that determine an individual's gender; in animals, the X and Y chromosomes
sex-linked describes a trait that is determined by a gene on the X chromosome
sexually transmitted disease (STD) an infectious disease that is spread during sex when a person comes into contact with another person's body fluids
sexual reproduction reproduction that involves (genes from) two parents
short tandem repeats (STRs) DNA sequences that show the greatest amount of polymorphism, or variation, between individuals
simple sugar a single sugar that has six carbon atoms; a monosaccharide
society a group of animals that lives together and shares responsibilities
speciation the evolution of new species
species a group of related organisms that can breed and produce fertile offspring
sperm the male gamete, or sex cell
spermatid an immature sperm cell that is produced by meiosis
spindle fibers that are formed during mitosis that help separate the chromatids
stamen the male organ of a flower; contains the filament and anther, which produces the pollen grains
starch a complex carbohydrate made up of many glucose molecules; used for energy storage in plants
sterile containing no viable bacteria or spores; incapable of reproduction
steroid hormones a group of hormones that is formed from cholesterol
stigma in plants, the part of the pistil (female organ) that traps pollen
stimulus (*plural,* **stimuli)** an event, change, or condition in the environment that causes an organism to make a response
stomate an opening on the surface of a leaf through which gas exchange occurs
style in flowers, the part of a pistil that connects the stigma and the ovary
succession the slow replacement of one ecological community by another
sustainable development economic gain that does not harm the environment
sweat glands glands in the skin that secrete perspiration to cool the body
symbiosis a close relationship between two or more different organisms that live together, often mutually beneficial
synapse the space between the end brush of one neuron and the dendrites of the next, across which the nerve impulse (message) travels
systemic pertaining to the whole body of an organism
systolic pressure the pressure of the blood when the heart is contracting, at its highest point

■ **taxonomy** the science of classifying organisms according to their shared traits
telophase the fourth and final stage of mitosis; when a new nuclear membrane forms around each set of chromosomes
temperature inversion occurs when a layer of cool air is trapped under a layer of warm air; traps air pollutants close to the ground
territory the area in which an animal feeds and breeds, and which it defends
tertiary consumer an organism on the fourth trophic level; obtains its energy by eating secondary consumers
testes the pair of male reproductive organs that produces the sperm cells
testosterone a steroid hormone; the main male sex hormone
tetrad in meiosis, the arrangement of four homologous chromosomes (two sets of sister chromatids) in which crossing-over occurs
theory of evolution idea that organisms change over time as a result of genetic variations that enable them to adapt to changing environments

thermal pollution a temperature increase in a body of water due to waste heat released by industry
thermoregulation the ability to maintain a constant internal body temperature
tides daily rise and fall of ocean water caused by the moon's gravitational pull
tissue a group of similar cells that work together to perform the same function
topsoil the nutrient-rich top layer of soil that can support crop growth
trachea the air tube that connects the pharynx to the bronchi; the windpipe
tracheae in insects, air tubes that connect the inside to outside for gas exchange
transcription the synthesis of RNA using DNA as a template
translation the synthesis of protein using the DNA code in RNA as a template
transpiration in plants, the loss of water vapor from the stomates of a leaf
trimester a period of three months, comprising one-third of a pregnancy
trophic level a feeding level in a food chain or in a food web
trophoblast the outer layer of a blastula that becomes part of the placenta
tumor a mass of cells that grows (divides) without restraint; often cancerous

■ **ultrasonography** the use of sound waves to form an image; used to view a fetus
umbilical cord in mammals, structure that connects the embryo to the placenta
urea a nitrogenous waste compound, dissolved in the urine, that is excreted
ureter a tube that transports urine from a kidney to the urinary bladder
urethra a tube that carries urine from the bladder to be excreted from the body
uric acid a form of nitrogenous waste that birds and insects excrete as a solid
urinary bladder an organ that stores urine before it is excreted
urine fluid waste that is produced by the kidney and excreted from the body
uterus in mammals, the reproductive organ that holds the developing embryo

■ **vaccination** an injection that prepares the immune system to better fight a specific disease in the future
vagina in mammals, the female structure that receives the sperm; the birth canal
vas deferens the tube in the male reproductive system that connects the epididymis with the urethra
vector a vehicle, mainly bacterial plasmids and viruses, used for transporting foreign DNA into a cell
vein a blood vessel that carries blood toward the heart
ventricle the thick-walled, lower chambers that push blood out of the heart
vestigial structure a structure with little or no function in an organism that did have a function in its ancestors
villi microscopic, fingerlike projections in the small intestine that increase the surface area for absorption of nutrients
virus a particle of genetic material that can only replicate within a host cell

■ **wellness** condition of how healthfully a person lives; a lifestyle that promotes good health

■ **xylem** tubes that carry water and nutrients from the roots up through a plant

■ **yolk sac** membrane in a terrestrial vertebrate embryo that surrounds the yolk

■ **zygote** the fertilized egg cell that is formed when two gametes' nuclei fuse